Introduction to Modern Statistics

Mine Çetinkaya-Rundel and Johanna Hardin

Contents

Preface   **8**

I   Introduction to data   **11**

1 Hello data   **12**
     1.1   Case study: Using stents to prevent strokes 12
     1.2   Data basics 14
     1.3   Chapter review 21
     1.4   Exercises 22

2 Study design   **31**
     2.1   Sampling principles and strategies 31
     2.2   Experiments 38
     2.3   Observational studies 41
     2.4   Chapter review 43
     2.5   Exercises 45

3 Applications: Data   **53**
     3.1   Case study: Passwords 53
     3.2   Interactive R tutorials 59
     3.3   R labs 59

II   Exploratory data analysis   **60**

4 Exploring categorical data   **61**
     4.1   Contingency tables and bar plots 61
     4.2   Visualizing two categorical variables 62
     4.3   Row and column proportions 64
     4.4   Pie charts 67
     4.5   Waffle charts 67
     4.6   Comparing numerical data across groups 68
     4.7   Chapter review 72
     4.8   Exercises 73

5 Exploring numerical data   **76**
     5.1   Scatterplots for paired data 76
     5.2   Dot plots and the mean 77
     5.3   Histograms and shape 81
     5.4   Variance and standard deviation 83
     5.5   Box plots, quartiles, and the median 85
     5.6   Robust statistics 88
     5.7   Transforming data 89
     5.8   Mapping data 91
     5.9   Chapter review 94

5.10 Exercises 95

6 Applications: Explore　104
　　6.1 Case study: Effective communication of exploratory results 104
　　6.2 Interactive R tutorials 110
　　6.3 R labs 110

III Regression modeling　111

7 Linear regression with a single predictor　112
　　7.1 Fitting a line, residuals, and correlation 112
　　7.2 Least squares regression 124
　　7.3 Outliers in linear regression 133
　　7.4 Chapter review 136
　　7.5 Exercises 137

8 Linear regression with multiple predictors　150
　　8.1 Indicator and categorical predictors 151
　　8.2 Many predictors in a model 154
　　8.3 Adjusted R-squared 157
　　8.4 Model selection 158
　　8.5 Chapter review 164
　　8.6 Exercises 165

9 Logistic regression　170
　　9.1 Discrimination in hiring 170
　　9.2 Modelling the probability of an event 172
　　9.3 Logistic model with many variables 174
　　9.4 Groups of different sizes 177
　　9.5 Chapter review 178
　　9.6 Exercises 179

10 Applications: Model　185
　　10.1 Case study: Houses for sale 185
　　10.2 Interactive R tutorials 192
　　10.3 R labs 192

IV Foundations of inference　193

11 Hypothesis testing with randomization　194
　　11.1 Sex discrimination case study 195
　　11.2 Opportunity cost case study 201
　　11.3 Hypothesis testing 206
　　11.4 Chapter review 209
　　11.5 Exercises 211

12 Confidence intervals with bootstrapping　217
　　12.1 Medical consultant case study 218
　　12.2 Tappers and listeners case study 224
　　12.3 Confidence intervals 225
　　12.4 Chapter review 227
　　12.5 Exercises 229

13 Inference with mathematical models　234
　　13.1 Central Limit Theorem 234
　　13.2 Normal Distribution 237
　　13.3 Quantifying the variability of a statistic 248
　　13.4 Case Study (test): Opportunity cost 250

13.5 Case study (test): Medical consultant 252
13.6 Case study (interval): Stents 253
13.7 Chapter review 257
13.8 Exercises 259

14 Decision Errors 261
14.1 Significance level 263
14.2 Two-sided hypotheses 263
14.3 Controlling the Type 1 Error rate 267
14.4 Power 268
14.5 Chapter review 269
14.6 Exercises 270

15 Applications: Foundations 272
15.1 Recap: Foundations 272
15.2 Case study: Malaria vaccine 275
15.3 Interactive R tutorials 279
15.4 R labs 279

V Statistical inference 280

16 Inference for a single proportion 281
16.1 Bootstrap test for a proportion 281
16.2 Mathematical model for a proportion 285
16.3 Chapter review 292
16.4 Exercises 293

17 Inference for comparing two proportions 306
17.1 Randomization test for the difference in proportions 306
17.2 Bootstrap confidence interval for the difference in proportions 308
17.3 Mathematical model for the difference in proportions 314
17.4 Chapter review 321
17.5 Exercises 322

18 Inference for two-way tables 333
18.1 Randomization test of independence 333
18.2 Mathematical model for test of independence 337
18.3 Chapter review 341
18.4 Exercises 342

19 Inference for a single mean 352
19.1 Bootstrap confidence interval for a mean 353
19.2 Mathematical model for a mean 359
19.3 Chapter review 370
19.4 Exercises 371

20 Inference for comparing two independent means 378
20.1 Randomization test for the difference in means 379
20.2 Bootstrap confidence interval for the difference in means 382
20.3 Mathematical model for testing the difference in means 384
20.4 Mathematical model for estimating the difference in means 387
20.5 Chapter review 390
20.6 Exercises 391

21 Inference for comparing paired means 398
21.1 Randomization test for the mean paired difference 399
21.2 Bootstrap confidence interval for the mean paired difference 401
21.3 Mathematical model for the mean paired difference 406
21.4 Chapter review 410

21.5 Exercises 411

22 Inference for comparing many means — 419
22.1 Case study: Batting 420
22.2 Randomization test for comparing many means 422
22.3 Mathematical model for test for comparing many means 427
22.4 Chapter review 431
22.5 Exercises 432

23 Applications: Infer — 439
23.1 Recap: Computational methods 439
23.2 Recap: Mathematical models 440
23.3 Case study: Redundant adjectives 442
23.4 Interactive R tutorials 448
23.5 R labs 448

VI Inferential modeling — 449

24 Inference for linear regression with a single predictor — 450
24.1 Case study: Sandwich store 450
24.2 Randomization test for the slope 453
24.3 Bootstrap confidence interval for the slope 455
24.4 Mathematical model for testing the slope 458
24.5 Mathematical model, interval for the slope 464
24.6 Checking model conditions 465
24.7 Chapter review 468
24.8 Exercises 469

25 Inference for linear regression with multiple predictors — 480
25.1 Multiple regression output from software 480
25.2 Multicollinearity 482
25.3 Cross-validation for prediction error 485
25.4 Chapter review 491
25.5 Exercises 492

26 Inference for logistic regression — 506
26.1 Model diagnostics 506
26.2 Multiple logistic regression output from software 508
26.3 Cross-validation for prediction error 509
26.4 Chapter review 513
26.5 Exercises 514

27 Applications: Model and infer — 519
27.1 Case study: Mario Kart 519
27.2 Interactive R tutorials 526
27.3 R labs 526

A Exercise solutions — 527

Copyright © 2021.

First Edition.

Version date: June 12, 2021.

This textbook and its supplements, including slides, labs, and interactive tutorials, may be downloaded for free at
openintro.org/book/ims.

This textbook is a derivative of *OpenIntro Statistics* 4th Edition and *Introduction to Statistics with Randomization and Simulation* 1st Edition by Diez, Barr, and Çetinkaya-Rundel, and it's available under a Creative Commons Attribution-ShareAlike 3.0 Unported United States License. License details are available at the Creative Commons website:
creativecommons.org.

Source files for this book may be found on GitHub at
github.com/openintrostat/ims.

Authors

Mine Çetinkaya-Rundel
Duke University, RStudio
mine@openintro.org

Mine Çetinkaya-Rundel is Associate Professor of the Practice position at the Department of Statistical Science at Duke University and Data Scientist and Professional Educator at RStudio. Mine's work focuses on innovation in statistics and data science pedagogy, with an emphasis on computing, reproducible research, student-centered learning, and open-source education as well as pedagogical approaches for enhancing retention of women and under-represented minorities in STEM. Mine works on integrating computation into the undergraduate statistics curriculum, using reproducible research methodologies and analysis of real and complex datasets. She also organizes ASA DataFest, an annual two-day competition in which teams of undergraduate students work to reveal insights into a rich and complex dataset. Mine has been working on the OpenIntro project since its founding and as part of this project she co-authored four open-source introductory statistics textbooks (including this one!). She is also the creator and maintainer of datasciencebox.org and she teaches the popular Statistics with R MOOC on Coursera.

Johanna Hardin
Pomona College
jo.hardin@pomona.edu

Jo Hardin is Professor of Mathematics and Statistics at Pomona College. She collaborates with molecular biologists to create novel statistical methods for analyzing high throughput data. She has also worked extensively in statistics and data science education, facilitating modern curricula for higher education instructors. She was a co-author on the 2014 ASA Curriculum Guidelines for Undergraduate Programs in Statistical Science, and she writes on the blog teachdatascience.com. The best part of her job is collaborating with undergraduate students. In her spare time, she loves reading, running, and breeding tortoises.

Preface

We hope readers will take away three ideas from this book in addition to forming a foundation of statistical thinking and methods.

1. Statistics is an applied field with a wide range of practical applications.
2. You don't have to be a math guru to learn from interesting, real data.
3. Data are messy, and statistical tools are imperfect. However, when you understand the strengths and weaknesses of these tools, you can use them to learn interesting things about the world.

Textbook overview

- **Part 1: Introduction to data.** Data structures, variables, summaries, graphics, and basic data collection and study design techniques.
- **Part 2: Exploratory data analysis.** Data visualization and summarisation, with particular emphasis on multivariable relationships.
- **Part 3: Regression modeling.** Modeling numerical and categorical outcomes with linear and logistic regression and using model results to describe relationships and made predictions.
- **Part 4: Foundations for inference.** Case studies are used to introduce the ideas of statistical inference with randomization tests, bootstrap intervals, and mathematical models.
- **Part 5: Statistical inference.** Further details of statistical inference using randomization tests, bootstrap intervals, and mathematical models for numerical and categorical data.
- **Part 6: Inferential modeling.** Extending inference techniques presented thus-far to linear and logistic regression settings and evaluating model performance.

Each part contains multiple chapters and ends with a case study. Building on the content covered in the part, the case study uses the tools and techniques to present a high level overview.

Each chapter ends with a review section which contains a chapter summary as well as a list of key terms introduced in the chapter. If you're not sure what some of these terms mean, we recommend you go back in the text and review their definitions. We purposefully present them in alphabetical order, instead of in order of appearance, so they will be a little more challenging to locate. However you should be able to easily spot them as **bolded text**.

Examples and exercises

Examples are provided to establish an understanding of how to apply methods.

EXAMPLE

This is an example. When a question is asked here, where can the answer be found?

The answer can be found here, in the solution section of the example!

When we think the reader is ready to try determining a solution on their own, we frame it as Guided Practice.

GUIDED PRACTICE

The reader may check or learn the answer to any Guided Practice problem by reviewing the full solution in a footnote.[1]

Exercises are also provided at the end of each chapter. Solutions are given for odd-numbered exercises in Appendix A.

Datasets and their sources

A large majority of the datasets used in the book can be found in various R packages. Each time a new dataset is introduced in the narrative, a reference to the package like the one below is provided. Many of these datasets are in the **openintro** R package that contains datasets used in OpenIntro's open-source textbooks.[2]

The `textbooks` data can be found in the **openintro** R package.

The datasets used throughout the book come from real sources like opinion polls and scientific articles, except for a handful of cases where we use toy data to highlight a particular feature or explain a particular concept. References for the sources of the real data are provided at the end of the book.

Computing with R

The narrative and the exercises in the book are computing language agnostic, however while it's possible to learn about modern statistics without computing, it's not possible to apply it. Therefore, we invite you to navigate the concepts you have learned in each chapter using the interactive R tutorials and the R labs that are included at the end of each chapter.

Interactive R tutorials

The self-paced and interactive R tutorials were developed using the learnr R package, and only an internet browser is needed to complete them.

Each chapter comes with a tutorial comprised of 4-8 lessons and listed like this.

Each of these lessons...

... is listed like this.

You can access the full list of tutorials supporting this book at https://openintrostat.github.io/ims-tutorials.

---

[1] Guided Practice problems are intended to stretch your thinking, and you can check yourself by reviewing the footnote solution for any Guided Practice.

[2] Mine Çetinkaya-Rundel, David Diez, Andrew Bray, Albert Kim, Ben Baumer, Chester Ismay and Christopher Barr (2020). openintro: Data Sets and Supplemental Functions from 'OpenIntro' Textbooks and Labs. R package version 2.0.0. https://github.com/OpenIntroStat/openintro

R labs

Once you feel comfortable with the material in the tutorials, we also encourage you to apply what you've learned via the computational labs that are also linked at the end of each chapter. The labs consist of data analysis case studies, and they require access to R and RStudio. The first lab includes installation instructions. If you'd rather not install the software locally, you can also try RStudio Cloud for free.

 Labs for each chapter are listed like this.

You can access the full list of labs supporting this book at
https://www.openintro.org/go?id=ims-r-labs.

OpenIntro, online resources, and getting involved

OpenIntro is an organization focused on developing free and affordable education materials.

We encourage anyone learning or teaching statistics to visit openintro.org and to get involved.

All OpenIntro resources are free and anyone is welcomed to use these online tools and resources with or without this textbook as a companion.

We value your feedback. If there is a part of the project you especially like or think needs improvement, we want to hear from you. For feedback on this specific book, you can open an issue on GitHub. You can also provide feedback on this book or any other OpenIntro resource via our contact form at openintro.org.

Acknowledgements

The *OpenIntro* project would not have been possible without the dedication and volunteer hours of all those involved, and we hope you will join us in extending a huge *thank you* to all those who volunteer with OpenIntro.

The authors would like to thank

- David Diez and Christopher Barr for their work on the 1st Edition of this book,
- Ben Baumer and Andrew Bray for their contribution rethinking how and which order we present this material as well as their work as original authors of the interactive tutorial content,
- Yanina Bellini Saibene, Florencia D'Andrea, and Roxana Noelia Villafañe for their work on creating the interactive tutorials in learnr,
- Will Gray for conceptual diagrams,
- Allison Theobold, Melinda Yager, and Randy Prium for their valuable feedback and review of the book,
- Colin Rundel for feedback on content and technical help with conversion from LaTeX to R Markdown,
- Christophe Dervieux for help with multi-output bookdown issues, and
- Müge Çetinkaya and Meenal Patel for their design vision.

We would like to also thank the developers of the open-source tools that make the development and authoring of this book possible, e.g., bookdown, tidyverse, and icons8.

We are also grateful to the many teachers, students, and other readers who have helped improve OpenIntro resources through their feedback.

PART I

INTRODUCTION TO DATA

Chapter 1

Hello data

 Scientists seek to answer questions using rigorous methods and careful observations. These observations – collected from the likes of field notes, surveys, and experiments – form the backbone of a statistical investigation and are called **data**. Statistics is the study of how best to collect, analyze, and draw conclusions from data. In this first chapter, we focus on both the properties of data and on the collection of data.

1.1 Case study: Using stents to prevent strokes

In this section we introduce a classic challenge in statistics: evaluating the efficacy of a medical treatment. Terms in this section, and indeed much of this chapter, will all be revisited later in the text. The plan for now is simply to get a sense of the role statistics can play in practice.

An experiment is designed to study the effectiveness of stents in treating patients at risk of stroke (Chimowitz et al., 2011). Stents are small mesh tubes that are placed inside narrow or weak arteries to assist in patient recovery after cardiac events and reduce the risk of an additional heart attack or death.

Many doctors have hoped that there would be similar benefits for patients at risk of stroke. We start by writing the principal question the researchers hope to answer:

Does the use of stents reduce the risk of stroke?

The researchers who asked this question conducted an experiment with 451 at-risk patients. Each volunteer patient was randomly assigned to one of two groups:

- **Treatment group.** Patients in the treatment group received a stent and medical management. The medical management included medications, management of risk factors, and help in lifestyle modification.
- **Control group.** Patients in the control group received the same medical management as the treatment group, but they did not receive stents.

Researchers randomly assigned 224 patients to the treatment group and 227 to the control group. In this study, the control group provides a reference point against which we can measure the medical impact of stents in the treatment group.

1.1. CASE STUDY: USING STENTS TO PREVENT STROKES

Researchers studied the effect of stents at two time points: 30 days after enrollment and 365 days after enrollment. The results of 5 patients are summarized in Table 1.1. Patient outcomes are recorded as stroke or no event, representing whether or not the patient had a stroke during that time period.

 The stent30 data and stent365 data can be found in the **openintro** R package.

Table 1.1: Results for five patients from the stent study.

patient	group	30 days	365 days
1	treatment	no event	no event
2	treatment	stroke	stroke
3	treatment	no event	no event
4	treatment	no event	no event
5	control	no event	no event

It would be difficult to answer a question on the impact of stents on the occurrence of strokes for **all** of the study patients using these *individual* observations. This question is better addressed by performing a statistical data analysis of *all* of the observations. Table 1.2 summarizes the raw data in a more helpful way. In this table, we can quickly see what happened over the entire study. For instance, to identify the number of patients in the treatment group who had a stroke within 30 days after the treatment, we look in the leftmost column (30 days), at the intersection of treatment and stroke: 33. To identify the number of control patients who did not have a stroke after 365 days after receiving treatment, we look at the rightmost column (365 days), at the intersection of control and no event: 199.

Table 1.2: Descriptive statistics for the stent study.

	30 days		365 days	
Group	Stroke	No event	Stroke	No event
Control	13	214	28	199
Treatment	33	191	45	179
Total	46	405	73	378

 GUIDED PRACTICE

Of the 224 patients in the treatment group, 45 had a stroke by the end of the first year. Using these two numbers, compute the proportion of patients in the treatment group who had a stroke by the end of their first year. (Note: answers to all Guided Practice exercises are provided in footnotes!)[1]

We can compute summary statistics from the table to give us a better idea of how the impact of the stent treatment differed between the two groups. A **summary statistic** is a single number summarizing data from a sample. For instance, the primary results of the study after 1 year could be described by two summary statistics: the proportion of people who had a stroke in the treatment and control groups.

- Proportion who had a stroke in the treatment (stent) group: $45/224 = 0.20 = 20\%$.
- Proportion who had a stroke in the control group: $28/227 = 0.12 = 12\%$.

---

[1]The proportion of the 224 patients who had a stroke within 365 days: $45/224 = 0.20$.

These two summary statistics are useful in looking for differences in the groups, and we are in for a surprise: an additional 8% of patients in the treatment group had a stroke! This is important for two reasons. First, it is contrary to what doctors expected, which was that stents would *reduce* the rate of strokes. Second, it leads to a statistical question: do the data show a "real" difference between the groups?

This second question is subtle. Suppose you flip a coin 100 times. While the chance a coin lands heads in any given coin flip is 50%, we probably won't observe exactly 50 heads. This type of variation is part of almost any type of data generating process. It is possible that the 8% difference in the stent study is due to this natural variation. However, the larger the difference we observe (for a particular sample size), the less believable it is that the difference is due to chance. So what we are really asking is the following: if in fact stents have no effect, how likely is it that we observe such a large difference?

While we don't yet have statistical tools to fully address this question on our own, we can comprehend the conclusions of the published analysis: there was compelling evidence of harm by stents in this study of stroke patients.

Be careful: Do not generalize the results of this study to all patients and all stents. This study looked at patients with very specific characteristics who volunteered to be a part of this study and who may not be representative of all stroke patients. In addition, there are many types of stents and this study only considered the self-expanding Wingspan stent (Boston Scientific). However, this study does leave us with an important lesson: we should keep our eyes open for surprises.

1.2 Data basics

Effective presentation and description of data is a first step in most analyses. This section introduces one structure for organizing data as well as some terminology that will be used throughout this book.

1.2.1 Observations, variables, and data matrices

Table 1.3 displays six rows of a dataset for 50 randomly sampled loans offered through Lending Club, which is a peer-to-peer lending company. This dataset will be referred to as `loan50`.

The `loan50` data can be found in the **openintro** R package.

Each row in the table represents a single loan. The formal name for a row is a **case** or **observational unit**. The columns represent characteristics of each loan, where each column is referred to as a **variable**. For example, the first row represents a loan of $22,000 with an interest rate of 10.90%, where the borrower is based in New Jersey (NJ) and has an income of $59,000.

GUIDED PRACTICE

What is the grade of the first loan in Table 1.3? And what is the home ownership status of the borrower for that first loan? Reminder: for these Guided Practice questions, you can check your answer in the footnote.[2]

In practice, it is especially important to ask clarifying questions to ensure important aspects of the data are understood. For instance, it is always important to be sure we know what each variable means and its units of measurement. Descriptions of the variables in the `loan50` dataset are given in Table 1.4.

---

[2]The loan's grade is B, and the borrower rents their residence.

1.2. DATA BASICS

Table 1.3: Six observations from the `loan50` dataset

	loan_amount	interest_rate	term	grade	state	total_income	homeownership
1	22,000	10.90	60	B	NJ	59,000	rent
2	6,000	9.92	36	B	CA	60,000	rent
3	25,000	26.30	36	E	SC	75,000	mortgage
4	6,000	9.92	36	B	CA	75,000	rent
5	25,000	9.43	60	B	OH	254,000	mortgage
6	6,400	9.92	36	B	IN	67,000	mortgage

Table 1.4: Variables and their descriptions for the `loan50` dataset.

Variable	Description
loan_amount	Amount of the loan received, in US dollars.
interest_rate	Interest rate on the loan, in an annual percentage.
term	The length of the loan, which is always set as a whole number of months.
grade	Loan grade, which takes a values A through G and represents the quality of the loan and its likelihood of being repaid.
state	US state where the borrower resides.
total_income	Borrower's total income, including any second income, in US dollars.
homeownership	Indicates whether the person owns, owns but has a mortgage, or rents.

The data in Table 1.3 represent a **data frame**, which is a convenient and common way to organize data, especially if collecting data in a spreadsheet. A data frame where each row is a unique case (observational unit), each column is a variable, and each cell is a single value is commonly referred to as **tidy data** Wickham (2014).

When recording data, use a tidy data frame unless you have a very good reason to use a different structure. This structure allows new cases to be added as rows or new variables as new columns and facilitates visualization, summarization, and other statistical analyses.

GUIDED PRACTICE

The grades for assignments, quizzes, and exams in a course are often recorded in a gradebook that takes the form of a data frame. How might you organize a course's grade data using a data frame? Describe the observational units and variables.[3]

GUIDED PRACTICE

We consider data for 3,142 counties in the United States, which includes the name of each county, the state where it resides, its population in 2017, the population change from 2010 to 2017, poverty rate, and nine additional characteristics. How might these data be organized in a data frame?[4]

---

[3] There are multiple strategies that can be followed. One common strategy is to have each student represented by a row, and then add a column for each assignment, quiz, or exam. Under this setup, it is easy to review a single line to understand the grade history of a student. There should also be columns to include student information, such as one column to list student names.

[4] Each county may be viewed as a case, and there are eleven pieces of information recorded for each case. A table with 3,142 rows and 14 columns could hold these data, where each row represents a county and each column represents a particular piece of information.

The data described in the Guided Practice above represents the county dataset, which is shown as a data frame in Table 1.5. The variables as well as the variables in the dataset that did not fit in Table 1.5 are described in Table 1.6.

Table 1.5: Six observations and six variables from the county dataset.

name	state	pop2017	pop_change	unemployment_rate	median_edu
Autauga County	Alabama	55,504	1.48	3.86	some_college
Baldwin County	Alabama	212,628	9.19	3.99	some_college
Barbour County	Alabama	25,270	-6.22	5.90	hs_diploma
Bibb County	Alabama	22,668	0.73	4.39	hs_diploma
Blount County	Alabama	58,013	0.68	4.02	hs_diploma
Bullock County	Alabama	10,309	-2.28	4.93	hs_diploma

Table 1.6: Variables and their descriptions for the county dataset.

Variable	Description
name	Name of county.
state	Name of state.
pop2000	Population in 2000.
pop2010	Population in 2010.
pop2017	Population in 2017.
pop_change	Population change from 2010 to 2017 (in percent).
poverty	Percent of population in poverty in 2017.
homeownership	Homeownership rate, 2006-2010.
multi_unit	Multi-unit rate: percent of housing units that are in multi-unit structures, 2006-2010.
unemployment_rate	Unemployment rate in 2017.
metro	Whether the county contains a metropolitan area, taking one of the values yes or no.
median_edu	Median education level (2013-2017), taking one of the values below_hs, hs_diploma, some_college, or bachelors.
per_capita_income	Per capita (per person) income (2013-2017).
median_hh_income	Median household income.
smoking_ban	Describes the type of county-level smoking ban in place in 2010, taking one of the values none, partial, or comprehensive.

 The county data can be found in the **usdata** R package.

1.2.2 Types of variables

Examine the unemployment_rate, pop2017, state, and median_edu variables in the county dataset. Each of these variables is inherently different from the other three, yet some share certain characteristics.

First consider unemployment_rate, which is said to be a **numerical** variable since it can take a wide range of numerical values, and it is sensible to add, subtract, or take averages with those values. On the other hand, we would not classify a variable reporting telephone area codes as numerical since the average, sum, and difference of area codes doesn't have any clear meaning. Instead, we would consider area codes as a categorical variable.

The pop2017 variable is also numerical, although it seems to be a little different than unemployment_rate. This variable of the population count can only take whole non-negative numbers (0, 1, 2, ...).

1.2. DATA BASICS

For this reason, the population variable is said to be **discrete** since it can only take numerical values with jumps. On the other hand, the unemployment rate variable is said to be **continuous**.

The variable state can take up to 51 values after accounting for Washington, DC: AL, AK, ..., and WY. Because the responses themselves are categories, state is called a **categorical** variable, and the possible values (states) are called the variable's **levels** (e.g., DC, AL, AK, etc.) .

Finally, consider the median_edu variable, which describes the median education level of county residents and takes values below_hs, hs_diploma, some_college, or bachelors in each county. This variable seems to be a hybrid: it is a categorical variable but the levels have a natural ordering. A variable with these properties is called an **ordinal** variable, while a regular categorical variable without this type of special ordering is called a **nominal** variable. To simplify analyses, any categorical variable in this book will be treated as a nominal (unordered) categorical variable.

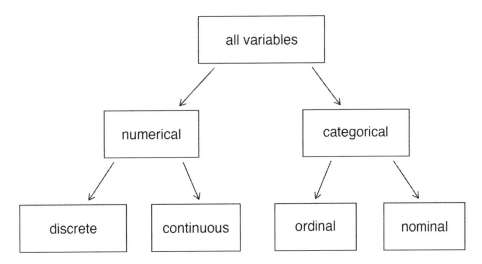

Figure 1.1: Breakdown of variables into their respective types.

 EXAMPLE

Data were collected about students in a statistics course. Three variables were recorded for each student: number of siblings, student height, and whether the student had previously taken a statistics course. Classify each of the variables as continuous numerical, discrete numerical, or categorical.

The number of siblings and student height represent numerical variables. Because the number of siblings is a count, it is discrete. Height varies continuously, so it is a continuous numerical variable. The last variable classifies students into two categories – those who have and those who have not taken a statistics course – which makes this variable categorical.

 GUIDED PRACTICE

An experiment is evaluating the effectiveness of a new drug in treating migraines. A group variable is used to indicate the experiment group for each patient: treatment or control. The num_migraines variable represents the number of migraines the patient experienced during a 3-month period. Classify each variable as either numerical or categorical?[5]

---

[5]The group variable can take just one of two group names, making it categorical. The num_migraines variable describes a count of the number of migraines, which is an outcome where basic arithmetic is sensible, which means this is numerical outcome; more specifically, since it represents a count, num_migraines is a discrete numerical variable.

1.2.3 Relationships between variables

Many analyses are motivated by a researcher looking for a relationship between two or more variables. A social scientist may like to answer some of the following questions:

> Does a higher than average increase in county population tend to correspond to counties with higher or lower median household incomes?
>
> If homeownership is lower than the national average in one county, will the percent of housing units that are in multi-unit structures in that county tend to be above or below the national average?
>
> How much can the median education level explain the median household income for counties in the US?

To answer these questions, data must be collected, such as the county dataset shown in Table 1.5. Examining **summary statistics** can provide numerical insights about the specifics of each of these questions. Alternatively, graphs can be used to visually explore the data, potentially providing more insight than a summary statistic.

Scatterplots are one type of graph used to study the relationship between two numerical variables. Figure 1.2 displays the relationship between the variables homeownership and multi_unit, which is the percent of housing units that are in multi-unit structures (e.g., apartments, condos). Each point on the plot represents a single county. For instance, the highlighted dot corresponds to County 413 in the county dataset: Chattahoochee County, Georgia, which has 39.4% of housing units that are in multi-unit structures and a homeownership rate of 31.3%. The scatterplot suggests a relationship between the two variables: counties with a higher rate of housing units that are in multi-unit structures tend to have lower homeownership rates. We might brainstorm as to why this relationship exists and investigate each idea to determine which are the most reasonable explanations.

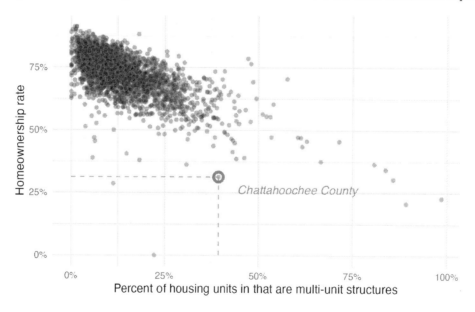

Figure 1.2: A scatterplot of homeownership versus the percent of housing units that are in multi-unit structures for US counties. The highlighted dot represents Chattahoochee County, Georgia, which has a multi-unit rate of 39.4% and a homeownership rate of 31.3%.

The multi-unit and homeownership rates are said to be associated because the plot shows a discernible pattern. When two variables show some connection with one another, they are called **associated** variables.

1.2. DATA BASICS

GUIDED PRACTICE

Examine the variables in the `loan50` dataset, which are described in Table 1.4. Create two questions about possible relationships between variables in `loan50` that are of interest to you.[6]

EXAMPLE

This example examines the relationship between the percent change in population from 2010 to 2017 and median household income for counties, which is visualized as a scatterplot in Figure 1.3. Are these variables associated?

The larger the median household income for a county, the higher the population growth observed for the county. While it isn't true that every county with a higher median household income has a higher population growth, the trend in the plot is evident. Since there is some relationship between the variables, they are associated.

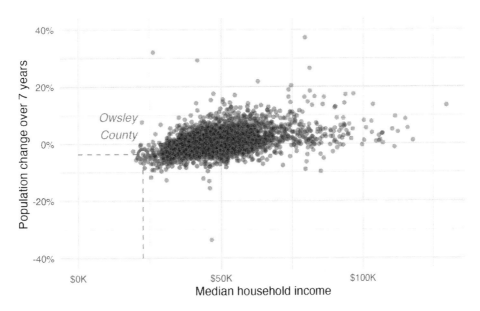

Figure 1.3: A scatterplot showing population chance against median household income. Owsley County of Kentucky, is highlighted, which lost 3.63% of its population from 2010 to 2017 and had median household income of $22,736.

Because there is a downward trend in Figure 1.2 – counties with more housing units that are in multi-unit structures are associated with lower homeownership – these variables are said to be **negatively associated**. A **positive association** is shown in the relationship between the `median_hh_income` and `pop_change` variables in Figure 1.3, where counties with higher median household income tend to have higher rates of population growth.

If two variables are not associated, then they are said to be **independent**. That is, two variables are independent if there is no evident relationship between the two.

---

[6]Two example questions: (1) What is the relationship between loan amount and total income? (2) If someone's income is above the average, will their interest rate tend to be above or below the average?

Associated or independent, not both.

A pair of variables are either related in some way (associated) or not (independent). No pair of variables is both associated and independent.

1.2.4 Explanatory and response variables

When we ask questions about the relationship between two variables, we sometimes also want to determine if the change in one variable causes a change in the other. Consider the following rephrasing of an earlier question about the county dataset:

> If there is an increase in the median household income in a county, does this drive an increase in its population?

In this question, we are asking whether one variable affects another. If this is our underlying belief, then *median household income* is the **explanatory variable** and the *population change* is the **response variable** in the hypothesized relationship.[7]

Explanatory and response variables.

When we suspect one variable might causally affect another, we label the first variable the explanatory variable and the second the response variable. We also use the terms **explanatory** and **response** to describe variables where the **response** might be predicted using the **explanatory** even if there is no causal relationship.

explanatory variable → *might affect* → response variable

For many pairs of variables, there is no hypothesized relationship, and these labels would not be applied to either variable in such cases.

Bear in mind that the act of labeling the variables in this way does nothing to guarantee that a causal relationship exists. A formal evaluation to check whether one variable causes a change in another requires an experiment.

1.2.5 Observational studies and experiments

There are two primary types of data collection: experiments and observational studies.

When researchers want to evaluate the effect of particular traits, treatments, or conditions, they conduct an **experiment**. For instance, we may suspect drinking a high-calorie energy drink will improve performance in a race. To check if there really is a causal relationship between the explanatory variable (whether the runner drank an energy drink or not) and the response variable (the race time), researchers identify a sample of individuals and split them into groups. The individuals in each group are *assigned* a treatment. When individuals are randomly assigned to a group, the experiment is called a **randomized experiment**. Random assignment organizes the participants in a study into groups that are roughly equal on all aspects, thus allowing us to control for any confounding variables that might affect the outcome (e.g., fitness level, racing experience, etc.). For example, each runner in the experiment could be randomly assigned, perhaps by flipping a coin, into one of two groups: the first group receives a **placebo** (fake treatment, in this case a no-calorie drink) and the second group receives the high-calorie energy drink. See the case study in Section 1.1 for another example of an experiment, though that study did not employ a placebo.

---

[7]In some disciplines, it's customary to refer to the explanatory variable as the **independent variable** and the response variable as the **dependent variable**. However, this becomes confusing since a *pair* of variables might be independent or dependent, so we avoid this language.

Researchers perform an **observational study** when they collect data in a way that does not directly interfere with how the data arise. For instance, researchers may collect information via surveys, review medical or company records, or follow a **cohort** of many similar individuals to form hypotheses about why certain diseases might develop. In each of these situations, researchers merely observe the data that arise. In general, observational studies can provide evidence of a naturally occurring association between variables, but they cannot by themselves show a causal connection as they don't offer a mechanism for controlling for confounding variables.

> **Association $\neq$ Causation.**
>
> In general, association does not imply causation. An advantage of a randomized experiment is that it is easier to establish causal relationships with such a study. The main reason for this is that observational studies do not control for confounding variables, and hence establishing causal relationships with observational studies requires advanced statistical methods (that are beyond the scope of this book). We will revisit this idea when we discuss experiments later in the book.

1.3 Chapter review

1.3.1 Summary

This chapter introduced you to the world of data. Data can be organized in many ways but tidy data, where each row represents an observation and each column represents a variable, lends itself most easily to statistical analysis. Many of the ideas from this chapter will be seen as we move on to doing full data analyses. In the next chapter you're going to learn about how we can design studies to collect the data we need to make conclusions with the desired scope of inference.

1.3.2 Terms

We introduced the following terms in the chapter. If you're not sure what some of these terms mean, we recommend you go back in the text and review their definitions. We are purposefully presenting them in alphabetical order, instead of in order of appearance, so they will be a little more challenging to locate. However you should be able to easily spot them as **bolded text**.

associated	experiment	ordinal
case	explanatory variable	placebo
categorical	independent	positive association
cohort	level	randomized experiment
continuous	negative association	response variable
data	nominal	summary statistic
data frame	numerical	tidy data
dependent	observational study	variable
discrete	observational unit	

1.4 Exercises

Answers to odd numbered exercises can be found in Appendix A.1.

1.1. **Marvel Cinematic Universe films.** The data frame below contains information on Marvel Cinematic Universe films through the Infinity saga (a movie storyline spanning from Ironman in 2008 to Endgame in 2019). Box office totals are given in millions of US Dollars. How many observations and how many variables does this data frame have?[8]

	Title	Length		Release Date	Opening Wknd US	Gross	
		Hrs	Mins			US	World
1	Iron Man	2	6	5/2/2008	98.62	319.03	585.8
2	The Incredible Hulk	1	52	6/12/2008	55.41	134.81	264.77
3	Iron Man 2	2	4	5/7/2010	128.12	312.43	623.93
4	Thor	1	55	5/6/2011	65.72	181.03	449.33
5	Captain America: The First Avenger	2	4	7/22/2011	65.06	176.65	370.57
...	...	...	...	...	...	...	...
23	Spiderman: Far from Home	2	9	7/2/2019	92.58	390.53	1131.93

1.2. **Cherry Blossom Run.** The data frame below contains information on runners in the 2017 Cherry Blossom Run, which is an annual road race that takes place in Washington, DC. Most runners participate in a 10-mile run while a smaller fraction take part in a 5k run or walk. How many observations and how many variables does this data frame have?[9]

	Bib	Name	Sex	Age	City / Country	Time		Pace	Event
						Net	Clock		
1	6	Hiwot G.	F	21	Ethiopia	3217	3217	321	10 Mile
2	22	Buze D.	F	22	Ethiopia	3232	3232	323	10 Mile
3	16	Gladys K.	F	31	Kenya	3276	3276	327	10 Mile
4	4	Mamitu D.	F	33	Ethiopia	3285	3285	328	10 Mile
5	20	Karolina N.	F	35	Poland	3288	3288	328	10 Mile
...	...	...	...	...	...	...	...	...	...
19961	25153	Andres E.	M	33	Woodbridge, VA	5287	5334	1700	5K

1.3. **Air pollution and birth outcomes, study components.** Researchers collected data to examine the relationship between air pollutants and preterm births in Southern California. During the study air pollution levels were measured by air quality monitoring stations. Specifically, levels of carbon monoxide were recorded in parts per million, nitrogen dioxide and ozone in parts per hundred million, and coarse particulate matter (PM_{10}) in $\mu g/m^3$. Length of gestation data were collected on 143,196 births between the years 1989 and 1993, and air pollution exposure during gestation was calculated for each birth. The analysis suggested that increased ambient PM_{10} and, to a lesser degree, CO concentrations may be associated with the occurrence of preterm births. (Ritz et al., 2000)

  a. Identify the main research question of the study.

  b. Who are the subjects in this study, and how many are included?

  c. What are the variables in the study? Identify each variable as numerical or categorical. If numerical, state whether the variable is discrete or continuous. If categorical, state whether the variable is ordinal.

---

[8] The mcu_films data used in this exercise can be found in the **openintro** R package.

[9] The run17 data used in this exercise can be found in the **cherryblossom** R package.

1.4. EXERCISES

1.4. **Cheaters, study components.** Researchers studying the relationship between honesty, age and self-control conducted an experiment on 160 children between the ages of 5 and 15. Participants reported their age, sex, and whether they were an only child or not. The researchers asked each child to toss a fair coin in private and to record the outcome (white or black) on a paper sheet, and said they would only reward children who report white. (Bucciol and Piovesan, 2011)

  a. Identify the main research question of the study.

  b. Who are the subjects in this study, and how many are included?

  c. The study's findings can be summarized as follows: *"Half the students were explicitly told not to cheat and the others were not given any explicit instructions. In the no instruction group probability of cheating was found to be uniform across groups based on child's characteristics. In the group that was explicitly told to not cheat, girls were less likely to cheat, and while rate of cheating didn't vary by age for boys, it decreased with age for girls."* How many variables were recorded for each subject in the study in order to conclude these findings? State the variables and their types.

1.5. **Gamification and statistics, study components.** Gamification is the application of game-design elements and game principles in non-game contexts. In educational settings, gamification is often implemented as educational activities to solve problems by using characteristics of game elements. Researchers investigating the effects of gamification on learning statistics conducted a study where they split college students in a statistics class into four groups: (1) no reading exercises and no gamification, (2) reading exercises but no gamification, (3) gamification but no reading exercises, and (4) gamification and reading exercises. Students in all groups also attended lectures. Students in the class were from two majors: Electrical and Computer Engineering (n = 279) and Business Administration (n = 86). After their assigned learning experience, each student took a final evaluation comprised of 30 multiple choice question and their score was measured as the number of questions they answered correctly. The researchers considered students' gender, level of studies (first through fourth year) and academic major. Other variables considered were expertise in the English language and use of personal computers and games, both of which were measured on a scale of 1 (beginner) to 5 (proficient). The study found that gamification had a positive effect on student learning compared to traditional teaching methods involving lectures and reading exercises. They also found that the effect was larger for females and Engineering students. (Legaki et al., 2020)

  a. Identify the main research question of the study.

  b. Who were the subjects in this study, and how many were included?

  c. What are the variables in the study? Identify each variable as numerical or categorical. If numerical, state whether the variable is discrete or continuous. If categorical, state whether the variable is ordinal.

1.6. **Stealers, study components.** In a study of the relationship between socio-economic class and unethical behavior, 129 University of California undergraduates at Berkeley were asked to identify themselves as having low or high social-class by comparing themselves to others with the most (least) money, most (least) education, and most (least) respected jobs. They were also presented with a jar of individually wrapped candies and informed that the candies were for children in a nearby laboratory, but that they could take some if they wanted. After completing some unrelated tasks, participants reported the number of candies they had taken. (Piff et al., 2012)

  a. Identify the main research question of the study.

  b. Who were the subjects in this study, and how many were included?

  c. The study found that students who were identified as upper-class took more candy than others. How many variables were recorded for each subject in the study in order to conclude these findings? State the variables and their types.

1.7. **Migraine and acupuncture.** A migraine is a particularly painful type of headache, which patients sometimes wish to treat with acupuncture. To determine whether acupuncture relieves migraine pain, researchers conducted a randomized controlled study where 89 individuals who identified as female diagnosed with migraine headaches were randomly assigned to one of two groups: treatment or control. Forty-three (43) patients in the treatment group received acupuncture that is specifically designed to treat migraines. Forty-six (46) patients in the control group received placebo acupuncture (needle insertion at non-acupoint locations). Twenty-four (24) hours after patients received acupuncture, they were asked if they were pain free. Results are summarized in the contingency table below. Also provided is a figure from the original paper displaying the appropriate area (M) versus the inappropriate area (S) used in the treatment of migraine attacks.[10] (Allais et al., 2011)

	Pain free?	
Group	No	Yes
Control	44	2
Treatment	33	10

a. What percent of patients in the treatment group were pain free 24 hours after receiving acupuncture?

b. What percent were pain free in the control group?

c. In which group did a higher percent of patients become pain free 24 hours after receiving acupuncture?

d. Your findings so far might suggest that acupuncture is an effective treatment for migraines for all people who suffer from migraines. However this is not the only possible conclusion. What is one other possible explanation for the observed difference between the percentages of patients that are pain free 24 hours after receiving acupuncture in the two groups?

e. What are the explanatory and response variables in this study?

1.8. **Sinusitis and antibiotics.** Researchers studying the effect of antibiotic treatment for acute sinusitis compared to symptomatic treatments randomly assigned 166 adults diagnosed with acute sinusitis to one of two groups: treatment or control. Study participants received either a 10-day course of amoxicillin (an antibiotic) or a placebo similar in appearance and taste. The placebo consisted of symptomatic treatments such as acetaminophen, nasal decongestants, etc. At the end of the 10-day period, patients were asked if they experienced improvement in symptoms. The distribution of responses is summarized below.[11] (Garbutt et al., 2012)

	Improvement	
Group	No	Yes
Control	16	65
Treatment	19	66

a. What percent of patients in the treatment group experienced improvement in symptoms?

b. What percent experienced improvement in symptoms in the control group?

c. In which group did a higher percentage of patients experience improvement in symptoms?

d. Your findings so far might suggest a real difference in the effectiveness of antibiotic and placebo treatments for improving symptoms of sinusitis. However this is not the only possible conclusion. What is one other possible explanation for the observed difference between the percentages patients who experienced improvement in symptoms?

e. What are the explanatory and response variables in this study?

---

[10] The `migraine` data used in this exercise can be found in the **openintro** R package.

[11] The `sinusitis` data used in this exercise can be found in the **openintro** R package.

1.9. **Daycare fines, study components.** Researchers tested the deterrence hypothesis which predicts that the introduction of a penalty will reduce the occurrence of the behavior subject to the fine, with the condition that the fine leaves everything else unchanged by instituting a fine for late pickup at daycare centers. For this study, they worked with 10 volunteer daycare centers that did not originally impose a fine to parents for picking up their kids late. They randomly selected 6 of these daycare centers and instituted a monetary fine (of a considerable amount) for picking up children late and then removed it. In the remaining 4 daycare centers no fine was introduced. The study period was divided into four: before the fine (weeks 1–4), the first 4 weeks with the fine (weeks 5-8), the last 8 weeks with fine (weeks 9–16), and the after fine period (weeks 17-20). Throughout the study, the number of kids who were picked up late was recorded each week for each daycare. The study found that the number of late-coming parents increased significantly when the fine was introduced, and no reduction occurred after the fine was removed.[12] (Gneezy and Rustichini, 2000)

center	week	group	late_pickups	study_period
1	1	test	8	before fine
1	2	test	8	before fine
1	3	test	7	before fine
1	4	test	6	before fine
1	5	test	8	first 4 weeks with fine
...	...	...	...	...
10	20	control	13	after fine

a. Is this an observational study or an experiment? Explain your reasoning.

b. What are the cases in this study and how many are included?

c. What is the response variable in the study and what type of variable is it?

d. What are the explanatory variables in the study and what types of variables are they?

1.10. **Efficacy of COVID-19 vaccine on adolescents, study components.** Results of a Phase 3 trial announced in March 2021 show that the Pfizer-BioNTech COVID-19 vaccine demonstrated 100% efficacy and robust antibody responses on 12 to 15 years old adolescents with or without prior evidence of SARS-CoV-2 infection. In this trial 2,260 adolescents were randomly assigned to two groups: one group got the vaccine (n = 1,131) and the other got a placebo (n = 1,129). While 18 cases of COVID-19 were observed in the placebo group, none were observed in the vaccine group.[13] (Pfizer, 2021)

a. Is this an observational study or an experiment? Explain your reasoning.

b. What are the cases in this study and how many are included?

c. What is the response variable in the study and what type of variable is it?

d. What are the explanatory variables in the study and what types of variables are they?

---

[12]The `daycare_fines` data used in this exercise can be found in the **openintro** R package.
[13]The `biontech_adolescents` data used in this exercise can be found in the **openintro** R package.

1.11. **Palmer penguins.** Data were collected on 344 penguins living on three islands (Torgersen, Biscoe, and Dream) in the Palmer Archipelago, Antarctica. In addition to which island each penguin lives on, the data contains information on the species of the penguin (*Adelie*, *Chinstrap*, or *Gentoo*), its bill length, bill depth, and flipper length (measured in millimeters), its body mass (measured in grams), and the sex of the penguin (female or male).[14] Bill length and depth are measured as shown in the image.[15] (Gorman et al., 2014a)

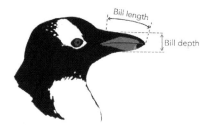

a. How many cases were included in the data?

b. How many numerical variables are included in the data? Indicate what they are, and if they are continuous or discrete.

c. How many categorical variables are included in the data, and what are they? List the corresponding levels (categories) for each.

1.12. **Smoking habits of UK residents.** A survey was conducted to study the smoking habits of 1,691 UK residents. Below is a data frame displaying a portion of the data collected in this survey. A blank cell indicates that data for that variable was not available for a given respondent.[16]

	sex	age	marital_status	gross_income	smoke	amount weekend	weekday
1	Female	61	Married	2,600 to 5,200	No		
2	Female	61	Divorced	10,400 to 15,600	Yes	5	4
3	Female	69	Widowed	5,200 to 10,400	No		
4	Female	50	Married	5,200 to 10,400	No		
5	Male	31	Single	10,400 to 15,600	Yes	10	20
...	...	...	...	...	...		
1691	Male	49	Divorced	Above 36,400	Yes	15	10

a. What does each row of the data frame represent?

b. How many participants were included in the survey?

c. Indicate whether each variable in the study is numerical or categorical. If numerical, identify as continuous or discrete. If categorical, indicate if the variable is ordinal.

---

[14] The penguins data used in this exercise can be found in the **palmerpenguins** R package.
[15] Artwork by Allison Horst.
[16] The smoking data used in this exercise can be found in the **openintro** R package.

1.13. **US Airports.** The visualization below shows the geographical distribution of airports in the contiguous United States and Washington, DC. This visualization was constructed based on a dataset where each observation is an airport.[17]

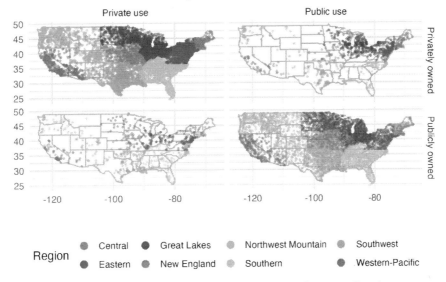

a. List the variables you believe were necessary to create this visualization.

b. Indicate whether each variable in the study is numerical or categorical. If numerical, identify as continuous or discrete. If categorical, indicate if the variable is ordinal.

1.14. **UN Votes.** The visualization below shows voting patterns in the United States, Canada, and Mexico in the United Nations General Assembly on a variety of issues. Specifically, for a given year between 1946 and 2019, it displays the percentage of roll calls in which the country voted yes for each issue. This visualization was constructed based on a dataset where each observation is a country/year pair.[18]

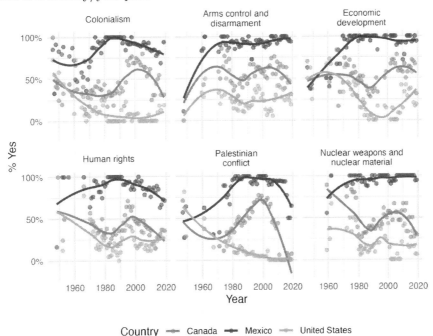

a. List the variables used in creating this visualization.

b. Indicate whether each variable in the study is numerical or categorical. If numerical, identify as continuous or discrete. If categorical, indicate if the variable is ordinal.

---

[17] The usairports data used in this exercise can be found in the **airports** R package.
[18] The data used in this exercise can be found in the **unvotes** R package.

1.15. **UK baby names.** The visualization below shows the number of baby girls born in the United Kingdom (comprised of England & Wales, Northern Ireland, and Scotland) who were given the name "Fiona" over the years.[19]

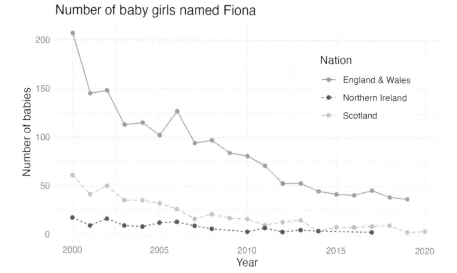

a. List the variables you believe were necessary to create this visualization.

b. Indicate whether each variable in the study is numerical or categorical. If numerical, identify as continuous or discrete. If categorical, indicate if the variable is ordinal.

1.16. **Shows on Netflix.** The visualization below shows the distribution of ratings of TV shows on Netflix (a streaming entertainment service) based on the decade they were released in and the country they were produced in. In the dataset, each observation is a TV show.[20]

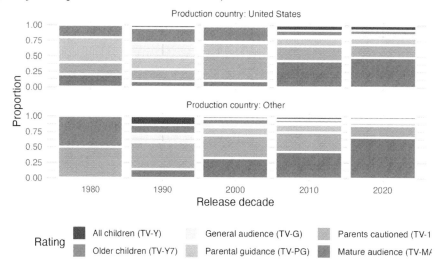

a. List the variables you believe were necessary to create this visualization.

b. Indicate whether each variable in the study is numerical or categorical. If numerical, identify as continuous or discrete. If categorical, indicate if the variable is ordinal.

---

[19]The ukbabynames data used in this exercise can be found in the **ukbabynames** R package.

[20]The netflix_titles data used in this exercise can be found in the **tidytuesdayR** R package.

1.17. **Stanford Open Policing.** The Stanford Open Policing project gathers, analyzes, and releases records from traffic stops by law enforcement agencies across the United States. Their goal is to help researchers, journalists, and policy makers investigate and improve interactions between police and the public. The following is an excerpt from a summary table created based off of the data collected as part of this project. (Pierson et al., 2020)

		Driver		Car	
County	State	Race / Ethnicity	Arrest rate	Stops / year	Search rate
Apache County	AZ	Black	0.016	266	0.077
Apache County	AZ	Hispanic	0.018	1008	0.053
Apache County	AZ	White	0.006	6322	0.017
Cochise County	AZ	Black	0.015	1169	0.047
Cochise County	AZ	Hispanic	0.01	9453	0.037
Cochise County	AZ	White	0.008	10826	0.024
...	...	...	...	...	...
Wood County	WI	Black	0.098	16	0.244
Wood County	WI	Hispanic	0.029	27	0.036
Wood County	WI	White	0.029	1157	0.033

a. What variables were collected on each individual traffic stop in order to create the summary table above?

b. State whether each variable is numerical or categorical. If numerical, state whether it is continuous or discrete. If categorical, state whether it is ordinal or not.

c. Suppose we wanted to evaluate whether vehicle search rates are different for drivers of different races. In this analysis, which variable would be the response variable and which variable would be the explanatory variable?

1.18. **Space launches.** The following summary table shows the number of space launches in the US by the type of launching agency and the outcome of the launch (success or failure).[21]

	1957 - 1999		2000-2018	
	Failure	Success	Failure	Success
Privste	13	295	10	562
State	281	3751	33	711
Startup	0	0	5	65

a. What variables were collected on each launch in order to create to the summary table above?

b. State whether each variable is numerical or categorical. If numerical, state whether it is continuous or discrete. If categorical, state whether it is ordinal or not.

c. Suppose we wanted to study how the success rate of launches vary between launching agencies and over time. In this analysis, which variable would be the response variable and which variable would be the explanatory variable?

---

[21] The data used in this exercise comes from the JSR Launch Vehicle Database, 2019 Feb 10 Edition.

1.19. **Pet names.** The city of Seattle, WA has an open data portal that includes pets registered in the city. For each registered pet, we have information on the pet's name and species. The following visualization plots the proportion of dogs with a given name versus the proportion of cats with the same name. The 20 most common cat and dog names are displayed. The diagonal line on the plot is the $x = y$ line; if a name appeared on this line, the name's popularity would be exactly the same for dogs and cats.[22]

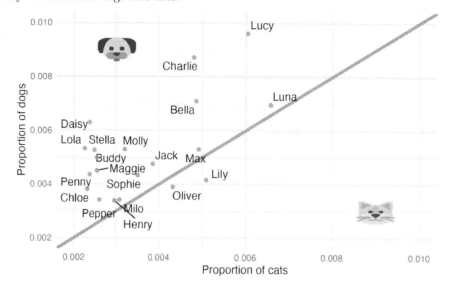

a. Are these data collected as part of an experiment or an observational study?

b. What is the most common dog name? What is the most common cat name?

c. What names are more common for cats than dogs?

d. Is the relationship between the two variables positive or negative? What does this mean in context of the data?

1.20. **Stressed out in an elevator.** In a study evaluating the relationship between stress and muscle cramps, half the subjects are randomly assigned to be exposed to increased stress by being placed into an elevator that falls rapidly and stops abruptly and the other half are left at no or baseline stress.

a. What type of study is this?

b. Can this study be used to conclude a causal relationship between increased stress and muscle cramps?

---

[22]The `seattlepets` data used in this exercise can be found in the **openintro** R package.

Chapter 2

Study design

 Before digging in to the details of working with data, we stop to think about how data come to be. That is, if the data are to be used to make broad and complete conclusions, then it is important to understand who or what the data represent. One important aspect of data provenance is sampling. Knowing how the observational units were selected from a larger entity will allow for generalizations back to the population from which the data were randomly selected. Additionally, by understanding the structure of the study, causal relationships can be separated from those relationships which are only associated. A good question to ask oneself before working with the data at all is, "How were these observations collected?". You will learn a lot about the data by understanding its source.

2.1 Sampling principles and strategies

The first step in conducting research is to identify topics or questions that are to be investigated. A clearly laid out research question is helpful in identifying what subjects or cases should be studied and what variables are important. It is also important to consider *how* data are collected so that the data are reliable and help achieve the research goals.

2.1.1 Populations and samples

Consider the following three research questions:

1. What is the average mercury content in swordfish in the Atlantic Ocean?
2. Over the last five years, what is the average time to complete a degree for Duke undergrads?
3. Does a new drug reduce the number of deaths in patients with severe heart disease?

Each research question refers to a target **population**. In the first question, the target population is all swordfish in the Atlantic ocean, and each fish represents a case. Often times, it is not feasible to collect data for every case in a population. Collecting data for an entire population is called a **census**. A census is difficult because it is too expensive to collect data for the entire population, but it might also be because it is difficult or impossible to identify the entire population of interest! Instead, a sample is taken. A **sample** is the data you have. Ideally, a sample is a small fraction of the population. For instance, 60 swordfish (or some other number) in the population might be selected, and this sample data may be used to provide an estimate of the population average and to answer the research question.

GUIDED PRACTICE

For the second and third questions above, identify the target population and what represents an individual case.[1]

2.1.2 Parameters and statistics

In the majority of statistical analysis procedures, the research question at hand boils down to understanding a numerical summary. The number (or set of numbers) may be a quantity you are already familiar with (like the average) or it may be something you learn through this text (like the slope and intercept from a least squares model, provided in Section 7.2).

A numerical summary can be calculated on either the sample of observation or the entire population. However, measuring every unit in the population is usually prohibitive (so the parameter is very rarely calculated). So, a "typical" numerical summary is calculated from a sample. Yet, we can still conceptualize calculating the average income of all adults in Argentina.

We use specific terms in order to differentiate when a number is being calculated on a sample of data (**statistic**) and when it is being calculated or considered for calculation on the entire population (**parameter**). The terms statistic and parameter are useful for communicating claims and models and will be used extensively in later chapters which delve into making inference on populations.

2.1.3 Anecdotal evidence

Consider the following possible responses to the three research questions:

1. A man on the news got mercury poisoning from eating swordfish, so the average mercury concentration in swordfish must be dangerously high.
2. I met two students who took more than 7 years to graduate from Duke, so it must take longer to graduate at Duke than at many other colleges.
3. My friend's dad had a heart attack and died after they gave him a new heart disease drug, so the drug must not work.

Each conclusion is based on data. However, there are two problems. First, the data only represent one or two cases. Second, and more importantly, it is unclear whether these cases are actually representative of the population. Data collected in this haphazard fashion are called **anecdotal evidence**.

Anecdotal evidence.

Be careful of data collected in a haphazard fashion. Such evidence may be true and verifiable, but it may only represent extraordinary cases and therefore not be a good representation of the population.

Anecdotal evidence typically is composed of unusual cases that we recall based on their striking characteristics. For instance, we are more likely to remember the two people we met who took 7

---

[1]The question *"Over the last five years, what is the average time to complete a degree for Duke undergrads?"* is only relevant to students who complete their degree; the average cannot be computed using a student who never finished their degree. Thus, only Duke undergrads who graduated in the last five years represent cases in the population under consideration. Each such student is an individual case. For the question *"Does a new drug reduce the number of deaths in patients with severe heart disease?"*, a person with severe heart disease represents a case. The population includes all people with severe heart disease.

2.1. SAMPLING PRINCIPLES AND STRATEGIES

Figure 2.1: In February 2010, some media pundits cited one large snow storm as evidence against global warming. As comedian Jon Stewart pointed out, "It is one storm, in one region, of one country."

years to graduate than the six others who graduated in four years. Instead of looking at the most unusual cases, we should examine a sample of many cases that better represent the population.

2.1.4 Sampling from a population

We might try to estimate the time to graduation for Duke undergraduates in the last five years by collecting a sample of graduates. All graduates in the last five years represent the *population*, and graduates who are selected for review are collectively called the *sample*. In general, we always seek to *randomly* select a sample from a population. The most basic type of random selection is equivalent to how raffles are conducted. For example, in selecting graduates, we could write each graduate's name on a raffle ticket and draw 10 tickets. The selected names would represent a random sample of 10 graduates.

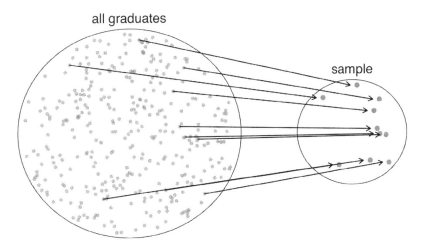

Figure 2.2: In this graphic, 10 graduates are randomly selected from the population to be included in the sample.

EXAMPLE

Suppose we ask a student who happens to be majoring in nutrition to select several graduates for the study. What kind of students do you think they might collect? Do you think their sample would be representative of all graduates?

Perhaps they would pick a disproportionate number of graduates from health-related fields. Or perhaps their selection would be a good representation of the population. When selecting samples by hand, we run the risk of picking a **biased** sample, even if our bias is unintended.

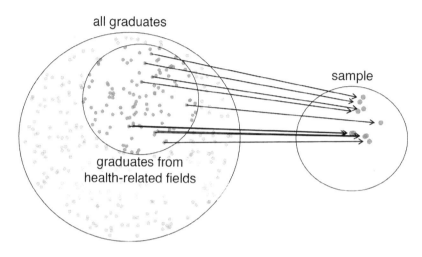

Figure 2.3: Asked to pick a sample of graduates, a nutrition major might inadvertently pick a disproportionate number of graduates from health-related majors.

If someone was permitted to pick and choose exactly which graduates were included in the sample, it is entirely possible that the sample would overrepresent that person's interests, which may be entirely unintentional. This introduces **bias** into a sample. Sampling randomly helps address this problem. The most basic random sample is called a **simple random sample**, and is equivalent to drawing names out of a hat to select cases. This means that each case in the population has an equal chance of being included and the cases in the sample are not related to each other.

The act of taking a simple random sample helps minimize bias. However, bias can crop up in other ways. Even when people are picked at random, e.g., for surveys, caution must be exercised if the **non-response rate** is high. For instance, if only 30% of the people randomly sampled for a survey actually respond, then it is unclear whether the results are **representative** of the entire population. This **non-response bias** can skew results.

Another common downfall is a **convenience sample**, where individuals who are easily accessible are more likely to be included in the sample. For instance, if a political survey is done by stopping people walking in the Bronx, this will not represent all of New York City. It is often difficult to discern what sub-population a convenience sample represents.

GUIDED PRACTICE

We can easily access ratings for products, sellers, and companies through websites. These ratings are based only on those people who go out of their way to provide a rating. If 50% of online reviews for a product are negative, do you think this means that 50% of buyers are dissatisfied with the product? Why or why not?[2]

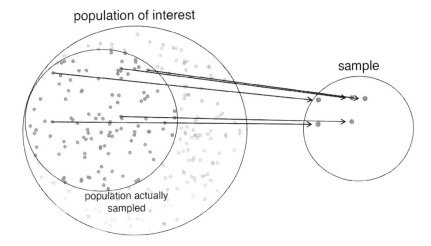

Figure 2.4: Due to the possibility of non-response, survey studies may only reach a certain group within the population. It is difficult, and often times impossible, to completely fix this problem.

2.1.5 Four sampling methods

Almost all statistical methods are based on the notion of implied randomness. If data are not collected in a random framework from a population, these statistical methods – the estimates and errors associated with the estimates – are not reliable. Here we consider four random sampling techniques: simple, stratified, cluster, and multistage sampling. Figures 2.5 and 2.6 provide graphical representations of these techniques.

Simple random sampling is probably the most intuitive form of random sampling. Consider the salaries of Major League Baseball (MLB) players, where each player is a member of one of the league's 30 teams. To take a simple random sample of 120 baseball players and their salaries, we could write the names of that season's several hundreds of players onto slips of paper, drop the slips into a bucket, shake the bucket around until we are sure the names are all mixed up, then draw out slips until we have the sample of 120 players. In general, a sample is referred to as "simple random" if each case in the population has an equal chance of being included in the final sample *and* knowing that a case is included in a sample does not provide useful information about which other cases are included.

Stratified sampling is a divide-and-conquer sampling strategy. The population is divided into groups called **strata**. The strata are chosen so that similar cases are grouped together, then a second sampling method, usually simple random sampling, is employed within each stratum. In the baseball salary example, each of the 30 teams could represent a strata, since some teams have a lot more money (up to 4 times as much!). Then we might randomly sample 4 players from each team for our sample of 120 players.

Stratified sampling is especially useful when the cases in each stratum are very similar with respect to the outcome of interest. The downside is that analyzing data from a stratified sample is a more complex task than analyzing data from a simple random sample. The analysis methods introduced in this book would need to be extended to analyze data collected using stratified sampling.

---

[2] Answers will vary. From our own anecdotal experiences, we believe people tend to rant more about products that fell below expectations than rave about those that perform as expected. For this reason, we suspect there is a negative bias in product ratings on sites like Amazon. However, since our experiences may not be representative, we also keep an open mind.

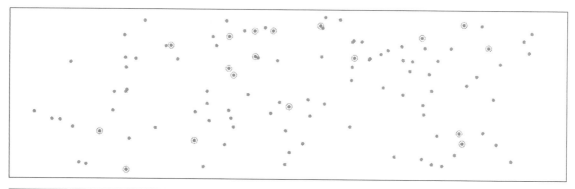

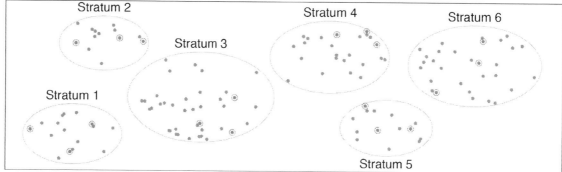

Figure 2.5: Examples of simple random and stratified sampling. In the top panel, simple random sampling was used to randomly select the 18 cases (denoted in red). In the bottom panel, stratified sampling was used: cases were first grouped into strata, then simple random sampling was employed to randomly select 3 cases within each stratum.

 EXAMPLE

Why would it be good for cases within each stratum to be very similar?

We might get a more stable estimate for the subpopulation in a stratum if the cases are very similar, leading to more precise estimates within each group. When we combine these estimates into a single estimate for the full population, that population estimate will tend to be more precise since each individual group estimate is itself more precise.

In a **cluster sample**, we break up the population into many groups, called **clusters**. Then we sample a fixed number of clusters and include all observations from each of those clusters in the sample. A **multistage sample** is like a cluster sample, but rather than keeping all observations in each cluster, we would collect a random sample within each selected cluster.

Sometimes cluster or multistage sampling can be more economical than the alternative sampling techniques. Also, unlike stratified sampling, these approaches are most helpful when there is a lot of case-to-case variability within a cluster but the clusters themselves don't look very different from one another. For example, if neighborhoods represented clusters, then cluster or multistage sampling work best when the neighborhoods are very diverse. A downside of these methods is that more advanced techniques are typically required to analyze the data, though the methods in this book can be extended to handle such data.

2.1. SAMPLING PRINCIPLES AND STRATEGIES

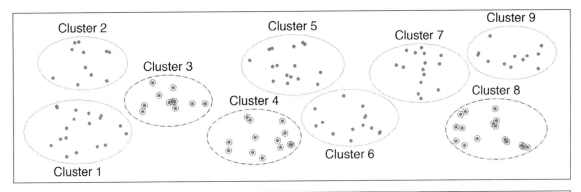

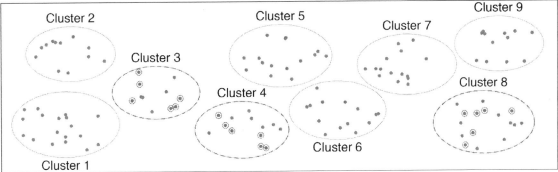

Figure 2.6: Examples of cluster and multistage sampling. In the top panel, cluster sampling was used: data were binned into nine clusters, three of these clusters were sampled, and all observations within these three cluster were included in the sample. In the bottom panel, multistage sampling was used, which differs from cluster sampling only in that we randomly select a subset of each cluster to be included in the sample rather than measuring every case in each sampled cluster.

 EXAMPLE

Suppose we are interested in estimating the malaria rate in a densely tropical portion of rural Indonesia. We learn that there are 30 villages in that part of the Indonesian jungle, each more or less similar to the next, but the distances between the villages are substantial. Our goal is to test 150 individuals for malaria. What sampling method should be employed?

A simple random sample would likely draw individuals from all 30 villages, which could make data collection extremely expensive. Stratified sampling would be a challenge since it is unclear how we would build strata of similar individuals. However, cluster sampling or multistage sampling seem like very good ideas. If we decided to use multistage sampling, we might randomly select half of the villages, then randomly select 10 people from each. This would probably reduce our data collection costs substantially in comparison to a simple random sample, and the cluster sample would still give us reliable information, even if we would need to analyze the data with slightly more advanced methods than we discuss in this book.

2.2 Experiments

Studies where the researchers assign treatments to cases are called **experiments**. When this assignment includes randomization, e.g., using a coin flip to decide which treatment a patient receives, it is called a **randomized experiment**. Randomized experiments are fundamentally important when trying to show a causal connection between two variables.

2.2.1 Principles of experimental design

1. **Controlling.** Researchers assign treatments to cases, and they do their best to **control** any other differences in the groups[3]. For example, when patients take a drug in pill form, some patients take the pill with only a sip of water while others may have it with an entire glass of water. To control for the effect of water consumption, a doctor may instruct every patient to drink a 12 ounce glass of water with the pill.

2. **Randomization.** Researchers randomize patients into treatment groups to account for variables that cannot be controlled. For example, some patients may be more susceptible to a disease than others due to their dietary habits. In this example dietary habit is a **confounding variable**[4], which is defined as a variable that is associated with both the explanatory and response variables. Randomizing patients into the treatment or control group helps even out such differences.

3. **Replication.** The more cases researchers observe, the more accurately they can estimate the effect of the explanatory variable on the response. In a single study, we **replicate** by collecting a sufficiently large sample. What is considered sufficiently large varies from experiment to experiment, but at a minimum we want to have multiple subjects (experimental units) per treatment group. Another way of achieving replication is replicating an entire study to verify an earlier finding. The term **replication crisis** refers to the ongoing methodological crisis in which past findings from scientific studies in several disciplines have failed to be replicated. **Pseudoreplication** occurs when individual observations under different treatments are heavily dependent on each other. For example, suppose you have 50 subjects in an experiment where you're taking blood pressure measurements at 10 time points throughout the course of the study. By the end, you will have $50 \times 10 = 500$ measurements. Reporting that you have 500 observations would be considered pseudoreplication, as the blood pressure measurements of a given individual are not independent of each other. Pseudoreplication often happens when the wrong entity is replicated, and the reported sample sizes are exaggerated.

4. **Blocking.** Researchers sometimes know or suspect that variables, other than the treatment, influence the response. Under these circumstances, they may first group individuals based on this variable into **blocks** and then randomize cases within each block to the treatment groups. This strategy is often referred to as **blocking**. For instance, if we are looking at the effect of a drug on heart attacks, we might first split patients in the study into low-risk and high-risk blocks, then randomly assign half the patients from each block to the control group and the other half to the treatment group, as shown in Figure 2.7. This strategy ensures that each treatment group has the same number of low-risk patients and also the same number of high-risk patients.

It is important to incorporate the first three experimental design principles into any study, and this book describes applicable methods for analyzing data from such experiments. Blocking is a slightly more advanced technique, and statistical methods in this book may be extended to analyze data collected using blocking.

---

[3]This is a different concept than a *control group*, which we discuss in the second principle and in Section 2.2.2.

[4]Also called a **lurking variable**, **confounding factor**, or a **confounder**.

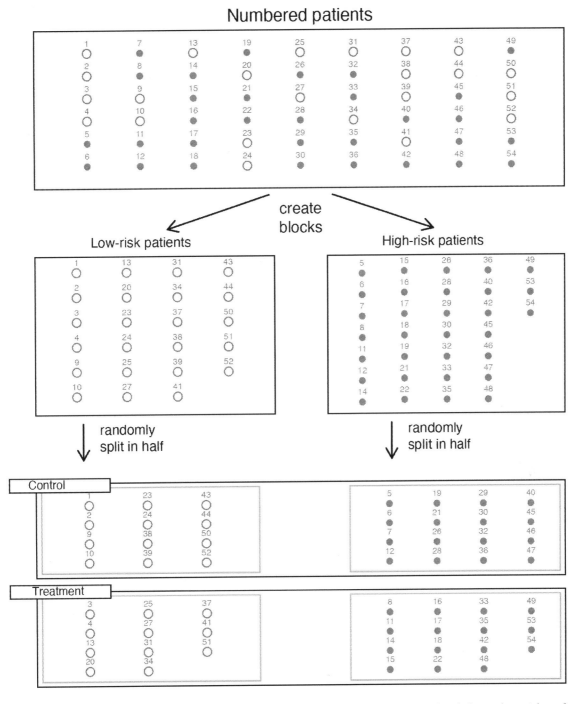

Figure 2.7: Blocking using a variable depicting patient risk. Patients are first divided into low-risk and high-risk blocks, then each block is evenly separated into the treatment groups using randomization. This strategy ensures an equal representation of patients in each treatment group from both the low-risk and high-risk categories.

2.2.2 Reducing bias in human experiments

Randomized experiments have long been considered to be the gold standard for data collection, but they do not ensure an unbiased perspective into the cause and effect relationship in all cases. Human studies are perfect examples where bias can unintentionally arise. Here we reconsider a study where a new drug was used to treat heart attack patients. In particular, researchers wanted to know if the drug reduced deaths in patients.

These researchers designed a randomized experiment because they wanted to draw causal conclusions about the drug's effect. Study volunteers[5] were randomly placed into two study groups. One group, the **treatment group**, received the drug. The other group, called the **control group**, did not receive any drug treatment.

Put yourself in the place of a person in the study. If you are in the treatment group, you are given a fancy new drug that you anticipate will help you. On the other hand, a person in the other group doesn't receive the drug and sits idly, hoping her participation doesn't increase her risk of death. These perspectives suggest there are actually two effects in this study: the one of interest is the effectiveness of the drug, and the second is an emotional effect of (not) taking the drug, which is difficult to quantify.

Researchers aren't usually interested in the emotional effect, which might bias the study. To circumvent this problem, researchers do not want patients to know which group they are in. When researchers keep the patients uninformed about their treatment, the study is said to be **blind**. But there is one problem: if a patient doesn't receive a treatment, they will know they're in the control group. A solution to this problem is to give a fake treatment to patients in the control group. This is called a **placebo**, and an effective placebo is the key to making a study truly blind. A classic example of a placebo is a sugar pill that is made to look like the actual treatment pill. However offering such a fake treatment may not be ethical in certain experiments. For example, in medical experiments, typically the control group must get the current standard of care. Often times, a placebo results in a slight but real improvement in patients. This effect has been dubbed the **placebo effect**.

The patients are not the only ones who should be blinded: doctors and researchers can unintentionally bias a study. When a doctor knows a patient has been given the real treatment, they might inadvertently give that patient more attention or care than a patient that they know is on the placebo. To guard against this bias, which again has been found to have a measurable effect in some instances, most modern studies employ a **double-blind** setup where doctors or researchers who interact with patients are, just like the patients, unaware of who is or is not receiving the treatment.[6]

GUIDED PRACTICE

Look back to the study in Section 1.1 where researchers were testing whether stents were effective at reducing strokes in at-risk patients. Is this an experiment? Was the study blinded? Was it double-blinded?[7]

GUIDED PRACTICE

For the study in Section 1.1, could the researchers have employed a placebo? If so, what would that placebo have looked like?[8]

---

[5]Human subjects are often called **patients, volunteers,** or **study participants**.

[6]There are always some researchers involved in the study who do know which patients are receiving which treatment. However, they do not interact with the study's patients and do not tell the blinded health care professionals who is receiving which treatment.

[7]The researchers assigned the patients into their treatment groups, so this study was an experiment. However, the patients could distinguish what treatment they received because a stent is a surgical procedure There is no equivalent surgical placebo, so this study was not blind. The study could not be double-blind since it was not blind.

[8]Ultimately, can we make patients think they got treated from a surgery? In fact, we can, and some experiments use a **sham surgery**. In a sham surgery, the patient does undergo surgery, but the patient does not receive the full treatment, though they will still get a placebo effect.

You may have many questions about the ethics of sham surgeries to create a placebo. These questions may have even arisen in your mind when in the general experiment context, where a possibly helpful treatment was withheld from individuals in the control group; the main difference is that a sham surgery tends to create additional risk, while withholding a treatment only maintains a person's risk.

There are always multiple viewpoints of experiments and placebos, and rarely is it obvious which is ethically "correct". For instance, is it ethical to use a sham surgery when it creates a risk to the patient? However, if we don't use sham surgeries, we may promote the use of a costly treatment that has no real effect; if this happens, money and other resources will be diverted away from other treatments that are known to be helpful. Ultimately, this is a difficult situation where we cannot perfectly protect both the patients who have volunteered for the study and the patients who may benefit (or not) from the treatment in the future.

2.3 Observational studies

Data where no treatment has been explicitly applied (or explicitly withheld) is called **observational data**. For instance, the loan data and county data described in Section 1.2 are both examples of observational data.

Making causal conclusions based on experiments is often reasonable, since we can randomly assign the explanatory variable(s), i.e., the treatments. However, making the same causal conclusions based on observational data can be treacherous and is not recommended. Thus, observational studies are generally only sufficient to show associations or form hypotheses that can be later checked with experiments.

GUIDED PRACTICE

Suppose an observational study tracked sunscreen use and skin cancer, and it was found that the more sunscreen someone used, the more likely the person was to have skin cancer. Does this mean sunscreen *causes* skin cancer?[9]

Some previous research tells us that using sunscreen actually reduces skin cancer risk, so maybe there is another variable that can explain this hypothetical association between sunscreen usage and skin cancer. One important piece of information that is absent is sun exposure. If someone is out in the sun all day, they are more likely to use sunscreen *and* more likely to get skin cancer. Exposure to the sun is unaccounted for in the simple observational investigation.

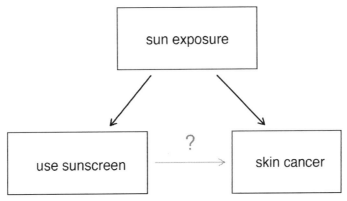

In this example, sun exposure is a confounding variable. The presence of confounding variables is what inhibits the ability for observational studies to make causal claims. While one method to justify making causal conclusions from observational studies is to exhaust the search for confounding variables, there is no guarantee that all confounding variables can be examined or measured.

---

[9] No. See the paragraph following the question!

GUIDED PRACTICE

Figure 1.2 shows a negative association between the homeownership rate and the percentage of housing units that are in multi-unit structures in a county. However, it is unreasonable to conclude that there is a causal relationship between the two variables. Suggest a variable that might explain the negative relationship.[10]

Observational studies come in two forms: prospective and retrospective studies. A **prospective study** identifies individuals and collects information as events unfold. For instance, medical researchers may identify and follow a group of patients over many years to assess the possible influences of behavior on cancer risk. One example of such a study is The Nurses' Health Study. Started in 1976 and expanded in 1989, the Nurses' Health Study has collected data on over 275,000 nurses and is still enrolling participants. This prospective study recruits registered nurses and then collects data from them using questionnaires. **Retrospective studies** collect data after events have taken place, e.g., researchers may review past events in medical records. Some datasets may contain both prospectively- and retrospectively-collected variables, such as medical studies which gather information on participants' lives before they enter the study and subsequently collect data on participants throughout the study.

---

[10] Answers will vary. Population density may be important. If a county is very dense, then this may require a larger percentage of residents to live in housing units that are in multi-unit structures. Additionally, the high density may contribute to increases in property value, making homeownership unfeasible for many residents.

2.4 Chapter review

2.4.1 Summary

A strong analyst will have a good sense of the types of data they are working with and how to visualize the data in order to gain a complete understanding of the variables. Equally important however, is an understanding of the data source. In this chapter, we have discussed randomized experiments and taking good, random, representative samples from a population. When we discuss inferential methods (starting in Chapter 11), the conclusions that can be drawn will be dependent on how the data were collected. Figure 2.8 summarizes the differences between random assignment of treatments and random samples.[11] Regularly revisiting Figure 2.8 will be important when making conclusions from a given data analysis.

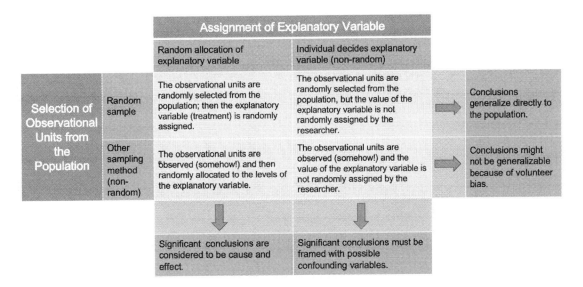

Figure 2.8: As we will see, analysis conclusions should be made carefully according to how the data were collected. Note that very few datasets come from the top left box because usually ethics require that randomly assignment of treatments can only be given to volunteers. Both representative (ideally random) sampling and experiments (random assignment of treatments) are important for how statistical conclusions can be made on populations.

2.4.2 Terms

We introduced the following terms in the chapter. If you're not sure what some of these terms mean, we recommend you go back in the text and review their definitions. We are purposefully presenting them in alphabetical order, instead of in order of appearance, so they will be a little more challenging to locate. However you should be able to easily spot them as **bolded text**.

---

[11] Derived from similar figures in Chance and Rossman (2018) and Ramsey and Schafer (2012).

anecdotal evidence	experiment	replicatation
bias	multistage sample	replication crisis
blind	non-response bias	representative
blocking	non-response rate	retrospective study
census	observational data	sample
cluster	parameter	sample bias
cluster sampling	placebo	simple random sample
confounding variable	placebo effect	simple random sampling
control	population	statistic
control group	prospective study	strata
convenience sample	pseudoreplication	stratified sampling
double-blind	randomized experiment	treatment group

2.5 Exercises

Answers to odd numbered exercises can be found in Appendix A.2.

2.1. **Parameters and statistics.** Identify which value represents the sample mean and which value represents the claimed population mean.

   a. American households spent an average of about $52 in 2007 on Halloween merchandise such as costumes, decorations and candy. To see if this number had changed, researchers conducted a new survey in 2008 before industry numbers were reported. The survey included 1,500 households and found that average Halloween spending was $58 per household.

   b. The average GPA of students in 2001 at a private university was 3.37. A survey on a sample of 203 students from this university yielded an average GPA of 3.59 a decade later.

2.2. **Sleeping in college.** A recent article in a college newspaper stated that college students get an average of 5.5 hrs of sleep each night. A student who was skeptical about this value decided to conduct a survey by randomly sampling 25 students. On average, the sampled students slept 6.25 hours per night. Identify which value represents the sample mean and which value represents the claimed population mean.

2.3. **Air pollution and birth outcomes, scope of inference.** Researchers collected data to examine the relationship between air pollutants and preterm births in Southern California. During the study air pollution levels were measured by air quality monitoring stations. Length of gestation data were collected on 143,196 births between the years 1989 and 1993, and air pollution exposure during gestation was calculated for each birth. (Ritz et al., 2000)

   a. Identify the population of interest and the sample in this study.

   b. Comment on whether or not the results of the study can be generalized to the population, and if the findings of the study can be used to establish causal relationships.

2.4. **Cheaters, scope of inference.** Researchers studying the relationship between honesty, age and self-control conducted an experiment on 160 children between the ages of 5 and 15. The researchers asked each child to toss a fair coin in private and to record the outcome (white or black) on a paper sheet, and said they would only reward children who report white. Half the students were explicitly told not to cheat and the others were not given any explicit instructions. Differences were observed in the cheating rates in the instruction and no instruction groups, as well as some differences across children's characteristics within each group. (Bucciol and Piovesan, 2011)

   a. Identify the population of interest and the sample in this study.

   b. Comment on whether or not the results of the study can be generalized to the population, and if the findings of the study can be used to establish causal relationships.

2.5. **Gamification and statistics, scope of inference.** Researchers investigating the effects of gamification (application of game-design elements and game principles in non-game contexts) on learning statistics randomly assigned 365 college students in a statistics course to one of four groups; one of these groups had no reading exercises and no gamification, one group had reading but no gamification, one group had gamification but no reading, and a final group had gamification and reading. Students in all groups also attended lectures. The study found that gamification had a positive impact on student learning compared to traditional teaching methods involving reading exercises. (Legaki et al., 2020)

   a. Identify the population of interest and the sample in this study.

   b. Comment on whether or not the results of the study can be generalized to the population, and if the findbngs of the study can be used to establish causal relationships.

2.6. **Stealers, scope of inference.** In a study of the relationship between socio-economic class and unethical behavior, 129 University of California undergraduates at Berkeley were asked to identify themselves as having low or high social-class by comparing themselves to others with the most (least) money, most (least) education, and most (least) respected jobs. They were also presented with a jar of individually wrapped candies and informed that the candies were for children in a nearby laboratory, but that they could take some if they wanted. After completing some unrelated tasks, participants reported the number of candies they had taken. It was found that those who were identified as upper-class took more candy than others. (Piff et al., 2012)

  a. Identify the population of interest and the sample in this study.

  b. Comment on whether or not the results of the study can be generalized to the population, and if the findings of the study can be used to establish causal relationships.

2.7. **Relaxing after work.** The General Social Survey asked the question, *"After an average work day, about how many hours do you have to relax or pursue activities that you enjoy?"* to a random sample of 1,155 Americans. The average relaxing time was found to be 1.65 hours. Determine which of the following is an observation, a variable, a sample statistic (value calculated based on the observed sample), or a population parameter.[12]

  a. An American in the sample.

  b. Number of hours spent relaxing after an average work day.

  c. 1.65.

  d. Average number of hours all Americans spend relaxing after an average work day.

2.8. **Cats on YouTube.** Suppose you want to estimate the percentage of videos on YouTube that are cat videos. It is impossible for you to watch all videos on YouTube so you use a random video picker to select 1000 videos for you. You find that 2% of these videos are cat videos. Determine which of the following is an observation, a variable, a sample statistic (value calculated based on the observed sample), or a population parameter.

  a. Percentage of all videos on YouTube that are cat videos.

  b. 2%.

  c. A video in your sample.

  d. Whether or not a video is a cat video.

2.9. **Course satisfaction across sections.** A large college class has 160 students. All 160 students attend the lectures together, but the students are divided into 4 groups, each of 40 students, for lab sections administered by different teaching assistants. The professor wants to conduct a survey about how satisfied the students are with the course, and he believes that the lab section a student is in might affect the student's overall satisfaction with the course.

  a. What type of study is this?

  b. Suggest a sampling strategy for carrying out this study.

2.10. **Housing proposal across dorms.** On a large college campus first-year students and sophomores live in dorms located on the eastern part of the campus and juniors and seniors live in dorms located on the western part of the campus. Suppose you want to collect student opinions on a new housing structure the college administration is proposing and you want to make sure your survey equally represents opinions from students from all years.

  a. What type of study is this?

  b. Suggest a sampling strategy for carrying out this study.

---

[12]The data used in this exercise comes from the General Social Survey, 2018.

2.11. **Internet use and life expectancy.** The following scatterplot was created as part of a study evaluating the relationship between estimated life expectancy at birth (as of 2014) and percentage of internet users (as of 2009) in 208 countries for which such data were available.[13]

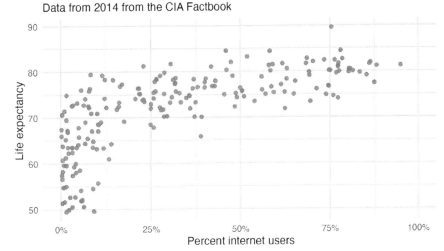

a. Describe the relationship between life expectancy and percentage of internet users.

b. What type of study is this?

c. State a possible confounding variable that might explain this relationship and describe its potential effect.

2.12. **Stressed out.** A study that surveyed a random sample of otherwise healthy high school students found that they are more likely to get muscle cramps when they are stressed. The study also noted that students drink more coffee and sleep less when they are stressed.

a. What type of study is this?

b. Can this study be used to conclude a causal relationship between increased stress and muscle cramps?

c. State possible confounding variables that might explain the observed relationship between increased stress and muscle cramps.

2.13. **Evaluate sampling methods.** A university wants to determine what fraction of its undergraduate student body support a new $25 annual fee to improve the student union. For each proposed method below, indicate whether the method is reasonable or not.

a. Survey a simple random sample of 500 students.

b. Stratify students by their field of study, then sample 10% of students from each stratum.

c. Cluster students by their ages (e.g., 18 years old in one cluster, 19 years old in one cluster, etc.), then randomly sample three clusters and survey all students in those clusters.

2.14. **Random digit dialing.** The Gallup Poll uses a procedure called random digit dialing, which creates phone numbers based on a list of all area codes in America in conjunction with the associated number of residential households in each area code. Give a possible reason the Gallup Poll chooses to use random digit dialing instead of picking phone numbers from the phone book.

---

[13]The cia_factbook data used in this exercise can be found in the **openintro** R package.

2.15. **Haters are gonna hate, study confirms.** A study published in the *Journal of Personality and Social Psychology* asked a group of 200 randomly sampled participants recruited online using Amazon's Mechanical Turk to evaluate how they felt about various subjects, such as camping, health care, architecture, taxidermy, crossword puzzles, and Japan in order to measure their attitude towards mostly independent stimuli. Then, they presented the participants with information about a new product: a microwave oven. This microwave oven does not exist, but the participants didn't know this, and were given three positive and three negative fake reviews. People who reacted positively to the subjects on the dispositional attitude measurement also tended to react positively to the microwave oven, and those who reacted negatively tended to react negatively to it. Researchers concluded that *"some people tend to like things, whereas others tend to dislike things, and a more thorough understanding of this tendency will lead to a more thorough understanding of the psychology of attitudes."* (Hepler and Albarracín, 2013)

   a. What are the cases?

   b. What is (are) the response variable(s) in this study?

   c. What is (are) the explanatory variable(s) in this study?

   d. Does the study employ random sampling? Explain. How could they have obtained participants?

   e. Is this an observational study or an experiment? Explain your reasoning.

   f. Can we establish a causal link between the explanatory and response variables?

   g. Can the results of the study be generalized to the population at large?

2.16. **Family size.** Suppose we want to estimate household size, where a *"household"* is defined as people living together in the same dwelling, and sharing living accommodations. If we select students at random at an elementary school and ask them what their family size is, will this be a good measure of household size? Or will our average be biased? If so, will it overestimate or underestimate the true value?

2.17. **Sampling strategies.** A statistics student who is curious about the relationship between the amount of time students spend on social networking sites and their performance at school decides to conduct a survey. Various research strategies for collecting data are described below. In each, name the sampling method proposed and any bias you might expect.

   a. They randomly sample 40 students from the study's population, give them the survey, ask them to fill it out and bring it back the next day.

   b. They give out the survey only to their friends, making sure each one of them fills out the survey.

   c. They post a link to an online survey on Facebook and ask their friends to fill out the survey.

   d. They randomly sample 5 classes and asks a random sample of students from those classes to fill out the survey.

2.18. **Reading the paper.** Below are excerpts from two articles published in the *NY Times*:

   a. An excerpt from an article titled *Risks: Smokers Found More Prone to Dementia* is below. Based on this study, can we conclude that smoking causes dementia later in life? Explain your reasoning. (Rabin, 2010)

   "Researchers analyzed data from 23,123 health plan members who participated in a voluntary exam and health behavior survey from 1978 to 1985, when they were 50-60 years old. 23 years later, about 25% of the group had dementia, including 1,136 with Alzheimer's disease and 416 with vascular dementia. After adjusting for other factors, the researchers concluded that pack-a-day smokers were 37% more likely than nonsmokers to develop dementia, and the risks went up with increased smoking; 44% for one to two packs a day; and twice the risk for more than two packs."

See next page for part b.

b. An excerpt from an article titled *The School Bully Is Sleepy* is below. A friend of yours who read the article says, *"The study shows that sleep disorders lead to bullying in school children."* Is this statement justified? If not, how best can you describe the conclusion that can be drawn from this study? (Parker-Pope, 2011)

"The University of Michigan study, collected survey data from parents on each child's sleep habits and asked both parents and teachers to assess behavioral concerns. About a third of the students studied were identified by parents or teachers as having problems with disruptive behavior or bullying. The researchers found that children who had behavioral issues and those who were identified as bullies were twice as likely to have shown symptoms of sleep disorders."

2.19. **Light and exam performance.** A study is designed to test the effect of light level on exam performance of students. The researcher believes that light levels might have different effects on people who wear glasses and people who don't, so they want to make sure both groups of people are equally represented in each treatment. The treatments are fluorescent overhead lighting, yellow overhead lighting, no overhead lighting (only desk lamps).

  a. What is the response variable?

  b. What is the explanatory variable? What are its levels?

  c. What is the blocking variable? What are its levels?

2.20. **Vitamin supplements.** To assess the effectiveness of taking large doses of vitamin C in reducing the duration of the common cold, researchers recruited 400 healthy volunteers from staff and students at a university. A quarter of the patients were assigned a placebo, and the rest were evenly divided between 1g Vitamin C, 3g Vitamin C, or 3g Vitamin C plus additives to be taken at onset of a cold for the following two days. All tablets had identical appearance and packaging. The nurses who handed the prescribed pills to the patients knew which patient received which treatment, but the researchers assessing the patients when they were sick did not. No significant differences were observed in any measure of cold duration or severity between the four groups, and the placebo group had the shortest duration of symptoms. (Audera et al., 2001)

  a. Was this an experiment or an observational study? Why?

  b. What are the explanatory and response variables in this study?

  c. Were the patients blinded to their treatment?

  d. Was this study double-blind?

  e. Participants are ultimately able to choose whether or not to use the pills prescribed to them. We might expect that not all of them will adhere and take their pills. Does this introduce a confounding variable to the study? Explain your reasoning.

2.21. **Light, noise, and exam performance.** A study is designed to test the effect of light level and noise level on exam performance of students. The researcher believes that light and noise levels might have different effects on people who wear glasses and people who don't, so they want to make sure both groups of people are equally represented in each treatment. The light treatments considered are fluorescent overhead lighting, yellow overhead lighting, no overhead lighting (only desk lamps). The noise treatments considered are no noise, construction noise, and human chatter noise.

  1. What type of study is this?

  2. How many factors are considered in this study? Identify them, and describe their levels.

  3. What is the role of the wearing glasses variable in this study?

2.22. **Music and learning.** You would like to conduct an experiment in class to see if students learn better if they study without any music, with music that has no lyrics (instrumental), or with music that has lyrics. Briefly outline a design for this study.

2.23. **Soda preference.** You would like to conduct an experiment in class to see if your classmates prefer the taste of regular Coke or Diet Coke. Briefly outline a design for this study.

2.24. **Exercise and mental health.** A researcher is interested in the effects of exercise on mental health and they propose the following study: use stratified random sampling to ensure representative proportions of 18-30, 31-40 and 41-55 year olds from the population. Next, randomly assign half the subjects from each age group to exercise twice a week, and instruct the rest not to exercise. Conduct a mental health exam at the beginning and at the end of the study, and compare the results.

    a. What type of study is this?

    b. What are the treatment and control groups in this study?

    c. Does this study make use of blocking? If so, what is the blocking variable?

    d. Does this study make use of blinding?

    e. Comment on whether or not the results of the study can be used to establish a causal relationship between exercise and mental health, and indicate whether or not the conclusions can be generalized to the population at large.

    f. Suppose you are given the task of determining if this proposed study should get funding. Would you have any reservations about the study proposal?

2.25. **Chia seeds and weight loss.** Chia Pets – those terra-cotta figurines that sprout fuzzy green hair – made the chia plant a household name. But chia has gained an entirely new reputation as a diet supplement. In one 2009 study, a team of researchers recruited 38 men and divided them randomly into two groups: treatment or control. They also recruited 38 women, and they randomly placed half of these participants into the treatment group and the other half into the control group. One group was given 25 grams of chia seeds twice a day, and the other was given a placebo. The subjects volunteered to be a part of the study. After 12 weeks, the scientists found no significant difference between the groups in appetite or weight loss. (Nieman et al., 2009)

    a. What type of study is this?

    b. What are the experimental and control treatments in this study?

    c. Has blocking been used in this study? If so, what is the blocking variable?

    d. Has blinding been used in this study?

    e. Comment on whether or not we can make a causal statement, and indicate whether or not we can generalize the conclusion to the population at large.

2.26. **City council survey.** A city council has requested a household survey be conducted in a suburban area of their city. The area is broken into many distinct and unique neighborhoods, some including large homes, some with only apartments, and others a diverse mixture of housing structures. For each part below, identify the sampling methods described, and describe the statistical pros and cons of the method in the city's context.

    a. Randomly sample 200 households from the city.

    b. Divide the city into 20 neighborhoods, and sample 10 households from each neighborhood.

    c. Divide the city into 20 neighborhoods, randomly sample 3 neighborhoods, and then sample all households from those 3 neighborhoods.

    d. Divide the city into 20 neighborhoods, randomly sample 8 neighborhoods, and then randomly sample 50 households from those neighborhoods.

    e. Sample the 200 households closest to the city council offices.

2.27. **Flawed reasoning.** Identify the flaw(s) in reasoning in the following scenarios. Explain what the individuals in the study should have done differently if they wanted to make such strong conclusions.

   a. Students at an elementary school are given a questionnaire that they are asked to return after their parents have completed it. One of the questions asked is, *"Do you find that your work schedule makes it difficult for you to spend time with your kids after school?"* Of the parents who replied, 85% said *"no"*. Based on these results, the school officials conclude that a great majority of the parents have no difficulty spending time with their kids after school.

   b. A survey is conducted on a simple random sample of 1,000 women who recently gave birth, asking them about whether or not they smoked during pregnancy. A follow-up survey asking if the children have respiratory problems is conducted 3 years later. However, only 567 of these women are reached at the same address. The researcher reports that these 567 women are representative of all mothers.

   c. An orthopedist administers a questionnaire to 30 of his patients who do not have any joint problems and finds that 20 of them regularly go running. He concludes that running decreases the risk of joint problems.

2.28. **Income and education in US counties.** The scatterplot below shows the relationship between per capita income (in thousands of dollars) and percent of population with a bachelor's degree in 3,142 counties in the US in 2019.[14]

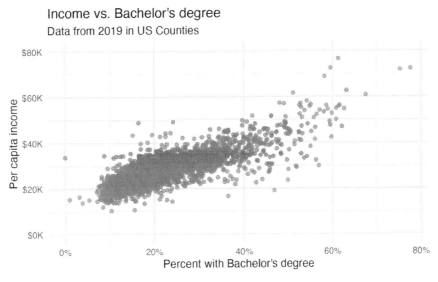

   a. What are the explanatory and response variables?

   b. Describe the relationship between the two variables. Make sure to discuss unusual observations, if any.

   c. Can we conclude that having a bachelor's degree increases one's income?

---

[14] The `county_complete` data used in this exercise can be found in the **openintro** R package.

2.29. **Eat well, feel better.** In a public health study on the effects of consumption of fruits and vegetables on psychological well-being in young adults, participants were randomly assigned to three groups: (1) diet-as-usual, (2) an ecological momentary intervention involving text message reminders to increase their fruits and vegetable consumption plus a voucher to purchase them, or (3) a fruit and vegetable intervention in which participants were given two additional daily servings of fresh fruits and vegetables to consume on top of their normal diet. Participants were asked to take a nightly survey on their smartphones. Participants were student volunteers at the University of Otago, New Zealand. At the end of the 14-day study, only participants in the third group showed improvements to their psychological well-being across the 14-days relative to the other groups. (Conner et al., 2017)

   a. What type of study is this?

   b. Identify the explanatory and response variables.

   c. Comment on whether the results of the study can be generalized to the population.

   d. Comment on whether the results of the study can be used to establish causal relationships.

   e. A newspaper article reporting on the study states, "The results of this study provide proof that giving young adults fresh fruits and vegetables to eat can have psychological benefits, even over a brief period of time." How would you suggest revising this statement so that it can be supported by the study?

2.30. **Screens, teens, and psychological well-being.** In a study of three nationally representative large-scale datasets from Ireland, the United States, and the United Kingdom (n = 17,247), teenagers between the ages of 12 to 15 were asked to keep a diary of their screen time and answer questions about how they felt or acted. The answers to these questions were then used to compute a psychological well-being score. Additional data were collected and included in the analysis, such as each child's sex and age, and on the mother's education, ethnicity, psychological distress, and employment. The study concluded that there is little clear-cut evidence that screen time decreases adolescent well-being. (Orben and Baukney-Przybylski, 2018)

   a. What type of study is this?

   b. Identify the explanatory variables.

   c. Identify the response variable.

   d. Comment on whether the results of the study can be generalized to the population, and why.

   e. Comment on whether the results of the study can be used to establish causal relationships.

Chapter 3

Applications: Data

3.1 Case study: Passwords

Stop for a second and think about how many passwords you've used so far today. You've probably used one to unlock your phone, one to check email, and probably at least one to log on to a social media account. Made a debit purchase? You've probably entered a password there too.

If you're reading this book, and particularly if you're reading it online, chances are you have had to create a password once or twice in your life. And if you are diligent about your safety and privacy, you've probably chosen passwords that would be hard for others to guess, or *crack*.

In this case study we introduce a dataset on passwords. The goal of the case study is to walk you through what a data scientist does when they first get a hold of a dataset as well as to provide some "foreshadowing" of concepts and techniques we'll introduce in the next few chapters on exploratory data analysis.

 The passwords data can be found in the **tidytuesdayR** R package.

Table 3.1 shows the first ten rows from the dataset, which are the ten most common passwords. Perhaps unsurprisingly, "password" tops the list, followed by "123456".

Table 3.1: Top ten rows of the passwords dataset.

rank	password	category	value	time_unit	offline_crack_sec	strength
1	password	password-related	6.91	years	2.170	8
2	123456	simple-alphanumeric	18.52	minutes	0.000	4
3	12345678	simple-alphanumeric	1.29	days	0.001	4
4	1234	simple-alphanumeric	11.11	seconds	0.000	4
5	qwerty	simple-alphanumeric	3.72	days	0.003	8
6	12345	simple-alphanumeric	1.85	minutes	0.000	4
7	dragon	animal	3.72	days	0.003	8
8	baseball	sport	6.91	years	2.170	4
9	football	sport	6.91	years	2.170	7
10	letmein	password-related	3.19	months	0.084	8

When you encounter a new dataset, taking a peek at the first few rows as we did in Table 3.1 is almost instinctual. It can often be helpful to take a look at the last few rows of the data as well to get a sense of the size of the data as well as potentially discover any characteristics that may not be apparent in the top few rows. Table 3.2 shows the bottom ten rows of the passwords dataset, which reveals that we are looking at a dataset of 500 passwords.

Table 3.2: Bottom ten rows of the passwords dataset.

rank	password	category	value	time_unit	offline_crack_sec	strength
491	natasha	name	3.19	months	0.084	7
492	sniper	cool-macho	3.72	days	0.003	8
493	chance	name	3.72	days	0.003	7
494	genesis	nerdy-pop	3.19	months	0.084	7
495	hotrod	cool-macho	3.72	days	0.003	7
496	reddog	cool-macho	3.72	days	0.003	6
497	alexande	name	6.91	years	2.170	9
498	college	nerdy-pop	3.19	months	0.084	7
499	jester	name	3.72	days	0.003	7
500	passw0rd	password-related	92.27	years	29.020	28

At this stage it's also useful to think about how these data were collected, as that will inform the scope of any inference you can make based on your analysis of the data.

GUIDED PRACTICE

Do these data come from an observational study or an experiment?[1]

GUIDED PRACTICE

There are 500 rows and 7 columns in the dataset. What does each row and each column represent?[2]

Once you've identified the rows and columns, it's useful to review the data dictionary to learn about what each column in the dataset represents. This is provided in Table 3.3.

Table 3.3: Variables and their descriptions for the passwords dataset.

Variable	Description
rank	Popularity in the database of released passwords.
password	Actual text of the password.
category	Category password falls into.
value	Time to crack by online guessing.
time_unit	Time unit to match with value.
offline_crack_sec	Time to crack offline in seconds.
strength	Strength of password, relative only to passwords in this dataset. Lower values indicate weaker passwords.

We now have a better sense of what each column represents, but we don't yet know much about the characteristics of each of the variables.

---

[1] This is an observational study. Researchers collected data on existing passwords in use and identified most common ones to put together this dataset.

[2] Each row represents a password and each column represents a variable which contains information on each password.

3.1. CASE STUDY: PASSWORDS

EXAMPLE

Determine whether each variable in the passwords dataset is numerical or categorical. For numerical variables, further classify them as continuous or discrete. For categorical variables, determine if the variable is ordinal.

The numerical variables in the dataset are `rank` (discrete), `value` (continuous), and `offline_crack_sec` (continuous). The categorical variables are `password`, `time_unit`. The strength variable is trickier to classify – we can think of it as discrete numerical or as an ordinal variable as it takes on numerical values, however it's used to categorize the passwords on an ordinal scale. One way of approaching this is thinking about whether the values the variable takes vary linearly, e.g., is the difference in strength between passwords with strength levels 8 and 9 the same as the difference with those with strength levels 9 and 10. If this is not necessarily the case, we would classify the variable as ordinal. Determining the classification of this variable requires understanding of how `strength` values were determined, which is a very typical workflow for working with data. Sometimes the data dictionary (presented in Table 3.3 isn't sufficient, and we need to go back to the data source and try to understand the data better before we can proceed with the analysis meaningfully.

Next, let's try to get to know each variable a little bit better. For categorical variables, this involves figuring out what their levels are and how commonly represented they are in the data. Figure 3.1 shows the distributions of the categorical variables in this dataset. We can see that password strengths of 0-10 are more common than higher values. The most common password category is name (e.g. michael, jennifer, jordan, etc.) and the least common is food (e.g., pepper, cheese, coffee, etc.). Many passwords can be cracked in the matter of days by online cracking with some taking as little as seconds and some as long as years to break. Each of these visualizations is a bar plot, which you will learn more about in Chapter 4.

Similarly, we can examine the distributions of the numerical variables as well. We already know that rank ranges between 1 and 500 in this dataset, based on Table 3.1 and Table 3.2. The value variable is slightly more complicated to consider since the numerical values in that column are meaningless without the time unit that accompanies them. Table 3.4 shows the minimum and maximum amount of time it takes to crack a password by online guessing. For example there are 11 passwords in the dataset that can be broken in a matter of seconds, and each of them take 11.11 seconds to break, since the minimum and the maximum of observations in this group are exactly equal to this value. And there are 65 passwords that take years to break, ranging from 2.56 years to 92.27 years.

Table 3.4: Minimum and maximum amount of time it takes to crack a password by online guessing as well as the number of observations that fall into each time unit category.

time_unit	n	min	max
seconds	11	11.11	11.11
minutes	51	1.85	18.52
hours	43	3.09	17.28
days	238	1.29	3.72
weeks	5	1.84	3.70
months	87	3.19	3.19
years	65	2.56	92.27

Even though passwords that take a large number of years to crack can seem like good options (see Table 3.5 for a list of them), now that you've seen them here (and the fact that they are in a dataset of 500 most common passwords), you should not use them as secure passwords!

Strengths, categories, and cracking time of 500 most common passwords

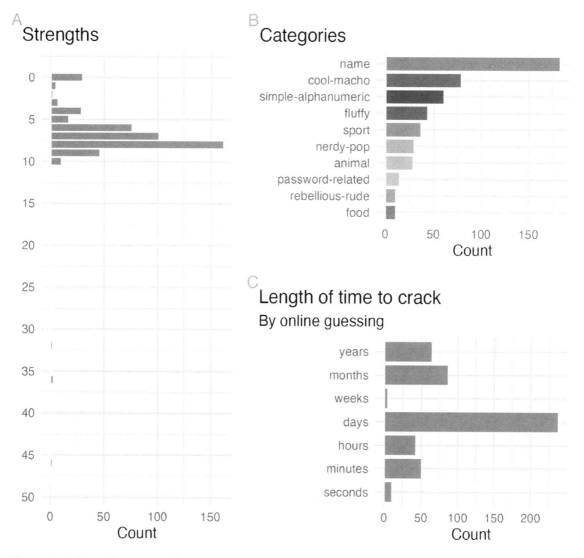

Figure 3.1: Distributions of the categorical variables in the `passwords` dataset. Plot A shows the distribution of password strengths, Plot B password categories, and Plot C length of time it takes to crack the passwords by online guessing.

Table 3.5: Passwords that take the longest amount of time to crack by online guessing.

rank	password	category	value	time_unit	offline_crack_sec	strength
26	trustno1	simple-alphanumeric	92.3	years	29.0	25
336	rush2112	nerdy-pop	92.3	years	29.0	48
406	jordan23	sport	92.3	years	29.3	34
500	passw0rd	password-related	92.3	years	29.0	28

The last numerical variable in the dataset is `offline_crack_sec`. Figure 3.2 shows the distribution of this variable, which reveals that all of these passwords can be cracked offline in under 30 seconds, with a large number of them being crackable in just a few seconds.

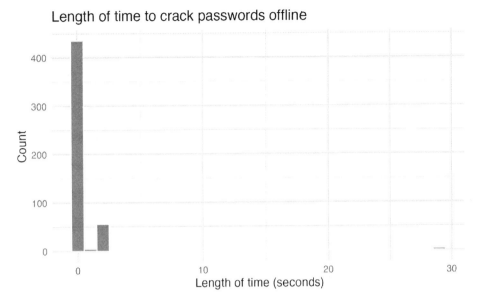

Figure 3.2: Histogram of the length of time it takes to crack passwords offline.

So far we examined the distributions of each individual variable, but it would be more interesting to explore relationships between multiple variables. Figure 3.3 shows the relationship between rank and strength of passwords by category, where more common passwords (those with higher rank) are plotted higher on the y-axis than those that are less common in this dataset. The stronger the password, the larger text it's represented with on the plot. While this visualization reveals some passwords that are less common, and stronger than others, we should reiterate that you should not use any of these passwords. And if you already do, it's time to go change it!

In this case study, we introduced you to the very first steps a data scientist takes when they start working with a new dataset. In the next few chapters we will introduce exploratory data analysis and you'll learn more about the various types of data visualizations and summary statistics you can make to get to know your data better.

Before you move on, we encourage you to think about whether the following questions can be answered with this dataset, and if yes, how you might go about answering them. It's okay if your answer is "I'm not sure", we simply want to get your exploratory juices flowing to prime you for what's to come!

1. What characteristics are associated with a strong vs. a weak password?
2. Do more popular passwords take shorter or longer to crack compared to less popular passwords?
3. Are passwords that start with letters or numbers more common among the list of top 500 most common passwords?

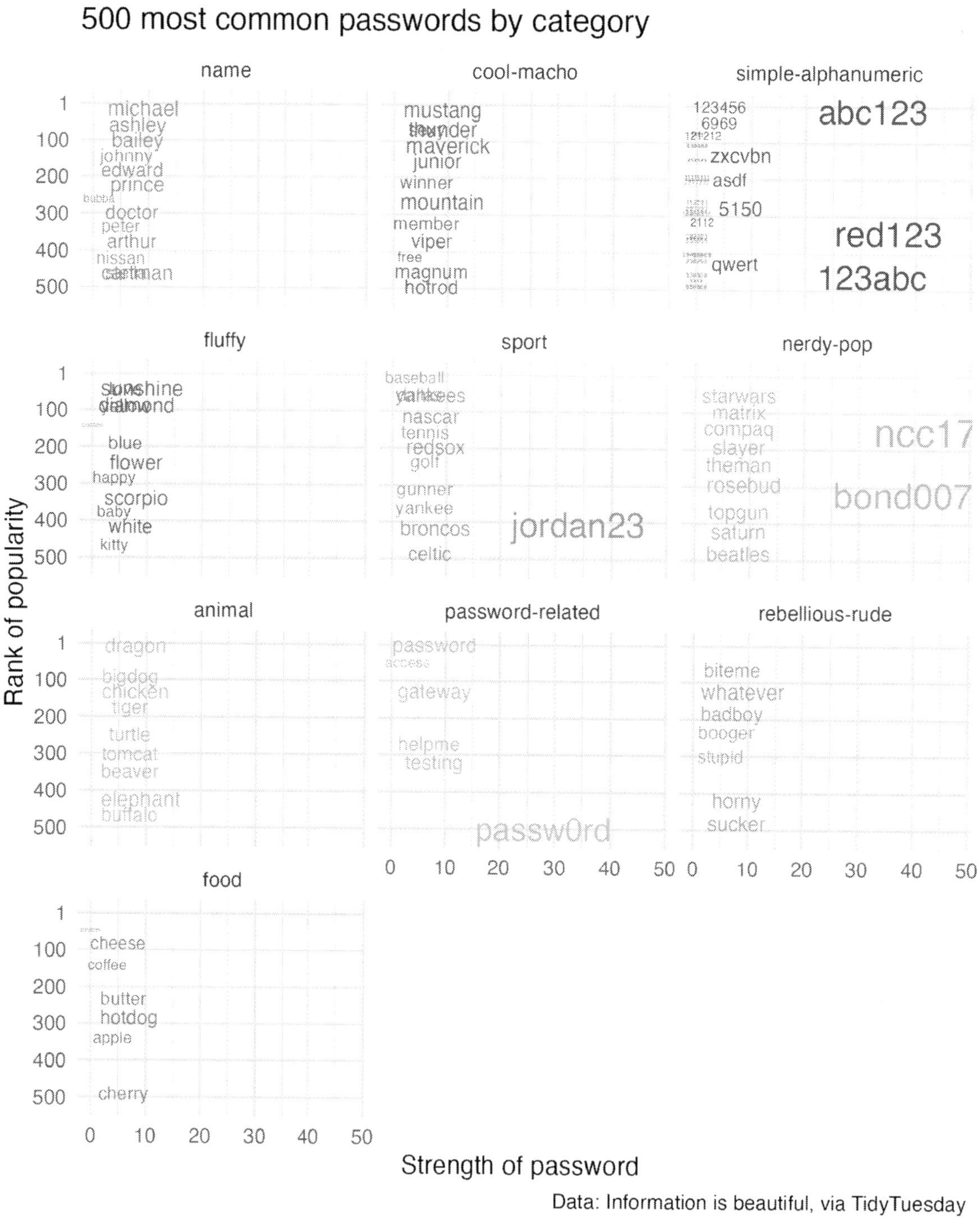

Figure 3.3: Rank vs. strength of 500 most common passwords by category.

3.2 Interactive R tutorials

Navigate the concepts you've learned in this chapter in R using the following self-paced tutorials. All you need is your browser to get started!

Tutorial 1: Introduction to data
https://openintrostat.github.io/ims-tutorials/01-data

Tutorial 1 - Lesson 1: Language of data
https://openintro.shinyapps.io/ims-01-data-01

Tutorial 1 - Lesson 2: Types of studies
https://openintro.shinyapps.io/ims-01-data-02

Tutorial 1 - Lesson 3: Sampling strategies and experimental design
https://openintro.shinyapps.io/ims-01-data-03

Tutorial 1 - Lesson 4: Case study
https://openintro.shinyapps.io/ims-01-data-04

You can also access the full list of tutorials supporting this book at https://openintrostat.github.io/ims-tutorials.

3.3 R labs

Further apply the concepts you've learned in this part in R with computational labs that walk you through a data analysis case study.

Intro to R - Birth rates
https://www.openintro.org/go?id=ims-r-lab-intro-to-r

You can also access the full list of labs supporting this book at https://www.openintro.org/go?id=ims-r-labs.

PART II

EXPLORATORY DATA ANALYSIS

Chapter 4

Exploring categorical data

This chapter focuses on exploring **categorical** data using summary statistics and visualizations. The summaries and graphs presented in this chapter are created using statistical software; however, since this might be your first exposure to the concepts, we take our time in this chapter to detail how to create them. Where possible, we present multivariate plots; plots that visualize the relationship between multiple variables. Mastery of the content presented in this chapter will be crucial for understanding the methods and techniques introduced in rest of the book.

In this chapter we will work with data on loans from Lending Club that you've previously seen in Chapter 1. The `loan50` dataset from Chapter 1 represents a sample from a larger loan dataset called `loans`. This larger dataset contains information on 10,000 loans made through Lending Club. We will examine the relationship between `homeownership`, which for the `loans` data can take a value of `rent`, `mortgage` (owns but has a mortgage), or `own`, and `app_type`, which indicates whether the loan application was made with a partner or whether it was an individual application.

The `loans_full_schema` data can be found in the **openintro** R package. Based on the data in this dataset we have modified the `homeownership` and `application_type` variables. We will refer to this modified dataset as `loans`.

4.1 Contingency tables and bar plots

Table 4.1 summarizes two variables: `application_type` and `homeownership`. A table that summarizes data for two categorical variables in this way is called a **contingency table**. Each value in the table represents the number of times a particular combination of variable outcomes occurred.

For example, the value 3496 corresponds to the number of loans in the dataset where the borrower rents their home and the application type was by an individual. Row and column totals are also included. The **row totals** provide the total counts across each row and the **column totals** down each column. We can also create a table that shows only the overall percentages or proportions for each combination of categories, or we can create a table for a single variable, such as the one shown in Table 4.2 for the `homeownership` variable.

A bar plot is a common way to display a single categorical variable. The left panel of Figure 4.1 shows a **bar plot** for the `homeownership` variable. In the right panel, the counts are converted into

Table 4.1: A contingency table for application type and homeownership.

application_type	homeownership			Total
	rent	mortgage	own	
joint	362	950	183	1495
individual	3496	3839	1170	8505
Total	3858	4789	1353	10000

Table 4.2: A table summarizing the frequencies for each value of the homeownership variable: mortgage, own, and rent.

homeownership	Count
rent	3858
mortgage	4789
own	1353
Total	10000

proportions, showing the proportion of observations that are in each level.

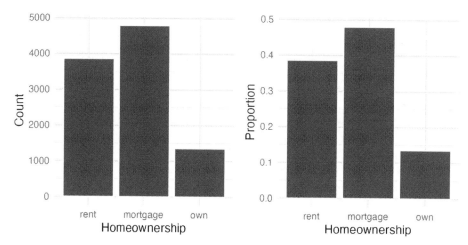

Figure 4.1: Two bar plots: the left panel shows the counts and the right panel shows the proportions of values of the homeownership variable.

4.2 Visualizing two categorical variables

4.2.1 Bar plots with two variables

We can display the distributions of two categorical variables on a bar plot concurrently. Such plots are generally useful for visualizing the relationship between two categorical variables. Figure 4.2 shows three such plots that visualize the relationship between homeownership and application_type variables. Plot A in Figure 4.2 is a **stacked bar plot**. This plot most clearly displays that loan applicants most commonly live in mortgaged homes. It is difficult to say, based on this plot alone, how different application types vary across the levels of homeownership. Plot B is a **dodged bar plot**. This plot most clearly displays that within each level of homeownership, individual applications are more common than joint applications. Finally, plot C is a **standardized bar plot** (also known as **filled bar plot**). This plot most clearly displays that joint applications are most common among loans for applicants who live in mortgaged homes, compared to renters and owners. This type of visualization is helpful in understanding the fraction of individual or joint loan applications for borrowers in each level of homeownership. Additionally, since the proportions of joint and individual loans vary across the groups, we can conclude that the two variables are associated for this sample.

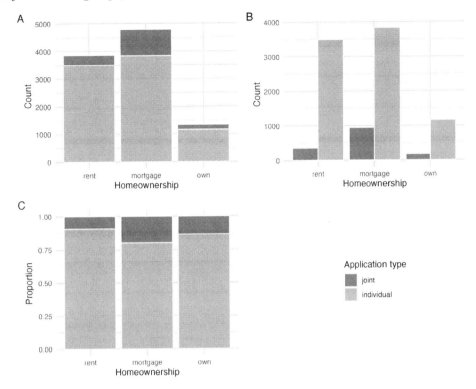

Figure 4.2: Three bar plots (stacked, dodged, and standardized) displaying homeownership and application type variables.

EXAMPLE

Examine the three bar plots in Figure 4.2. When is the stacked, dodged, or standardized bar plot the most useful?

The stacked bar plot is most useful when it's reasonable to assign one variable as the explanatory variable (here homeownership) and the other variable as the response (here application_type), since we are effectively grouping by one variable first and then breaking it down by the others.

Dodged bar plots are more agnostic in their display about which variable, if any, represents the explanatory and which the response variable. It is also easy to discern the number of cases in each of the six different group combinations. However, one downside is that it tends to require more horizontal space; the narrowness of Plot B compared to the other two in Figure 4.2 makes the plot feel a bit cramped. Additionally, when two groups are of very different sizes, as we see in the group own relative to either of the other two groups, it is difficult to discern if there is an association between the variables.

The standardized stacked bar plot is helpful if the primary variable in the stacked bar plot is relatively imbalanced, e.g., the category has only a third of the observations in the category, making the simple stacked bar plot less useful for checking for an association. The major downside of the standardized version is that we lose all sense of how many cases each of the bars represents.

4.2.2 Mosaic plots

A **mosaic plot** is a visualization technique suitable for contingency tables that resembles a standardized stacked bar plot with the benefit that we still see the relative group sizes of the primary variable as well.

To get started in creating our first mosaic plot, we'll break a square into columns for each category of the variable, with the result shown in Plot A of Figure 4.3. Each column represents a level of `homeownership`, and the column widths correspond to the proportion of loans in each of those categories. For instance, there are fewer loans where the borrower is an owner than where the borrower has a mortgage. In general, mosaic plots use box *areas* to represent the number of cases in each category.

Plot B in Figure 4.3 displays the relationship between homeownership and application type. Each column is split proportionally to the number of loans from individual and joint borrowers. For example, the second column represents loans where the borrower has a mortgage, and it was divided into individual loans (upper) and joint loans (lower). As another example, the bottom segment of the third column represents loans where the borrower owns their home and applied jointly, while the upper segment of this column represents borrowers who are homeowners and filed individually. We can again use this plot to see that the `homeownership` and `application_type` variables are associated, since some columns are divided in different vertical locations than others, which was the same technique used for checking an association in the standardized stacked bar plot.

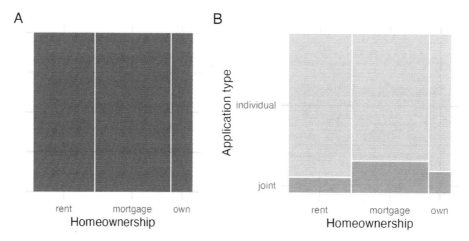

Figure 4.3: The mosaic plots: one for homeownership alone and the other displaying the relationship between homeownership and application type.

In Figure 4.3, we chose to first split by the homeowner status of the borrower. However, we could have instead first split by the application type, as in Figure 4.4. Like with the bar plots, it's common to use the explanatory variable to represent the first split in a mosaic plot, and then for the response to break up each level of the explanatory variable, if these labels are reasonable to attach to the variables under consideration.

4.3 Row and column proportions

In the previous sections we inspected visualizations of two categorical variables in bar plots and mosaic plots. However, we have not discussed how the values in the bar and mosaic plots that show proportions are calculated. In this section we will investigate fractional breakdown of one variable in another and we can modify our contingency table to provide such a view. Table 4.3 shows **row proportions** for Table 4.1, which are computed as the counts divided by their row totals. The value 3496 at the intersection of individual and rent is replaced by $3496/8505 = 0.411$, i.e., 3496 divided by its row total, 8505. So what does 0.411 represent? It corresponds to the proportion of individual applicants who rent.

4.3. ROW AND COLUMN PROPORTIONS

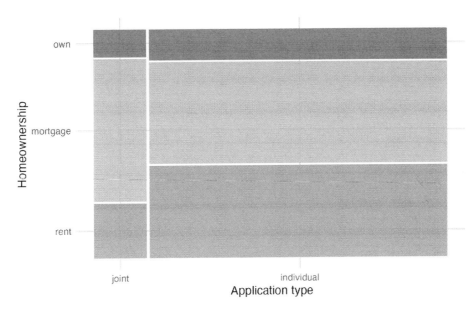

Figure 4.4: Mosaic plot where loans are grouped by homeownership after they have been divided into individual and joint application types.

Table 4.3: A contingency table with row proportions for the application type and homeownership variables.

application_type	homeownership			Total
	rent	mortgage	own	
joint	0.242	0.635	0.122	1
individual	0.411	0.451	0.138	1

A contingency table of the **column proportions** is computed in a similar way, where each is computed as the count divided by the corresponding column total. Table 4.4 shows such a table, and here the value 0.906 indicates that 90.6% of renters applied as individuals for the loan. This rate is higher compared to loans from people with mortgages (80.2%) or who own their home (86.5%). Because these rates vary between the three levels of homeownership (rent, mortgage, own), this provides evidence that app_type and homeownership variables may be associated.

Table 4.4: A contingency table with column proportions for the application type and homeownership variables.

application_type	homeownership		
	rent	mortgage	own
joint	0.094	0.198	0.135
individual	0.906	0.802	0.865
Total	1.000	1.000	1.000

Row and column proportions can also be thought of as **conditional proportions** as they tell us about the proportion of observations in a given level of a categorical variable conditional on the level of another categorical variable.

We could also have checked for an association between application_type and homeownership in Table 4.3 using row proportions. When comparing these row proportions, we would look down columns to see if the fraction of loans where the borrower rents, has a mortgage, or owns varied across the application types.

GUIDED PRACTICE

What does 0.451 represent in Table 4.3? What does 0.802 represent in Table 4.4?[1]

GUIDED PRACTICE

What does 0.122 represent in Table 4.3? What does 0.135 represent in Table 4.4?[2]

EXAMPLE

Data scientists use statistics to build email spam filters. By noting specific characteristics of an email, a data scientist may be able to classify some emails as spam or not spam with high accuracy. One such characteristic is whether the email contains no numbers, small numbers, or big numbers. Another characteristic is the email format, which indicates whether or not an email has any HTML content, such as bolded text. We'll focus on email format and spam status using the dataset; these variables are summarized in a contingency table in Table 4.5. Which would be more helpful to someone hoping to classify email as spam or regular email for this table: row or column proportions?

A data scientist would be interested in how the proportion of spam changes within each email format. This corresponds to column proportions: the proportion of spam in plain text emails and the proportion of spam in HTML emails.

If we generate the column proportions, we can see that a higher fraction of plain text emails are spam (209/1195 = 17.5%) than compared to HTML emails (158/2726 = 5.8%). This information on its own is insufficient to classify an email as spam or not spam, as over 80% of plain text emails are not spam. Yet, when we carefully combine this information with many other characteristics, we stand a reasonable chance of being able to classify some emails as spam or not spam with confidence. This example points out that row and column proportions are not equivalent. Before settling on one form for a table, it is important to consider each to ensure that the most useful table is constructed. However, sometimes it simply isn't clear which, if either, is more useful.

The email data can be found in the **openintro** R package.

Table 4.5: A contingency table for spam and format

spam	HTML	text	Total
not spam	2568	986	3554
spam	158	209	367
Total	2726	1195	3921

---

[1] 0.451 represents the proportion of individual applicants who have a mortgage. 0.802 represents the fraction of applicants with mortgages who applied as individuals.

[2] 0.122 represents the fraction of joint borrowers who own their home. 0.135 represents the home-owning borrowers who had a joint application for the loan.

EXAMPLE

Look back to Table 4.3 and Table 4.4. Are there any obvious scenarios where one might be more useful than the other?

None that we think are obvious! What is distinct about the email example is that the two loan variables don't have a clear explanatory-response variable relationship that we might hypothesize. Usually it is most useful to "condition" on the explanatory variable. For instance, in the email example, the email format was seen as a possible explanatory variable of whether the message was spam, so we would find it more interesting to compute the relative frequencies (proportions) for each email format.

4.4 Pie charts

A **pie chart** is shown in Figure 4.5 alongside a bar plot representing the same information. Pie charts can be useful for giving a high-level overview to show how a set of cases break down. However, it is also difficult to decipher certain details in a pie chart. For example, it's not immediately obvious that there are more loans where the borrower has a mortgage than rent when looking at the pie chart, while this detail is very obvious in the bar plot.

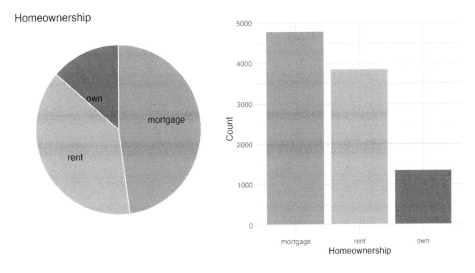

Figure 4.5: A pie chart and bar plot of homeownership.

Pie charts can work well when the goal is to visualize a categorical variable with very few levels, and especially if each level represents a simple fraction (e.g., one-half, one-quarter, etc.). However they can be quite difficult to read when they are used to visualize a categorical variable with many levels. For example, the pie chart and the bar plot in Figure 4.6 both represent the distribution of loan grades (A through G). In this case, it is far easier to compare the counts of each loan grade using the bar plot than the pie chart.

4.5 Waffle charts

Another useful technique of visualizing categorical data is a **waffle chart**. Waffle charts can be used to communicate the proportion of the data that falls into each level of a categorical variable. Just like with pie charts, they work best when the number of levels represented is low. However, unlike pie charts, they can make it easier to compare proportions that represent non-simple fractions. Figure

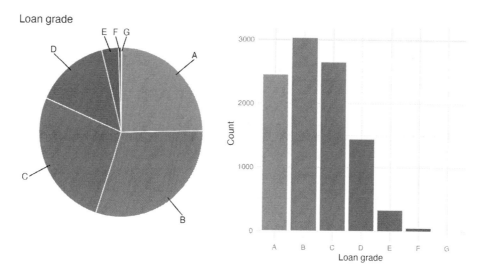

Figure 4.6: A pie chart and bar plot of loan grades.

4.7 displays two examples of waffle charts: one for the distribution of homeownership and the other for the distribution of loan status.

Figure 4.7: Plot A: Waffle chart of homeownership, with levels rent, morgage, and own. Plot B: Waffle chart of loan status, with levels current, fully paid, in grade period, and late.

4.6 Comparing numerical data across groups

Some of the more interesting investigations can be considered by examining numerical data across groups. In this section we will expand on a few methods we've already seen to make plots for numerical data from multiple groups on the same graph as well as introduce a few new methods for comparing numerical data across groups.

We will revisit the county dataset and compare the median household income for counties that gained population from 2010 to 2017 versus counties that had no gain. While we might like to make a causal connection between income and population growth, remember that these are observational data and so such an interpretation would be, at best, half-baked.

4.6. COMPARING NUMERICAL DATA ACROSS GROUPS

We have data on 3142 counties in the United States. We are missing 2017 population data from 3 of them, and of the remaining 3139 counties, in 1541 the population increased from 2010 to 2017 and in the remaining 1598 the population decreased. Table 4.6 shows a sample of 5 observations from each group.

Table 4.6: The median household income from a random sample of 5 counties with population gain between 2010 to 2017 and another random sample of 5 counties with no population gain.

State	County	Population change (%)	Gain / No gain	Median household income
Colorado	Custer County	14.28	gain	41330
Georgia	Murray County	1.35	gain	41617
Georgia	Pickens County	7.41	gain	61542
Texas	Wharton County	2.12	gain	50145
Washington	Grays Harbor County	2.30	gain	45483
Alabama	Conecuh County	-3.40	no gain	30434
Illinois	McDonough County	-4.32	no gain	42911
Iowa	Iowa County	-1.08	no gain	58077
Michigan	Genesee County	-1.95	no gain	45231
Wyoming	Campbell County	-3.76	no gain	80178

Color can be used to split histograms for numerical variables by levels of a categorical variable. An example of this is shown in Plot A of Figure 4.8. The **side-by-side box plot** is another traditional tool for comparing across groups. An example is shown in Plot B of Figure 4.8, where there are two box plots, one for each group, placed into one plotting window and drawn on the same scale.

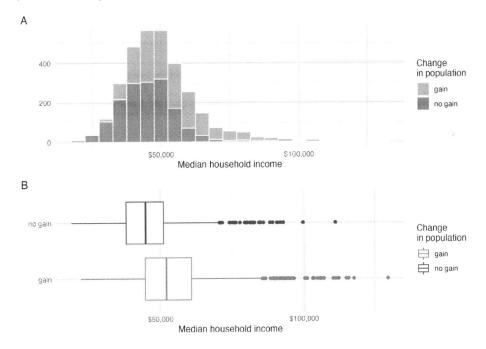

Figure 4.8: Histograms (Plot A) and side by-side box plots (Plot B) for median household income, where counties are split by whether there was a population gain or not.

GUIDED PRACTICE

Use the plots in Figure 4.8 to compare the incomes for counties across the two groups. What do you notice about the approximate center of each group? What do you notice about the variability between groups? Is the shape relatively consistent between groups? How many *prominent* modes are there for each group?[3]

 GUIDED PRACTICE

What components of each plot in Figure 4.8 do you find most useful?[4]

Another useful visualization for comparing numerical data across groups is a **ridge plot**, which combines density plots for various groups drawn on the same scale in a single plotting window. Figure 4.9 displays a ridge plot for the distribution of median household income in counties, split by whether there was a population gain or not.

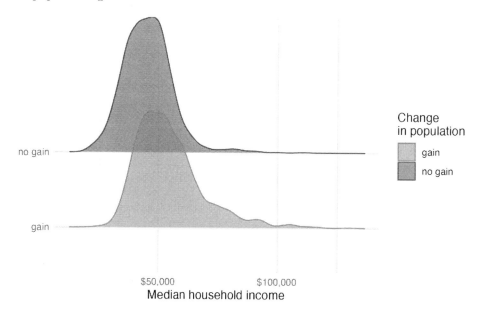

Figure 4.9: Ridge plot for median household income, where counties are split by whether there was a population gain or not.

 GUIDED PRACTICE

What components of the ridge plot in Figure 4.9 do you find most useful compared to those in Figure 4.8?[5]

One last visualization technique we'll highlight for comparing numerical data across groups is **faceting**. In this technique we split (facet) the graphical display of the data across plotting windows based

---

[3]Answers may vary a little. The counties with population gains tend to have higher income (median of about $45,000) versus counties without a gain (median of about $40,000). The variability is also slightly larger for the population gain group. This is evident in the IQR, which is about 50% bigger in the *gain* group. Both distributions show slight to moderate right skew and are unimodal. The box plots indicate there are many observations far above the median in each group, though we should anticipate that many observations will fall beyond the whiskers when examining any dataset that contain more than a few hundred data points.

[4]Answers will vary. The side-by-side box plots are especially useful for comparing centers and spreads, while the hollow histograms are more useful for seeing distribution shape, skew, modes, and potential anomalies.

[5]The ridge plot give us a better sense of the shape, and especially modality, of the data.

on groups. Plot A in Figure 4.10 displays the same information as Plot A in Figure 4.8, however here the distributions of median household income for counties with and without population gain are faceted across two plotting windows. We preserve the same scale on the x and y axes for easier comparison. An advantage of this approach is that it extends to splitting the data across levels of two categorical variables, which allows for displaying relationships between three variables. In Plot B in Figure 4.10 we have now split the data into four groups using the `pop_change` and `metro` variables:

- top left represents counties that are *not* in a `metropolitan` area with population gain,
- top right represents counties that are in a `metropolitan` area with population gain,
- bottom left represents counties that are *not* in a `metropolitan` area without population gain, and finally
- bottom right represents counties that are in a `metropolitan` area without population gain.

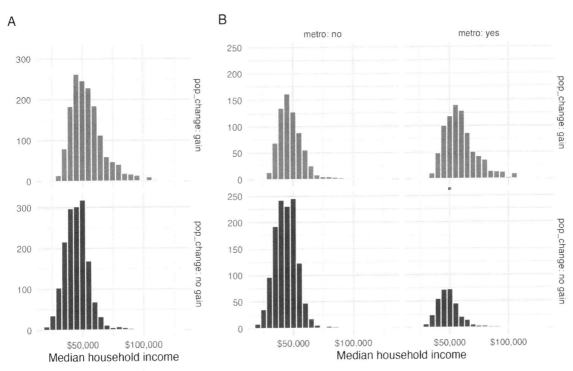

Figure 4.10: Distribution of median income in counties using faceted histograms: Plot A facets by whether there was a population gain or not and Plot B facets by both population gain and whether the county is in a metropolitan area.

We can continue building up on this visualization to add one more variable, `median_edu`, which is the median education level in the county. In Figure 4.11, we represent median education level using color, where pink (solid line) represents counties where the median education level is high school diploma, yellow (dashed line) is some college degree, and red (dotted line) is Bachelor's.

 GUIDED PRACTICE

Based on Figure 4.11, what can you say about how median household income in counties vary depending on population gain/no gain, metropolitan area/not, and median degree?[6]

---

[6]The ridge plot give us a better sense of the shape, and especially modality, of the data.

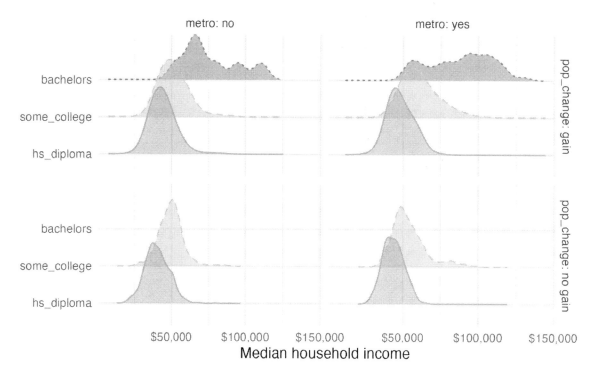

Figure 4.11: Distribution of median income in counties using a ridge plot, faceted by whether the county had a population gain or not as well as whether the county is in a metropolitan area and colored by the median education level in the county.

4.7 Chapter review

4.7.1 Summary

Fluently working with categorical variables is an important skill for data analysts. In this chapter we've introduced different visualizations and numerical summaries applied to categorical variables. The graphical visualizations are even more descriptive when two variables are presented simultaneously. We presented bar plots, mosaic plots, pie charts, and estimations of conditional proportions.

4.7.2 Terms

We introduced the following terms in the chapter. If you're not sure what some of these terms mean, we recommend you go back in the text and review their definitions. We are purposefully presenting them in alphabetical order, instead of in order of appearance, so they will be a little more challenging to locate. However you should be able to easily spot them as **bolded text**.

column proportions	faceted plot	row totals
column totals	filled bar plot	side-by-side box plot
conditional proportions	mosaic plot	stacked bar plot
contingency table	ridge plot	standardized bar plot
dodged bar plot	row proportions	

4.8 Exercises

Answers to odd numbered exercises can be found in Appendix A.4.

4.1. **Antibiotic use in children.** The bar plot and the pie chart below show the distribution of pre-existing medical conditions of children involved in a study on the optimal duration of antibiotic use in treatment of tracheitis, which is an upper respiratory infection.[7]

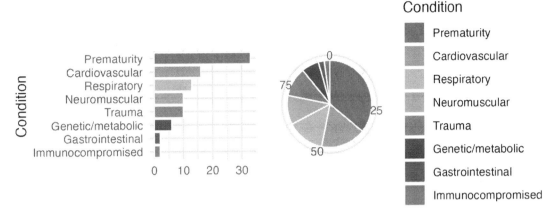

a. What features are apparent in the bar plot but not in the pie chart?

b. What features are apparent in the pie chart but not in the bar plot?

c. Which graph would you prefer to use for displaying these categorical data?

4.2. **Views on immigration.** Nine-hundred and ten (910) randomly sampled registered voters from Tampa, FL were asked if they thought workers who have illegally entered the US should be (i) allowed to keep their jobs and apply for US citizenship, (ii) allowed to keep their jobs as temporary guest workers but not allowed to apply for US citizenship, or (iii) lose their jobs and have to leave the country. The results of the survey by political ideology are shown below.[8]

Response	Conservative	Liberal	Moderate	Total
Apply for citizenship	57	101	120	278
Guest worker	121	28	113	262
Leave the country	179	45	126	350
Not sure	15	1	4	20
Total	372	175	363	910

a. What percent of these Tampa, FL voters identify themselves as conservatives?

b. What percent of these Tampa, FL voters are in favor of the citizenship option?

c. What percent of these Tampa, FL voters identify themselves as conservatives and are in favor of the citizenship option?

d. What percent of these Tampa, FL voters who identify themselves as conservatives are also in favor of the citizenship option? What percent of moderates share this view? What percent of liberals share this view?

e. Do political ideology and views on immigration appear to be associated? Explain your reasoning.

f. Conjecture other possible variables that might explain the potential relationship between these two variables.

---

[7] The antibiotics data used in this exercise can be found in the **openintro** R package.

[8] The immigration data used in this exercise can be found in the **openintro** R package.

4.3. **Black Lives Matter.** A Washington Post-Schar School poll conducted in the United States in June 2020, among a random national sample of 1,006 adults, asked respondents whether they support or oppose protests following George Floyd's killing that have taken place in cities across the US. The survey also collected information on the age of the respondents. (Washington Post, 2020) The results are summarized in the stacked bar plot below.

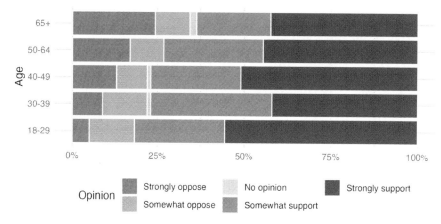

a. Based on the stacked bar plot, do views on the protests and age appear to be associated? Explain your reasoning.

b. Conjecture other possible variables that might explain the potential association between these two variables.

4.4. **Raise taxes.** A random sample of registered voters nationally were asked whether they think it's better to raise taxes on the rich or raise taxes on the poor. The survey also collected information on the political party affiliation of the respondents. (Public Policy Polling, 2015)

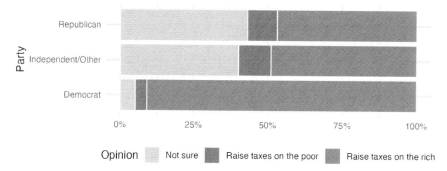

a. Based on the stacked bar plot shown above, do views on raising taxes and political affiliation appear to be associated? Explain your reasoning.

b. Conjecture other possible variables that might explain the potential association between these two variables.

4.5. **Heart transplant data display.** The Stanford University Heart Transplant Study was conducted to determine whether an experimental heart transplant program increased lifespan. Each patient entering the program was officially designated a heart transplant candidate, meaning that he was gravely ill and might benefit from a new heart. Patients were randomly assigned into treatment and control groups. Patients in the treatment group received a transplant, and those in the control group did not. The visualization below displays two different versions of the data.[9] (Turnbull et al., 1974)

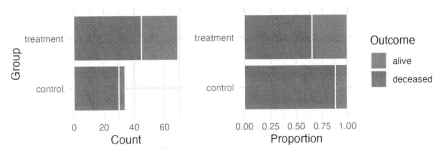

a. Provide one aspect of the two group comparison that is easier to see from the stacked bar plot (left)?

b. Provide one aspect of the two group comparison that is easeir to see from the standardized bar plot (right)?

c. For the Heart Transplant Study which of those aspects would be more important to display? That is, which bar plot would be better as a data visualization?

4.6. **Shipping holiday gifts data display.** A local news survey asked 500 randomly sampled Los Angeles residents which shipping carrier they prefer to use for shipping holiday gifts. The table below shows the distribution of responses by age group as well as the expected counts for each cell (shown in italics).

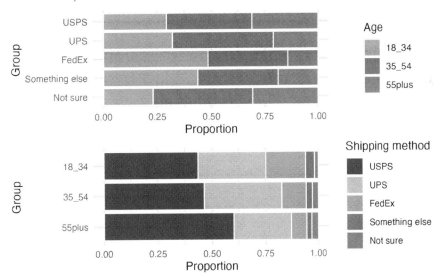

a. Which graph (top or bottom) would you use to understand the shipping choices of people of different ages?

b. Which graph (top or bottom) would you use to understand the age distribution across different types of shipping choices?

c. A new shipping company would like to market to people over the age of 55. Who will be their biggest competitor?

d. FedEx would like to reach out to grow their market share so as to balance the age demographics of FedEx users. To what age group should FedEx market?

---

[9] The `heart_transplant` data used in this exercise can be found in the **openintro** R package.

Chapter 5

Exploring numerical data

This chapter focuses on exploring **numerical** data using summary statistics and visualizations. The summaries and graphs presented in this chapter are created using statistical software; however, since this might be your first exposure to the concepts, we take our time in this chapter to detail how to create them. Mastery of the content presented in this chapter will be crucial for understanding the methods and techniques introduced in rest of the book.

Consider the `loan_amount` variable from the `loan50` dataset, which represents the loan size for each of 50 loans in the dataset.

This variable is numerical since we can sensibly discuss the numerical difference of the size of two loans. On the other hand, area codes and zip codes are not numerical, but rather they are categorical variables.

Throughout this chapter, we will apply numerical methods using the `loan50` and `county` datasets, which were introduced in Section 1.2. If you'd like to review the variables from either dataset, see Tables 1.4 and 1.6.

The `county` data can be found in the **usdata** R package and the `loan50` data can be found in the **openintro** R package.

5.1 Scatterplots for paired data

A **scatterplot** provides a case-by-case view of data for two numerical variables. In Figure 1.2, a scatterplot was used to examine the homeownership rate against the percentage of housing units that are in multi-unit structures (e.g., apartments) in the `county` dataset. Another scatterplot is shown in Figure 5.1, comparing the total income of a borrower `total_income` and the amount they borrowed `loan_amount` for the `loan50` dataset. In any scatterplot, each point represents a single case. Since there are 50 cases in `loan50`, there are 50 points in Figure 5.1.

Looking at Figure 5.1, we see that there are many borrowers with income below $100,000 on the left side of the graph, while there are a handful of borrowers with income above $250,000.

5.2. DOT PLOTS AND THE MEAN

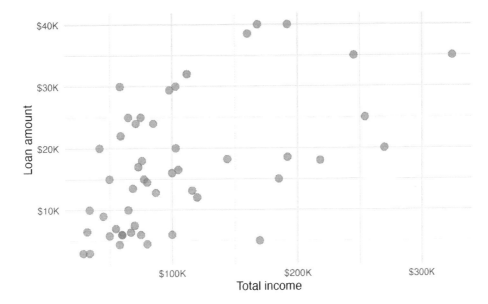

Figure 5.1: A scatterplot of loan amount versus total income for the `loan50` dataset.

EXAMPLE

Figure 5.2 shows a plot of median household income against the poverty rate for 3142 counties in the US. What can be said about the relationship between these variables?

The relationship is evidently **nonlinear**, as highlighted by the dashed line. This is different from previous scatterplots we have seen, which indicate very little, if any, curvature in the trend.

GUIDED PRACTICE

What do scatterplots reveal about the data, and how are they useful?[1]

GUIDED PRACTICE

Describe two variables that would have a horseshoe-shaped association in a scatterplot (∩ or ⌢).[2]

5.2 Dot plots and the mean

Sometimes we are interested in the distribution of a single variable. In these cases, a dot plot provides the most basic of displays. A **dot plot** is a one-variable scatterplot; an example using the interest rate of 50 loans is shown in Figure 5.3.

---

[1] Answers may vary. Scatterplots are helpful in quickly spotting associations relating variables, whether those associations come in the form of simple trends or whether those relationships are more complex.

[2] Consider the case where your vertical axis represents something "good" and your horizontal axis represents something that is only good in moderation. Health and water consumption fit this description: we require some water to survive, but consume too much and it become toxic and can kill a person.

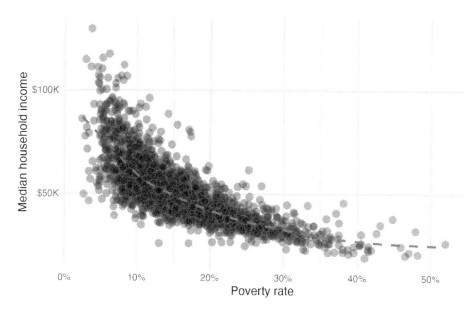

Figure 5.2: A scatterplot of the median household income against the poverty rate for the county dataset. Data are from 2017. A statistical model has also been fit to the data and is shown as a dashed line.

The **mean**, often called the **average** is a common way to measure the center of a **distribution** of data. To compute the mean interest rate, we add up all the interest rates and divide by the number of observations.

The sample mean is often labeled $\bar{x}$. The letter x is being used as a generic placeholder for the variable of interest and the bar over the x communicates we're looking at the average interest rate, which for these 50 loans is 11.57%. It's useful to think of the mean as the balancing point of the distribution, and it's shown as a triangle in Figure 5.3.

Mean.

The sample mean can be calculated as the sum of the observed values divided by the number of observations:

$$\bar{x} = \frac{x_1 + x_2 + \cdots + x_n}{n}$$

GUIDED PRACTICE

Examine the equation for the mean. What does x_1 correspond to? And x_2? Can you infer a general meaning to what x_i might represent?[3]

GUIDED PRACTICE

What was n in this sample of loans?[4]

---

[3]x_1 corresponds to the interest rate for the first loan in the sample, x_2 to the second loan's interest rate, and x_i corresponds to the interest rate for the ith loan in the dataset. For example, if $i = 4$, then we're examining x_4, which refers to the fourth observation in the dataset.

[4]The sample size was $n = 50$.

5.2. DOT PLOTS AND THE MEAN

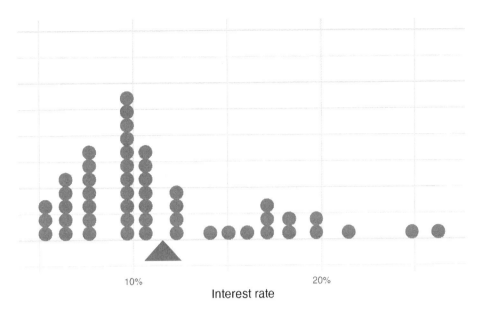

Figure 5.3: A dot plot of interest rate for the `loan50` dataset. The rates have been rounded to the nearest percent in this plot, and the distribution's mean is shown as a red triangle.

The `loan50` dataset represents a sample from a larger population of loans made through Lending Club. We could compute a mean for the entire population in the same way as the sample mean. However, the population mean has a special label: μ. The symbol μ is the Greek letter *mu* and represents the average of all observations in the population. Sometimes a subscript, such as $_x$, is used to represent which variable the population mean refers to, e.g., μ_x. Oftentimes it is too expensive to measure the population mean precisely, so we often estimate μ using the sample mean, $\bar{x}$.

 The Greek letter μ is pronounced *mu*, listen to the pronunciation here.

 EXAMPLE

Although we don't have an ability to *calculate* the average interest rate across all loans in the populations, we can *estimate* the population value using the sample data. Based on the sample of 50 loans, what would be a reasonable estimate of μ_x, the mean interest rate for all loans in the full dataset?

The sample mean, 11.57, provides a rough estimate of μ_x. While it is not perfect, this is our single best guess **point estimate** of the average interest rate of all the loans in the population under study. In Chapter 11 and beyond, we will develop tools to characterize the accuracy of point estimates, like the sample mean. As you might have guessed, point estimates based on larger samples tend to be more accurate than those based on smaller samples.

The mean is useful because it allows us to rescale or standardize a metric into something more easily interpretable and comparable. Suppose we would like to understand if a new drug is more effective at treating asthma attacks than the standard drug. A trial of 1,500 adults is set up, where 500 receive the new drug, and 1000 receive a standard drug in the control group. Results of this trial are summarized in Table 5.1.

Comparing the raw counts of 200 to 300 asthma attacks would make it appear that the new drug is better, but this is an artifact of the imbalanced group sizes. Instead, we should look at the average number of asthma attacks per patient in each group:

Table 5.1: Results of a trial of 1500 adults that suffer from asthma.

	New drug	Standard drug
Number of patients	500	1000
Total asthma attacks	200	300

- New drug: 200/500 = 0.4 asthma attacks per patient
- Standard drug: 300/1000 = 0.3 asthma attacks per patient

The standard drug has a lower average number of asthma attacks per patient than the average in the treatment group.

EXAMPLE

Come up with another example where the mean is useful for making comparisons.

Emilio opened a food truck last year where he sells burritos, and his business has stabilized over the last 3 months. Over that 3 month period, he has made $11,000 while working 625 hours. Emilio's average hourly earnings provides a useful statistic for evaluating whether his venture is, at least from a financial perspective, worth it:

$$\frac{\$11000}{625 \text{ hours}} = \$17.60 \text{ per hour}$$

By knowing his average hourly wage, Emilio now has put his earnings into a standard unit that is easier to compare with many other jobs that he might consider.

EXAMPLE

Suppose we want to compute the average income per person in the US. To do so, we might first think to take the mean of the per capita incomes across the 3,142 counties in the county dataset. What would be a better approach?

The county dataset is special in that each county actually represents many individual people. If we were to simply average across the income variable, we would be treating counties with 5,000 and 5,000,000 residents equally in the calculations. Instead, we should compute the total income for each county, add up all the counties' totals, and then divide by the number of people in all the counties. If we completed these steps with the county data, we would find that the per capita income for the US is $30,861. Had we computed the *simple* mean of per capita income across counties, the result would have been just $26,093!

This example used what is called a **weighted mean**. For more information on this topic, check out the following online supplement regarding weighted means.

5.3 Histograms and shape

Dot plots show the exact value for each observation. They are useful for small datasets but can become hard to read with larger samples. Rather than showing the value of each observation, we prefer to think of the value as belonging to a *bin*. For example, in the `loan50` dataset, we created a table of counts for the number of loans with interest rates between 5.0% and 7.5%, then the number of loans with rates between 7.5% and 10.0%, and so on. Observations that fall on the boundary of a bin (e.g., 10.00%) are allocated to the lower bin. The tabulation is shown in Table 5.2, and the binned counts are plotted as bars in Figure 5.4 into what is called a **histogram**. Note that the histogram resembles a more heavily binned version of the stacked dot plot shown in Figure 5.3.

Table 5.2: Counts for the binned interest rate data.

Interest rate	Count
(5% - 7.5%]	11
(7.5% - 10%]	15
(10% - 12.5%]	8
(12.5% - 15%]	4
(15% - 17.5%]	5
(17.5% - 20%]	4
(20% - 22.5%]	1
(22.5% - 25%]	1
(25% - 27.5%]	1

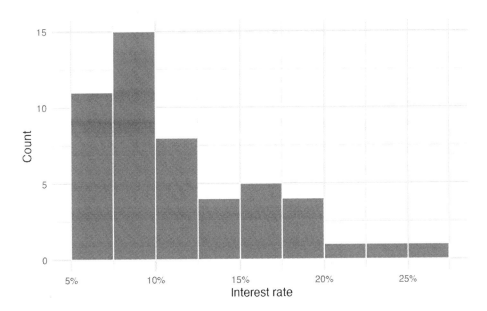

Figure 5.4: A histogram of interest rate. This distribution is strongly skewed to the right.

Histograms provide a view of the **data density**. Higher bars represent where the data are relatively more common. For instance, there are many more loans with rates between 5% and 10% than loans with rates between 20% and 25% in the dataset. The bars make it easy to see how the density of the data changes relative to the interest rate.

Histograms are especially convenient for understanding the shape of the data distribution. Figure 5.4 suggests that most loans have rates under 15%, while only a handful of loans have rates above 20%. When the distribution of a variable trails off to the right in this way and has a longer right **tail**, the shape is said to be **right skewed**.[5]

---

[5]Other ways to describe data that are right skewed: skewed to the right, skewed to the high end, or skewed to the positive end.

Variables with the reverse characteristic – a long, thinner tail to the left – are said to be **left skewed**. We also say that such a distribution has a long left tail. Variables that show roughly equal trailing off in both directions are called **symmetric**.

When data trail off in one direction, the distribution has a **long tail**. If a distribution has a long left tail, it is left skewed. If a distribution has a long right tail, it is right skewed.

GUIDED PRACTICE

Besides the mean (since it was labeled), what can you see in the dot plot in Figure 5.3 that you cannot see in the histogram in Figure 5.4?[6]

In addition to looking at whether a distribution is skewed or symmetric, histograms can be used to identify modes. A **mode** is represented by a prominent peak in the distribution. There is only one prominent peak in the histogram of `interest_rate`.

A definition of *mode* sometimes taught in math classes is the value with the most occurrences in the dataset. However, for many real-world datasets, it is common to have *no* observations with the same value in a dataset, making this definition impractical in data analysis.

Figure 5.5 shows histograms that have one, two, or three prominent peaks. Such distributions are called **unimodal**, **bimodal**, and **multimodal**, respectively. Any distribution with more than two prominent peaks is called multimodal. Notice that there was one prominent peak in the unimodal distribution with a second less prominent peak that was not counted since it only differs from its neighboring bins by a few observations.

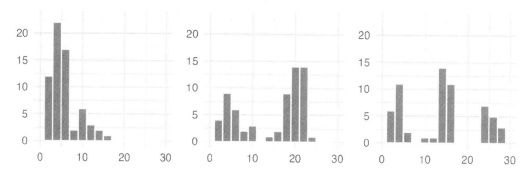

Figure 5.5: Counting only prominent peaks, the distributions are (left to right) unimodal, bimodal, and multimodal. Note that the left plot is unimodal because we are counting prominent peaks, not just any peak.

EXAMPLE

Figure 5.4 reveals only one prominent mode in the interest rate. Is the distribution unimodal, bimodal, or multimodal?[7]

---

[6]The interest rates for individual loans.
[7]Remember that *uni* stands for 1 (think *uni*cycles), and *bi* stands for 2 (think *bi*cycles).

> **GUIDED PRACTICE**
>
> Height measurements of young students and adult teachers at a K-3 elementary school were taken. How many modes would you expect in this height dataset?[8].

Looking for modes isn't about finding a clear and correct answer about the number of modes in a distribution, which is why *prominent* is not rigorously defined in this book. The most important part of this examination is to better understand your data.

5.4 Variance and standard deviation

The mean was introduced as a method to describe the center of a variable, and **variability** in the data is also important. Here, we introduce two measures of variability: the variance and the standard deviation. Both of these are very useful in data analysis, even though their formulas are a bit tedious to calculate by hand. The standard deviation is the easier of the two to comprehend, as it roughly describes how far away the typical observation is from the mean.

We call the distance of an observation from its mean its **deviation**. Below are the deviations for the 1st, 2nd, 3rd, and 50th observations in the `interest_rate` variable:

$$x_1 - \bar{x} = 10.9 - 11.57 = -0.67$$
$$x_2 - \bar{x} = 9.92 - 11.57 = -1.65$$
$$x_3 - \bar{x} = 26.3 - 11.57 = 14.73$$
$$\vdots$$
$$x_{50} - \bar{x} = 6.08 - 11.57 = -5.49$$

If we square these deviations and then take an average, the result is equal to the sample **variance**, denoted by s^2:

$$\begin{aligned} s^2 &= \frac{(-0.67)^2 + (-1.65)^2 + (14.73)^2 + \cdots + (-5.49)^2}{50 - 1} \\ &= \frac{0.45 + 2.72 + \cdots + 30.14}{49} \\ &= 25.52 \end{aligned}$$

We divide by $n - 1$, rather than dividing by n, when computing a sample's variance. There's some mathematical nuance here, but the end result is that doing this makes this statistic slightly more reliable and useful.

Notice that squaring the deviations does two things. First, it makes large values relatively much larger. Second, it gets rid of any negative signs.

---

[8]There might be two height groups visible in the dataset: one of the students and one of the adults. That is, the data are probably bimodal.

Standard deviation.

The sample standard deviation can be calculated as the square root of the sum of the squared distance of each value from the mean divided by the number of observations minus one:

$$s = \sqrt{\frac{\sum_{i=1}^{n}(x_i - \bar{x})^2}{n-1}}$$

The **standard deviation** is defined as the square root of the variance:

$$s = \sqrt{25.52} = 5.05$$

While often omitted, a subscript of x may be added to the variance and standard deviation, i.e., s_x^2 and s_x, if it is useful as a reminder that these are the variance and standard deviation of the observations represented by $x_1, x_2, \ldots, x_n$.

 Variance and standard deviation.

The variance is the average squared distance from the mean. The standard deviation is the square root of the variance. The standard deviation is useful when considering how far the data are distributed from the mean.

The standard deviation represents the typical deviation of observations from the mean. Often about 68% of the data will be within one standard deviation of the mean and about 95% will be within two standard deviations. However, these percentages are not strict rules.

Like the mean, the population values for variance and standard deviation have special symbols: σ^2 for the variance and σ for the standard deviation.

 The Greek letter σ is pronounced *sigma*, listen to the pronunciation here.

 GUIDED PRACTICE

A good description of the shape of a distribution should include modality and whether the distribution is symmetric or skewed to one side. Using Figure 5.7 as an example, explain why such a description is important.[9]

---

[9] Figure 5.7 shows three distributions that look quite different, but all have the same mean, variance, and standard deviation. Using modality, we can distinguish between the first plot (bimodal) and the last two (unimodal). Using skewness, we can distinguish between the last plot (right skewed) and the first two. While a picture, like a histogram, tells a more complete story, we can use modality and shape (symmetry/skew) to characterize basic information about a distribution.

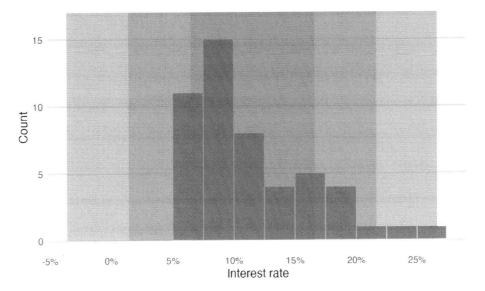

Figure 5.6: For the interest rate variable, 34 of the 50 loans (68%) had interest rates within 1 standard deviation of the mean, and 48 of the 50 loans (96%) had rates within 2 standard deviations. Usually about 68% of the data are within 1 standard deviation of the mean and 95% within 2 standard deviations, though this is far from a hard rule.)

EXAMPLE

Describe the distribution of the interest_rate variable using the histogram in Figure 5.4. The description should incorporate the center, variability, and shape of the distribution, and it should also be placed in context. Also note any especially unusual cases.

The distribution of interest rates is unimodal and skewed to the high end. Many of the rates fall near the mean at 11.57%, and most fall within one standard deviation (5.05%) of the mean. There are a few exceptionally large interest rates in the sample that are above 20%.

In practice, the variance and standard deviation are sometimes used as a means to an end, where the "end" is being able to accurately estimate the uncertainty associated with a sample statistic. For example, in Chapter 13 the standard deviation is used in calculations that help us understand how much a sample mean varies from one sample to the next.

5.5 Box plots, quartiles, and the median

A **box plot** summarizes a dataset using five statistics while also identifying unusual observations. Figure 5.8 provides a dot plot alongside a box plot of the interest_rate variable from the loan50 dataset.

The dark line inside the box represents the **median**, which splits the data in half. 50% of the data fall below this value and 50% fall above it. Since in the loan50 dataset there are 50 observations (an even number), the median is defined as the average of the two observations closest to the 50th percentile. Table 5.3 shows all interest rates, arranged in ascending order. We can see that the 25th and the 26th values are both 9.93, which corresponds to the dark line in the box plot in Figure 5.8.

When there are an odd number of observations, there will be exactly one observation that splits the data into two halves, and in such a case that observation is the median (no average needed).

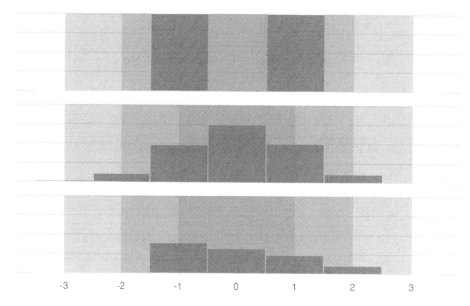

Figure 5.7: Three very different population distributions with the same mean (0) and standard deviation (1).

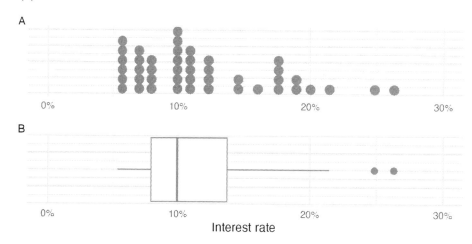

Figure 5.8: Plot A shows a dot plot and Plot B shows a box plot of the distribution of interest rates from the `loan50` dataset.

Table 5.3: Interest rates from the `loan50` dataset, arranged in ascending order.

	1	2	3	4	5	6	7	8	9	10
1	5.31	5.31	5.32	6.08	6.08	6.08	6.71	6.71	7.34	7.35
10	7.35	7.96	7.96	7.96	7.97	9.43	9.43	9.44	9.44	9.44
20	9.92	9.92	9.92	9.92	9.93	9.93	10.42	10.42	10.90	10.90
30	10.91	10.91	10.91	11.98	12.62	12.62	12.62	14.08	15.04	16.02
40	17.09	17.09	17.09	18.06	18.45	19.42	20.00	21.45	24.85	26.30

 Median: the number in the middle.

If the data are ordered from smallest to largest, the **median** is the observation right in the middle. If there are an even number of observations, there will be two values in the middle, and the median is taken as their average.

The second step in building a box plot is drawing a rectangle to represent the middle 50% of the data.

5.5. BOX PLOTS, QUARTILES, AND THE MEDIAN

The length of the the box is called the **interquartile range**, or **IQR** for short. It, like the standard deviation, is a measure of variability in data. The more variable the data, the larger the standard deviation and IQR tend to be. The two boundaries of the box are called the **first quartile** (the 25th percentile, i.e., 25% of the data fall below this value) and the **third quartile** (the 75th percentile, i.e., 75% of the data fall below this value), and these are often labeled Q_1 and Q_3, respectively.

Interquartile range (IQR).

The IQR interquartile range is the length of the box in a box plot. It is computed as $IQR = Q_3 - Q_1$, where Q_1 and Q_3 are the 25th and 75th percentiles, respectively.

A α **percentile** is a number with α% of the observations below and $100 - \alpha$% of the observations above. For example, the 90th percentile of SAT scores is the value of the SAT score with 90% of students below that value and 10% of students above that value.

GUIDED PRACTICE

What percent of the data fall between Q_1 and the median? What percent is between the median and Q_3?[10]

Extending out from the box, the **whiskers** attempt to capture the data outside of the box. The whiskers of a box plot reach to the minimum and the maximum values in the data, unless there are points that are considered unusually high or unusually low, which are identified as potential **outliers** by the box plot. These are labeled with a dot on the box plot. The purpose of labeling the outlying points – instead of extending the whiskers to the minimum and maximum observed values – is to help identify any observations that appear to be unusually distant from the rest of the data. There are a variety of formulas for determining whether a particular data point is considered an outlier, and different statistical software use different formulas. A commonly used formula is that any observation beyond $1.5 \times IQR$ away from the first or the third quartile is considered an outlier. In a sense, the box is like the body of the box plot and the whiskers are like its arms trying to reach the rest of the data, up to the outliers.

Outliers are extreme.

An **outlier** is an observation that appears extreme relative to the rest of the data. Examining data for outliers serves many useful purposes, including

- identifying strong skew in the distribution,
- identifying possible data collection or data entry errors, and
- providing insight into interesting properties of the data.

Keep in mind, however, that some datasets have a naturally long skew and outlying points do **not** represent any sort of problem in the dataset.

GUIDED PRACTICE

Using the box plot in Figure 5.8, estimate the values of the Q_1, Q_3, and IQR for `interest_rate` in the `loan50` dataset.[11]

---

[10]Since Q_1 and Q_3 capture the middle 50% of the data and the median splits the data in the middle, 25% of the data fall between Q_1 and the median, and another 25% falls between the median and Q_3.

[11]These visual estimates will vary a little from one person to the next: $Q_1 \approx 8$%, $Q_3 \approx 14$%, IQR $\approx 14 - 8 = 6$%.

5.6 Robust statistics

How are the **sample statistics** of the interest_rate dataset affected by the observation, 26.3%? What would have happened if this loan had instead been only 15%? What would happen to these summary statistics if the observation at 26.3% had been even larger, say 35%? The three conjectured scenarios are plotted alongside the original data in Figure 5.9, and sample statistics are computed under each scenario in Table 5.4.

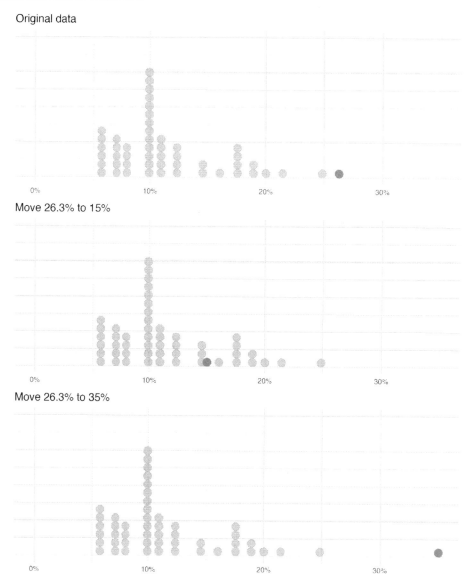

Figure 5.9: Dot plots of the original interest rate data and two modified datasets.

Table 5.4: A comparison of how the median, IQR, mean, and standard deviation change as the value of an extereme observation from the original interest data changes.

Scenario	Robust		Not robust	
	Median	IQR	Mean	SD
Original data	9.93	5.75	11.6	5.05
Move 26.3% to 15%	9.93	5.75	11.3	4.61
Move 26.3% to 35%	9.93	5.75	11.7	5.68

5.7. TRANSFORMING DATA

GUIDED PRACTICE

Which is more affected by extreme observations, the mean or median? Is the standard deviation or IQR more affected by extreme observations?[12]

The median and IQR are called **robust statistics** because extreme observations have little effect on their values: moving the most extreme value generally has little influence on these statistics. On the other hand, the mean and standard deviation are more heavily influenced by changes in extreme observations, which can be important in some situations.

EXAMPLE

The median and IQR did not change under the three scenarios in Table 5.4. Why might this be the case?

The median and IQR are only sensitive to numbers near Q_1, the median, and Q_3. Since values in these regions are stable in the three datasets, the median and IQR estimates are also stable.

GUIDED PRACTICE

The distribution of loan amounts in the `loan50` dataset is right skewed, with a few large loans lingering out into the right tail. If you were wanting to understand the typical loan size, should you be more interested in the mean or median?[13]

5.7 Transforming data

When data are very strongly skewed, we sometimes transform them so they are easier to model. Figure 5.10 shows two right skewed distributions: distribution of the percentage of unemployed people and the distribution of the population in all counties in the United States. The distribution of population is more strongly skewed than the distribution of unemployed, hence the log transformation results in a much bigger change in the shape of the distribution.

EXAMPLE

Consider the histogram of county populations shown in Plot C of Figure 5.10, which shows extreme skew. What characteristics of the plot keep it from being useful?

Nearly all of the data fall into the left-most bin, and the extreme skew obscures many of the potentially interesting details at the low values.

There are some standard transformations that may be useful for strongly right skewed data where much of the data is positive but clustered near zero. A **transformation** is a rescaling of the data using a function. For instance, a plot of the logarithm (base 10) of unemployment rates and county

---

[12] Mean is affected more than the median. Standard deviation is affected more than the IQR.
[13] If we are looking to simply understand what a typical individual loan looks like, the median is probably more useful. However, if the goal is to understand something that scales well, such as the total amount of money we might need to have on hand if we were to offer 1,000 loans, then the mean would be more useful.

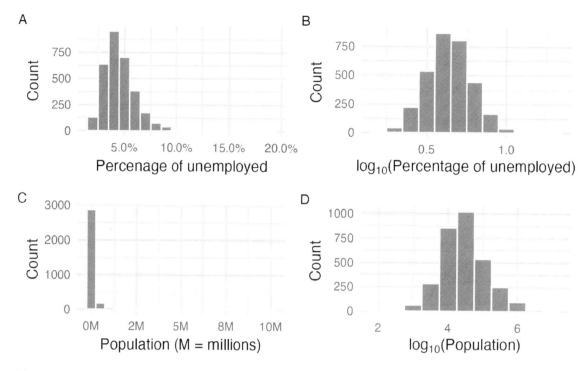

Figure 5.10: Plot A: A histogram of the percentage of unemployed in all US counties. Plot B: A histogram of $\log_{10}$-transformed unemployed percentages. Plot C: A histogram of population in all US counties. Plot D: A histogram of $\log_{10}$-transformed populations. For Plots B and D, the x-value corresponds to the power of 10, e.g., 1 on the x-axis corresponds to $10^1 = 10$ and 5 on the x-axis corresponds to $10^5 = 100{,}000$. Data are from 2017.

populations results in the new histograms on the right in Figure 5.10. The transformed data are symmetric, and any potential outliers appear much less extreme than in the original data set. By reigning in the outliers and extreme skew, transformations often make it easier to build statistical models for the data.

Transformations can also be applied to one or both variables in a scatterplot. A scatterplot of the population change from 2010 to 2017 against the population in 2010 is shown in Figure 5.11. In this first scatterplot, it's hard to decipher any interesting patterns because the population variable is so strongly skewed (left plot). However, if we apply a $\log_{10}$ transformation to the population variable, as shown in Figure 5.11, a positive association between the variables is revealed (right plot). In fact, we may be interested in fitting a trend line to the data when we explore methods around fitting regression lines in Chapter 7.

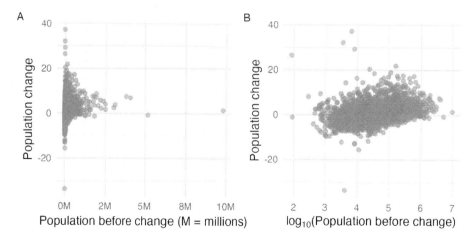

Figure 5.11: Plot A: Scatterplot of population change against the population before the change. Plot B: A scatterplot of the same data but where the population size has been log-transformed.

5.8 Mapping data

Transformations other than the logarithm can be useful, too. For instance, the square root ($\sqrt{\text{original observation}}$) and inverse ($\frac{1}{\text{original observation}}$) are commonly used by data scientists. Common goals in transforming data are to see the data structure differently, reduce skew, assist in modeling, or straighten a nonlinear relationship in a scatterplot.

5.8 Mapping data

The county dataset offers many numerical variables that we could plot using dot plots, scatterplots, or box plots, but they can miss the true nature of the data as geographic. When we encounter geographic data, we should create an **intensity map**, where colors are used to show higher and lower values of a variable. Figures 5.12 and 5.13 show intensity maps for poverty rate in percent (poverty), unemployment rate in percent (unemployment_rate), homeownership rate in percent (homeownership), and median household income in $1000s (median_hh_income). The color key indicates which colors correspond to which values. The intensity maps are not generally very helpful for getting precise values in any given county, but they are very helpful for seeing geographic trends and generating interesting research questions or hypotheses.

EXAMPLE

What interesting features are evident in the poverty and unemployment rate intensity maps?

Poverty rates are evidently higher in a few locations. Notably, the deep south shows higher poverty rates, as does much of Arizona and New Mexico. High poverty rates are evident in the Mississippi flood plains a little north of New Orleans and also in a large section of Kentucky.

The unemployment rate follows similar trends, and we can see correspondence between the two variables. In fact, it makes sense for higher rates of unemployment to be closely related to poverty rates. One observation that stands out when comparing the two maps: the poverty rate is much higher than the unemployment rate, meaning while many people may be working, they are not making enough to break out of poverty.

GUIDED PRACTICE

What interesting features are evident in the median household income intensity map in Figure 5.13?[14]

---

[14] Answers will vary. There is some correspondence between high earning and metropolitan areas, where we can see darker spots (higher median household income), though there are several exceptions. You might look for large cities you are familiar with and try to spot them on the map as dark spots.

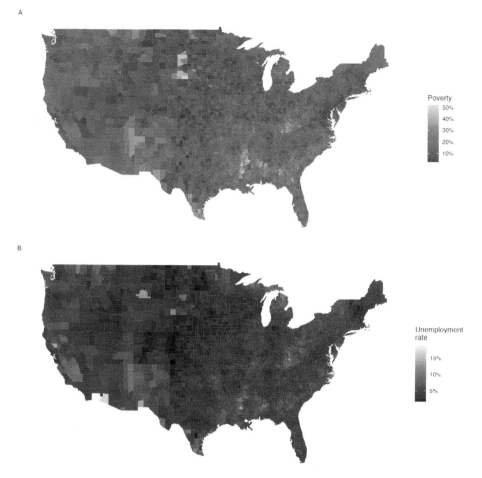

Figure 5.12: Plot A: Intensity map of poverty rate (percent). Plot B: Intensity map of the unemployment rate (percent).

5.8. MAPPING DATA

Figure 5.13: Plot A: Intensity map of homeownership rate (percent). Plot B: Intensity map of median household income (in thousands of USD).

5.9 Chapter review

5.9.1 Summary

Fluently working with numerical variables is an important skill for data analysts. In this chapter we've introduced different visualizations and numerical summaries applied to numeric variables. The graphical visualizations are even more descriptive when two variables are presented simultaneously. We presented scatterplots, dot plots, histograms, and box plots. Numerical variables can be summarized using the mean, median, quartiles, standard deviation, and variance.

5.9.2 Terms

We introduced the following terms in the chapter. If you're not sure what some of these terms mean, we recommend you go back in the text and review their definitions. We are purposefully presenting them in alphabetical order, instead of in order of appearance, so they will be a little more challenging to locate. However you should be able to easily spot them as **bolded text**.

average	IQR	scatterplot
bimodal	left skewed	standard deviation
box plot	mean	symmetric
data density	median	tail
deviation	multimodal	third quartile
distribution	nonlinear	transformation
dot plot	outlier	unimodal
first quartile	percentile	variability
histogram	point estimate	variance
intensity map	right skewed	weighted mean
interquartile range	robust statistics	whiskers

5.10 Exercises

Answers to odd numbered exercises can be found in Appendix A.5.

5.1. **Mammal life spans.** Data were collected on life spans (in years) and gestation lengths (in days) for 62 mammals. A scatterplot of life span versus length of gestation is shown below.[15] (Allison and Cicchetti, 1975)

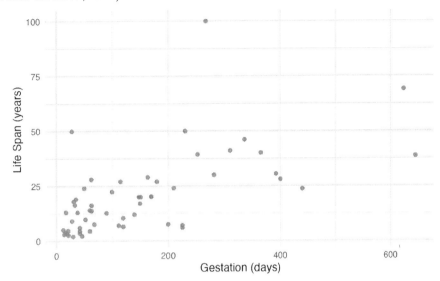

  a. What type of an association is apparent between life span and length of gestation?

  b. What type of an association would you expect to see if the axes of the plot were reversed, i.e., if we plotted length of gestation versus life span?

  c. Are life span and length of gestation independent? Explain your reasoning.

5.2. **Associations.** Indicate which of the plots show (a) a positive association, (b) a negative association, or (c) no association. Also determine if the positive and negative associations are linear or nonlinear. Each part may refer to more than one plot.

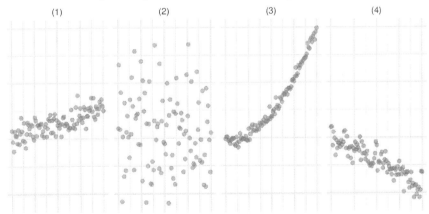

5.3. **Reproducing bacteria.** Suppose that there is only sufficient space and nutrients to support one million bacterial cells in a petri dish. You place a few bacterial cells in this petri dish, allow them to reproduce freely, and record the number of bacterial cells in the dish over time. Sketch a plot representing the relationship between number of bacterial cells and time.

5.4. **Office productivity.** Office productivity is relatively low when the employees feel no stress about their work or job security. However, high levels of stress can also lead to reduced employee productivity. Sketch a plot to represent the relationship between stress and productivity.

---

[15] The `mammals` data used in this exercise can be found in the **openintro** R package.

5.5. **Make-up exam.** In a class of 25 students, 24 of them took an exam in class and 1 student took a make-up exam the following day. The professor graded the first batch of 24 exams and found an average score of 74 points with a standard deviation of 8.9 points. The student who took the make-up the following day scored 64 points on the exam.

   a. Does the new student's score increase or decrease the average score?

   b. What is the new average?

   c. Does the new student's score increase or decrease the standard deviation of the scores?

5.6. **Infant mortality.** The infant mortality rate is defined as the number of infant deaths per 1,000 live births. This rate is often used as an indicator of the level of health in a country. The relative frequency histogram below shows the distribution of estimated infant death rates for 224 countries for which such data were available in 2014.[16]

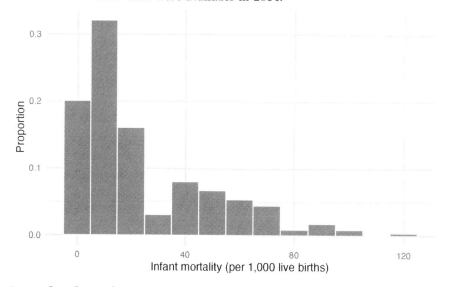

   a. Estimate Q1, the median, and Q3 from the histogram.

   b. Would you expect the mean of this dataset to be smaller or larger than the median? Explain your reasoning.

5.7. **Days off at a mining plant.** Workers at a particular mining site receive an average of 35 days paid vacation, which is lower than the national average. The manager of this plant is under pressure from a local union to increase the amount of paid time off. However, he does not want to give more days off to the workers because that would be costly. Instead he decides he should fire 10 employees in such a way as to raise the average number of days off that are reported by his employees. In order to achieve this goal, should he fire employees who have the most number of days off, least number of days off, or those who have about the average number of days off?

5.8. **Medians and IQRs.** For each part, compare distributions A and B based on their medians and IQRs. You do not need to calculate these statistics; simply state how the medians and IQRs compare. Make sure to explain your reasoning. *Hint:* It may be useful to sketch dot plots of the distributions.

   a. **A:** 3, 5, 6, 7, 9; **B:** 3, 5, 6, 7, 20

   b. **A:** 3, 5, 6, 7, 9; **B:** 3, 5, 7, 8, 9

   c. **A:** 1, 2, 3, 4, 5; **B:** 6, 7, 8, 9, 10

   d. **A:** 0, 10, 50, 60, 100; **B:** 0, 100, 500, 600, 1000

---

[16]The `cia_factbook` data used in this exercise can be found in the **openintro** R package.

5.10. EXERCISES

5.9. **Means and SDs.** For each part, compare distributions A and B based on their means and standard deviations. You do not need to calculate these statistics; simply state how the means and the standard deviations compare. Make sure to explain your reasoning. *Hint:* It may be useful to sketch dot plots of the distributions.

   a. **A:** 3, 5, 5, 5, 8, 11, 11, 11, 13; **B:** 3, 5, 5, 5, 8, 11, 11, 11, 20

   b. **A:** -20, 0, 0, 0, 15, 25, 30, 30; **B:** -40, 0, 0, 0, 15, 25, 30, 30

   c. **A:** 0, 2, 4, 6, 8, 10; **B:** 20, 22, 24, 26, 28, 30

   d. **A:** 100, 200, 300, 400, 500; **B:** 0, 50, 300, 550, 600

5.10. **Histograms and box plots.** Describe (in words) the distribution in the histograms below and match them to the box plots.

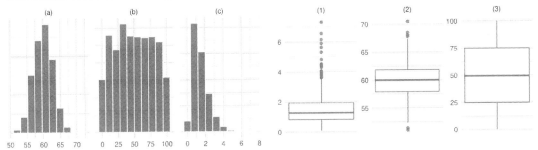

5.11. **Air quality.** Daily air quality is measured by the air quality index (AQI) reported by the Environmental Protection Agency. This index reports the pollution level and what associated health effects might be a concern. The index is calculated for five major air pollutants regulated by the Clean Air Act and takes values from 0 to 300, where a higher value indicates lower air quality. AQI was reported for a sample of 91 days in 2011 in Durham, NC. The histogram below shows the distribution of the AQI values on these days.[17]

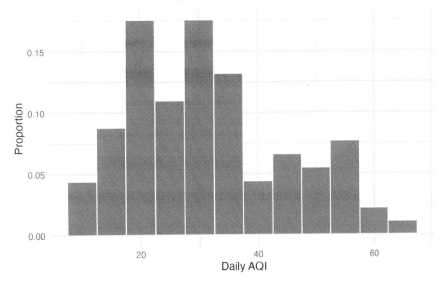

   a. Estimate the median AQI value of this sample.

   b. Would you expect the mean AQI value of this sample to be higher or lower than the median? Explain your reasoning.

   c. Estimate Q1, Q3, and IQR for the distribution.

   d. Would any of the days in this sample be considered to have an unusually low or high AQI? Explain your reasoning.

---

[17] The pm25_2011_durham data used in this exercise can be found in the **openintro** R package.

5.12. **Median vs. mean.** Estimate the median for the 400 observations shown in the histogram, and note whether you expect the mean to be higher or lower than the median.

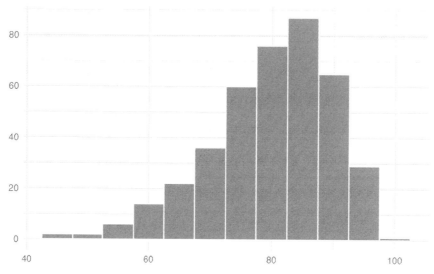

5.13. **Histograms vs. box plots.** Compare the two plots below. What characteristics of the distribution are apparent in the histogram and not in the box plot? What characteristics are apparent in the box plot but not in the histogram?

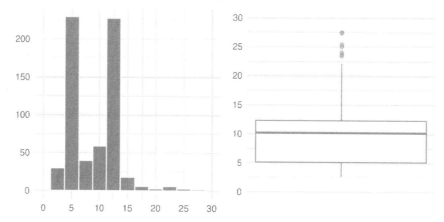

5.14. **Facebook friends.** Facebook data indicate that 50% of Facebook users have 100 or more friends, and that the average friend count of users is 190. What do these findings suggest about the shape of the distribution of number of friends of Facebook users? (Backstrom, 2011)

5.15. **Distributions and appropriate statistics.** For each of the following, state whether you expect the distribution to be symmetric, right skewed, or left skewed. Also specify whether the mean or median would best represent a typical observation in the data, and whether the variability of observations would be best represented using the standard deviation or IQR. Explain your reasoning.

   a. Number of pets per household.

   b. Distance to work, i.e., number of miles between work and home.

   c. Heights of adult males.

   d. Age at death.

   e. Exam grade on an easy test.

5.16. **Distributions and appropriate statistics.** For each of the following, state whether you expect the distribution to be symmetric, right skewed, or left skewed. Also specify whether the mean or median would best represent a typical observation in the data, and whether the variability of observations would be best represented using the standard deviation or IQR. Explain your reasoning.

   a. Housing prices in a country where 25% of the houses cost below $350,000, 50% of the houses cost below $450,000, 75% of the houses cost below $1,000,000, and there are a meaningful number of houses that cost more than $6,000,000.

   b. Housing prices in a country where 25% of the houses cost below $300,000, 50% of the houses cost below $600,000, 75% of the houses cost below $900,000, and very few houses that cost more than $1,200,000.

   c. Number of alcoholic drinks consumed by college students in a given week. Assume that most of these students don't drink since they are under 21 years old, and only a few drink excessively.

   d. Annual salaries of the employees at a Fortune 500 company where only a few high level executives earn much higher salaries than all the other employees.

   e. Gestation time in humans where 25% of the babies are born by 38 weeks of gestation, 50% of the babies are born by 39 weeks, 75% of the babies are born by 40 weeks, and the maximum gestation length is 46 weeks.

5.17. **TV watchers.** College students in a statistics class were asked how many hours of television they watch per week, including online streaming services. This sample yielded an average of 8.28 hours, with a standard deviation of 7.18 hours. Is the distribution of number of hours students watch television weekly symmetric? If not, what shape would you expect this distribution to have? Explain your reasoning.

5.18. **Exam scores.** The average on a history exam (scored out of 100 points) was 85, with a standard deviation of 15. Is the distribution of the scores on this exam symmetric? If not, what shape would you expect this distribution to have? Explain your reasoning.

5.19. **Midrange.** The *midrange* of a distribution is defined as the average of the maximum and the minimum of that distribution. Is this statistic robust to outliers and extreme skew? Explain your reasoning.

5.20. **Oscar winners.** The first Oscar awards for best actor and best actress were given out in 1929. The histograms below show the age distribution for all of the best actor and best actress winners from 1929 to 2019. Summary statistics for these distributions are also provided. Compare the distributions of ages of best actor and actress winners.[18]

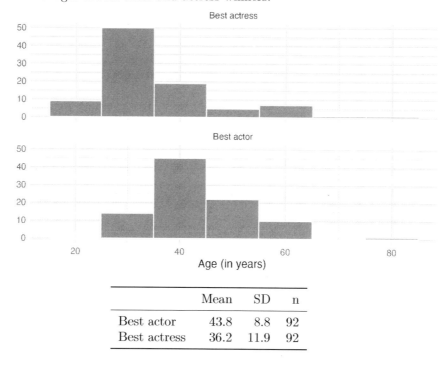

	Mean	SD	n
Best actor	43.8	8.8	92
Best actress	36.2	11.9	92

5.21. **Stats scores.** The final exam scores of twenty introductory statistics students, arranged in ascending order, as as follows: 57, 66, 69, 71, 72, 73, 74, 77, 78, 78, 79, 79, 81, 81, 82, 83, 83, 88, 89, 94. Suppose students who score above the 75th percentile on the final exam get an A in the class. How many students will get an A in this class?

5.22. **Income at the coffee shop.** The first histogram below shows the distribution of the yearly incomes of 40 patrons at a college coffee shop. Suppose two new people walk into the coffee shop: one making $225,000 and the other $250,000. The second histogram shows the new income distribution. Summary statistics are also provided, rounded to the nearest whole number.

	n	Min	Q1	Median	Mean	Max	SD
Before	40	$60,679	$60,818	$65,238	$65,089	$69,885	$2,122
After	42	$60,679	$60,838	$65,352	$73,299	$250,000	$37,321

---

[18]The oscars data used in this exercise can be found in the **openintro** R package.

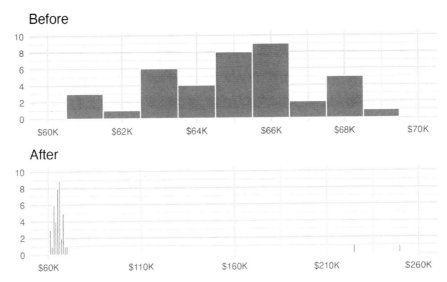

a. Would the mean or the median best represent what we might think of as a typical income for the 42 patrons at this coffee shop? What does this say about the robustness of the two measures?

b. Would the standard deviation or the IQR best represent the amount of variability in the incomes of the 42 patrons at this coffee shop? What does this say about the robustness of the two measures?

5.23. **A new statistic.** The statistic $\frac{\bar{x}}{median}$ can be used as a measure of skewness. Suppose we have a distribution where all observations are greater than 0, $x_i > 0$. What is the expected shape of the distribution under the following conditions? Explain your reasoning.

a. $\frac{\bar{x}}{median} = 1$

b. $\frac{\bar{x}}{median} < 1$

c. $\frac{\bar{x}}{median} > 1$

5.24. **Commute times.** The US census collects data on the time it takes Americans to commute to work, among many other variables. The histogram below shows the distribution of average commute times in 3,142 US counties in 2017. Also shown below is a spatial intensity map of the same data.[19]

a. Describe the numerical distribution and comment on whether or not a log transformation may be advisable for these data.

b. Describe the spatial distribution of commuting times using the map.

---

[19] The `county_complete` data used in this exercise can be found in the **usdata** R package.

5.25. **Hispanic population.** The US census collects data on race and ethnicity of Americans, among many other variables. The histogram below shows the distribution of the percentage of the population that is Hispanic in 3,142 counties in the US in 2010. Also shown is a histogram of logs of these values.[20]

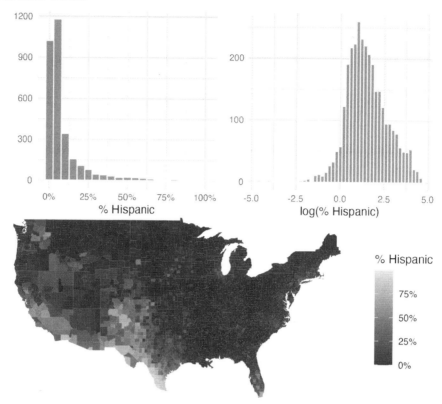

a. Describe the numerical distribution and comment on why we might want to use log-transformed values in analyzing or modeling these data.

b. What features of the distribution of the Hispanic population in US counties are apparent in the map but not in the histogram? What features are apparent in the histogram but not the map?

c. Is one visualization more appropriate or helpful than the other? Explain your reasoning.

---

[20]The `county_complete` data used in this exercise can be found in the **usdata** R package.

5.26. **NYC marathon winners.** The histogram and box plots below show the distribution of finishing times for male and female winners of the New York City Marathon between 1970 and 2020.[21]

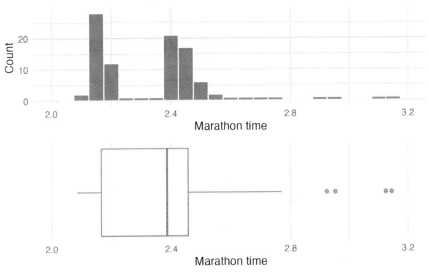

a. What features of the distribution are apparent in the histogram and not the box plot? What features are apparent in the box plot but not in the histogram?

b. What may be the reason for the bimodal distribution? Explain.

c. Compare the distribution of marathon times for men and women based on the box plot shown below.

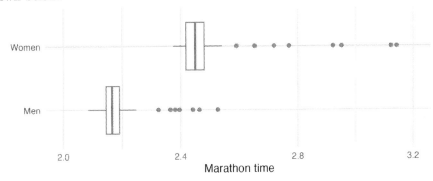

d. The time series plot shown below is another way to look at these data. Describe what is visible in this plot but not in the others.

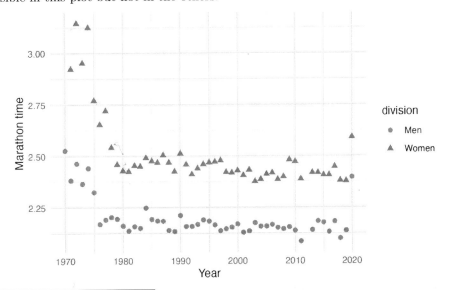

---

[21] The `nyc_marathon` data used in this exercise can be found in the **openintro** R package.

Chapter 6

Applications: Explore

6.1 Case study: Effective communication of exploratory results

Graphs can powerfully communicate ideas directly and quickly. We all know, after all, that "a picture is worth 1000 words." Unfortunately, however, there are times when an image conveys a message which is inaccurate or misleading.

This chapter focuses on how graphs can best be utilized to present data accurately and effectively. Along with data modeling, creative visualization is somewhat of an art. However, even with an art, there are recommended guiding principles. We provide a few best practices for creating data visualizations.

6.1.1 Keep it simple

When creating a graphic, keep in mind what it is that you'd like your reader to see. Colors should be used to group items or differentiate levels in meaningful ways. Colors can be distracting when they are only used to brighten up the plot.

Consider a manufacturing company who has summarized their costs into five different categories. In the two graphics provided in Figure 6.1, notice that the magnitudes in the pie chart are difficult for the eye to compare. That is, can your eye tell how different "Buildings and administration" is from "Workplace materials" when looking at the slices of pie? Additionally, the colors in the pie chart do not mean anything and are therefore distracting. Lastly, the three-dimensional aspect of the image does not improve the reader's ability to understand the data presented.

As an alternative, a bar plot has been provided. Notice how much easier it is to identify the magnitude of the differences across categories while not being distracted by other aspects of the image. Typically, a bar plot will be easier for the reader to digest than a pie chart, especially if the categorical data being plotted has more than just a few levels.

6.1.2 Use color to draw attention

There are many reasons why you might choose to add **color** to your plots. An important principle to keep in mind is to use color to draw attention. Of course, you should still think about how visually pleasing your visualization is, and if you're adding color for making it visually pleasing without drawing attention to a particular feature, that might be fine. However you should be critical of default coloring and explicitly decide whether to include color and how. Notice that in Plot B in Figure 6.2 the coloring is done in such a way to draw the reader's attention to one particular piece of information. The default coloring in Plot A can be distracting and makes the reader question, for

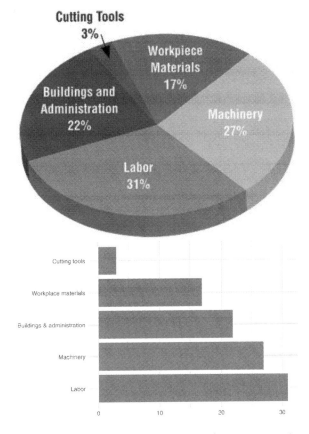

Figure 6.1: A pie chart (with added irrelevant features) as compared to a simple bar plot.

example, is there something similar about the red and purple bars? Also note that not everyone sees color the same way, often it's useful to add color and one more feature (e.g., pattern) so that you can refer to the features you're drawing attention to in multiple ways.

6.1.3 Tell a story

For many graphs, an important aspect is the inclusion of information which is not provided in the dataset that is being plotted. The external information serves to contextualize the data and helps communicate the narrative of the research. In Figure 6.3, the graph on the right is **annotated** with information about the start of the university's fiscal year which contextualizes the information provided by the data. Sometimes the additional information may be a diagonal line given by $y = x$, points above the line quickly show the reader which values have a y coordinate larger than the x coordinate; points below the line show the opposite.

6.1.4 Order matters

Most software programs have built in methods for some of the plot details. For example, the default option for the software program used in this text, R, is to order the bars in a bar plot alphabetically. As seen in Figure 6.4, the alphabetical ordering isn't particularly meaningful for describing the data. Sometimes it makes sense to **order** the bars from tallest to shortest (or vice versa). But in this case, the best ordering is probably the one in which the questions were asked. An ordering which doesn't make sense in the context of the problem (e.g., alphabetically here), can mislead the reader who might take a quick glance at the axes and not read the bar labels carefully.

In September 2019, YouGov survey asked 1,639 Great Britain adults the following question[1]:

---

[1] Source: YouGov Survey Results, retrieved Oct 7, 2019.

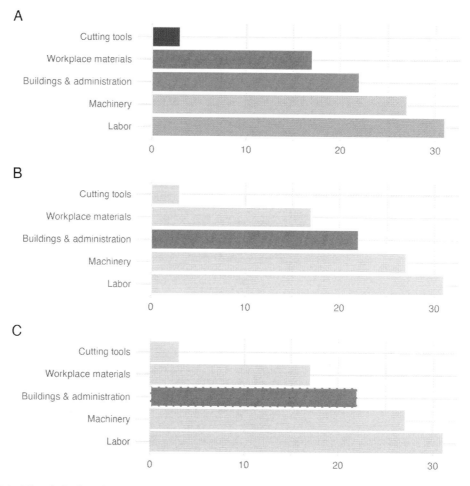

Figure 6.2: The default coloring in the first bar plot does nothing for the understanding of the data. In the second plot, the color draws attention directly to the bar on Buildings and Administration.

How well or badly do you think the government are doing at handling Britain's exit from the European Union?

- Very well
- Fairly well
- Fairly badly
- Very badly
- Don't know

6.1.5 Make the labels as easy to read as possible

The Brexit survey results were additionally broken down by region in Great Britain. The stacked bar plot allows for comparison of Brexit opinion across the five regions. In Figure 6.5 the bars are vertical in Plot A and horizontal in Plot B. While the quantitative information in the two graphics is identical, flipping the graph and creating horizontal bars provides more space for the **axis labels**. The easier the categories are to read, the more the reader will learn from the visualization. Remember, the goal is to convey as much information as possible in a succinct and clear manner.

6.1.6 Pick a purpose

Every graphical decision should be made with a **purpose**. As previously mentioned, sticking with default options is not always best for conveying the narrative of your data story. Stacked bar plots tell one part of a story. Depending on your research question, they may not tell the part of the story most important to the research. Figure 6.6 provides three different ways of representing the same

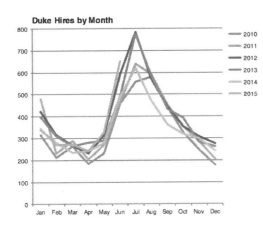 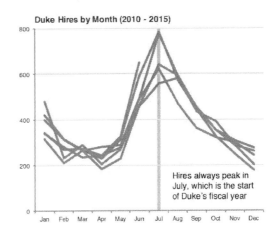

Figure 6.3: Credit: Angela Zoss and Eric Monson, Duke Data Visualization Services

information. If the most important comparison across regions is proportion, you might prefer Plot A. If the most important comparison across regions also considers the total number of individuals in the region, you might prefer Plot B. If a separate bar plot for each region makes the point you'd like, use Plot C, which has been **faceted** by region.

Plot C in Figure 6.6 also provides full titles and a succinct URL with the data source. Other deliberate decisions to consider include using informative labels and avoiding redundancy.

6.1.7 Select meaningful colors

One last consideration for building graphs is to consider color choices. Default or rainbow colors are not always the choice which will best distinguish the level of your variables. Much research has been done to find color combinations which are distinct and also which are clear for differently sighted individuals. The cividis scale works well with ordinal data. (Nuñez et al., 2018) Figure 6.7 shows the same plot with two different colorings.

In this chapter different representations are contrasted to demonstrate best practices in creating graphs. The fundamental principle is that your graph should provide maximal information succinctly and clearly. Labels should be clear and oriented horizontally for the reader. Don't forget titles and, if possible, include the source of the data.

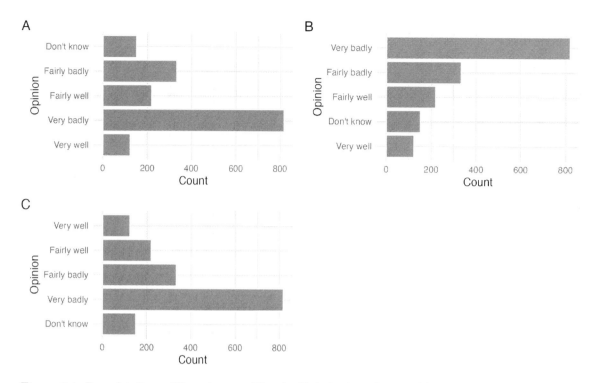

Figure 6.4: Bar plot three different ways. Plot A: Alphabetic ordering of levels, Plot B: Bars ordered in descending order of frequency, Plot C: Bars ordered in the same order as they were presented in the survey question.

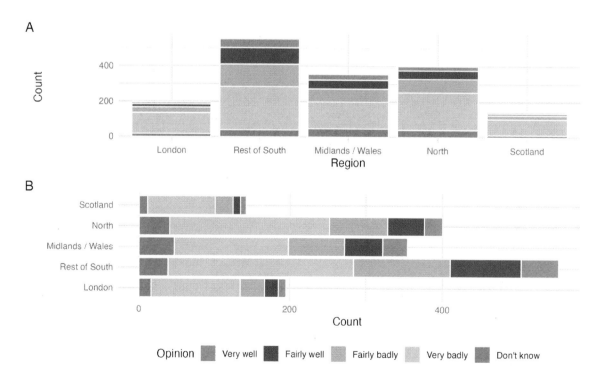

Figure 6.5: Stacked bar plots vertically and horizontally. The horizontal orientation makes the region labels easier to read.

6.1. CASE STUDY: EFFECTIVE COMMUNICATION OF EXPLORATORY RESULTS 109

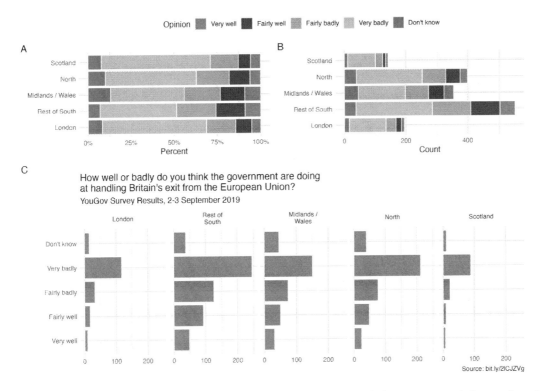

Figure 6.6: Three different representations of the two variables including survey opinion and region. Use the graphic that best conveys the data narrative at hand.

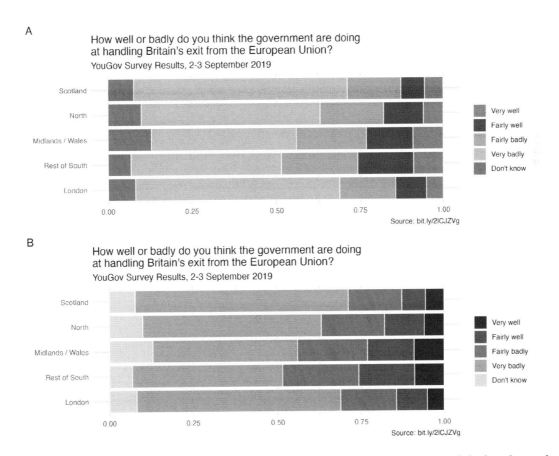

Figure 6.7: Identical bar plots with two different coloring options. Plot A uses a default color scale, Plot B uses colors from the cividis scale.

6.2 Interactive R tutorials

Navigate the concepts you've learned in this chapter in R using the following self-paced tutorials. All you need is your browser to get started!

Tutorial 2: Exploratory data analysis
https://openintrostat.github.io/ims-tutorials/02-explore

Tutorial 2 - Lesson 1: Visualizing categorical data
https://openintrostat.github.io/ims-tutorials/02-explore-01

Tutorial 2 - Lesson 2: Visualizing numerical data
https://openintro.shinyapps.io/ims-02-explore-02

Tutorial 2 - Lesson 3: Summarizing with statistics
https://openintro.shinyapps.io/ims-02-explore-03

Tutorial 2 - Lesson 4: Case study
https://openintro.shinyapps.io/ims-02-explore-04

You can also access the full list of tutorials supporting this book at https://openintrostat.github.io/ims-tutorials.

6.3 R labs

Further apply the concepts you've learned in this part in R with computational labs that walk you through a data analysis case study.

Intro to data - Flight delays
https://www.openintro.org/go?id=ims-r-lab-intro-to-data

You can also access the full list of labs supporting this book at https://www.openintro.org/go?id=ims-r-labs.

PART III

REGRESSION MODELING

Chapter 7

Linear regression with a single predictor

Linear regression is a very powerful statistical technique. Many people have some familiarity with regression models just from reading the news, where straight lines are overlaid on scatterplots. Linear models can be used for prediction or to evaluate whether there is a linear relationship between a numerical variable on the horizontal axis and the average of the numerical variable on the vertical axis.

7.1 Fitting a line, residuals, and correlation

When considering linear regression, it's helpful to think deeply about the line fitting process. In this section, we define the form of a linear model, explore criteria for what makes a good fit, and introduce a new statistic called *correlation*.

7.1.1 Fitting a line to data

Figure 7.1 shows two variables whose relationship can be modeled perfectly with a straight line. The equation for the line is $y = 5 + 64.96x$. Consider what a perfect linear relationship means: we know the exact value of y just by knowing the value of x. A perfect linear relationship is unrealistic in almost any natural process. For example, if we took family income (x), this value would provide some useful information about how much financial support a college may offer a prospective student (y). However, the prediction would be far from perfect, since other factors play a role in financial support beyond a family's finances.

Linear regression is the statistical method for fitting a line to data where the relationship between two variables, x and y, can be modeled by a straight line with some error:

$$y = b_0 + b_1\ x + e$$

The values b_0 and b_1 represent the model's intercept and slope, respectively, and the error is represented by e. These values are calculated based on the data, i.e., they are sample statistics. If the observed data is a random sample from a target population that we are interested in making inferences about, these values are considered to be point estimates for the population parameters β_0

7.1. FITTING A LINE, RESIDUALS, AND CORRELATION

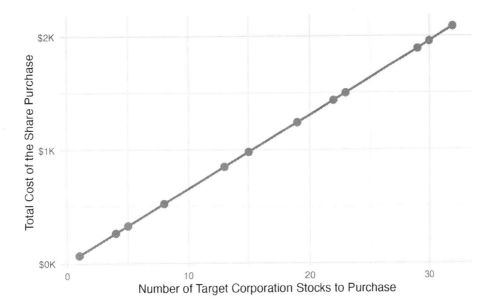

Figure 7.1: Requests from twelve separate buyers were simultaneously placed with a trading company to purchase Target Corporation stock (ticker TGT, December 28th, 2018), and the total cost of the shares were reported. Because the cost is computed using a linear formula, the linear fit is perfect.

and β_1. We will discuss how to make inferences about parameters of a linear model based on sample statistics in Chapter 24.

 The Greek letter β is pronounced *beta*, listen to the pronunciation here.

When we use x to predict y, we usually call x the **predictor** variable and we call y the **outcome**. We also often drop the e term when writing down the model since our main focus is often on the prediction of the average outcome.

It is rare for all of the data to fall perfectly on a straight line. Instead, it's more common for data to appear as a *cloud of points*, such as those examples shown in Figure 7.2. In each case, the data fall around a straight line, even if none of the observations fall exactly on the line. The first plot shows a relatively strong downward linear trend, where the remaining variability in the data around the line is minor relative to the strength of the relationship between x and y. The second plot shows an upward trend that, while evident, is not as strong as the first. The last plot shows a very weak downward trend in the data, so slight we can hardly notice it. In each of these examples, we will have some uncertainty regarding our estimates of the model parameters, β_0 and β_1. For instance, we might wonder, should we move the line up or down a little, or should we tilt it more or less? As we move forward in this chapter, we will learn about criteria for line-fitting, and we will also learn about the uncertainty associated with estimates of model parameters.

There are also cases where fitting a straight line to the data, even if there is a clear relationship between the variables, is not helpful. One such case is shown in Figure 7.3 where there is a very clear relationship between the variables even though the trend is not linear. We discuss nonlinear trends in this chapter and the next, but details of fitting nonlinear models are saved for a later course.

7.1.2 Using linear regression to predict possum head lengths

Brushtail possums are marsupials that live in Australia, and a photo of one is shown in Figure 7.4. Researchers captured 104 of these animals and took body measurements before releasing the animals back into the wild. We consider two of these measurements: the total length of each possum, from head to tail, and the length of each possum's head.

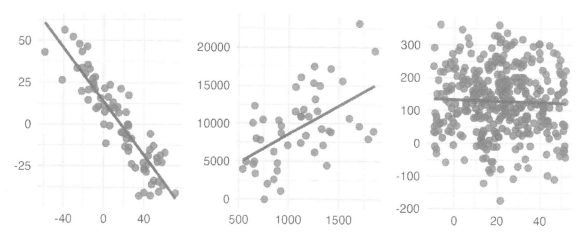

Figure 7.2: Three datasets where a linear model may be useful even though the data do not all fall exactly on the line.

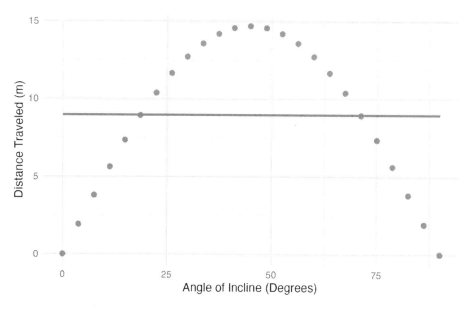

Figure 7.3: The best fitting line for these data is flat, which is not a useful way to describe the non-linear relationship. These data are from a physics experiment.

 The possum data can be found in the **openintro** R package.

Figure 7.5 shows a scatterplot for the head length (mm) and total length (cm) of the possums. Each point represents a single possum from the data. The head and total length variables are associated: possums with an above average total length also tend to have above average head lengths. While the relationship is not perfectly linear, it could be helpful to partially explain the connection between these variables with a straight line.

We want to describe the relationship between the head length and total length variables in the possum dataset using a line. In this example, we will use the total length as the predictor variable, x, to predict a possum's head length, y. We could fit the linear relationship by eye, as in Figure 7.6.

The equation for this line is

$$\hat{y} = 41 + 0.59x$$

A "hat" on y is used to signify that this is an estimate. We can use this line to discuss properties of

Figure 7.4: The common brushtail possum of Australia. Photo by Greg Schecter, flic.kr/p/9BAFbR, CC BY 2.0 license.

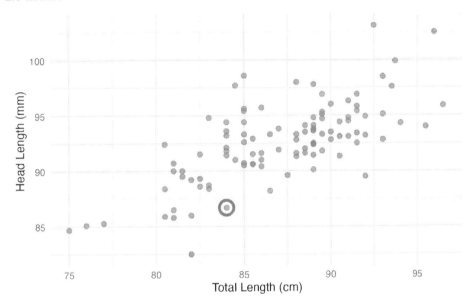

Figure 7.5: A scatterplot showing head length against total length for 104 brushtail possums. A point representing a possum with head length 86.7 mm and total length 84 cm is highlighted.

possums. For instance, the equation predicts a possum with a total length of 80 cm will have a head length of

$$\hat{y} = 41 + 0.59 \times 80 = 88.2$$

The estimate may be viewed as an average: the equation predicts that possums with a total length of 80 cm will have an average head length of 88.2 mm. Absent further information about an 80 cm possum, the prediction for head length that uses the average is a reasonable estimate.

There may be other variables that could help us predict the head length of a possum besides its length. Perhaps the relationship would be a little different for male possums than female possums, or perhaps it would differ for possums from one region of Australia versus another region. Plot A in Figure 7.7 shows the relationship between total length and head length of brushtail possums, taking into consideration their sex. Male possums (represented by blue triangles) seem to be larger in terms of total length and head length than female possums (represented by red circles). Plot B in Figure 7.7 shows the same relationship, taking into consideration their age. It's harder to tell if age changes the relationship between total length and head length for these possums.

In Chapter 8, we'll learn about how we can include more than one predictor in our model. Before we get there, we first need to better understand how to best build a linear model with one predictor.

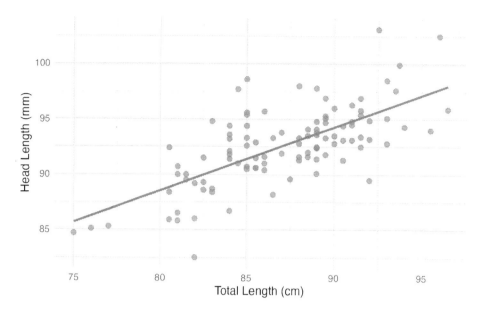

Figure 7.6: A reasonable linear model was fit to represent the relationship between head length and total length.

7.1.3 Residuals

Residuals are the leftover variation in the data after accounting for the model fit:

$$\text{Data} = \text{Fit} + \text{Residual}$$

Each observation will have a residual, and three of the residuals for the linear model we fit for the possum data are shown in Figure 7.8. If an observation is above the regression line, then its residual, the vertical distance from the observation to the line, is positive. Observations below the line have negative residuals. One goal in picking the right linear model is for these residuals to be as small as possible.

Figure 7.8 is almost a replica of Figure 7.6, with three points from the data highlighted. The observation marked by a red circle has a small, negative residual of about -1; the observation marked by a gray diamond has a large positive residual of about +7; and the observation marked by a pink triangle has a moderate negative residual of about -4. The size of a residual is usually discussed in terms of its absolute value. For example, the residual for the observation marked by a pink triangle is larger than that of the observation marked by a red circle because $|-4|$ is larger than $|-1|$.

 Residual: Difference between observed and expected.

The residual of the ith observation (x_i, y_i) is the difference of the observed outcome (y_i) and the outcome we would predict based on the model fit $(\hat{y}_i)$:

$$e_i = y_i - \hat{y}_i$$

We typically identify $\hat{y}_i$ by plugging x_i into the model.

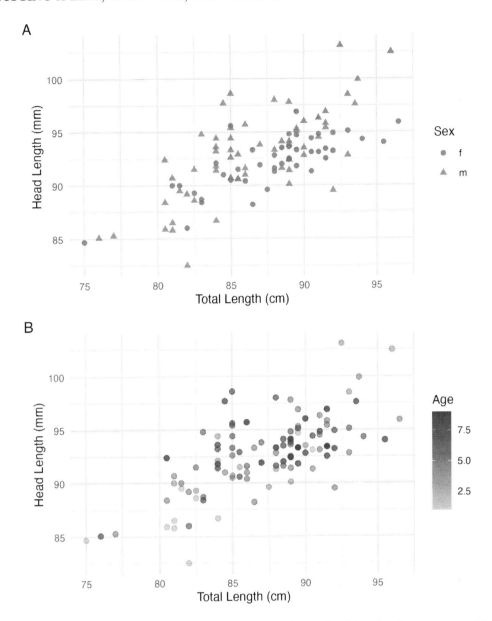

Figure 7.7: Relationship between total length and head length of brushtail possums, taking into consideration their sex (Plot A) or age (Plot B).

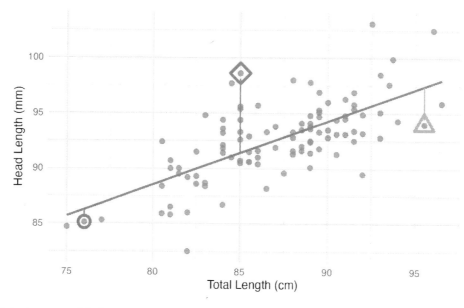

Figure 7.8: A reasonable linear model was fit to represent the relationship between head length and total length, with three points highlighted.

EXAMPLE

The linear fit shown in Figure 7.8 is given as $\hat{y} = 41 + 0.59x$. Based on this line, formally compute the residual of the observation (76.0, 85.1). This observation is marked by a red circle in Figure 7.8. Check it against the earlier visual estimate, -1.

We first compute the predicted value of the observation marked by a red circle based on the model:

$$\hat{y} = 41 + 0.59x = 41 + 0.59 \times 76.0 = 85.84$$

Next we compute the difference of the actual head length and the predicted head length:

$$e = y - \hat{y} = 85.1 - 85.84 = -0.74$$

The model's error is $e = -0.74$ mm, which is very close to the visual estimate of -1 mm. The negative residual indicates that the linear model overpredicted head length for this particular possum.

GUIDED PRACTICE

If a model underestimates an observation, will the residual be positive or negative? What about if it overestimates the observation?[1]

---

[1] If a model underestimates an observation, then the model estimate is below the actual. The residual, which is the actual observation value minus the model estimate, must then be positive. The opposite is true when the model overestimates the observation: the residual is negative.

7.1. FITTING A LINE, RESIDUALS, AND CORRELATION

GUIDED PRACTICE

Compute the residuals for the observation marked by a blue diamond, $(85.0, 98.6)$, and the observation marked by a pink triangle, $(95.5, 94.0)$, in the figure using the linear relationship $\hat{y} = 41 + 0.59x$.[2]

Residuals are helpful in evaluating how well a linear model fits a dataset. We often display them in a scatterplot such as the one shown in Figure 7.9 for the regression line in Figure 7.8. The residuals are plotted with their predicted outcome variable value as the horizontal coordinate, and the vertical coordinate as the residual. For instance, the point $(85.0, 98.6)$ (marked by the blue diamond) had a predicted value of 91.4 mm and had a residual of 7.45 mm, so in the residual plot it is placed at $(91.4, 7.45)$. Creating a residual plot is sort of like tipping the scatterplot over so the regression line is horizontal, as indicated by the dashed line.

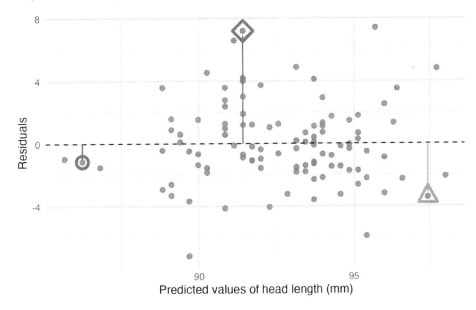

Figure 7.9: Residual plot for the model predicting head length from total length for brushtail possums.

---

[2] Gray diamond: $\hat{y} = 41 + 0.59x = 41 + 0.59 \times 85.0 = 91.15 \rightarrow e = y - \hat{y} = 98.6 - 91.15 = 7.45$. This is close to the earlier estimate of 7. pink triangle: $\hat{y} = 41 + 0.59x = 97.3 \rightarrow e = -3.3$. This is also close to the estimate of -4.

EXAMPLE

One purpose of residual plots is to identify characteristics or patterns still apparent in data after fitting a model. Figure 7.10 shows three scatterplots with linear models in the first row and residual plots in the second row. Can you identify any patterns remaining in the residuals?

In the first dataset (first column), the residuals show no obvious patterns. The residuals appear to be scattered randomly around the dashed line that represents 0.

The second dataset shows a pattern in the residuals. There is some curvature in the scatterplot, which is more obvious in the residual plot. We should not use a straight line to model these data. Instead, a more advanced technique should be used to model the curved relationship, such as the variable transformations discussed in Section 5.7.

The last plot shows very little upwards trend, and the residuals also show no obvious patterns. It is reasonable to try to fit a linear model to the data. However, it is unclear whether there is evidence that the slope parameter is different from zero. The point estimate of the slope parameter, labeled b_1, is not zero, but we might wonder if this could just be due to chance. We will address this sort of scenario in Chapter 24.

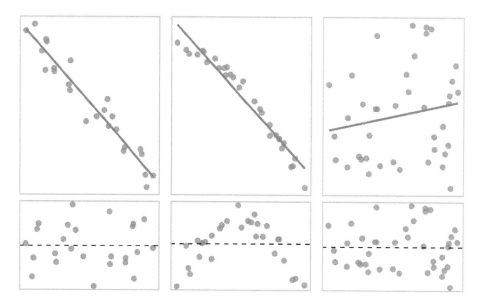

Figure 7.10: Sample data with their best fitting lines (top row) and their corresponding residual plots (bottom row).

7.1.4 Describing linear relationships with correlation

We've seen plots with strong linear relationships and others with very weak linear relationships. It would be useful if we could quantify the strength of these linear relationships with a statistic.

Correlation: strength of a linear relationship.

Correlation which always takes values between -1 and 1, describes the strength and direction of the linear relationship between two variables. We denote the correlation by r.

The correlation value has no units and will not be affected by a linear change in the units (e.g., going from inches to centimeters).

We can compute the correlation using a formula, just as we did with the sample mean and standard deviation. The formula for correlation, however, is rather complex[3], and like with other statistics, we generally perform the calculations on a computer or calculator.

Figure 7.11 shows eight plots and their corresponding correlations. Only when the relationship is perfectly linear is the correlation either -1 or 1. If the relationship is strong and positive, the correlation will be near +1. If it is strong and negative, it will be near -1. If there is no apparent linear relationship between the variables, then the correlation will be near zero.

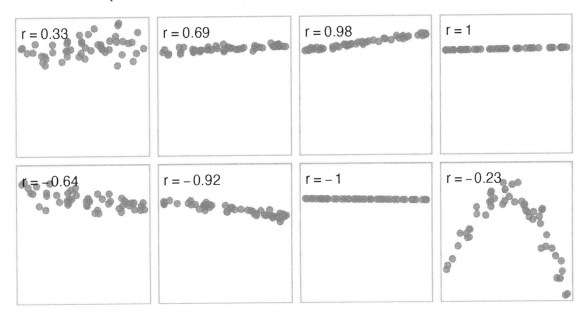

Figure 7.11: Sample scatterplots and their correlations. The first row shows variables with a positive relationship, represented by the trend up and to the right. The second row shows variables with a negative trend, where a large value in one variable is associated with a lower value in the other.

The correlation is intended to quantify the strength of a linear trend. Nonlinear trends, even when strong, sometimes produce correlations that do not reflect the strength of the relationship; see three such examples in Figure 7.12.

---

[3]Formally, we can compute the correlation for observations $(x_1, y_1), (x_2, y_2), \ldots, (x_n, y_n)$ using the formula

$$r = \frac{1}{n-1} \sum_{i=1}^{n} \frac{x_i - \bar{x}}{s_x} \frac{y_i - \bar{y}}{s_y}$$

where $\bar{x}$, $\bar{y}$, s_x, and s_y are the sample means and standard deviations for each variable.

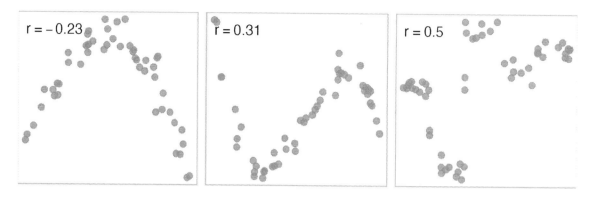

Figure 7.12: Sample scatterplots and their correlations. In each case, there is a strong relationship between the variables. However, because the relationship is not linear, the correlation is relatively weak.

 GUIDED PRACTICE

No straight line is a good fit for any of the datasets represented in Figure 7.12. Try drawing nonlinear curves on each plot. Once you create a curve for each, describe what is important in your fit.[4]

---

[4]We'll leave it to you to draw the lines. In general, the lines you draw should be close to most points and reflect overall trends in the data.

7.1. FITTING A LINE, RESIDUALS, AND CORRELATION

EXAMPLE

The scatterplots below display the relationships between various crop yields in countries. In the plots, each point represents a different country. The x and y variables represent the proportion of total yield in the last 50 years which is due to that crop type.

Order the six scatterplots from strongest negative to strongest positive linear relationship.

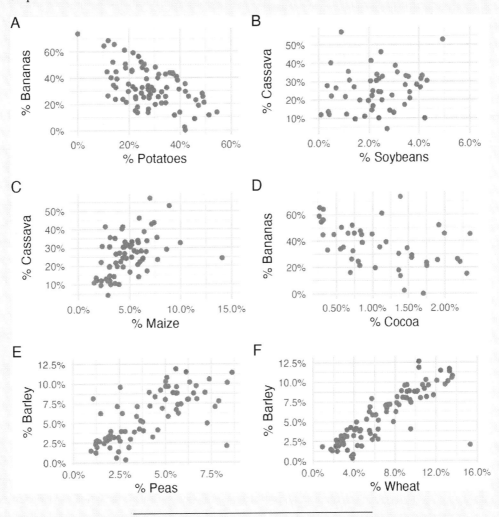

The order of most negative correlation to most positive correlation is:

$$A \to D \to B \to C \to E \to F$$

- Plot A - bananas vs. potatoes: -0.54
- Plot B - cassava vs. soybeans: 0.16
- Plot C - cassava vs. maize: 0.46
- Plot D - cocoa vs. bananas: -0.44
- Plot E - peas vs. barley: 0.69
- Plot F - wheat vs. barley: 0.85

One important aspect of the correlation is that it's *unitless*. That is, unlike a measurement of the slope of a line (see the next section) which provides an increase in the y-coordinate for a one unit increase in the x-coordinate (in units of the x and y variable), there are no units associated with the correlation of x and y. Figure 7.13 shows the relationship between weights and heights of 507 physically active individuals. In Plot A, weight is measured in kilograms (kg) and height in centimeters (cm). In Plot B, weight has been converted to pounds (lbs) and height to inches (in). The correlation coefficient ($r = 0.72$) is also noted on both plots. We can see that the shape of the relationship has not changed, and neither has the correlation coefficient. The only visual change to the plot is the axis *labeling* of the points.

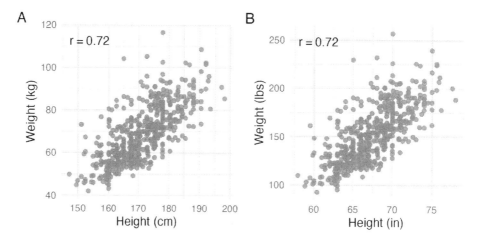

Figure 7.13: Two scatterplots, both displaying the relationship between weights and heights of 507 physically healthy adults. In Plot A, the units are kilograms and centimeters. In Plot B, the units are pounds and inches. Also noted on both plots is the correlation coefficient, $r = 0.72$.

7.2 Least squares regression

Fitting linear models by eye is open to criticism since it is based on an individual's preference. In this section, we use *least squares regression* as a more rigorous approach to fitting a line to a scatterplot.

7.2.1 Gift aid for freshman at Elmhurst College

This section considers a dataset on family income and gift aid data from a random sample of fifty students in the freshman class of Elmhurst College in Illinois. Gift aid is financial aid that does not need to be paid back, as opposed to a loan. A scatterplot of these data is shown in Figure 7.14 along with a linear fit. The line follows a negative trend in the data; students who have higher family incomes tended to have lower gift aid from the university.

 GUIDED PRACTICE

Is the correlation positive or negative in Figure 7.14?[5]

7.2.2 An objective measure for finding the best line

We begin by thinking about what we mean by the "best" line. Mathematically, we want a line that has small residuals. But beyond the mathematical reasons, hopefully it also makes sense intuitively that whatever line we fit, the residuals should be small (i.e., the points should be close to the line). The first option that may come to mind is to minimize the sum of the residual magnitudes:

---

[5]Larger family incomes are associated with lower amounts of aid, so the correlation will be negative. Using a computer, the correlation can be computed: -0.499.

7.2. LEAST SQUARES REGRESSION

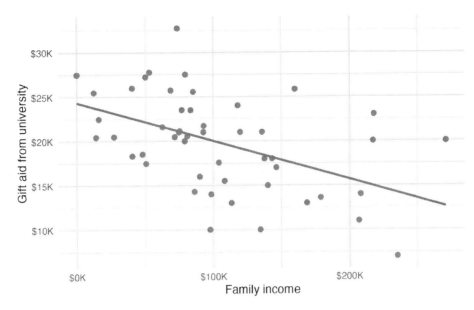

Figure 7.14: Gift aid and family income for a random sample of 50 freshman students from Elmhurst College.

$$|e_1| + |e_2| + \cdots + |e_n|$$

which we could accomplish with a computer program. The resulting dashed line shown in Figure 7.15 demonstrates this fit can be quite reasonable.

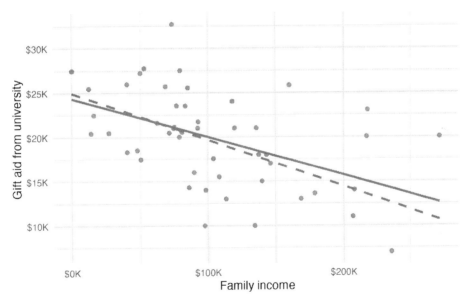

Figure 7.15: Gift aid and family income for a random sample of 50 freshman students from Elmhurst College. The dashed line represents the line that minimizes the sum of the absolute value of residuals, the solid line represents the line that minimizes the sum of squared residuals, i.e., the least squares line.

However, a more common practice is to choose the line that minimizes the sum of the squared residuals:

$$e_1^2 + e_2^2 + \cdots + e_n^2$$

The line that minimizes this least squares criterion is represented as the solid line in Figure 7.15 and is commonly called the **least squares line**. The following are three possible reasons to choose the least squares option instead of trying to minimize the sum of residual magnitudes without any squaring:

1. It is the most commonly used method.
2. Computing the least squares line is widely supported in statistical software.
3. In many applications, a residual twice as large as another residual is more than twice as bad. For example, being off by 4 is usually more than twice as bad as being off by 2. Squaring the residuals accounts for this discrepancy.
4. The analyses which link the model to inference about a population are most straightforward when the line is fit through least squares.

The first two reasons are largely for tradition and convenience; the third and fourth reasons explain why the least squares criterion is typically most helpful when working with real data.[6]

7.2.3 Finding and interpreting the least squares line

For the Elmhurst data, we could write the equation of the least squares regression line as

$$\widehat{\texttt{aid}} = \beta_0 + \beta_1 \times \texttt{family\_income}$$

Here the equation is set up to predict gift aid based on a student's family income, which would be useful to students considering Elmhurst. These two values, β_0 and β_1, are the parameters of the regression line.

The parameters are estimated using the observed data. In practice, this estimation is done using a computer in the same way that other estimates, like a sample mean, can be estimated using a computer or calculator.

The dataset where these data are stored is called `elmhurst`. The first 5 rows of this dataset are given in Table 7.1.

Table 7.1: First five rows of the `elmhurst` dataset.

family_income	gift_aid	price_paid
92.92	21.7	14.28
0.25	27.5	8.53
53.09	27.8	14.25
50.20	27.2	8.78
137.61	18.0	24.00

We can see that family income is recorded in a variable called `family_income` and gift aid from university is recorded in a variable called `gift_aid`. For now, we won't worry about the `price_paid` variable. We should also note that these data are from the 2011-2012 academic year, and all monetary amounts are given in $1,000s, i.e., the family income of the first student in the data shown in Table 7.1 is $92,920 and they received a gift aid of $21,700. (The data source states that all numbers have been rounded to the nearest whole dollar.)

Statistical software is usually used to compute the least squares line and the typical output generated as a result of fitting regression models looks like the one shown in Table 7.2. For now we will focus on the first column of the output, which lists b_0 and b_1. In Chapter 24 we will dive deeper into the remaining columns which give us information on how accurate and precise these values of intercept and slope that are calculated from a sample of 50 students are in estimating the population parameters of intercept and slope for *all* students.

The model output tells us that the intercept is approximately 24.319 and the slope on `family_income` is approximately -0.043.

---

[6]There are applications where the sum of residual magnitudes may be more useful, and there are plenty of other criteria we might consider. However, this book only applies the least squares criterion.

7.2. LEAST SQUARES REGRESSION

Table 7.2: Summary of least squares fit for the Elmhurst data.

term	estimate	std.error	statistic	p.value
(Intercept)	24.32	1.29	18.83	<0.0001
family_income	-0.04	0.01	-3.98	2e-04

But what do these values mean? Interpreting parameters in a regression model is often one of the most important steps in the analysis.

EXAMPLE

The intercept and slope estimates for the Elmhurst data are $b_0 = 24.319$ and $b_1 = -0.043$. What do these numbers really mean?

Interpreting the slope parameter is helpful in almost any application. For each additional $1,000 of family income, we would expect a student to receive a net difference of $1,000 \times (-0.0431) = -\$43.10$ in aid on average, i.e., $43.10 *less*. Note that a higher family income corresponds to less aid because the coefficient of family income is negative in the model. We must be cautious in this interpretation: while there is a real association, we cannot interpret a causal connection between the variables because these data are observational. That is, increasing a particular student's family income may not cause the student's aid to drop. (Although it would be reasonable to contact the college and ask if the relationship is causal, i.e., if Elmhurst College's aid decisions are partially based on students' family income.)

The estimated intercept $b_0 = 24.319$ describes the average aid if a student's family had no income, $24,319. The meaning of the intercept is relevant to this application since the family income for some students at Elmhurst is $0. In other applications, the intercept may have little or no practical value if there are no observations where x is near zero.

Interpreting parameters estimated by least squares.

The slope describes the estimated difference in the predicted average outcome of y if the predictor variable x happened to be one unit larger. The intercept describes the average outcome of y if $x = 0$ *and* the linear model is valid all the way to $x = 0$ (values of $x = 0$ are not observed or relevant in many applications).

If you would like to learn more about using R to fit linear models, see Section 10.2 for the interactive R tutorials. An alternative way of calculating the values of intercept and slope of a least squares line is manual calculations using formulas. While manual calculations are not commonly used by practicing statisticians and data scientists, it is useful to work through the first time you're learning about the least squares line and modeling in general. Calculating the values by hand leverages two properties of the least squares line:

1. The slope of the least squares line can be estimated by

$$b_1 = \frac{s_y}{s_x} r$$

where r is the correlation between the two variables, and s_x and s_y are the sample standard deviations of the predictor and outcome, respectively.

2. If $\bar{x}$ is the sample mean of the predictor variable and $\bar{y}$ is the sample mean of the outcome variable, then the point $(\bar{x}, \bar{y})$ falls on the least squares line.

Table 7.3 shows the sample means for the family income and gift aid as $101,780 and $19,940, respectively. We could plot the point $(102, 19.9)$ on Figure 7.14 to verify it falls on the least squares line (the solid line).

Table 7.3: Summary statistics for family income and gift aid.

Family income, x		Gift aid, y		
mean	sd	mean	sd	r
102	63.2	19.9	5.46	-0.499

Next, we formally find the point estimates b_0 and b_1 of the parameters β_0 and β_1.

EXAMPLE

Using the summary statistics in Table 7.3, compute the slope for the regression line of gift aid against family income.

Compute the slope using the summary statistics from Table 7.3:

$$b_1 = \frac{s_y}{s_x} r = \frac{5.46}{63.2}(-0.499) = -0.0431$$

You might recall the form of a line from math class, which we can use to find the model fit, including the estimate of b_0. Given the slope of a line and a point on the line, (x_0, y_0), the equation for the line can be written as

$$y - y_0 = slope \times (x - x_0)$$

Identifying the least squares line from summary statistics.

To identify the least squares line from summary statistics:

- Estimate the slope parameter, $b_1 = (s_y/s_x)r$.
- Note that the point $(\bar{x}, \bar{y})$ is on the least squares line, use $x_0 = \bar{x}$ and $y_0 = \bar{y}$ with the point-slope equation: $y - \bar{y} = b_1(x - \bar{x})$.
- Simplify the equation, we get $y = \bar{y} - b_1\bar{x} + b_1 x$, which reveals that $b_0 = \bar{y} - b_1\bar{x}$.

7.2. LEAST SQUARES REGRESSION

EXAMPLE

Using the point $(102, 19.9)$ from the sample means and the slope estimate $b_1 = -0.0431$, find the least-squares line for predicting aid based on family income.

Apply the point-slope equation using $(102, 19.9)$ and the slope $b_1 = -0.0431$:

$$y - y_0 = b_1(x - x_0)$$
$$y - 19.9 = -0.0431(x - 102)$$

Expanding the right side and then adding 19.9 to each side, the equation simplifies:

$$\widehat{\texttt{aid}} = 24.3 - 0.0431 \times \texttt{family\_income}$$

Here we have replaced y with $\widehat{\texttt{aid}}$ and x with $\texttt{family\_income}$ to put the equation in context. The final least squares equation should always include a "hat" on the variable being predicted, whether it is a generic "y" or a named variable like "aid".

EXAMPLE

Suppose a high school senior is considering Elmhurst College. Can they simply use the linear equation that we have estimated to calculate her financial aid from the university?

She may use it as an estimate, though some qualifiers on this approach are important. First, the data all come from one freshman class, and the way aid is determined by the university may change from year to year. Second, the equation will provide an imperfect estimate. While the linear equation is good at capturing the trend in the data, no individual student's aid will be perfectly predicted (as can be seen from the individual data points in the cloud around the line).

7.2.4 Extrapolation is treacherous

> *When those blizzards hit the East Coast this winter, it proved to my satisfaction that global warming was a fraud. That snow was freezing cold. But in an alarming trend, temperatures this spring have risen. Consider this: On February 6 it was 10 degrees. Today it hit almost 80. At this rate, by August it will be 220 degrees. So clearly folks the climate debate rages on.*[7]

Stephen Colbert April 6th, 2010

Linear models can be used to approximate the relationship between two variables. However, like any model, they have real limitations. Linear regression is simply a modeling framework. The truth is almost always much more complex than a simple line. For example, we do not know how the data outside of our limited window will behave.

---

[7] http://www.cc.com/video-clips/l4nkoq

EXAMPLE

Use the model $\widehat{\texttt{aid}} = 24.3 - 0.0431 \times \texttt{family\_income}$ to estimate the aid of another freshman student whose family had income of $1 million.

We want to calculate the aid for a family with $1 million income. Note that in our model this will be represented as 1,000 since the data are in $1,000s.

$$24.3 - 0.0431 \times 1000 = -18.8$$

The model predicts this student will have -$18,800 in aid (!). However, Elmhurst College does not offer *negative aid* where they select some students to pay extra on top of tuition to attend.

Applying a model estimate to values outside of the realm of the original data is called **extrapolation**. Generally, a linear model is only an approximation of the real relationship between two variables. If we extrapolate, we are making an unreliable bet that the approximate linear relationship will be valid in places where it has not been analyzed.

7.2.5 Describing the strength of a fit

We evaluated the strength of the linear relationship between two variables earlier using the correlation, r. However, it is more common to explain the strength of a linear fit using R^2, called **R-squared**. If provided with a linear model, we might like to describe how closely the data cluster around the linear fit.

The R^2 of a linear model describes the amount of variation in the outcome variable that is explained by the least squares line. For example, consider the Elmhurst data, shown in Figure 7.14. The variance of the outcome variable, aid received, is about $s^2_{aid} \approx 29.8$ million (calculated from the data, some of which is shown in Table 7.1). However, if we apply our least squares line, then this model reduces our uncertainty in predicting aid using a student's family income. The variability in the residuals describes how much variation remains after using the model: $s^2_{RES} \approx 22.4$ million. In short, there was a reduction of

$$\frac{s^2_{aid} - s^2_{RES}}{s^2_{aid}} = \frac{29800 - 22400}{29800} = \frac{7500}{29800} \approx 0.25,$$

or about 25%, of the outcome variable's variation by using information about family income for predicting aid using a linear model. It turns out that R^2 corresponds exactly to the squared value of the correlation:

$$r = -0.499 \rightarrow R^2 = 0.25$$

GUIDED PRACTICE

If a linear model has a very strong negative relationship with a correlation of -0.97, how much of the variation in the outcome is explained by the predictor?[8]

R^2 is also called the **coefficient of determination**.

---

[8] About $R^2 = (-0.97)^2 = 0.94$ or 94% of the variation in the outcome variable is explained by the linear model.

7.2. LEAST SQUARES REGRESSION

Coefficient of determination: proportion of variability in the outcome variable explained by the model.

Since r is always between -1 and 1, R^2 will always be between 0 and 1. This statistic is called the **coefficient of determination**, and it measures the proportion of variation in the outcome variable, y, that can be explained by the linear model with predictor x.

More generally, R^2 can be calculated as a ratio of a measure of variability around the line divided by a measure of total variability.

Sums of squares to measure variability in y.

We can measure the variability in the y values by how far they tend to fall from their mean, $\bar{y}$. We define this value as the **total sum of squares**, calculated using the formula below, where y_i represents each y value in the sample, and $\bar{y}$ represents the mean of the y values in the sample.

$$SST = (y_1 - \bar{y})^2 + (y_2 - \bar{y})^2 + \cdots + (y_n - \bar{y})^2.$$

Left-over variability in the y values if we know x can be measured by the **sum of squared errors**, or sum of squared residuals, calculated using the formula below, where $\hat{y}_i$ represents the predicted value of y_i based on the least squares regression.[9],

$$SSE = (y_1 - \hat{y}_1)^2 + (y_2 - \hat{y}_2)^2 + \cdots + (y_n - \hat{y}_n)^2$$
$$= e_1^2 + e_2^2 + \cdots + e_n^2$$

The coefficient of determination can then be calculated as

$$R^2 = \frac{SST - SSE}{SST} = 1 - \frac{SSE}{SST}$$

EXAMPLE

Among 50 students in the `elmhurst` dataset, the total variability in gift aid is $SST = 1461$[10] The sum of squared residuals is $SSE = 1098$. Find R^2.

Since we know SSE and SST, we can calculate R^2 as

$$R^2 = 1 - \frac{SSE}{SST} = 1 - \frac{1098}{1461} = 0.25,$$

the same value we found when we squared the correlation: $R^2 = (-0.499)^2 = 0.25$.

7.2.6 Categorical predictors with two levels

Categorical variables are also useful in predicting outcomes. Here we consider a categorical predictor with two levels (recall that a *level* is the same as a *category*). We'll consider Ebay auctions for a

---

[9]The difference $SST - SSE$ is called the **regression sum of squares**, SSR, and can also be calculated as $SSR = (\hat{y}_1 - \bar{y})^2 + (\hat{y}_2 - \bar{y})^2 + \cdots + (\hat{y}_n - \bar{y})^2$. SSR represents the variation in y that was accounted for in our model.

[10]SST can be calculated by finding the sample variance of the outcome variable, s^2 and multiplying by $n - 1$.

video game, *Mario Kart* for the Nintendo Wii, where both the total price of the auction and the condition of the game were recorded. Here we want to predict total price based on game condition, which takes values used and new.

The mariokart data can be found in the **openintro** R package.

A plot of the auction data is shown in Figure 7.16. Note that the original dataset contains some Mario Kart games being sold at prices above $100 but for this analysis we have limited our focus to the 141 Mario Kart games that were sold below $100.

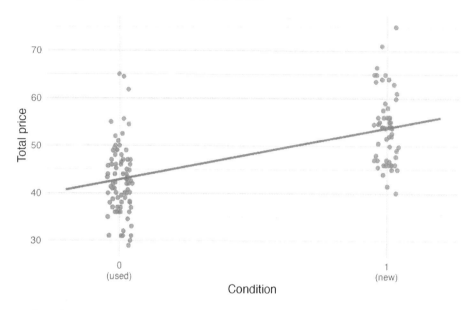

Figure 7.16: Total auction prices for the video game Mario Kart, divided into used ($x = 0$) and new ($x = 1$) condition games. The least squares regression line is also shown.

To incorporate the game condition variable into a regression equation, we must convert the categories into a numerical form. We will do so using an **indicator variable** called condnew, which takes value 1 when the game is new and 0 when the game is used. Using this indicator variable, the linear model may be written as

$$\widehat{\texttt{price}} = b_0 + b_1 \times \texttt{condnew}$$

The parameter estimates are given in Table 7.4.

Table 7.4: Least squares regression summary for the final auction price against the condition of the game.

term	estimate	std.error	statistic	p.value
(Intercept)	42.9	0.81	52.67	<0.0001
condnew	10.9	1.26	8.66	<0.0001

Using values from Table 7.4, the model equation can be summarized as

$$\widehat{\texttt{price}} = 42.87 + 10.9 \times \texttt{condnew}$$

EXAMPLE

Interpret the two parameters estimated in the model for the price of Mario Kart in eBay auctions.

The intercept is the estimated price when `condnew` has a value 0, i.e., when the game is in used condition. That is, the average selling price of a used version of the game is $42.9. The slope indicates that, on average, new games sell for about $10.9 more than used games.

Interpreting model estimates for categorical predictors.

The estimated intercept is the value of the outcome variable for the first category (i.e., the category corresponding to an indicator value of 0). The estimated slope is the average change in the outcome variable between the two categories.

Note that, fundamentally, the intercept and slope interpretations don't change when modeling categorical variables with two levels. However, when the predictor variable is binary, the coefficient estimates (b_0 and b_1) are directly interpretable with respect to the dataset at hand.

We'll elaborate further on modeling categorical predictors in Chapter 8, where we examine the influence of many predictor variables simultaneously using multiple regression.

7.3 Outliers in linear regression

In this section, we identify criteria for determining which outliers are important and influential. Outliers in regression are observations that fall far from the cloud of points. These points are especially important because they can have a strong influence on the least squares line.

EXAMPLE

There are three plots shown in Figure 7.17 along with the corresponding least squares line and residual plots. For each scatterplot and residual plot pair, identify the outliers and note how they influence the least squares line. Recall that an outlier is any point that doesn't appear to belong with the vast majority of the other points.

- A: There is one outlier far from the other points, though it only appears to slightly influence the line.

- B: There is one outlier on the right, though it is quite close to the least squares line, which suggests it wasn't very influential.

- C: There is one point far away from the cloud, and this outlier appears to pull the least squares line up on the right; examine how the line around the primary cloud doesn't appear to fit very well.

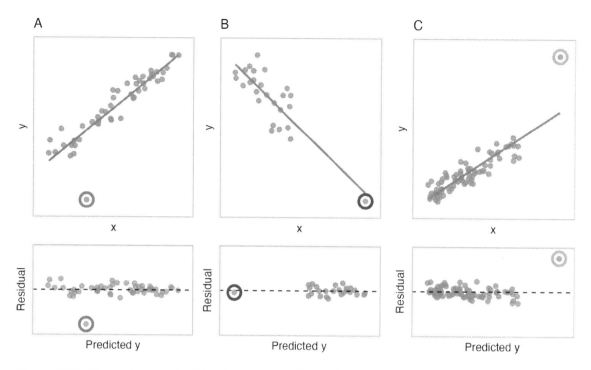

Figure 7.17: Three plots, each with a least squares line and corresponding residual plot. Each dataset has at least one outlier.

 EXAMPLE

There are three plots shown in Figure 7.18 along with the least squares line and residual plots. As you did in previous exercise, for each scatterplot and residual plot pair, identify the outliers and note how they influence the least squares line. Recall that an outlier is any point that doesn't appear to belong with the vast majority of the other points.

- D: There is a primary cloud and then a small secondary cloud of four outliers. The secondary cloud appears to be influencing the line somewhat strongly, making the least square line fit poorly almost everywhere. There might be an interesting explanation for the dual clouds, which is something that could be investigated.

- E: There is no obvious trend in the main cloud of points and the outlier on the right appears to largely (and problematically) control the slope of the least squares line.

- F: There is one outlier far from the cloud. However, it falls quite close to the least squares line and does not appear to be very influential.

Examine the residual plots in Figures 7.17 and 7.18. In Plots C, D, and E, you will probably find that there are a few observations which are both away from the remaining points along the x-axis and not in the trajectory of the trend in the rest of the data. In these cases, the outliers influenced the slope of the least squares lines. In Plot E, the bulk of the data show no clear trend, but if we fit a line to these data, we impose a trend where there isn't really one.

7.3. OUTLIERS IN LINEAR REGRESSION

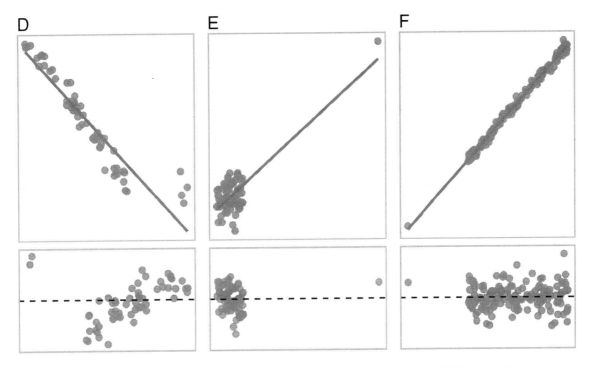

Figure 7.18: Three plots, each with a least squares line and residual plot. All datasets have at least one outlier.

 Leverage.

Points that fall horizontally away from the center of the cloud tend to pull harder on the line, so we call them points with **high leverage** or **leverage points**.

Points that fall horizontally far from the line are points of high leverage; these points can strongly influence the slope of the least squares line. If one of these high leverage points does appear to actually invoke its influence on the slope of the line – as in Plots C, D, and E of Figures 7.17 and 7.18 – then we call it an **influential point**. Usually we can say a point is influential if, had we fitted the line without it, the influential point would have been unusually far from the least squares line.

 Types of outliers.

A point (or a group of points) that stands out from the rest of the data is called an outlier. Outliers that fall horizontally away from the center of the cloud of points are called leverage points. Outliers that influence on the slope of the line are called influential points.

It is tempting to remove outliers. Don't do this without a very good reason. Models that ignore exceptional (and interesting) cases often perform poorly. For instance, if a financial firm ignored the largest market swings – the "outliers" – they would soon go bankrupt by making poorly thought-out investments.

7.4 Chapter review

7.4.1 Summary

Throughout this chapter, the nuances of the linear model have been described. You have learned how to create a linear model with explanatory variables that are numerical (e.g., total possum length) and those that are categorical (e.g., whether or not a video game was new). The residuals in a linear model are an important metric used to understand how well a model fits; high leverage points, influential points, and other types of outliers can impact the fit of a model. Correlation is a measure of the strength and direction of the linear relationship of two variables, without specifying which variable is the explanatory and which is the outcome. Future chapters will focus on generalizing the linear model from the sample of data to claims about the population of interest.

7.4.2 Terms

We introduced the following terms in the chapter. If you're not sure what some of these terms mean, we recommend you go back in the text and review their definitions. We are purposefully presenting them in alphabetical order, instead of in order of appearance, so they will be a little more challenging to locate. However you should be able to easily spot them as **bolded text**.

coefficient of determination	influential point	predictor
correlation	least squares line	R-squared
extrapolation	leverage point	residuals
high leverage	outcome	sum of squared error
indicator variable	outlier	total sum of squares

7.5 Exercises

Answers to odd numbered exercises can be found in Appendix A.7.

7.1. **Visualize the residuals.** The scatterplots shown below each have a superimposed regression line. If we were to construct a residual plot (residuals versus x) for each, describe in words what those plots would look like.

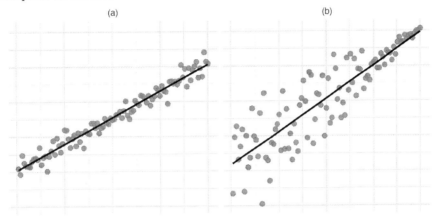

7.2. **Trends in the residuals.** Shown below are two plots of residuals remaining after fitting a linear model to two different sets of data. For each plot, describe important features and determine if a linear model would be appropriate for these data. Explain your reasoning.

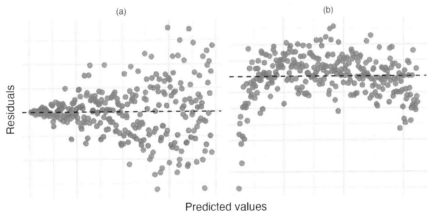

7.3. **Identify relationships, I.** For each of the six plots, identify the strength of the relationship (e.g., weak, moderate, or strong) in the data and whether fitting a linear model would be reasonable.

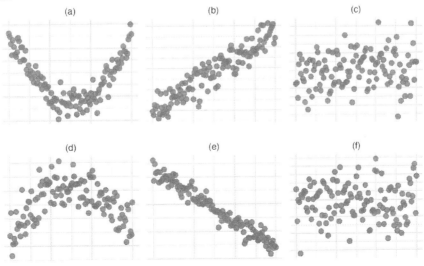

7.4. **Identify relationships, II.** For each of the six plots, identify the strength of the relationship (e.g., weak, moderate, or strong) in the data and whether fitting a linear model would be reasonable.

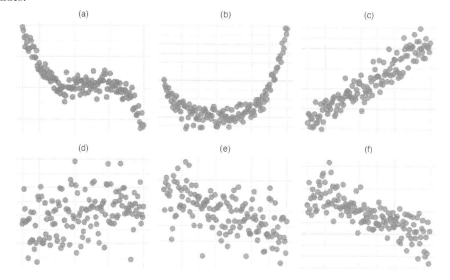

7.5. **Midterms and final.** The two scatterplots below show the relationship between the overall course average and two midterm exams (Exam 1 and Exam 2) recorded for 233 students during several years for a statistics course at a university.[11]

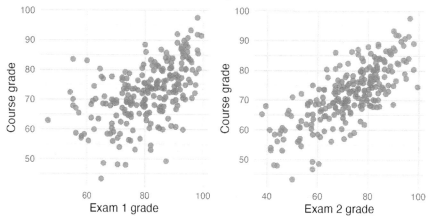

a. Based on these graphs, which of the two exams has the strongest correlation with the course grade? Explain.

b. Can you think of a reason why the correlation between the exam you chose in part (a) and the course grade is higher?

---

[11] The `exam_grades` data used in this exercise can be found in the **openintro** R package.

7.6. **Partners' ages and heights.** The Great Britain Office of Population Census and Surveys collected data on a random sample of 170 married couples in Britain, recording the age (in years) and heights (converted here to inches) of the partners. The scatterplot on the left shows the heights of the partners plotted against each other and the plot on the right shows the ages of the partners plotted against each other.[12]

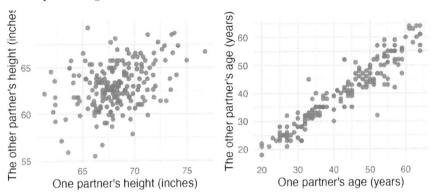

a. Describe the relationship between partners' ages.

b. Describe the relationship between partners' heights.

c. Which plot shows a stronger correlation? Explain your reasoning.

d. Data on heights were originally collected in centimeters, and then converted to inches. Does this conversion affect the correlation between partners' heights?

7.7. **Match the correlation, I.** Match each correlation to the corresponding scatterplot.[13]

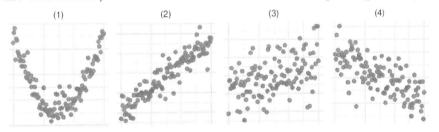

a. $r = -0.7$

b. $r = 0.45$

c. $r = 0.06$

d. $r = 0.92$

7.8. **Match the correlation, II.** Match each correlation to the corresponding scatterplot.[14]

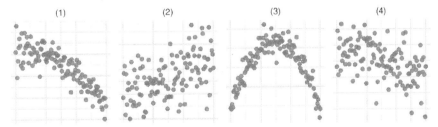

a. $r = 0.49$

b. $r = -0.48$

c. $r = -0.03$

d. $r = -0.85$

---

[12]The husbands_wives data used in this exercise can be found in the **openintro** R package.

[13]The corr_match data used in this exercise can be found in the **openintro** R package.

[14]The corr_match data used in this exercise can be found in the **openintro** R package.

7.9. **Body measurements, correlation.** Researchers studying anthropometry collected body and skeletal diameter measurements, as well as age, weight, height and sex for 507 physically active individuals. The scatterplot below shows the relationship between height and shoulder girth (circumference of shoulders measured over deltoid muscles), both measured in centimeters.[15] (Heinz et al., 2003)

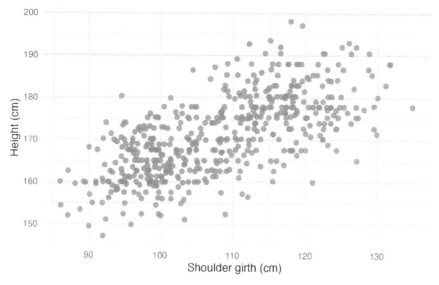

a. Describe the relationship between shoulder girth and height.

b. How would the relationship change if shoulder girth was measured in inches while the units of height remained in centimeters?

7.10. **Compare correlations.** Eduardo and Rosie are both collecting data on number of rainy days in a year and the total rainfall for the year. Eduardo records rainfall in inches and Rosie in centimeters. How will their correlation coefficients compare?

7.11. **The Coast Starlight, correlation.** The Coast Starlight Amtrak train runs from Seattle to Los Angeles. The scatterplot below displays the distance between each stop (in miles) and the amount of time it takes to travel from one stop to another (in minutes).[16]

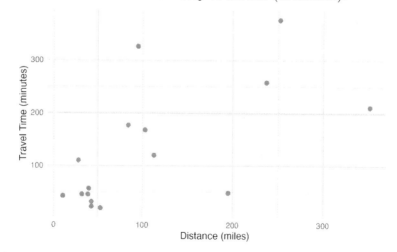

a. Describe the relationship between distance and travel time.

b. How would the relationship change if travel time was instead measured in hours, and distance was instead measured in kilometers?

c. Correlation between travel time (in miles) and distance (in minutes) is $r = 0.636$. What is the correlation between travel time (in kilometers) and distance (in hours)?

---

[15]The `bdims` data used in this exercise can be found in the **openintro** R package.

[16]The `coast_starlight` data used in this exercise can be found in the **openintro** R package.

7.12. **Crawling babies, correlation.** A study conducted at the University of Denver investigated whether babies take longer to learn to crawl in cold months, when they are often bundled in clothes that restrict their movement, than in warmer months. Infants born during the study year were split into twelve groups, one for each birth month. We consider the average crawling age of babies in each group against the average temperature when the babies are six months old (that's when babies often begin trying to crawl). Temperature is measured in degrees Fahrenheit (F) and age is measured in weeks.[17] (Benson, 1993)

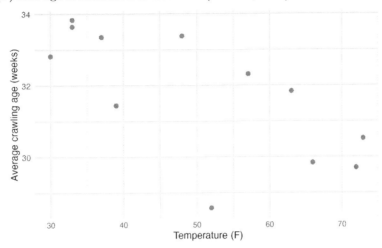

a. Describe the relationship between temperature and crawling age.

b. How would the relationship change if temperature was measured in degrees Celsius (C) and age was measured in months?

c. The correlation between temperature in F and age in weeks was $r = -0.70$. If we converted the temperature to C and age to months, what would the correlation be?

7.13. **Partners' ages.** What would be the correlation between the ages of partners if people always dated others who are

a. 3 years younger than themselves?

b. 2 years older than themselves?

c. half as old as themselves?

7.14. **Graduate degrees and salaries.** What would be the correlation between the annual salaries of people with and without a graduate degree at a company if for a certain type of position someone with a graduate degree always made

a. $5,000 more than those without a graduate degree?

b. 25% more than those without a graduate degree?

c. 15% less than those without a graduate degree?

7.15. **Units of regression.** Consider a regression predicting the number of calories (cal) from width (cm) for a sample of square shaped chocolate brownies. What are the units of the correlation coefficient, the intercept, and the slope?

7.16. **Which is higher?** Determine if (I) or (II) is higher or if they are equal: *"For a regression line, the uncertainty associated with the slope estimate, b_1, is higher when (I) there is a lot of scatter around the regression line or (II) there is very little scatter around the regression line."* Explain your reasoning.

7.17. **Over-under, I.** Suppose we fit a regression line to predict the shelf life of an apple based on its weight. For a particular apple, we predict the shelf life to be 4.6 days. The apple's residual is -0.6 days. Did we over or under estimate the shelf-life of the apple? Explain your reasoning.

---

[17] The `babies_crawl` data used in this exercise can be found in the **openintro** R package.

7.18. **Over-under, II.** Suppose we fit a regression line to predict the number of incidents of skin cancer per 1,000 people from the number of sunny days in a year. For a particular year, we predict the incidence of skin cancer to be 1.5 per 1,000 people, and the residual for this year is 0.5. Did we over or under estimate the incidence of skin cancer? Explain your reasoning.

7.19. **Starbucks, calories, and protein.** The scatterplot below shows the relationship between the number of calories and amount of protein (in grams) Starbucks food menu items contain. Since Starbucks only lists the number of calories on the display items, we might be interested in predicting the amount of protein a menu item has based on its calorie content.[18]

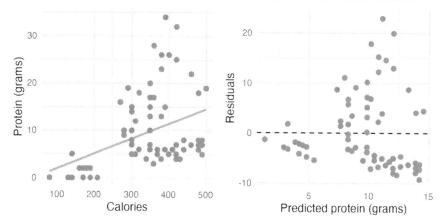

a. Describe the relationship between number of calories and amount of protein (in grams) that Starbucks food menu items contain.

b. In this scenario, what are the predictor and outcome variables?

c. Why might we want to fit a regression line to these data?

d. What does the residuals vs. predicted plot tell us about the variability in our prediction errors based on this model for items with lower vs. higher predicted protein?

7.20. **Starbucks, calories, and carbs.** The scatterplot below shows the relationship between the number of calories and amount of carbohydrates (in grams) Starbucks food menu items contain. Since Starbucks only lists the number of calories on the display items, we might be interested in predicting the amount of carbs a menu item has based on its calorie content.[19]

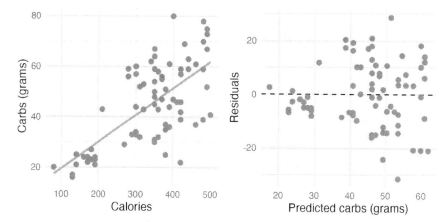

a. Describe the relationship between number of calories and amount of carbohydrates (in grams) that Starbucks food menu items contain.

b. In this scenario, what are the predictor and outcome variables?

c. Why might we want to fit a regression line to these data?

d. What does the residuals vs. predicted plot tell us about the variability in our prediction errors based on this model for items with lower vs. higher predicted carbs?

---

[18]The `starbucks` data used in this exercise can be found in the **openintro** R package.
[19]The `starbucks` data used in this exercise can be found in the **openintro** R package.

7.21. **The Coast Starlight, regression.** The Coast Starlight Amtrak train runs from Seattle to Los Angeles. The scatterplot below displays the distance between each stop (in miles) and the amount of time it takes to travel from one stop to another (in minutes). The mean travel time from one stop to the next on the Coast Starlight is 129 mins, with a standard deviation of 113 minutes. The mean distance traveled from one stop to the next is 108 miles with a standard deviation of 99 miles. The correlation between travel time and distance is 0.636.[20]

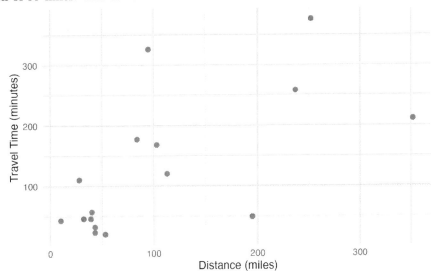

a. Write the equation of the regression line for predicting travel time.

b. Interpret the slope and the intercept in this context.

c. Calculate R^2 of the regression line for predicting travel time from distance traveled for the Coast Starlight, and interpret R^2 in the context of the application.

d. The distance between Santa Barbara and Los Angeles is 103 miles. Use the model to estimate the time it takes for the Starlight to travel between these two cities.

e. It actually takes the Coast Starlight about 168 mins to travel from Santa Barbara to Los Angeles. Calculate the residual and explain the meaning of this residual value.

f. Suppose Amtrak is considering adding a stop to the Coast Starlight 500 miles away from Los Angeles. Would it be appropriate to use this linear model to predict the travel time from Los Angeles to this point?

---

[20] The coast_starlight data used in this exercise can be found in the **openintro** R package.

7.22. **Body measurements, regression.** Researchers studying anthropometry collected body and skeletal diameter measurements, as well as age, weight, height and sex for 507 physically active individuals. The scatterplot below shows the relationship between height and shoulder girth (circumference of shoulders measured over deltoid muscles), both measured in centimeters. The mean shoulder girth is 107.20 cm with a standard deviation of 10.37 cm. The mean height is 171.14 cm with a standard deviation of 9.41 cm. The correlation between height and shoulder girth is 0.67.[21] (Heinz et al., 2003)

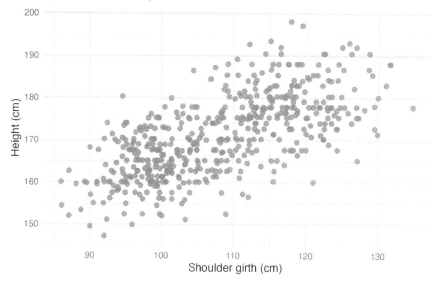

a. Write the equation of the regression line for predicting height.

b. Interpret the slope and the intercept in this context.

c. Calculate R^2 of the regression line for predicting height from shoulder girth, and interpret it in the context of the application.

d. A randomly selected student from your class has a shoulder girth of 100 cm. Predict the height of this student using the model.

e. The student from part (d) is 160 cm tall. Calculate the residual, and explain what this residual means.

f. A one year old has a shoulder girth of 56 cm. Would it be appropriate to use this linear model to predict the height of this child?

---

[21] The **bdims** data used in this exercise can be found in the **openintro** R package.

7.23. **Poverty and unemployment.** The following scatterplot shows the relationship between percent of population below the poverty level (poverty) from unemployment rate among those ages 20-64 (unemployment_rate) in counties in the US, as provided by data from the 2019 American Community Survey. The regression output for the model for predicting poverty from unemployment_rate is also provided.[22]

term	estimate	std.error	statistic	p.value
(Intercept)	4.60	0.349	13.2	<0.0001
unemployment_rate	2.05	0.062	33.1	<0.0001

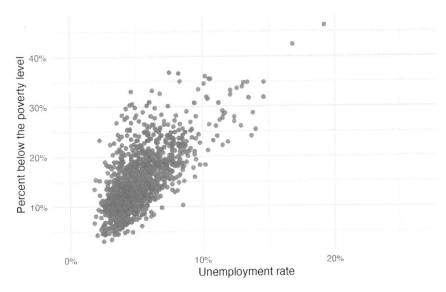

a. Write out the linear model.

b. Interpret the intercept.

c. Interpret the slope.

d. For this model R^2 is 46%. Interpret this value.

e. Calculate the correlation coefficient.

---

[22] The county_2019 data used in this exercise can be found in the **usdata** R package.

7.24. **Cats weights.** The following regression output is for predicting the heart weight (Hwt, in g) of cats from their body weight (Bwt, in kg). The coefficients are estimated using a dataset of 144 domestic cats.[23]

term	estimate	std.error	statistic	p.value
(Intercept)	-0.357	0.692	-0.515	0.6072
Bwt	4.034	0.250	16.119	<0.0001

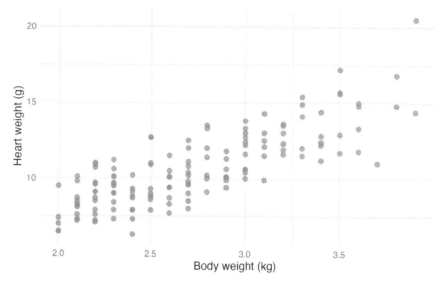

a. Write out the linear model.

b. Interpret the intercept.

c. Interpret the slope.

d. The R^2 of this model is 1%. Interpret R^2.

e. Calculate the correlation coefficient.

7.25. **Outliers, I.** Identify the outliers in the scatterplots shown below, and determine what type of outliers they are. Explain your reasoning.

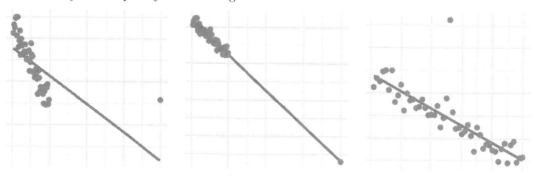

---

[23]The cats data used in this exercise can be found in the **MASS** R package.

7.26. **Outliers, II.** Identify the outliers in the scatterplots shown below and determine what type of outliers they are. Explain your reasoning.

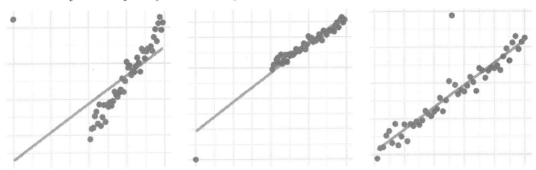

7.27. **Urban homeowners, outliers.** The scatterplot below shows the percent of families who own their home vs. the percent of the population living in urban areas. There are 52 observations, each corresponding to a state in the US. Puerto Rico and District of Columbia are also included.[24]

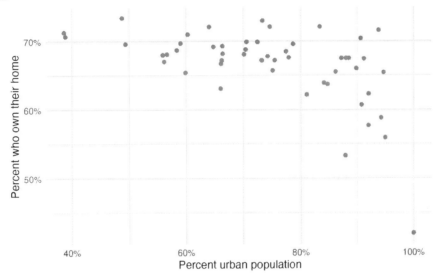

a. Describe the relationship between the percent of families who own their home and the percent of the population living in urban areas.

b. The outlier at the bottom right corner is District of Columbia, where 100% of the population is considered urban. What type of an outlier is this observation?

---

[24] The `urban_owner` data used in this exercise can be found in the **usdata** R package.

7.28. **Crawling babies, outliers.** A study conducted at the University of Denver investigated whether babies take longer to learn to crawl in cold months, when they are often bundled in clothes that restrict their movement, than in warmer months. The plot below shows the relationship between average crawling age of babies born in each month and the average temperature in the month when the babies are six months old. The plot reveals a potential outlying month when the average temperature is about 53F and average crawling age is about 28.5 weeks. Does this point have high leverage? Is it an influential point?[25] (Benson, 1993)

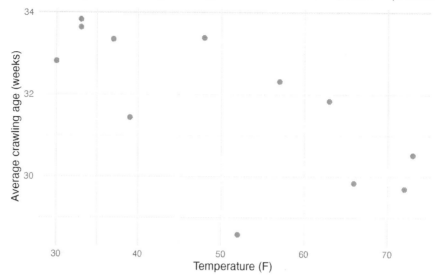

7.29. **True / False.** Determine if the following statements are true or false. If false, explain why.

a. A correlation coefficient of -0.90 indicates a stronger linear relationship than a correlation of 0.5.

b. Correlation is a measure of the association between any two variables.

7.30. **Cherry trees.** The scatterplots below show the relationship between height, diameter, and volume of timber in 31 felled black cherry trees. The diameter of the tree is measured 4.5 feet above the ground.[26]

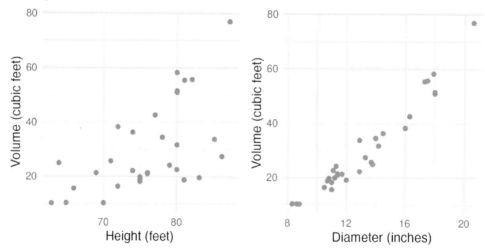

a. Describe the relationship between volume and height of these trees.

b. Describe the relationship between volume and diameter of these trees.

c. Suppose you have height and diameter measurements for another black cherry tree. Which of these variables would be preferable to use to predict the volume of timber in this tree using a simple linear regression model? Explain your reasoning.

---

[25]The `babies_crawl` data used in this exercise can be found in the **openintro** R package.
[26]The `trees` data used in this exercise can be found in the **datasets** R package.

7.31. **Match the correlation, III.** Match each correlation to the corresponding scatterplot.[27]

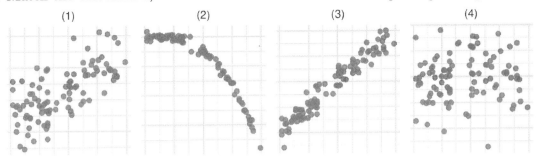

a. r = 0.69

b. r = 0.09

c. r = -0.91

d. r = 0.97

7.32. **Helmets and lunches.** The scatterplot shows the relationship between socioeconomic status measured as the percentage of children in a neighborhood receiving reduced-fee lunches at school (lunch) and the percentage of bike riders in the neighborhood wearing helmets (helmet). The average percentage of children receiving reduced-fee lunches is 30.833% with a standard deviation of 26.724% and the average percentage of bike riders wearing helmets is 30.883% with a standard deviation of 16.948%.

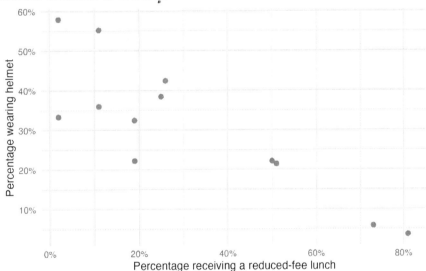

a. If the R^2 for the least-squares regression line for these data is 72%, what is the correlation between lunch and helmet?

b. Calculate the slope and intercept for the least-squares regression line for these data.

c. Interpret the intercept of the least-squares regression line in the context of the application.

d. Interpret the slope of the least-squares regression line in the context of the application.

5. What would the value of the residual be for a neighborhood where 40% of the children receive reduced-fee lunches and 40% of the bike riders wear helmets? Interpret the meaning of this residual in the context of the application.

---

[27]The corr_match data used in this exercise can be found in the **openintro** R package.

Chapter 8

Linear regression with multiple predictors

Building on the ideas of one predictor variable in a linear regression model (from Chapter 7), a multiple linear regression model is now fit to two or more predictor variables. By considering how different explanatory variables interact, we can uncover complicated relationships between the predictor variables and the response variable. One challenge to working with multiple variables is that it is sometimes difficult to know which variables are most important to include in the model. Model building is an extensive topic, and we scratch the surface here by defining and utilizing the adjusted R^2 value.

Multiple regression extends single predictor variable regression to the case that still has one response but many predictors (denoted x_1, x_2, x_3, ...). The method is motivated by scenarios where many variables may be simultaneously connected to an output.

We will consider data about loans from the peer-to-peer lender, Lending Club, which is a dataset we first encountered in Chapter 1. The loan data includes terms of the loan as well as information about the borrower. The outcome variable we would like to better understand is the interest rate assigned to the loan. For instance, all other characteristics held constant, does it matter how much debt someone already has? Does it matter if their income has been verified? Multiple regression will help us answer these and other questions.

The dataset includes results from 10,000 loans, and we'll be looking at a subset of the available variables, some of which will be new from those we saw in earlier chapters. The first six observations in the dataset are shown in Table 8.1, and descriptions for each variable are shown in Table 8.2. Notice that the past bankruptcy variable (`bankruptcy`) is an indicator variable, where it takes the value 1 if the borrower had a past bankruptcy in their record and 0 if not. Using an indicator variable in place of a category name allows for these variables to be directly used in regression. Two of the other variables are categorical (`verified_income` and `issue_month`), each of which can take one of a few different non-numerical values; we'll discuss how these are handled in the model in Section 8.1.

8.1. INDICATOR AND CATEGORICAL PREDICTORS

 The `loans_full_schema` data can be found in the **openintro** R package. Based on the data in this dataset we have created two new variables: `credit_util` which is calculated as the total credit utilized divided by the total credit limit and `bankruptcy` which turns the number of bankruptcies to an indicator variable (0 for no bankruptcies and 1 for at least 1 bankruptcy). We will refer to this modified dataset as `loans`.

Table 8.1: First six rows of the `loans` dataset.

interest_rate	verified_income	debt_to_income	credit_util	bankruptcy	term	credit_checks	issue_month
14.07	Verified	18.01	0.548	0	60	6	Mar-2018
12.61	Not Verified	5.04	0.150	1	36	1	Feb-2018
17.09	Source Verified	21.15	0.661	0	36	4	Feb-2018
6.72	Not Verified	10.16	0.197	0	36	0	Jan-2018
14.07	Verified	57.96	0.755	0	36	7	Mar-2018
6.72	Not Verified	6.46	0.093	0	36	6	Jan-2018

Table 8.2: Variables and their descriptions for the `loans` dataset.

Variable	Description
`interest_rate`	Interest rate on the loan, in an annual percentage.
`verified_income`	Categorical variable describing whether the borrower's income source and amount have been verified, with levels 'Verified', 'Source Verified', and 'Not Verified'.
`debt_to_income`	Debt-to-income ratio, which is the percentage of total debt of the borrower divided by their total income.
`credit_util`	Of all the credit available to the borrower, what fraction are they utilizing. For example, the credit utilization on a credit card would be the card's balance divided by the card's credit limit.
`bankruptcy`	An indicator variable for whether the borrower has a past bankruptcy in their record. This variable takes a value of '1' if the answer is *yes* and '0' if the answer is *no*.
`term`	The length of the loan, in months.
`issue_month`	The month and year the loan was issued, which for these loans is always during the first quarter of 2018.
`credit_checks`	Number of credit checks in the last 12 months. For example, when filing an application for a credit card, it is common for the company receiving the application to run a credit check.

8.1 Indicator and categorical predictors

Let's start by fitting a linear regression model for interest rate with a single predictor indicating whether or not a person has a bankruptcy in their record:

$$\widehat{\texttt{interest\_rate}} = 12.34 + 0.74 \times \texttt{bankruptcy}$$

Results of this model are shown in Table 8.3.

Table 8.3: Summary of a linear model for predicting `interest_rate` based on whether the borrower has a bankruptcy in their record. Degrees of freedom for this model is 9998.

term	estimate	std.error	statistic	p.value
(Intercept)	12.34	0.05	231.49	<0.0001
bankruptcy1	0.74	0.15	4.82	<0.0001

EXAMPLE

Interpret the coefficient for the past bankruptcy variable in the model.

The variable takes one of two values: 1 when the borrower has a bankruptcy in their history and 0 otherwise. A slope of 0.74 means that the model predicts a 0.74% higher interest rate for those borrowers with a bankruptcy in their record. (See Section 7.2.6 for a review of the interpretation for two-level categorical predictor variables.)

Suppose we had fit a model using a 3-level categorical variable, such as `verified_income`. The output from software is shown in Table 8.4. This regression output provides multiple rows for the variable. Each row represents the relative difference for each level of `verified_income`. However, we are missing one of the levels: Not Verified. The missing level is called the **reference level** and it represents the default level that other levels are measured against.

Table 8.4: Summary of a linear model for predicting `interest_rate` based on whether the borrower's income source and amount has been verified. This predictor has three levels, which results in 2 rows in the regression output.

term	estimate	std.error	statistic	p.value
(Intercept)	11.10	0.08	137.2	<0.0001
verified_incomeSource Verified	1.42	0.11	12.8	<0.0001
verified_incomeVerified	3.25	0.13	25.1	<0.0001

EXAMPLE

How would we write an equation for this regression model?

The equation for the regression model may be written as a model with two predictors:

$$\widehat{interest\_rate} = 11.10 \\ + 1.42 \times \texttt{verified\_income}_{\text{Source Verified}} \\ + 3.25 \times \texttt{verified\_income}_{\text{Verified}}$$

We use the notation $\texttt{variable}_{\text{level}}$ to represent indicator variables for when the categorical variable takes a particular value. For example, $\texttt{verified\_income}_{\text{Source Verified}}$ would take a value of 1 if it was for a borrower that was source verified, and it would take a value of 0 otherwise. Likewise, $\texttt{verified\_income}_{\text{Verified}}$ would take a value of 1 if it was for a borrower that was verified, and 0 if it took any other value.

The notation $\texttt{variable}_{\text{level}}$ may feel a bit confusing. Let's figure out how to use the equation for

8.1. INDICATOR AND CATEGORICAL PREDICTORS

each level of the `verified_income` variable.

EXAMPLE

Using the model for predicting interest rate from income verification type, compute the average interest rate for borrowers whose income source and amount are both *unverified*.

When `verified_income` takes a value of `Not Verified`, then both indicator functions in the equation for the linear model are set to 0:

$$\widehat{\texttt{interest\_rate}} = 11.10 + 1.42 \times 0 + 3.25 \times 0 = 11.10$$

The average interest rate for these borrowers is 11.1%. Because the level does not have its own coefficient and it is the reference value, the indicators for the other levels for this variable all drop out.

EXAMPLE

Using the model for predicting interest rate from income verification type, compute the average interest rate for borrowers whose income source and amount are both *source verified*.

When `verified_income` takes a value of `Source Verified`, then the corresponding variable takes a value of 1 while the other is 0:

$$\widehat{\texttt{interest\_rate}} = 11.10 + 1.42 \times 1 + 3.25 \times 0 = 12.52$$

The average interest rate for these borrowers is 12.52%.

GUIDED PRACTICE

Compute the average interest rate for borrowers whose income source and amount are both verified.[1]

Predictors with several categories.

When fitting a regression model with a categorical variable that has k levels where $k > 2$, software will provide a coefficient for $k - 1$ of those levels. For the last level that does not receive a coefficient, this is the reference level, and the coefficients listed for the other levels are all considered relative to this reference level.

---

[1] When `verified_income` takes a value of `Verified`, then the corresponding variable takes a value of 1 while the other is 0: $11.10 + 1.42 \times 0 + 3.25 \times 1 = 14.35$. The average interest rate for these borrowers is 14.35%.

GUIDED PRACTICE

Interpret the coefficients from the model above.[2]

The higher interest rate for borrowers who have verified their income source or amount is surprising. Intuitively, we'd think that a loan would look *less* risky if the borrower's income has been verified. However, note that the situation may be more complex, and there may be confounding variables that we didn't account for. For example, perhaps lenders require borrowers with poor credit to verify their income. That is, verifying income in our dataset might be a signal of some concerns about the borrower rather than a reassurance that the borrower will pay back the loan. For this reason, the borrower could be deemed higher risk, resulting in a higher interest rate. (What other confounding variables might explain this counter-intuitive relationship suggested by the model?)

GUIDED PRACTICE

How much larger of an interest rate would we expect for a borrower who has verified their income source and amount vs a borrower whose income source has only been verified?[3]

8.2 Many predictors in a model

The world is complex, and it can be helpful to consider many factors at once in statistical modeling. For example, we might like to use the full context of borrowers to predict the interest rate they receive rather than using a single variable. This is the strategy used in **multiple regression**. While we remain cautious about making any causal interpretations using multiple regression on observational data, such models are a common first step in gaining insights or providing some evidence of a causal connection.

We want to construct a model that accounts not only for any past bankruptcy or whether the borrower had their income source or amount verified, but simultaneously accounts for all the variables in the `loans` dataset: `verified_income`, `debt_to_income`, `credit_util`, `bankruptcy`, `term`, `issue_month`, and `credit_checks`.

$$\widehat{\texttt{interest\_rate}} = b_0$$
$$+ b_1 \times \texttt{verified\_income}_{\text{Source Verified}}$$
$$+ b_2 \times \texttt{verified\_income}_{\text{Verified}}$$
$$+ b_3 \times \texttt{debt\_to\_income}$$
$$+ b_4 \times \texttt{credit\_util}$$
$$+ b_5 \times \texttt{bankruptcy}$$
$$+ b_6 \times \texttt{term}$$
$$+ b_9 \times \texttt{credit\_checks}$$
$$+ b_7 \times \texttt{issue\_month}_{\text{Jan-2018}}$$
$$+ b_8 \times \texttt{issue\_month}_{\text{Mar-2018}}$$

---

[2] Each of the coefficients gives the incremental interest rate for the corresponding level relative to the `Not Verified` level, which is the reference level. For example, for a borrower whose income source and amount have been verified, the model predicts that they will have a 3.25% higher interest rate than a borrower who has not had their income source or amount verified.

[3] Relative to the `Not Verified` category, the `Verified` category has an interest rate of 3.25% higher, while the `Source Verified` category is only 1.42% higher. Thus, `Verified` borrowers will tend to get an interest rate about 3.25 higher than `Source Verified` borrowers.

8.2. MANY PREDICTORS IN A MODEL

This equation represents a holistic approach for modeling all of the variables simultaneously. Notice that there are two coefficients for `verified_income` and also two coefficients for `issue_month`, since both are 3-level categorical variables.

We calculate $b_0, b_1, b_2, \cdots, b_9$ the same way as we did in the case of a model with a single predictor – we select values that minimize the sum of the squared residuals:

$$SSE = e_1^2 + e_2^2 + \cdots + e_{10000}^2 = \sum_{i=1}^{10000} e_i^2 = \sum_{i=1}^{10000} (y_i - \hat{y}_i)^2$$

where y_i and $\hat{y}_i$ represent the observed interest rates and their estimated values according to the model, respectively. 10,000 residuals are calculated, one for each observation. Note that these values are sample statistics and in the case where the observed data is a random sample from a target population that we are interested in making inferences about, they are estimates of the population parameters $\beta_0, \beta_1, \beta_2, \cdots, \beta_9$. We will discuss inference based on linear models in Chapter 25, for now we will focus on calculating sample statistics b_i.

We typically use a computer to minimize the sum of squares and compute point estimates, as shown in the sample output in Table 8.5. Using this output, we identify b_i, just as we did in the one-predictor case.

Table 8.5: Output for the regression model, where interest rate is the outcome and the variables listed are the predictors. Degrees of freedom for this model is 9990.

term	estimate	std.error	statistic	p.value
(Intercept)	1.89	0.21	9.01	<0.0001
verified_incomeSource Verified	1.00	0.10	10.06	<0.0001
verified_incomeVerified	2.56	0.12	21.87	<0.0001
debt_to_income	0.02	0.00	7.43	<0.0001
credit_util	4.90	0.16	30.25	<0.0001
bankruptcy1	0.39	0.13	2.96	0.0031
term	0.15	0.00	38.89	<0.0001
credit_checks	0.23	0.02	12.52	<0.0001
issue_monthJan-2018	0.05	0.11	0.42	0.6736
issue_monthMar-2018	-0.04	0.11	-0.39	0.696

Multiple regression model.

A multiple regression model is a linear model with many predictors. In general, we write the model as

$$\hat{y} = b_0 + b_1 x_1 + b_2 x_2 + \cdots + b_k x_k$$

when there are k predictors. We always calculate b_i using statistical software.

EXAMPLE

Write out the regression model using the regression output from Table 8.5. How many predictors are there in this model?

The fitted model for the interest rate is given by:

$$\widehat{\text{interest\_rate}} = 1.89$$
$$+ 1.00 \times \text{verified\_income}_{\text{Source Verified}}$$
$$+ 2.56 \times \text{verified\_income}_{\text{Verified}}$$
$$+ 0.02 \times \text{debt\_to\_income}$$
$$+ 4.90 \times \text{credit\_util}$$
$$+ 0.39 \times \text{bankruptcy}$$
$$+ 0.15 \times \text{term}$$
$$+ 0.23 \times \text{credit\_checks}$$
$$+ 0.05 \times \text{issue\_month}_{\text{Jan-2018}}$$
$$- 0.04 \times \text{issue\_month}_{\text{Mar-2018}}$$

If we count up the number of predictor coefficients, we get the *effective* number of predictors in the model; there are nine of those. Notice that the categorical predictor counts as two, once for each of the two levels shown in the model. In general, a categorical predictor with p different levels will be represented by $p - 1$ terms in a multiple regression model. A total of seven variables were used as predictors to fit this model: `verified_income`, `debt_to_income`, `credit_util`, `bankruptcy`, `term`, `credit_checks`, `issue_month`.

GUIDED PRACTICE

Interpret the coefficient of the variable `credit_checks`.[4]

GUIDED PRACTICE

Compute the residual of the first observation in Table 8.1 on page using the full model.[5]

---

[4] All else held constant, for each additional inquiry into the applicant's credit during the last 12 months, we would expect the interest rate for the loan to be higher, on average, by 0.23 points.

[5] To compute the residual, we first need the predicted value, which we compute by plugging values into the equation from earlier. For example, `verified_income`$_{\text{Source Verified}}$ takes a value of 0, `verified_income`$_{\text{Verified}}$ takes a value of 1 (since the borrower's income source and amount were verified), was 18.01, and so on. This leads to a prediction of $\widehat{\text{interest\_rate}}_1 = 18.09$. The observed interest rate was 14.07%, which leads to a residual of $e_1 = 14.07 - 18.09 = -4.02$.

EXAMPLE

We calculated a slope coefficient of 0.74 for `bankruptcy` in Section 8.1 while the coefficient is 0.386 here. Why is there a difference between the coefficient values between the models with single and multiple predictors?

If we examined the data carefully, we would see that some predictors are correlated. For instance, when we modeled the relationship of the outcome `interest_rate` and predictor `bankruptcy` using linear regression, we were unable to control for other variables like whether the borrower had their income verified, the borrower's debt-to-income ratio, and other variables. That original model was constructed in a vacuum and did not consider the full context of everything that is considered when an interest rate is decided. When we include all of the variables, underlying and unintentional bias that was missed by not including these other variables is reduced or eliminated. Of course, bias can still exist from other confounding variables.

The previous example describes a common issue in multiple regression: correlation among predictor variables. We say the two predictor variables are collinear (pronounced as *co-linear*) when they are correlated, and this **multicollinearity** complicates model estimation. While it is impossible to prevent multicollinearity from arising in observational data, experiments are usually designed to prevent predictors from being multicollinearity.

GUIDED PRACTICE

The estimated value of the intercept is 1.925, and one might be tempted to make some interpretation of this coefficient, such as, it is the model's predicted price when each of the variables take value zero: income source is not verified, the borrower has no debt (debt-to-income and credit utilization are zero), and so on. Is this reasonable? Is there any value gained by making this interpretation?[6]

8.3 Adjusted R-squared

We first used R^2 in Section 7.2.5 to determine the amount of variability in the response that was explained by the model:

$$R^2 = 1 - \frac{\text{variability in residuals}}{\text{variability in the outcome}} = 1 - \frac{Var(e_i)}{Var(y_i)}$$

where e_i represents the residuals of the model and y_i the outcomes. This equation remains valid in the multiple regression framework, but a small enhancement can make it even more informative when comparing models.

GUIDED PRACTICE

The variance of the residuals for the model given in the earlier Guided Practice is 18.53, and the variance of the total price in all the auctions is 25.01. Calculate R^2 for this model.[7]

---

[6] Many of the variables do take a value 0 for at least one data point, and for those variables, it is reasonable. However, one variable never takes a value of zero: `term`, which describes the length of the loan, in months. If `term` is set to zero, then the loan must be paid back immediately; the borrower must give the money back as soon as they receive it, which means it is not a real loan. Ultimately, the interpretation of the intercept in this setting is not insightful.

This strategy for estimating R^2 is acceptable when there is just a single variable. However, it becomes less helpful when there are many variables. The regular R^2 is a biased estimate of the amount of variability explained by the model when applied to model with more than one predictor. To get a better estimate, we use the adjusted R^2.

Adjusted R-squared as a tool for model assessment.

The **adjusted R-squared** is computed as

$$R^2_{adj} = 1 - \frac{s^2_{residuals}/(n-k-1)}{s^2_{outcome}/(n-1)}$$
$$= 1 - \frac{s^2_{residuals}}{s^2_{outcome}} \times \frac{n-1}{n-k-1}$$

where n is the number of observations used to fit the model and k is the number of predictor variables in the model. Remember that a categorical predictor with p levels will contribute $p-1$ to the number of variables in the model.

Because k is never negative, the adjusted R^2 will be smaller – often times just a little smaller – than the unadjusted R^2. The reasoning behind the adjusted R^2 lies in the **degrees of freedom** associated with each variance, which is equal to $n-k-1$ in the multiple regression context. If we were to make predictions for *new data* using our current model, we would find that the unadjusted R^2 would tend to be slightly overly optimistic, while the adjusted R^2 formula helps correct this bias.

GUIDED PRACTICE

There were n = 10,000 auctions in the dataset and $k = 9$ predictor variables in the model. Use n, k, and the variances from the earlier Guided Practice to calculate R^2_{adj} for the interest rate model.[8]

GUIDED PRACTICE

Suppose you added another predictor to the model, but the variance of the errors $Var(e_i)$ didn't go down. What would happen to the R^2? What would happen to the adjusted R^2?[9]

Adjusted R^2 could also have been used in Chapter 7 where we introduced regression models with a single predictor. However, when there is only $k = 1$ predictors, adjusted R^2 is very close to regular R^2, so this nuance isn't typically important when the model has only one predictor.

8.4 Model selection

The best model is not always the most complicated. Sometimes including variables that are not evidently important can actually reduce the accuracy of predictions. In this section, we discuss model selection strategies, which will help us eliminate variables from the model that are found to be

---

[7] $R^2 = 1 - \frac{18.53}{25.01} = 0.2591$.

[8] $R^2_{adj} = 1 - \frac{18.53}{25.01} \times \frac{10000-1}{1000-9-1} = 0.2584$. While the difference is very small, it will be important when we fine tune the model in the next section.

[9] The unadjusted R^2 would stay the same and the adjusted R^2 would go down.

less important. It's common (and hip, at least in the statistical world) to refer to models that have undergone such variable pruning as **parsimonious**.

In practice, the model that includes all available predictors is often referred to as the **full model**. The full model may not be the best model, and if it isn't, we want to identify a smaller model that is preferable.

8.4.1 Stepwise selection

Two common strategies for adding or removing variables in a multiple regression model are called backward elimination and forward selection. These techniques are often referred to as **stepwise selection** strategies, because they add or delete one variable at a time as they "step" through the candidate predictors.

Backward elimination starts with the full model (the model that includes all potential predictor variables. Variables are eliminated one-at-a-time from the model until we cannot improve the model any further.

Forward selection is the reverse of the backward elimination technique. Instead of eliminating variables one-at-a-time, we add variables one-at-a-time until we cannot find any variables that improve the model any further.

An important consideration in implementing either of these stepwise selection strategies is the criterion used to decide whether to eliminate or add a variable. One commonly used decision criterion is adjusted R^2. When using adjusted R^2 as the decision criterion, we seek to eliminate or add variables depending on whether they lead to the largest improvement in adjusted R^2 and we stop when adding or elimination of another variable does not lead to further improvement in adjusted R^2.

Adjusted R^2 describes the strength of a model fit, and it is a useful tool for evaluating which predictors are adding value to the model, where *adding value* means they are (likely) improving the accuracy in predicting future outcomes.

Let's consider two models, which are shown in Table 8.6 and Table 8.7. The first table summarizes the full model since it includes all predictors, while the second does not include the `issue_month` variable.

Table 8.6: The fit for the full regression model, including the adjusted R^2.

term	estimate	std.error	statistic	p.value
(Intercept)	1.89	0.21	9.01	<0.0001
verified_incomeSource Verified	1.00	0.10	10.06	<0.0001
verified_incomeVerified	2.56	0.12	21.87	<0.0001
debt_to_income	0.02	0.00	7.43	<0.0001
credit_util	4.90	0.16	30.25	<0.0001
bankruptcy1	0.39	0.13	2.96	0.0031
term	0.15	0.00	38.89	<0.0001
credit_checks	0.23	0.02	12.52	<0.0001
issue_monthJan-2018	0.05	0.11	0.42	0.6736
issue_monthMar-2018	-0.04	0.11	-0.39	0.696

Adjusted R-sq = 0.2597
df = 9964

Table 8.7: The fit for the regression model after dropping issue month, including the adjusted R^2.

term	estimate	std.error	statistic	p.value
(Intercept)	1.90	0.20	9.56	<0.0001
verified_incomeSource Verified	1.00	0.10	10.05	<0.0001
verified_incomeVerified	2.56	0.12	21.86	<0.0001
debt_to_income	0.02	0.00	7.44	<0.0001
credit_util	4.90	0.16	30.25	<0.0001
bankruptcy1	0.39	0.13	2.96	0.0031
term	0.15	0.00	38.89	<0.0001
credit_checks	0.23	0.02	12.52	<0.0001

Adjusted R-sq = 0.2598
df = 9966

EXAMPLE

Which of the two models is better?

We compare the adjusted R^2 of each model to determine which to choose. Since the second model has a higher R^2_{adj} compared to the first model, we prefer the second model to the first.

Will the model without issue_month be better than the model with issue_month? We cannot know for sure, but based on the adjusted R^2, this is our best assessment.

8.4. MODEL SELECTION

EXAMPLE

Results corresponding to the full model for the `loans` data are shown in Table 8.6. How should we proceed under the backward elimination strategy?

Our baseline adjusted R^2 from the full model is 0.2597, and we need to determine whether dropping a predictor will improve the adjusted R^2. To check, we fit models that each drop a different predictor, and we record the adjusted R^2:

- Excluding `verified_income`: 0.2238
- Excluding `debt_to_income`: 0.2557
- Excluding `credit_util`: 0.1916
- Excluding `bankruptcy`: 0.2589
- Excluding `term`: 0.1468
- Excluding `credit_checks`: 0.2484
- Excluding `issue_month`: 0.2598

The model without `issue_month` has the highest adjusted R^2 of 0.2598, higher than the adjusted R^2 for the full model. Because eliminating `issue_month` leads to a model with a higher adjusted R^2, we drop `issue_month` from the model.

Since we eliminated a predictor from the model in the first step, we see whether we should eliminate any additional predictors. Our baseline adjusted R^2 is now $R^2_{adj} = 0.2598$. We now fit new models, which consider eliminating each of the remaining predictors in addition to `issue_month`:

- Excluding `issue_month` and `verified_income`: 0.22395
- Excluding `issue_month` and `debt_to_income`: 0.25579
- Excluding `issue_month` and `credit_util`: 0.19174
- Excluding `issue_month` and `bankruptcy`: 0.25898
- Excluding `issue_month` and `term`: 0.14692
- Excluding `issue_month` and `credit_checks`: 0.24801

None of these models lead to an improvement in adjusted R^2, so we do not eliminate any of the remaining predictors. That is, after backward elimination, we are left with the model that keeps all predictors except `issue_month`, which we can summarize using the coefficients from Table 8.7.

$$\widehat{\texttt{interest\_rate}} = 1.90$$
$$+ 1.00 \times \texttt{verified\_income}_{\texttt{Source only}}$$
$$+ 2.56 \times \texttt{verified\_income}_{\texttt{Verified}}$$
$$+ 0.02 \times \texttt{debt\_to\_income}$$
$$+ 4.90 \times \texttt{credit\_util}$$
$$+ 0.39 \times \texttt{bankruptcy}$$
$$+ 0.15 \times \texttt{term}$$
$$+ 0.23 \times \texttt{credit\_check}$$

EXAMPLE

Construct a model for predicting `interest_rate` from the `loans` data using forward selection.

We start with the model that includes no predictors. Then we fit each of the possible models with just one predictor. Then we examine the adjusted R^2 for each of these models:

- Including `verified_income`: 0.05926
- Including `debt_to_income`: 0.01946
- Including `credit_util`: 0.06452
- Including `bankruptcy`: 0.00222
- Including `term`: 0.12855
- Including `credit_checks`: -0.0001
- Including `issue_month`: 0.01711

In this first step, we compare the adjusted R^2 against a baseline model that has no predictors. The no-predictors model always has $R^2_{adj} = 0$. The model with one predictor that has the largest adjusted R^2 is the model with the `term` predictor, and because this adjusted R^2 is larger than the adjusted R^2 from the model with no predictors ($R^2_{adj} = 0$), we will add this variable to our model.

We repeat the process again, this time considering 2-predictor models where one of the predictors is `term` and with a new baseline of $R^2_{adj} = 0.12855$:

- Including `term` and `verified_income`: 0.16851
- Including `term` and `debt_to_income`: 0.14368
- Including `term` and `credit_util`: 0.20046
- Including `term` and `bankruptcy`: 0.13070
- Including `term` and `credit_checks`: 0.12840
- Including `term` and `issue_month`: 0.14294

Adding `credit_util` yields the highest increase in adjusted R^2 and has a higher adjusted R^2 (0.20046) than the baseline (0.12855). Thus, we will also add `credit_util` to the model as a predictor.

Since we have again added a predictor to the model, we have a new baseline adjusted R^2 of 0.20046. We can continue on and see whether it would be beneficial to add a third predictor:

- Including `term`, `credit_util`, and `verified_income`: 0.24183
- Including `term`, `credit_util`, and `debt_to_income`: 0.20810
- Including `term`, `credit_util`, and `bankruptcy`: 0.20169
- Including `term`, `credit_util`, and `credit_checks`: 0.20031
- Including `term`, `credit_util`, and `issue_month`: 0.21629

The model including `verified_income` has the largest increase in adjusted R^2 (0.24183) from the baseline (0.20046), so we add `verified_income` to the model as a predictor as well.

We continue on in this way, next adding `debt_to_income`, then `credit_checks`, and `bankruptcy`. At this point, we come again to the `issue_month` variable: adding this as a predictor leads to $R^2_{adj} = 0.25843$, while keeping all the other predictors but excluding `issue_month` has a higher $R^2_{adj} = 0.25854$. This means we do not add `issue_month` to the model as a predictor. In this example, we have arrived at the same model that we identified from backward elimination.

Stepwise selection strategies.

Backward elimination begins with the model having the largest number of predictors and eliminates variables one-by-one until we are satisfied that all remaining variables are important to the model. Forward selection starts with no variables included in the model, then it adds in variables according to their importance until no other important variables are found. Notice that, for both methods, we have always chosen to retain the model with the largest adjusted R^2 value, even if the difference is less than half a percent (e.g., 0.2597 versus 0.2598). One could argue that the difference between these two models is negligible, as they both explain nearly the same amount of variability in the `interest_rate`. These negligible differences are an important aspect to model selection. It is highly advised that *before* you begin the model selection process, you decide what a "meaningful" difference in adjusted R^2 is for the context of your data. Maybe this difference is 1% or maybe it is 5%. This "threshold" is what you will then use to decide if one model is "better" than another model. Using meaningful thresholds in model selection requires more critical thinking about what the adjusted R^2 values mean.

Additionally, backward elimination and forward selection sometimes arrive at different final models. This is because the decision for whether to include a given variable or not depends on the other variables that are already in the model. With forward selection, you start with a model that includes no variables and add variables one at a time. In backward elimination, you start with a model that includes all of the variables and remove variables one at a time. How much a given variable changes the percentage of the variability in the outcome that is explained by the model depends on what other variables are in the model. This is especially important if the predictor variables are correlated with each other.

There is no "one size fits all" model selection strategy, which is why there are so many different model selection methods. We hope you walk away from this exploration understanding how stepwise selection is carried out and the considerations that should be made when using stepwise selection with regression models.

8.4.2 Other model selection strategies

Stepwise selection using adjusted R^2 as the decision criteria is one of many commonly used model selection strategies. Stepwise selection can also be carried out with decision criteria other than adjusted R^2, such as p-values, which you'll learn about in Chapters 24 onwards, or AIC (Akaike information criterion) or BIC (Bayesian information criterion), which you might learn about in more advanced courses.

Alternatively, one could choose to include or exclude variables from a model based on expert opinion or due to research focus. In fact, many statisticians discourage the use of stepwise regression alone for model selection and advocate, instead, for a more thoughtful approach that carefully considers the research focus and features of the data.

8.5 Chapter review

8.5.1 Summary

With real data, there is often a need to describe how multiple variables can be modeled together. In this chapter, we have presented one approach using multiple linear regression. Each coefficient represents a one unit increase of that predictor variable on the response variable *given* the rest of the predictor variables in the model. Working with and interpreting multivariable models can be tricky, especially when the predictor variables show multicollinearity. There is often no perfect or "right" final model, but using the adjusted R^2 value is one way to identify important predictor variables for a final regression model. In later chapters we will generalize multiple linear regression models to a larger population of interest from which the dataset was generated.

8.5.2 Terms

We introduced the following terms in the chapter. If you're not sure what some of these terms mean, we recommend you go back in the text and review their definitions. We are purposefully presenting them in alphabetical order, instead of in order of appearance, so they will be a little more challenging to locate. However you should be able to easily spot them as **bolded text**.

adjusted R-squared	full model	reference level
backward elimination	multicollinearity	stepwise selection
degrees of freedom	multiple regression	
forward selection	parsimonious	

8.6 Exercises

Answers to odd numbered exercises can be found in Appendix A.8.

8.1. **High correlation, good or bad?** Two friends, Frances and Annika, are in disagreement about whether high correlation values are *always* good in the context of regression. Frances claims that it's desirable for all variables in the dataset to be highly correlated to each other when building linear models. Annika claims that while it's desirable for each of the predictors to be highly correlated with the outcome, it is not desirable for the predictors to be highly correlated with each other. Who is right: Frances, Annika, both, or neither? Explain your reasoning using appropriate terminology.

8.2. **Dealing with categorical predictors.** Two friends, Elliott and Adrian, want to build a model predicting typing speed (average number of words typed per minute) from whether the person wears glasses or not. Before building the model they want to conduct some exploratory analysis to evaluate the strength of the association between these two variables, but they're in disagreement about how to evaluate how strongly a categorical predictor is associated with a numerical outcome. Elliott claims that it is not possible to calculate a correlation coefficient to summarize the relationship between a categorical predictor and a numerical outcome, however they're not sure what a better alternative is. Adrian claims that you can recode a binary predictor as a 0/1 variable (assign one level to be 0 and the other to be 1), thus converting it to a numerical variable. According to Adrian, you can then calculate the correlation coefficient between the predictor and the outcome. Who is right: Elliott or Adrian? If you pick Elliott, can you suggest a better alternative for evaluating the association between the categorical predictor and the numerical outcome?

8.3. **Training for the 5K.** Nico signs up for a 5K (a 5,000 metre running race) 30 days prior to the race. They decide to run a 5K every day to train for it, and each day they record the following information: `days_since_start` (number of days since starting training), `days_till_race` (number of days left until the race), `mood` (poor, good, awesome), `tiredness` (1-not tired to 10-very tired), and `time` (time it takes to run 5K, recorded as mm:ss). Top few rows of the data they collect is shown below.

days_since_start	days_till_race	mood	tiredness	time
1	29	good	3	25:45
2	28	poor	5	27:13
3	27	awesome	4	24:13
...	...	...	...	...

Using these data Nico wants to build a model predicting `time` from the other variables. Should they include all variables shown above in their model? Why or why not?

8.4. **Multiple regression fact checking.** Determine which of the following statements are true and false. For each statement that is false, explain why it is false.

a. If predictors are collinear, then removing one variable will have no influence on the point estimate of another variable's coefficient.

b. Suppose a numerical variable x has a coefficient of $b_1 = 2.5$ in the multiple regression model. Suppose also that the first observation has $x_1 = 7.2$, the second observation has a value of $x_1 = 8.2$, and these two observations have the same values for all other predictors. Then the predicted value of the second observation will be 2.5 higher than the prediction of the first observation based on the multiple regression model.

c. If a regression model's first variable has a coefficient of $b_1 = 5.7$, then if we are able to influence the data so that an observation will have its x_1 be 1 larger than it would otherwise, the value y_1 for this observation would increase by 5.7.

8.5. **Baby weights and smoking.** US Department of Health and Human Services, Centers for Disease Control and Prevention collect information on births recorded in the country. The data used here are a random sample of 1,000 births from 2014. Here, we study the relationship between smoking and weight of the baby. The variable smoke is coded 1 if the mother is a smoker, and 0 if not. The summary table below shows the results of a linear regression model for predicting the average birth weight of babies, measured in pounds, based on the smoking status of the mother.[10] (ICPSR, 2014)

term	estimate	std.error	statistic	p.value
(Intercept)	7.270	0.0435	167.22	<0.0001
habitsmoker	-0.593	0.1275	-4.65	<0.0001

a. Write the equation of the regression model.

b. Interpret the slope in this context, and calculate the predicted birth weight of babies born to smoker and non-smoker mothers.

8.6. **Baby weights and mature moms.** The following is a model for predicting baby weight from whether the mom is classified as a mature mom (35 years or older at the time of pregnancy). (ICPSR, 2014)

term	estimate	std.error	statistic	p.value
(Intercept)	7.354	0.103	71.02	<0.0001
matureyounger mom	-0.185	0.113	-1.64	0.102

a. Write the equation of the regression model.

b. Interpret the slope in this context, and calculate the predicted birth weight of babies born to mature and younger mothers.

8.7. **Movie returns, prediction.** A model was fit to predict return-on-investment (ROI) on movies based on release year and genre (Adventure, Action, Drama, Horror, and Comedy). The model output is shown below. (FiveThirtyEight, 2015)

term	estimate	std.error	statistic	p.value
(Intercept)	-156.04	169.15	-0.92	0.3565
release_year	0.08	0.08	0.94	0.348
genreAdventure	0.30	0.74	0.40	0.6914
genreComedy	0.57	0.69	0.83	0.4091
genreDrama	0.37	0.62	0.61	0.5438
genreHorror	8.61	0.86	9.97	<0.0001

a. For a given release year, which genre of movies are predicted, on average, to have the highest predicted return on investment?

b. The adjusted R^2 of this model is 10.71%. Adding the production budget of the movie to the model increases the adjusted R^2 to 10.84%. Should production budget be added to the model?

---

[10]The births14 data used in this exercise can be found in the **openintro** R package.

8.8. **Movie returns by genre.** A model was fit to predict return-on-investment (ROI) on movies based on release year and genre (Adventure, Action, Drama, Horror, and Comedy). The plots below show the predicted ROI vs. actual ROI for each of the genres separately. Do these figures support the comment in the FiveThirtyEight.com article that states, "The return-on-investment potential for horror movies is absurd." Note that the x-axis range varies for each plot. (FiveThirtyEight, 2015)

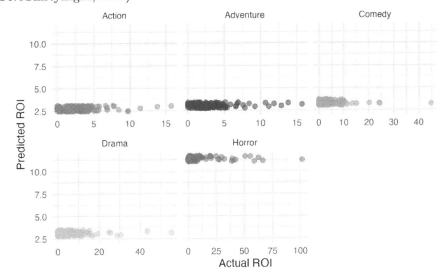

8.9. **Predicting baby weights.** A more realistic approach to modeling baby weights is to consider all possibly related variables at once. Other variables of interest include length of pregnancy in weeks (weeks), mother's age in years (mage), the sex of the baby (sex), smoking status of the mother (habit), and the number of hospital (visits) visits during pregnancy. Below are three observations from this data set.

weight	weeks	mage	sex	visits	habit
6.96	37	34	male	14	nonsmoker
8.86	41	31	female	12	nonsmoker
7.51	37	36	female	10	nonsmoker

The summary table below shows the results of a regression model for predicting the average birth weight of babies based on all of the variables presented above.

term	estimate	std.error	statistic	p.value
(Intercept)	-3.82	0.57	-6.73	<0.0001
weeks	0.26	0.01	18.93	<0.0001
mage	0.02	0.01	2.53	0.0115
sexmale	0.37	0.07	5.30	<0.0001
visits	0.02	0.01	2.09	0.0373
habitsmoker	-0.43	0.13	-3.41	7e-04

a. Write the equation of the regression model that includes all of the variables.

b. Interpret the slopes of weeks and habit in this context.

c. If we fit a model predicting baby weight from only habit (whether the mom smokes), we observe a difference in the slope coefficient for habit in this small model and the slope coefficient for habit in the larger model. Why might there be a difference?

d. Calculate the residual for the first observation in the data set.

8.10. **Palmer penguins, predicting body mass.** Researchers studying a community of Antarctic penguins collected body measurement (bill length, bill depth, and flipper length measured in millimeters and body mass, measured in grams), species (*Adelie*, *Chinstrap*, or *Gentoo*), and sex (female or male) data on 344 penguins living on three islands (Torgersen, Biscoe, and Dream) in the Palmer Archipelago, Antarctica.[11] The summary table below shows the results of a linear regression model for predicting body mass (which is more difficult to measure) from the other variables in the dataset. (Gorman et al., 2014a)

term	estimate	std.error	statistic	p.value
(Intercept)	-1461.0	571.3	-2.6	0.011
bill_length_mm	18.2	7.1	2.6	0.0109
bill_depth_mm	67.2	19.7	3.4	7e-04
flipper_length_mm	16.0	2.9	5.5	<0.0001
sexmale	389.9	47.8	8.1	<0.0001
speciesChinstrap	-251.5	81.1	-3.1	0.0021
speciesGentoo	1014.6	129.6	7.8	<0.0001

a. Write the equation of the regression model.

b. Interpret each one of the slopes in this context.

c. Calculate the residual for a male Adelie penguin that weighs 3750 grams with the following body measurements: bill_length_mm = 39.1, bill_depth_mm = 18.7, flipper_length_mm = 181. Does the model overpredict or underpredict this penguin's weight?

d. The R^2 of this model is 87.5%. Interpret this value in context of the data and the model.

8.11. **Baby weights, backwards elimination.** Let's consider a model that predicts weight of newborns using several predictors: whether the mother is considered mature, number of weeks of gestation, number of hospital visits during pregnancy, weight gained by the mother during pregnancy, sex of the baby, and whether the mother smoke cigarettes during pregnancy (habit). (ICPSR, 2014)

The adjusted R^2 of the full model is 0.326. We remove each variable one by one, refit the model, and record the adjusted R^2. Which, if any, variable should be removed from the model?

- Drop mature: 0.321
- Drop weeks: 0.061
- Drop visits: 0.326
- Drop gained: 0.327
- Drop sex: 0.301

Which, if any, variable should be removed from the model first?

---

[11] The penguins data used in this exercise can be found in the **palmerpenguins** R package.

8.6. EXERCISES

8.12. **Palmer penguins, backwards elimination.** The following full model is built to predict the weights of three species (*Adelie*, *Chinstrap*, or *Gentoo*) of penguins living in the Palmer Archipelago, Antarctica. (Gorman et al., 2014a)

term	estimate	std.error	statistic	p.value
(Intercept)	-1461.0	571.3	-2.6	0.011
bill_length_mm	18.2	7.1	2.6	0.0109
bill_depth_mm	67.2	19.7	3.4	7e-04
flipper_length_mm	16.0	2.9	5.5	<0.0001
sexmale	389.9	47.8	8.1	<0.0001
speciesChinstrap	-251.5	81.1	-3.1	0.0021
speciesGentoo	1014.6	129.6	7.8	<0.0001

The adjusted R^2 of the full model is 0.9. In order to evaluate whether any of the predictors can be dropped from the model without losing predictive performance of the model, the researchers dropped one variable at a time, refit the model, and recorded the adjusted R^2 of the smaller model. These values are given below.

- Drop `bill_length_mm`: 0.87
- Drop `bill_depth_mm`: 0.869
- Drop `flipper_length_mm`: 0.861
- Drop `sex`: 0.845
- Drop `species`: 0.821

Which, if any, variable should be removed from the model first?

8.13. **Baby weights, forward selection.** Using information on the mother and the sex of the baby (which can be determined prior to birth), we want to build a model that predicts the birth weight of babies. In order to do so, we will evaluate six candidate predictors: whether the mother is considered `mature`, number of `weeks` of gestation, number of hospital `visits` during pregnancy, weight `gained` by the mother during pregnancy, `sex` of the baby, and whether the mother smoke cigarettes during pregnancy (`habit`). And we will make a decision about including them in the model using forward selection and adjusted R^2. Below are the six models we evaluate and their adjusted R^2 values. (ICPSR, 2014)

- Predict `weight` from `mature`: 0.002
- Predict `weight` from `weeks`: 0.3
- Predict `weight` from `visits`: 0.034
- Predict `weight` from `gained`: 0.021
- Predict `weight` from `sex`: 0.018
- Predict `weight` from `habit`: 0.021

Which variable should be added to the model first?

8.14. **Palmer penguins, forward selection.** Using body measurement and other relevant data on three species (*Adelie*, *Chinstrap*, or *Gentoo*) of penguins living in the Palmer Archipelago, Antarctica, we want to predict their body mass. In order to do so, we will evaluate five candidate predictors and make a decision about including them in the model using forward selection and adjusted R^2. Below are the five models we evaluate and their adjusted R^2 values:

- Predict body mass from `bill_length_mm`: 0.352
- Predict body mass from `bill_depth_mm`: 0.22
- Predict body mass from `flipper_length_mm`: 0.758
- Predict body mass from `sex`: 0.178
- Predict body mass from `species`: 0.668

Which variable should be added to the model first?

Chapter 9

Logistic regression

 In this chapter we introduce **logistic regression** as a tool for building models when there is a categorical response variable with two levels, e.g., yes and no. Logistic regression is a type of **generalized linear model (GLM)** for response variables where regular multiple regression does not work very well. GLMs can be thought of as a two-stage modeling approach. We first model the response variable using a probability distribution, such as the binomial or Poisson distribution. Second, we model the parameter of the distribution using a collection of predictors and a special form of multiple regression. Ultimately, the application of a GLM will feel very similar to multiple regression, even if some of the details are different.

9.1 Discrimination in hiring

We will consider experiment data from a study that sought to understand the effect of race and sex on job application callback rates (Bertrand and Mullainathan, 2003). To evaluate which factors were important, job postings were identified in Boston and Chicago for the study, and researchers created many fake resumes to send off to these jobs to see which would elicit a callback.[1] The researchers enumerated important characteristics, such as years of experience and education details, and they used these characteristics to randomly generate fake resumes. Finally, they randomly assigned a name to each resume, where the name would imply the applicant's sex and race.

 The `resume` data can be found in the **openintro** R package.

The first names that were used and randomly assigned in this experiment were selected so that they would predominantly be recognized as belonging to Black or White individuals; other races were not considered in this study. While no name would definitively be inferred as pertaining to a Black individual or to a White individual, the researchers conducted a survey to check for racial association

---

[1] We did omit discussion of some structure in the data for the analysis presented: the experiment design included blocking, where typically four resumes were sent to each job: one for each inferred race/sex combination (as inferred based on the first name). We did not worry about this blocking aspect, since accounting for the blocking would *reduce* the standard error without notably changing the point estimates for the `race` and `sex` variables versus the analysis performed in the section. That is, the most interesting conclusions in the study are unaffected even when completing a more sophisticated analysis.

9.1. DISCRIMINATION IN HIRING

of the names; names that did not pass this survey check were excluded from usage in the experiment. You can find the full set of names that did pass the survey test and were ultimately used in the study in Table 9.1. For example, Lakisha was a name that their survey indicated would be interpreted as a Black woman, while Greg was a name that would generally be interpreted to be associated with a White male.

Table 9.1: List of all 36 unique names along with the commonly inferred race and sex associated with these names.

first_name	race	sex	first_name	race	sex	first_name	race	sex
Aisha	Black	female	Hakim	Black	male	Laurie	White	female
Allison	White	female	Jamal	Black	male	Leroy	Black	male
Anne	White	female	Jay	White	male	Matthew	White	male
Brad	White	male	Jermaine	Black	male	Meredith	White	female
Brendan	White	male	Jill	White	female	Neil	White	male
Brett	White	male	Kareem	Black	male	Rasheed	Black	male
Carrie	White	female	Keisha	Black	female	Sarah	White	female
Darnell	Black	male	Kenya	Black	female	Tamika	Black	female
Ebony	Black	female	Kristen	White	female	Tanisha	Black	female
Emily	White	female	Lakisha	Black	female	Todd	White	male
Geoffrey	White	male	Latonya	Black	female	Tremayne	Black	male
Greg	White	male	Latoya	Black	female	Tyrone	Black	male

The response variable of interest is whether or not there was a callback from the employer for the applicant, and there were 8 attributes that were randomly assigned that we'll consider, with special interest in the race and sex variables. Race and sex are in protected classes the United States, meaning they are not legally permitted factors for hiring or employment decisions. The full set of attributes considered is provided in Table 26.1.

Table 9.2: Descriptions of nine variables from the `resume` dataset. Many of the variables are indicator variables, meaning they take the value 1 if the specified characteristic is present and 0 otherwise.

variable	description
received_callback	Specifies whether the employer called the applicant following submission of the application for the job.
job_city	City where the job was located: Boston or Chicago.
college_degree	An indicator for whether the resume listed a college degree.
years_experience	Number of years of experience listed on the resume.
honors	Indicator for the resume listing some sort of honors, e.g. employee of the month.
military	Indicator for if the resume listed any military experience.
has_email_address	Indicator for if the resume listed an email address for the applicant.
race	Race of the applicant, implied by their first name listed on the resume.
sex	Sex of the applicant (limited to only and in this study), implied by the first name listed on the resume.

All of the attributes listed on each resume were randomly assigned. This means that no attributes that might be favorable or detrimental to employment would favor one demographic over another on these resumes. Importantly, due to the experimental nature of this study, we can infer causation between these variables and the callback rate, if substantial differences are found. Our analysis will allow us to compare the practical importance of each of the variables relative to each other.

9.2 Modelling the probability of an event

Logistic regression is a generalized linear model where the outcome is a two-level categorical variable. The outcome, Y_i, takes the value 1 (in our application, this represents a callback for the resume) with probability p_i and the value 0 with probability $1 - p_i$. Because each observation has a slightly different context, e.g., different education level or a different number of years of experience, the probability p_i will differ for each observation. Ultimately, it is this **probability** that we model in relation to the predictor variables: we will examine which resume characteristics correspond to higher or lower callback rates.

Notation for a logistic regression model.

The outcome variable for a GLM is denoted by Y_i, where the index i is used to represent observation i. In the resume application, Y_i will be used to represent whether resume i received a callback ($Y_i = 1$) or not ($Y_i = 0$).

The predictor variables are represented as follows: $x_{1,i}$ is the value of variable 1 for observation i, $x_{2,i}$ is the value of variable 2 for observation i, and so on.

$$transformation(p_i) = b_0 + b_1 x_{1,i} + b_2 x_{2,i} + \cdots + b_k x_{k,i}$$

We want to choose a **transformation** in the equation that makes practical and mathematical sense. For example, we want a transformation that makes the range of possibilities on the left hand side of the equation equal to the range of possibilities for the right hand side; if there was no transformation for this equation, the left hand side could only take values between 0 and 1, but the right hand side could take values outside of this range.

A common transformation for p_i is the **logit transformation**, which may be written as

$$logit(p_i) = \log_e \left(\frac{p_i}{1 - p_i} \right)$$

The logit transformation is shown in Figure 9.1. Below, we rewrite the equation relating Y_i to its predictors using the logit transformation of p_i:

$$\log_e \left(\frac{p_i}{1 - p_i} \right) = b_0 + b_1 x_{1,i} + b_2 x_{2,i} + \cdots + b_k x_{k,i}$$

In our resume example, there are 8 predictor variables, so $k = 8$. While the precise choice of a logit function isn't intuitive, it is based on theory that underpins generalized linear models, which is beyond the scope of this book. Fortunately, once we fit a model using software, it will start to feel like we're back in the multiple regression context, even if the interpretation of the coefficients is more complex.

To convert from values on the logistic regression scale to the probability scale, we need to back transform and then solve for p_i:

9.2. MODELLING THE PROBABILITY OF AN EVENT

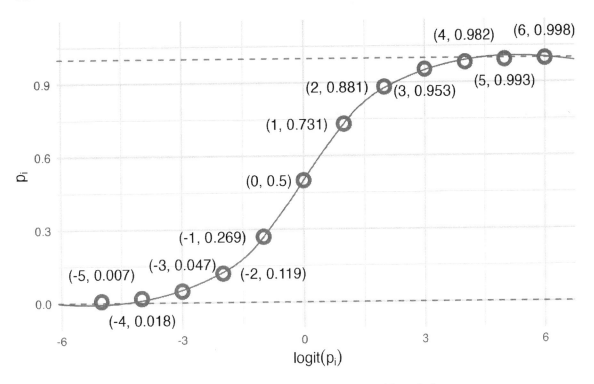

Figure 9.1: Values of p_i against values of $logit(p_i)$.

$$\log_e\left(\frac{p_i}{1-p_i}\right) = b_0 + b_1 x_{1,i} + \cdots + b_k x_{k,i}$$

$$\frac{p_i}{1-p_i} = e^{b_0 + b_1 x_{1,i} + \cdots + b_k x_{k,i}}$$

$$p_i = (1-p_i)\, e^{b_0 + b_1 x_{1,i} + \cdots + b_k x_{k,i}}$$

$$p_i = e^{b_0 + b_1 x_{1,i} + \cdots + b_k x_{k,i}} - p_i \times e^{b_0 + b_1 x_{1,i} + \cdots + b_k x_{k,i}}$$

$$p_i + p_i\, e^{b_0 + b_1 x_{1,i} + \cdots + b_k x_{k,i}} = e^{b_0 + b_1 x_{1,i} + \cdots + b_k x_{k,i}}$$

$$p_i(1 + e^{b_0 + b_1 x_{1,i} + \cdots + b_k x_{k,i}}) = e^{b_0 + b_1 x_{1,i} + \cdots + b_k x_{k,i}}$$

$$p_i = \frac{e^{b_0 + b_1 x_{1,i} + \cdots + b_k x_{k,i}}}{1 + e^{b_0 + b_1 x_{1,i} + \cdots + b_k x_{k,i}}}$$

As with most applied data problems, we substitute the point estimates for the parameters (the b_i) so that we can make use of this formula.

EXAMPLE

We start by fitting a model with a single predictor: honors. This variable indicates whether the applicant had any type of honors listed on their resume, such as employee of the month. A logistic regression model was fit using statistical software and the following model was found:

$$\log_e\left(\frac{\widehat{p}_i}{1-\widehat{p}_i}\right) = -2.4998 + 0.8668 \times \texttt{honors}$$

a. If a resume is randomly selected from the study and it does not have any honors listed, what is the probability it resulted in a callback?
b. What would the probability be if the resume did list some honors?

---

a. If a randomly chosen resume from those sent out is considered, and it does not list honors, then honors takes the value of 0 and the right side of the model equation equals -2.4998. Solving for p_i: $\frac{e^{-2.4998}}{1+e^{-2.4998}} = 0.076$. Just as we labeled a fitted value of y_i with a "hat" in single-variable and multiple regression, we do the same for this probability: $\hat{p}_i = 0.076$.
b. If the resume had listed some honors, then the right side of the model equation is $-2.4998 + 0.8668 \times 1 = -1.6330$, which corresponds to a probability $\hat{p}_i = 0.163$. Notice that we could examine -2.4998 and -1.6330 in Figure 9.1 to estimate the probability before formally calculating the value.

While knowing whether a resume listed honors provides some signal when predicting whether or not the employer would call, we would like to account for many different variables at once to understand how each of the different resume characteristics affected the chance of a callback.

9.3 Logistic model with many variables

We used statistical software to fit the logistic regression model with all 8 predictors described in Table 26.1. Like multiple regression, the result may be presented in a summary table, which is shown in Table 9.3.

Table 9.3: Summary table for the full logistic regression model for the resume callback example.

term	estimate	std.error	statistic	p.value
(Intercept)	-2.66	0.18	-14.64	<0.0001
job_cityChicago	-0.44	0.11	-3.85	1e-04
college_degree1	-0.07	0.12	-0.55	0.5821
years_experience	0.02	0.01	1.96	0.0503
honors1	0.77	0.19	4.14	<0.0001
military1	-0.34	0.22	-1.59	0.1127
has_email_address1	0.22	0.11	1.93	0.0541
raceWhite	0.44	0.11	4.10	<0.0001
sexman	-0.18	0.14	-1.32	0.1863

Just like multiple regression, we could trim some variables from the model. Here we'll use a statistic called **Akaike information criterion (AIC)**, which is analogous to how we used adjusted R^2 in multiple regression. AIC is a popular model selection method used in many disciplines, and is praised for its emphasis on model uncertainty and parsimony. AIC selects a "best" model by ranking models

9.3. LOGISTIC MODEL WITH MANY VARIABLES

from best to worst according to their AIC values. In the calculation of a model's AIC, a penalty is given for including additional variables. This penalty for added model complexity attempts to strike a balance between underfitting (too few variables in the model) and overfitting (too many variables in the model). When using AIC for model selection, models with a lower AIC value are considered to be "better." Remember that when using adjusted R^2 we select models with higher values instead. It is important to note that AIC provides information about the quality of a model relative to other models, but does not provide information about the overall quality of a model.

We will look for models with a lower AIC using a backward elimination strategy. After using this criteria, the variable `college_degree` is eliminated, giving the smaller model summarized in Table 9.4, which is what we'll rely on for the remainder of this section.

Table 9.4: Summary table for the logistic regression model for the resume callback example, where variable selection has been performed using AIC.

term	estimate	std.error	statistic	p.value
(Intercept)	-2.72	0.16	-17.51	<0.0001
job_cityChicago	-0.44	0.11	-3.83	1e-04
years_experience	0.02	0.01	2.02	0.043
honors1	0.76	0.19	4.12	<0.0001
military1	-0.34	0.22	-1.60	0.1105
has_email_address1	0.22	0.11	1.97	0.0494
raceWhite	0.44	0.11	4.10	<0.0001
sexman	-0.20	0.14	-1.45	0.1473

EXAMPLE

The `race` variable had taken only two levels: `Black` and `White`. Based on the model results, what does the coefficient of this variable say about callback decisions?

The coefficient shown corresponds to the level of `White`, and it is positive. This positive coefficient reflects a positive gain in callback rate for resumes where the candidate's first name implied they were White. The model results suggest that prospective employers favor resumes where the first name is typically interpreted to be White.

The coefficient of race$_{\text{White}}$ in the full model in Table 9.3, is nearly identical to the model shown in Table 9.4. The predictors in this experiment were thoughtfully laid out so that the coefficient estimates would typically not be much influenced by which other predictors were in the model, which aligned with the motivation of the study to tease out which effects were important to getting a callback. In most observational data, it's common for point estimates to change a little, and sometimes a lot, depending on which other variables are included in the model.

EXAMPLE

Use the model summarized in Table 9.4 to estimate the probability of receiving a callback for a job in Chicago where the candidate lists 14 years experience, no honors, no military experience, includes an email address, and has a first name that implies they are a White male.

We can start by writing out the equation using the coefficients from the model:

$$log_e\left(\frac{p}{1-p}\right)$$
$$= -2.7162 - 0.4364 \times \texttt{job\_city}_{\texttt{Chicago}}$$
$$+ 0.0206 \times \texttt{years\_experience}$$
$$+ 0.7634 \times \texttt{honors} - 0.3443 \times \texttt{military} + 0.2221 \times \texttt{email}$$
$$+ 0.4429 \times \texttt{race}_{\texttt{White}} - 0.1959 \times \texttt{sex}_{\texttt{man}}$$

Now we can add in the corresponding values of each variable for this individual:

$$log_e\left(\frac{\widehat{p}}{1-\widehat{p}}\right)$$
$$= -2.7162 - 0.4364 \times 1 + 0.0206 \times 14$$
$$+ 0.7634 \times 0 - 0.3443 \times 0 + 0.2221 \times 1$$
$$+ 0.4429 \times 1 - 0.1959 \times 1 = -2.3955$$

We can now back-solve for $\widehat{p}$: the chance such an individual will receive a callback is about $\frac{e^{-2.3955}}{1+e^{-2.3955}} = 8.35\%$.

EXAMPLE

Compute the probability of a callback for an individual with a name commonly inferred to be from a Black male but who otherwise has the same characteristics as the one described in the previous example.

We can complete the same steps for an individual with the same characteristics who is Black, where the only difference in the calculation is that the indicator variable $\texttt{race}_{\texttt{White}}$ will take a value of 0. Doing so yields a probability of 0.0553. Let's compare the results with those of the previous example..

In practical terms, an individual perceived as White based on their first name would need to apply to $\frac{1}{0.0835} \approx 12$ jobs on average to receive a callback, while an individual perceived as Black based on their first name would need to apply to $\frac{1}{0.0553} \approx 18$ jobs on average to receive a callback. That is, applicants who are perceived as Black need to apply to 50% more employers to receive a callback than someone who is perceived as White based on their first name for jobs like those in the study.

What we've quantified in this section is alarming and disturbing. However, one aspect that makes this racism so difficult to address is that the experiment, as well-designed as it is, cannot send us much signal about which employers are discriminating. It is only possible to say that discrimination is happening, even if we cannot say which particular callbacks — or non-callbacks — represent discrimination. Finding strong evidence of racism for individual cases is a persistent challenge in enforcing anti-discrimination laws.

9.4 Groups of different sizes

Any form of discrimination is concerning, and this is why we decided it was so important to discuss this topic using data. The resume study also only examined discrimination in a single aspect: whether a prospective employer would call a candidate who submitted their resume. There was a 50% higher barrier for resumes simply when the candidate had a first name that was perceived to be from a Black individual. It's unlikely that discrimination would stop there.

EXAMPLE

Let's consider a sex-imbalanced company that consists of 20% women and 80% men, and we'll suppose that the company is very large, consisting of perhaps 20,000 employees. (A more thoughtful example would include more inclusive gender identities.) Suppose when someone goes up for promotion at this company, 5 of their colleagues are randomly chosen to provide feedback on their work.

Now let's imagine that 10% of the people in the company are prejudiced against the other sex. That is, 10% of men are prejudiced against women, and similarly, 10% of women are prejudiced against men.

Who is discriminated against more at the company, men or women?

Let's suppose we took 100 men who have gone up for promotion in the past few years. For these men, $5 \times 100 = 500$ random colleagues will be tapped for their feedback, of which about 20% will be women (100 women). Of these 100 women, 10 are expected to be biased against the man they are reviewing. Then, of the 500 colleagues reviewing them, men will experience discrimination by about 2% of their colleagues when they go up for promotion.

Let's do a similar calculation for 100 women who have gone up for promotion in the last few years. They will also have 500 random colleagues providing feedback, of which about 400 (80%) will be men. Of these 400 men, about 40 (10%) hold a bias against women. Of the 500 colleagues providing feedback on the promotion packet for these women, 8% of the colleagues hold a bias against the women.

This example highlights something profound: even in a hypothetical setting where each demographic has the same degree of prejudice against the other demographic, the smaller group experiences the negative effects more frequently. Additionally, if we would complete a handful of examples like the one above with different numbers, we'd learn that the greater the imbalance in the population groups, the more the smaller group is disproportionately impacted.[2]

Of course, there are other considerable real-world omissions from the hypothetical example. For example, studies have found instances where people from an oppressed group also discriminate against others within their own oppressed group. As another example, there are also instances where a majority group can be oppressed, with apartheid in South Africa being one such historic example. Ultimately, discrimination is complex, and there are many factors at play beyond the mathematics property we observed in the previous example.

We close this chapter on this serious topic, and we hope it inspires you to think about the power of reasoning with data. Whether it is with a formal statistical model or by using critical thinking skills to structure a problem, we hope the ideas you have learned will help you do more and do better in life.

---

[2] If a proportion p of a company are women and the rest of the company consists of men, then under the hypothetical situation the ratio of rates of discrimination against women versus men would be given by $(1-p)/p$; this ratio is always greater than 1 when $p < 0.5$.

9.5 Chapter review

9.5.1 Summary

Logistic and linear regression models have many similarities. The strongest of which is the linear combination of the explanatory variables which is used to form predictions related to the response variable. However, with logistic regression, the response variable is binary and therefore a prediction is given on the probability of a successful event. Logistic model fit and variable selection can be carried out in similar ways as multiple linear regression.

9.5.2 Terms

We introduced the following terms in the chapter. If you're not sure what some of these terms mean, we recommend you go back in the text and review their definitions. We are purposefully presenting them in alphabetical order, instead of in order of appearance, so they will be a little more challenging to locate. However you should be able to easily spot them as **bolded text**.

Akaike information criterion	logistic regression	probability of an event
generalized linear model	logit transformation	transformation

9.6 Exercises

Answers to odd numbered exercises can be found in Appendix A.9.

9.1. **True / False.** Determine which of the following statements are true and false. For each statement that is false, explain why it is false.

   a. In logistic regression we fit a line to model the relationship between the predictor(s) and the binary outcome.

   b. In logistic regression, we expect the residuals to be even scattered on either side of zero, just like with linear regression.

   c. In logistic regression, the outcome variable is binary but the predictor variable(s) can be either binary or continuous.

9.2. **Logistic regression fact checking.** Determine which of the following statements are true and false. For each statement that is false, explain why it is false.

   a. Suppose we consider the first two observations based on a logistic regression model, where the first variable in observation 1 takes a value of $x_1 = 6$ and observation 2 has $x_1 = 4$. Suppose we realized we made an error for these two observations, and the first observation was actually $x_1 = 7$ (instead of 6) and the second observation actually had $x_1 = 5$ (instead of 4). Then the predicted probability from the logistic regression model would increase the same amount for each observation after we correct these variables.

   b. When using a logistic regression model, it is impossible for the model to predict a probability that is negative or a probability that is greater than 1.

   c. Because logistic regression predicts probabilities of outcomes, observations used to build a logistic regression model need not be independent.

   d. When fitting logistic regression, we typically complete model selection using adjusted R^2.

9.3. **Possum classification, model selection.** The common brushtail possum of the Australia region is a bit cuter than its distant cousin, the American opossum (see Figure 7.4. We consider 104 brushtail possums from two regions in Australia, where the possums may be considered a random sample from the population. The first region is Victoria, which is in the eastern half of Australia and traverses the southern coast. The second region consists of New South Wales and Queensland, which make up eastern and northeastern Australia.[3]

We use logistic regression to differentiate between possums in these two regions. The outcome variable, called pop, takes value 1 when a possum is from Victoria and 0 when it is from New South Wales or Queensland. We consider five predictors: sex (an indicator for a possum being male), head_l (head length), skull_w (skull width), total_l (total length), and tail_l (tail length). Each variable is summarized in a histogram. The full logistic regression model and a reduced model after variable selection are summarized in the tables below.

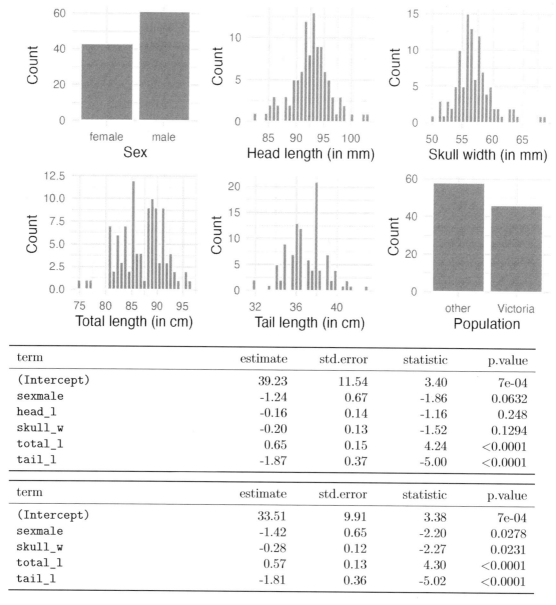

term	estimate	std.error	statistic	p.value
(Intercept)	39.23	11.54	3.40	7e-04
sexmale	-1.24	0.67	-1.86	0.0632
head_l	-0.16	0.14	-1.16	0.248
skull_w	-0.20	0.13	-1.52	0.1294
total_l	0.65	0.15	4.24	<0.0001
tail_l	-1.87	0.37	-5.00	<0.0001

term	estimate	std.error	statistic	p.value
(Intercept)	33.51	9.91	3.38	7e-04
sexmale	-1.42	0.65	-2.20	0.0278
skull_w	-0.28	0.12	-2.27	0.0231
total_l	0.57	0.13	4.30	<0.0001
tail_l	-1.81	0.36	-5.02	<0.0001

a. Examine each of the predictors. Are there any outliers that are likely to have a very large influence on the logistic regression model?

b. The summary table for the full model indicates that at least one variable should be eliminated when using the p-value approach for variable selection: head_l. The second component of the table summarizes the reduced model following variable selection. Explain why the remaining estimates change between the two models.

---

[3]The possum data used in this exercise can be found in the **openintro** R package.

9.4. **Challenger disaster and model building.** On January 28, 1986, a routine launch was anticipated for the Challenger space shuttle. Seventy-three seconds into the flight, disaster happened: the shuttle broke apart, killing all seven crew members on board. An investigation into the cause of the disaster focused on a critical seal called an O-ring, and it is believed that damage to these O-rings during a shuttle launch may be related to the ambient temperature during the launch. The table below summarizes observational data on O-rings for 23 shuttle missions, where the mission order is based on the temperature at the time of the launch. temperature gives the temperature in Fahrenheit, damaged represents the number of damaged O-rings, and undamaged represents the number of O-rings that were not damaged.[4]

mission	1	2	3	4	5	6	7	8	9	10	11	12
temperature	53	57	58	63	66	67	67	67	68	69	70	70
damaged	5	1	1	1	0	0	0	0	0	0	1	0
undamaged	1	5	5	5	6	6	6	6	6	6	5	6

mission	13	14	15	16	17	18	19	20	21	22	23
temperature	70	70	72	73	75	75	76	76	78	79	81
damaged	1	0	0	0	0	1	0	0	0	0	0
undamaged	5	6	6	6	6	5	6	6	6	6	6

term	estimate	std.error	statistic	p.value
(Intercept)	11.66	3.30	3.54	4e-04
temperature	-0.22	0.05	-4.07	<0.0001

a. Each column of the table above represents a different shuttle mission. Examine these data and describe what you observe with respect to the relationship between temperatures and damaged O-rings.

b. Failures have been coded as 1 for a damaged O-ring and 0 for an undamaged O-ring, and a logistic regression model was fit to these data. The regression output for this model is given above. Describe the key components of the output in words.

c. Write out the logistic model using the point estimates of the model parameters.

d. Based on the model, do you think concerns regarding O-rings are justified? Explain.

9.5. **Possum classification, prediction.** A logistic regression model was proposed for classifying common brushtail possums into their two regions. The outcome variable took value 1 if the possum was from Victoria and 0 otherwise.

term	estimate	std.error	statistic	p.value
(Intercept)	33.51	9.91	3.38	7e-04
sexmale	-1.42	0.65	-2.20	0.0278
skull_w	-0.28	0.12	-2.27	0.0231
total_l	0.57	0.13	4.30	<0.0001
tail_l	-1.81	0.36	-5.02	<0.0001

a. Write out the form of the model. Also identify which of the variables are positively associated with the outcome of living in Victoria, when controlling for other variables.

b. Suppose we see a brushtail possum at a zoo in the US, and a sign says the possum had been captured in the wild in Australia, but it doesn't say which part of Australia. However, the sign does indicate that the possum is male, its skull is about 63 mm wide, its tail is 37 cm long, and its total length is 83 cm. What is the reduced model's computed probability that this possum is from Victoria? How confident are you in the model's accuracy of this probability calculation?

---

[4]The orings data used in this exercise can be found in the **openintro** R package.

9.6. **Challenger disaster and prediction.** On January 28, 1986, a routine launch was anticipated for the Challenger space shuttle. Seventy-three seconds into the flight, disaster happened: the shuttle broke apart, killing all seven crew members on board. An investigation into the cause of the disaster focused on a critical seal called an O-ring, and it is believed that damage to these O-rings during a shuttle launch may be related to the ambient temperature during the launch. The investigation found that the ambient temperature at the time of the shuttle launch was closely related to the damage of O-rings, which are a critical component of the shuttle.

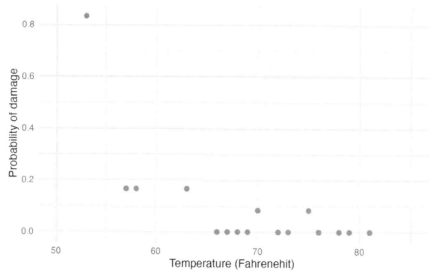

a. The data provided in the previous exercise are shown in the plot. The logistic model fit to these data may be written as

$$\log\left(\frac{\hat{p}}{1-\hat{p}}\right) = 11.6630 - 0.2162 \times \texttt{temperature}$$

where $\hat{p}$ is the model-estimated probability that an O-ring will become damaged. Use the model to calculate the probability that an O-ring will become damaged at each of the following ambient temperatures: 51, 53, and 55 degrees Fahrenheit. The model-estimated probabilities for several additional ambient temperatures are provided below, where subscripts indicate the temperature:

$$\hat{p}_{57} = 0.341 \quad \hat{p}_{59} = 0.251 \quad \hat{p}_{61} = 0.179 \quad \hat{p}_{63} = 0.124$$
$$\hat{p}_{65} = 0.084 \quad \hat{p}_{67} = 0.056 \quad \hat{p}_{69} = 0.037 \quad \hat{p}_{71} = 0.024$$

b. Add the model-estimated probabilities from part (a) on the plot, then connect these dots using a smooth curve to represent the model-estimated probabilities.

c. Describe any concerns you may have regarding applying logistic regression in this application, and note any assumptions that are required to accept the model's validity.

9.7. **Spam filtering, model selection.** Spam filters are built on principles similar to those used in logistic regression. Using characteristics of individual emails, we fit a probability that each message is spam or not spam. We have several email variables for this problem, and we won't describe what each variable means here for the sake of brevity, but each is either a numerical or indicator variable.[5]

term	estimate	std.error	statistic	p.value
(Intercept)	-0.69	0.09	-7.42	<0.0001
to_multiple1	-2.82	0.31	-9.05	<0.0001
cc	0.03	0.02	1.41	0.1585
attach	0.28	0.08	3.44	6e-04
dollar	-0.08	0.02	-3.45	6e-04
winneryes	1.72	0.34	5.09	<0.0001
inherit	0.32	0.15	2.10	0.0355
password	-0.79	0.30	-2.64	0.0083
format1	-1.50	0.13	-12.01	<0.0001
re_subj1	-1.92	0.38	-5.10	<0.0001
exclaim_subj	0.26	0.23	1.14	0.2531
sent_email1	-16.67	293.19	-0.06	0.9547

The AIC of the full model is 1863.5. We remove each variable one by one, refit the model, and record the updated AIC. Which, if any, variable should be removed from the model?

a. For variable selection, we fit the full model, which includes all variables, and then we also fit each model where we've dropped exactly one of the variables. In each of these reduced models, the AIC value for the model is reported below. Based on these results, which variable, if any, should we drop as part of model selection? Explain.

- None Dropped: 1863.5
- Drop `to_multiple`: 2023.5
- Drop `cc`: 1863.2
- Drop `attach`: 1871.9
- Drop `dollar`: 1879.7
- Drop `winner`: 1885
- Drop `inherit`: 1865.5
- Drop `password`: 1879.3
- Drop `format`: 2008.9
- Drop `re_subj`: 1904.6
- Drop `exclaim_subj`: 1862.8
- Drop `sent_email`: 1958.2

b. Consider the subsequent model selection stage (where the variable from part (a) has been removed, and we are considering removal of a second variable). Here again we've computed the AIC for each leave-one-variable-out model. Based on the results, which variable, if any, should we drop as part of model selection? Explain.

- None Dropped: 1862.8
- Drop `to_multiple`: 2021.5
- Drop `cc`: 1862.4
- Drop `attach`: 1871.2
- Drop `dollar`: 1877.8
- Drop `winner`: 1885.2
- Drop `inherit`: 1864.8
- Drop `password`: 1878.4
- Drop `format`: 2007
- Drop `re_subj`: 1904.3
- Drop `sent_email`: 1957.3

See next page for part c.

---

[5]The `email` data used in this exercise can be found in the **openintro** R package.

c. Consider one more step in the process. Now we've removed two the subsequent model selection stage (where the variable from part (a) has been removed, and we are considering removal of a second variable). Here again we've computed the AIC for each leave-one-variable-out model. Based on the results, which variable, if any, should we drop as part of model selection? Explain.

- None Dropped: 1862.4
- Drop `to_multiple`: 2019.6
- Drop `attach`: 1871.2
- Drop `dollar`: 1877.7
- Drop `winner`: 1885
- Drop `inherit`: 1864.5
- Drop `password`: 1878.2
- Drop `format`: 2007.4
- Drop `re_subj`: 1902.9
- Drop `sent_email`: 1957.6

9.8. **Spam filtering, prediction.** Recall running a logistic regression to aid in spam classification for individual emails. In this exercise, we've taken a small set of the variables and fit a logistic model with the following output:

term	estimate	std.error	statistic	p.value
`(Intercept)`	-0.81	0.09	-9.34	<0.0001
`to_multiple1`	-2.64	0.30	-8.68	<0.0001
`winneryes`	1.63	0.32	5.11	<0.0001
`format1`	-1.59	0.12	-13.28	<0.0001
`re_subj1`	-3.05	0.36	-8.40	<0.0001

a. Write down the model using the coefficients from the model fit.

b. Suppose we have an observation where `to_multiple` = 0, `winner` = 1, `format` = 0, and `re_subj` = 0. What is the predicted probability that this message is spam?

c. Put yourself in the shoes of a data scientist working on a spam filter. For a given message, how high must the probability a message is spam be before you think it would be reasonable to put it in a *spambox* (which the user is unlikely to check)? What tradeoffs might you consider? Any ideas about how you might make your spam-filtering system even better from the perspective of someone using your email service?

Chapter 10

Applications: Model

10.1 Case study: Houses for sale

Take a walk around your neighborhood and you'll probably see a few houses for sale. If you find a house for sale, you can probably go online and look up its price. You'll quickly note that the prices seem a bit arbitrary – the homeowners get to decide what the amount they want to list their house for, and many criteria factor into this decision, e.g., what do comparable houses ("comps" in real estate speak) sell for, how quickly they need to sell the house, etc.

In this case study we'll formalize the process of figuring out how much to list a house for by using data on current home sales In November of 2020, information on 98 houses in the Duke Forest neighborhood of Durham, NC were scraped from Zillow. The homes were all recently sold at the time of data collection, and the goal of the project was to build a model for predicting the sale price based on a particular home's characteristics. The first four homes are shown in Table 10.1, and descriptions for each variable are shown in Table 10.2.

 The `duke_forest` data can be found in the **openintro** R package.

Table 10.1: Top four rows of the data describing homes for sale in the Duke Forest neighborhood of Durham, NC.

price	bed	bath	area	year_built	lot
1,520,000	3	4	6,040	1,972	0.97
1,030,000	5	4	4,475	1,969	1.38
420,000	2	3	1,745	1,959	0.51
680,000	4	3	2,091	1,961	0.84

Table 10.2: Variables and their descriptions for the `duke_forest` dataset.

Variable	Description
price	Sale price, in USD
bed	Number of bedrooms
bath	Number of bathrooms
area	Area of home, in square feet
year_built	Year the home was built
cooling	Cooling system: central or other
lot	Area of the entire property, in acres

10.1.1 Correlating with `price`

As mentioned, the goal of the data collection was to build a model for the sale price of homes. While using multiple predictor variables is likely preferable to using only one variable, we start by learning about the variables themselves and their relationship to price. Figure 10.1 shows scatterplots describing price as a function of each of the predictor variables. All of the variables seem to be positively associated with price (higher values of the variable are matched with higher price values).

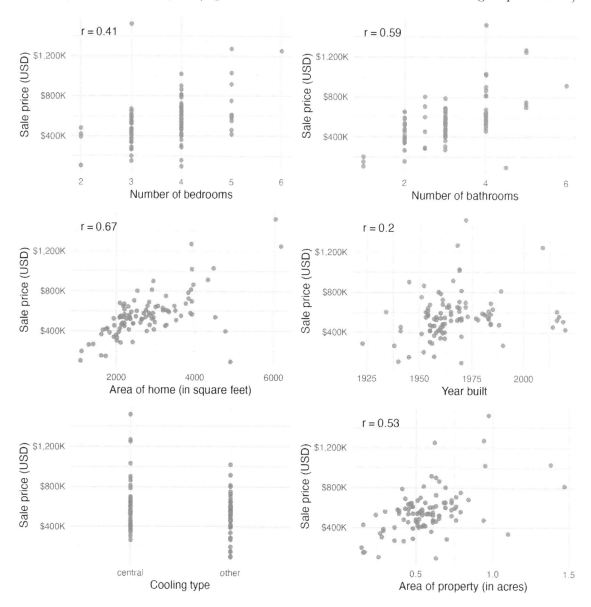

Figure 10.1: Scatter plots describing six different predictor variables' relationship with the price of a home.

   GUIDED PRACTICE

In Figure 10.1 there does not appear to be a correlation value calculated for the predictor variable, `cooling`. Why not? Can the variable still be used in the linear model? Explain.[1]

---

[1] The correlation coefficient can only be calculated to describe the relationship between two numerical variables. The predictor variable `cooling` is categorical, not numerical. It *can*, however, be used in the linear model as a binary indicator variable coded, for example, with a 1 for central and 0 for other.

10.1. CASE STUDY: HOUSES FOR SALE

EXAMPLE

In Figure 10.1 which variable seems to be most informative for predicting house price? Provide two reasons for your answer.

The area of the home is the variable which is most highly correlated with `price`. Additionally, the scatterplot for `price` vs. `area` seems to show a strong linear relationship between the two variables. Note that the correlation coefficient and the scatterplot linearity will often give the same conclusion. However, recall that the correlation coefficient is very sensitive to outliers, so it is always wise to look at the scatterplot even when the variables are highly correlated.

10.1.2 Modeling `price` with `area`

A linear model was fit to predict `price` from `area`. The resulting model information is given in Table 10.3.

Table 10.3: Summary of least squares fit for price on area.

term	estimate	std.error	statistic	p.value
(Intercept)	116,652	53,302	2.19	0.0311
area	159	18	8.78	<0.0001

Adjusted R-sq = 0.4394
df = 96

GUIDED PRACTICE

Interpret the value of $b_1 = 159$ in the context of the problem.[2]

GUIDED PRACTICE

Using the output in Table 10.3, write out the model for predicting `price` from `area`.[3]

The residuals from the linear model can be to assess whether or not a linear model is appropriate. Figure 10.2 plots the residuals $e_i = y_i - \hat{y}_i$ on the y-axis and the fitted (or predicted) values $\hat{y}_i$ on the x-axis.

GUIDED PRACTICE

What aspect(s) of the residual plot indicate that a linear model is appropriate? What aspect(s) of the residual plot seem concerning when fitting a linear model?[4]

---

[2] For each additional square foot of house, we would expect such houses to cost, on average, $159 more.
[3] $\widehat{\text{price}} = 116,652 - 159 \times \text{area}$
[4] The residual plot shows that the relationship between `area` and the average `price` of a home is indeed linear. However, the residuals are quite large for expensive homes. The large residuals indicate potential outliers or increasing variability, either of which could warrant more involved modeling techniques than are presented in this text.

Figure 10.2: Residuals versus predicted values for the model predicting sale price from area of home.

10.1.3 Modeling `price` with multiple variables

It seems as though the predictions of home price might be more accurate if more than one predictor variable was used in the linear model. Table 10.4 displays the output from a linear model of `price` regressed on `area`, `bed`, `bath`, `year_built`, `cooling`, and `lot`.

Table 10.4: Summary of least squares fit for price on multiple predictor variables.

term	estimate	std.error	statistic	p.value
(Intercept)	-2,826,650	1,790,616	-1.58	0.1179
area	102	23	4.42	<0.0001
bed	-13,692	25,928	-0.53	0.5987
bath	41,076	24,662	1.67	0.0993
year_built	1,459	914	1.60	0.1139
coolingother	-84,065	30,338	-2.77	0.0068
lot	356,141	75,940	4.69	<0.0001

Adjusted R-sq = 0.5896
df = 90

 EXAMPLE

Using Table 10.4, write out the linear model of price on the six predictor variables.

$$\widehat{\text{price}} = -2,826,650 \\ + 102 \times \text{area} - 13,692 \times \text{bed} \\ + 41,076 \times \text{bath} + 1,459 \times \text{year\_built} \\ + 84,065 \times \text{cooling}_{\text{central}} + 356,141 \times \text{lot}$$

10.1. CASE STUDY: HOUSES FOR SALE

GUIDED PRACTICE

The value of the estimated coefficient on cooling$_{\text{central}}$ is $b_5 = 84,065$. Interpret the value of b_5 in the context of the problem.[5]

A friend suggests that maybe you don't need all six variables to have a good model for price. You consider taking a variable out, but you aren't sure which one to remove.

EXAMPLE

Results corresponding to the full model for the housing data are shown in Table 10.4. How should we proceed under the backward elimination strategy?

Our baseline adjusted R^2 from the full model is 0.59, and we need to determine whether dropping a predictor will improve the adjusted R^2. To check, we fit models that each drop a different predictor, and we record the adjusted R^2:

- Excluding area: 0.506
- Excluding bed: 0.593
- Excluding bath: 0.582
- Excluding year_built: 0.583
- Excluding cooling: 0.559
- Excluding lot: 0.489

The model without bed has the highest adjusted R^2 of 0.593, higher than the adjusted R^2 for the full model. Because eliminating bed leads to a model with a higher adjusted R^2, we drop bed from the model.

It might seem counter-intuitive to exclude information on number of bedrooms from the model. After all, we would expect homes with more bedrooms to cost more, and we can see a clear relationship between number of bedrooms and sale price in Figure 10.1. However note that area is still in the model, and it's quite likely that the area of the home and the number of bedrooms are highly associated. Therefore, the model already has information on "how much space is available in the house" with the inclusion of area alone.

Since we eliminated a predictor from the model in the first step, we see whether we should eliminate any additional predictors. Our baseline adjusted R^2 is now 0.593. We fit another set of new models, which consider eliminating each of the remaining predictors in addition to bed:

- Excluding bed and area: 0.51
- Excluding bed and bath: 0.586
- Excluding bed and year_built: 0.586
- Excluding bed and cooling: 0.563
- Excluding bed and lot: 0.493

None of these models lead to an improvement in adjusted R^2, so we do not eliminate any of the remaining predictors.

---

[5]The coefficient indicates that if all the other variables are kept constant, homes with central air conditioning cost $84,065 more, on average.

That is, after backward elimination, we are left with the model that keeps all predictors except bed, which we can summarize using the coefficients from Table 10.5.

Table 10.5: Summary of least squares fit for price on multiple predictor variables, excluding number of bedrooms.

term	estimate	std.error	statistic	p.value
(Intercept)	-2,868,784	1,781,736	-1.61	0.1108
area	99	22	4.44	<0.0001
bath	36,228	22,799	1.59	0.1155
year_built	1,466	910	1.61	0.1107
coolingother	-83,856	30,215	-2.78	0.0067
lot	357,119	75,617	4.72	<0.0001

Adjusted R-sq = 0.5929
df = 91

Then, the linear model for predicting sale price based on this model is as follows:

$$\widehat{\text{price}} = -2,868,784 + 99 \times \text{area}$$
$$+ 36,228 \times \text{bath} + 1,466 \times \text{year\_built}$$
$$+ 83,856 \times \text{cooling}_{\text{central}} + 357,119 \times \text{lot}$$

EXAMPLE

The residual plot for the model with all of the predictor variables except bed is given in Figure 10.3. How do the residuals in Figure 10.3 compare to the residuals in Figure 10.2?

The residuals, for the most part, are randomly scattered around 0. However there is one extreme outlier with a residual of -$750,000, a house whose actual sale price is a lot lower than its predicted price. Also, we observe again that the residuals are quite large for expensive homes.

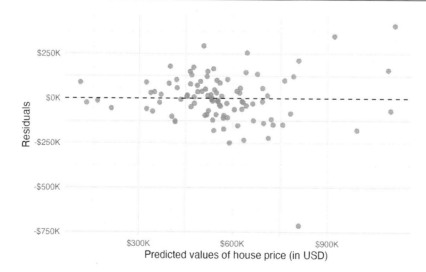

Figure 10.3: Residuals versus predicted values for the model predicting sale price from all predictors except for number of bedrooms.

10.1. CASE STUDY: HOUSES FOR SALE

GUIDED PRACTICE

Consider a house with 1,803 square feet, 2.5 bathrooms, 0.145 acres, built in 1941, that has central air conditioning. What is the predicted price of the home?[6]

GUIDED PRACTICE

If you later learned that the house (with a predicted price of $298,152) had recently sold for $804,133, would you think the model was terrible? What if you learned that the house was in California?[7]

---

[6]$\widehat{\text{price}} = -2,868,784 + 99 \times 1803$
$+ 36,228 \times 2.5 + 1,466 \times 1941$
$+ 83,856 \times 1 + 357,119 \times 0.145$
$= \$298,152$.

[7]A residual of $505,981 is reasonably big. Note that the large residuals (except a few homes) in Figure 10.3 are closer to $250,000 (about half as big). After we learn that the house is in California, we realize that the model shouldn't be applied to the new home at all! The original data are from Durham, NC, and models based on the Durham, NC data should be used only to explore patterns in prices for homes in Durham, NC.

10.2 Interactive R tutorials

Navigate the concepts you've learned in this chapter in R using the following self-paced tutorials. All you need is your browser to get started!

Tutorial 3: Regression modeling
https://openintrostat.github.io/ims-tutorials/03-model

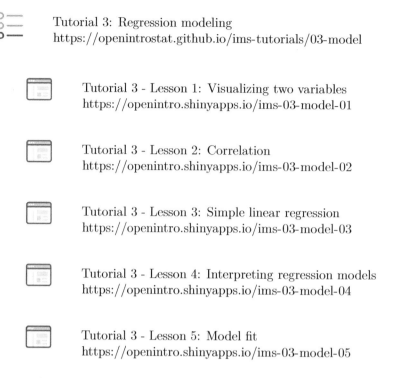

Tutorial 3 - Lesson 1: Visualizing two variables
https://openintro.shinyapps.io/ims-03-model-01

Tutorial 3 - Lesson 2: Correlation
https://openintro.shinyapps.io/ims-03-model-02

Tutorial 3 - Lesson 3: Simple linear regression
https://openintro.shinyapps.io/ims-03-model-03

Tutorial 3 - Lesson 4: Interpreting regression models
https://openintro.shinyapps.io/ims-03-model-04

Tutorial 3 - Lesson 5: Model fit
https://openintro.shinyapps.io/ims-03-model-05

You can also access the full list of tutorials supporting this book at
https://openintrostat.github.io/ims-tutorials.

10.3 R labs

Further apply the concepts you've learned in this part in R with computational labs that walk you through a data analysis case study.

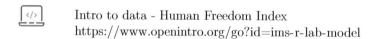

Intro to data - Human Freedom Index
https://www.openintro.org/go?id=ims-r-lab-model

You can also access the full list of labs supporting this book at
https://www.openintro.org/go?id=ims-r-labs.

PART IV

FOUNDATIONS OF INFERENCE

Chapter 11

Hypothesis testing with randomization

 Statistical inference is primarily concerned with understanding and quantifying the uncertainty of parameter estimates. While the equations and details change depending on the setting, the foundations for inference are the same throughout all of statistics.

We start with two case studies designed to motivate the process of making decisions about research claims. We formalize the process through the introduction of the **hypothesis testing framework**, which allows us to formally evaluate claims about the population.

Throughout the book so far, you have worked with data in a variety of contexts. You have learned how to summarize and visualize the data as well as how to model multiple variables at the same time. Sometimes the dataset at hand represents the entire research question. But more often than not, the data have been collected to answer a research question about a larger group of which the data are a (hopefully) representative subset.

You may agree that there is almost always variability in data – one dataset will not be identical to a second dataset even if they are both collected from the same population using the same methods. However, quantifying the variability in the data is neither obvious nor easy to do, i.e., answering the question "*how* different is one dataset from another?" is not trivial.

First, a note on notation. We generally use p to denote a population proportion and $\hat{p}$ to a sample proportion. Similarly, we generally use μ to denote a population mean and $\bar{x}$ to denote a sample mean.

11.1 Sex discrimination case study

EXAMPLE

Suppose your professor splits the students in your class into two groups: students who sit on the left side of the classroom and students who sit on the right side of the classroom. If $\hat{p}_L$ represents the proportion of students who prefer to read books on screen who sit on the left side of the classroom and $\hat{p}_R$ represents the proportion of students who prefer to read books on screen who sit on the right side of the classroom, would you be surprised if $\hat{p}_L$ did not *exactly* equal $\hat{p}_R$?

While the proportions $\hat{p}_L$ and $\hat{p}_R$ would probably be close to each other, it would be unusual for them to be exactly the same. We would probably observe a small difference due to *chance*.

GUIDED PRACTICE

If we don't think the side of the room a person sits on in class is related to whether they prefer to read books on screen, what assumption are we making about the relationship between these two variables?[1]

Studying randomness of this form is a key focus of statistics. Throughout this chapter, and those that follow, we provide three different approaches for quantifying the variability inherent in data: randomization, bootstrapping, and mathematical models. Using the methods provided in this chapter, we will be able to draw conclusions beyond the dataset at hand to research questions about larger populations that the samples come from.

The first type of variability we will explore comes from experiments where the explanatory variable (or treatment) is randomly assigned to the observational units. As you learned in Chapter 1, a randomized experiment can be used to assess whether or not one variable (the explanatory variable) causes changes in a second variable (the response variable). Every dataset has some variability in it, so to decide whether the variability in the data is due to (1) the causal mechanism (the randomized explanatory variable in the experiment) or instead (2) natural variability inherent to the data, we set up a sham randomized experiment as a comparison. That is, we assume that each observational unit would have gotten the exact same response value regardless of the treatment level. By reassigning the treatments many many times, we can compare the actual experiment to the sham experiment. If the actual experiment has more extreme results than any of the sham experiments, we are led to believe that it is the explanatory variable which is causing the result and not just variability inherent to the data. Using a few different case studies, let's look more carefully at this idea of a **randomization test**.

11.1 Sex discrimination case study

We consider a study investigating sex discrimination in the 1970s, which is set in the context of personnel decisions within a bank. The research question we hope to answer is, "Are individuals who identify as female discriminated against in promotion decisions made by their managers who identify as male?" (Rosen and Jerdee, 1974)

The `sex_discrimination` data can be found in the **openintro** R package.

This study considered sex roles, and only allowed for options of "male" and "female". We should

---
[1]We would be assuming that these two variables are **independent**.

note that the identities being considered are not gender identities and also that the study allowed only for a binary classification of sex.

11.1.1 Observed data

The participants in this study were 48 bank supervisors who identified as male, attending a management institute at the University of North Carolina in 1972. They were asked to assume the role of the personnel director of a bank and were given a personnel file to judge whether the person should be promoted to a branch manager position. The files given to the participants were identical, except that half of them indicated the candidate identified as male and the other half indicated the candidate identified as female. These files were randomly assigned to the subjects.

GUIDED PRACTICE

Is this an observational study or an experiment? How does the type of study impact what can be inferred from the results?[2]

For each supervisor both the sex associated with the assigned file and the promotion decision were recorded. Using the results of the study summarized in Table 11.1, we would like to evaluate if individuals who identify as female are unfairly discriminated against in promotion decisions. In this study, a smaller proportion of female identifying applications were promoted than males (0.583 versus 0.875), but it is unclear whether the difference provides *convincing evidence* that individuals who identify as female are unfairly discriminated against.

Table 11.1: Summary results for the sex discrimination study.

sex	decision		Total
	promoted	not promoted	
male	21	3	24
female	14	10	24
Total	35	13	48

The data are visualized in Figure 11.1 as a set of cards. Note that each card denotes a personnel file (an observation from our dataset) and the colors indicate the decision: red for promoted and white for not promoted. Additionally, the observations are broken up into groups of male and female identifying groups.

---

[2]The study is an experiment, as subjects were randomly assigned a "male" file or a "female" file (remember, all the files were actually identical in content). Since this is an experiment, the results can be used to evaluate a causal relationship between the sex of a candidate and the promotion decision.

11.1. SEX DISCRIMINATION CASE STUDY

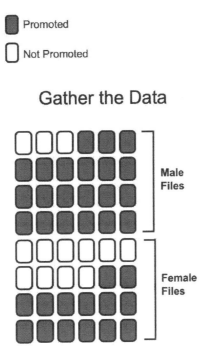

Figure 11.1: The sex discrimination study can be thought of as 48 red and white cards.

 EXAMPLE

Statisticians are sometimes called upon to evaluate the strength of evidence. When looking at the rates of promotion in this study, why might we be tempted to immediately conclude that individuals identifying as female are being discriminated against?

The large difference in promotion rates (58.3% for female personnel versus 87.5% for male personnel) suggest there might be discrimination against women in promotion decisions. However, we cannot yet be sure if the observed difference represents discrimination or is just due to random chance when there is no discrimination occurring. Since we wouldn't expect the sample proportions to be *exactly* equal, even if the truth was that the promotion decisions were independent of sex, we can't rule out random chance as a possible explanation when simply comparing the sample proportions.

The previous example is a reminder that the observed outcomes in the sample may not perfectly reflect the true relationships between variables in the underlying population. Table 11.1 shows there were 7 fewer promotions for female identifying personnel than for the male personnel, a difference in promotion rates of 29.2% $\left(\frac{21}{24} - \frac{14}{24} = 0.292\right)$. This observed difference is what we call a **point estimate** of the true difference. The point estimate of the difference in promotion rate is large, but the sample size for the study is small, making it unclear if this observed difference represents discrimination or whether it is simply due to chance when there is no discrimination occurring. Chance can be thought of as the claim due to natural variability; discrimination can be thought of as the claim the researchers set out to demonstrate. We label these two competing claims, H_0 and H_A:

- H_0 : **Null hypothesis**. The variables `sex` and `decision` are independent. They have no relationship, and the observed difference between the proportion of males and females who were promoted, 29.2%, was due to the natural variability inherent in the population.
- H_A : **Alternative hypothesis**. The variables `sex` and `decision` are *not* independent. The difference in promotion rates of 29.2% was not due to natural variability, and equally qualified female personnel are less likely to be promoted than male personnel.

Hypothesis testing.

These hypotheses are part of what is called a **hypothesis test**. A hypothesis test is a statistical technique used to evaluate competing claims using data. Often times, the null hypothesis takes a stance of *no difference* or *no effect*. This hypothesis assumes that any differences seen are due to the variability inherent in the population and could have occurred by random chance.

If the null hypothesis and the data notably disagree, then we will reject the null hypothesis in favor of the alternative hypothesis.

There are many nuances to hypothesis testing, so don't worry if you aren't a master of hypothesis testing at the end of this section. We'll discuss these ideas and details many times in this chapter as well as in the chapters that follow.

What would it mean if the null hypothesis, which says the variables sex and decision are unrelated, was true? It would mean each banker would decide whether to promote the candidate without regard to the sex indicated on the personnel file. That is, the difference in the promotion percentages would be due to the natural variability in how the files were randomly allocated to different bankers, and this randomization just happened to give rise to a relatively large difference of 29.2%.

Consider the alternative hypothesis: bankers were influenced by which sex was listed on the personnel file. If this was true, and especially if this influence was substantial, we would expect to see some difference in the promotion rates of male and female candidates. If this sex bias was against female candidates, we would expect a smaller fraction of promotion recommendations for female personnel relative to the male personnel.

We will choose between the two competing claims by assessing if the data conflict so much with H_0 that the null hypothesis cannot be deemed reasonable. If data and the null claim seem to be at odds with one another, and the data seem to support H_A, then we will reject the notion of independence and conclude that the data provide evidence of discrimination.

11.1.2 Variability of the statistic

Table 11.1 shows that 35 bank supervisors recommended promotion and 13 did not. Now, suppose the bankers' decisions were independent of the sex of the candidate. Then, if we conducted the experiment again with a different random assignment of sex to the files, differences in promotion rates would be based only on random fluctuation in promotion decisions. We can actually perform this **randomization**, which simulates what would have happened if the bankers' decisions had been independent of sex but we had distributed the file sexes differently.[3]

In the **simulation**, we thoroughly shuffle the 48 personnel files, 35 labelled promoted and 13 labelled not promoted, together and we deal files into two new stacks. Note that by keeping 35 promoted and 13 not promoted, we are assuming that 35 of the bank managers would have promoted the individual whose content is contained in the file **independent** of the sex indicated on their file. We will deal 24 files into the first stack, which will represent the 24 "female" files. The second stack will also have 24 files, and it will represent the 24 "male" files. Figure 11.2 highlights both the shuffle and the reallocation to the sham sex groups.

Then, as we did with the original data, we tabulate the results and determine the fraction of personnel files designated as "male" and "female" who were promoted.

Since the randomization of files in this simulation is independent of the promotion decisions, any difference in promotion rates is due to chance. Table 11.2 show the results of one such simulation.

---

[3]The test procedure we employ in this section is sometimes referred to as a **permutation test**. The difference between the two is how the explanatory variable was assigned. Permutation tests are used for observational studies, where the explanatory variable was not randomly assigned..

11.1. SEX DISCRIMINATION CASE STUDY

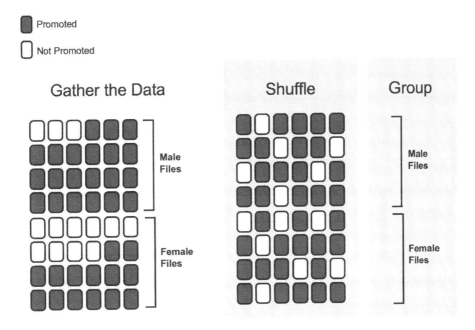

Figure 11.2: The sex discrimination data is shuffled and reallocated to new groups of male and female files.

Table 11.2: Simulation results, where the difference in promotion rates between male and female is purely due to random chance.

sex	decision		Total
	promoted	not promoted	
male	18	6	24
female	17	7	24
Total	35	13	48

 GUIDED PRACTICE

What is the difference in promotion rates between the two simulated groups in Table 11.2 ? How does this compare to the observed difference 29.2% from the actual study?[4]

Figure 11.3 shows that the difference in promotion rates is much larger in the original data than it is in the simulated groups (0.292 > 0.042). The quantity of interest throughout this case study has been the difference in promotion rates. We call the summary value the **statistic** of interest (or often the **test statistic**). When we encounter different data structures, the statistic is likely to change (e.g., we might calculate an average instead of a proportion), but we will always want to understand how the statistic varies from sample to sample.

11.1.3 Observed statistic vs. null statistics

We computed one possible difference under the null hypothesis in Guided Practice, which represents one difference due to chance when the null hypothesis is assumed to be true. While in this first simulation, we physically dealt out files, it is much more efficient to perform this simulation using

---
[4]$18/24 - 17/24 = 0.042$ or about 4.2% in favor of the male personnel. This difference due to chance is much smaller than the difference observed in the actual groups.

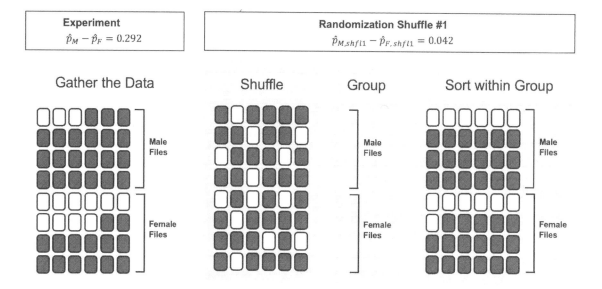

Figure 11.3: We summarize the randomized data to produce one estimate of the difference in proportions given no sex discrimination. Note that the sort step is only used to make it easier to visually calculate the simulated sample proportions.

a computer. Repeating the simulation on a computer, we get another difference due to chance under the same assumption: -0.042. And another: 0.208. And so on until we repeat the simulation enough times that we have a good idea of the shape of the *distribution of differences* under the null hypothesis. Figure 11.4 shows a plot of the differences found from 100 simulations, where each dot represents a simulated difference between the proportions of male and female files recommended for promotion.

Note that the distribution of these simulated differences in proportions is centered around 0. Under the null hypothesis our simulations made no distinction between male and female personnel files. Thus, a center of 0 makes sense: we should expect differences from chance alone to fall around zero with some random fluctuation for each simulation.

EXAMPLE

How often would you observe a difference of at least 29.2% (0.292) according to Figure 11.4? Often, sometimes, rarely, or never?

It appears that a difference of at least 29.2% under the null hypothesis would only happen about 2% of the time according to Figure 11.4. Such a low probability indicates that observing such a large difference from chance alone is rare.

The difference of 29.2% is a rare event if there really is no impact from listing sex in the candidates' files, which provides us with two possible interpretations of the study results:

- If H_0, the **Null hypothesis** is true: Sex has no effect on promotion decision, and we observed a difference that is so large that it would only happen rarely.

- If H_A, the **Alternative hypothesis** is true: Sex has an effect on promotion decision, and what we observed was actually due to equally qualified female candidates being discriminated against in promotion decisions, which explains the large difference of 29.2%.

When we conduct formal studies, we reject a null position (the idea that the data are a result of

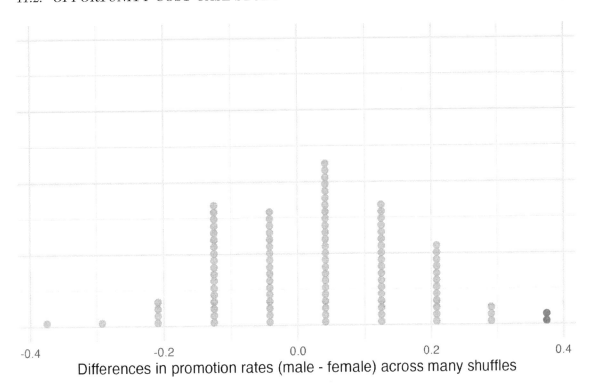

Figure 11.4: A stacked dot plot of differences from 100 simulations produced under the null hypothesis, H_0, where the simulated sex and decision are independent. Two of the 100 simulations had a difference of at least 29.2%, the difference observed in the study, and are shown as solid blue dots.

chance only) if the data strongly conflict with that null position.[5] In our analysis, we determined that there was only a $\approx 2\%$ probability of obtaining a sample where $\geq 29.2\%$ more male candidates than female candidates get promoted under the null hypothesis, so we conclude that the data provide strong evidence of sex discrimination against female candidates by the male supervisors. In this case, we reject the null hypothesis in favor of the alternative.

Statistical inference is the practice of making decisions and conclusions from data in the context of uncertainty. Errors do occur, just like rare events, and the dataset at hand might lead us to the wrong conclusion. While a given dataset may not always lead us to a correct conclusion, statistical inference gives us tools to control and evaluate how often these errors occur. Before getting into the nuances of hypothesis testing, let's work through another case study.

11.2 Opportunity cost case study

How rational and consistent is the behavior of the typical American college student? In this section, we'll explore whether college student consumers always consider the following: money not spent now can be spent later.

In particular, we are interested in whether reminding students about this well-known fact about money causes them to be a little thriftier. A skeptic might think that such a reminder would have no

---

[5]This reasoning does not generally extend to anecdotal observations. Each of us observes incredibly rare events every day, events we could not possibly hope to predict. However, in the non-rigorous setting of anecdotal evidence, almost anything may appear to be a rare event, so the idea of looking for rare events in day-to-day activities is treacherous. For example, we might look at the lottery: there was only a 1 in 176 million chance that the Mega Millions numbers for the largest jackpot in history (October 23, 2018) would be (5, 28, 62, 65, 70) with a Mega ball of (5), but nonetheless those numbers came up! However, no matter what numbers had turned up, they would have had the same incredibly rare odds. That is, *any set of numbers we could have observed would ultimately be incredibly rare.* This type of situation is typical of our daily lives: each possible event in itself seems incredibly rare, but if we consider every alternative, those outcomes are also incredibly rare. We should be cautious not to misinterpret such anecdotal evidence.

impact. We can summarize the two different perspectives using the null and alternative hypothesis framework.

- H_0 : **Null hypothesis.** Reminding students that they can save money for later purchases will not have any impact on students' spending decisions.
- H_A : **Alternative hypothesis.** Reminding students that they can save money for later purchases will reduce the chance they will continue with a purchase.

In this section, we'll explore an experiment conducted by researchers that investigates this very question for students at a university in the southwestern United States. (Frederick et al., 2009)

11.2.1 Observed data

One-hundred and fifty students were recruited for the study, and each was given the following statement:

> *Imagine that you have been saving some extra money on the side to make some purchases, and on your most recent visit to the video store you come across a special sale on a new video. This video is one with your favorite actor or actress, and your favorite type of movie (such as a comedy, drama, thriller, etc.). This particular video that you are considering is one you have been thinking about buying for a long time. It is available for a special sale price of $14.99. What would you do in this situation? Please circle one of the options below.*[6]

Half of the 150 students were randomized into a control group and were given the following two options:

(A) Buy this entertaining video.

(B) Not buy this entertaining video.

The remaining 75 students were placed in the treatment group, and they saw a slightly modified option (B):

(A) Buy this entertaining video.

(B) Not buy this entertaining video. Keep the $14.99 for other purchases.

Would the extra statement reminding students of an obvious fact impact the purchasing decision? Table 11.3 summarizes the study results.

 The `opportunity_cost` data can be found in the **openintro** R package.

Table 11.3: Summary results of the opportunity cost study.

group	buy video	not buy video	Total
control	56	19	75
treatment	41	34	75
Total	97	53	150

(decision column spans buy video and not buy video)

It might be a little easier to review the results using a visualization. Figure 11.5 shows that a higher proportion of students in the treatment group chose not to buy the video compared to those in the control group.

---

[6]This context might feel strange if physical video stores predate you. If you're curious about what those were like, look up "Blockbuster".

11.2. OPPORTUNITY COST CASE STUDY

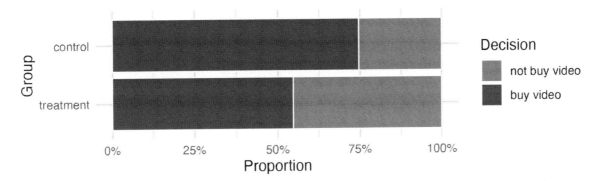

Figure 11.5: Stacked bar plot of results of the opportunity cost study.

Another useful way to review the results from Table 11.3 is using row proportions, specifically considering the proportion of participants in each group who said they would buy or not buy the video. These summaries are given in Table 11.4.

Table 11.4: The opportunity cost data are summarized using row proportions. Row proportions are particularly useful here since we can view the proportion of *buy* and *not buy* decisions in each group.

	decision		
group	buy video	not buy video	Total
control	0.747	0.253	1
treatment	0.547	0.453	1

We will define a **success** in this study as a student who chooses not to buy the video.[7] Then, the value of interest is the change in video purchase rates that results by reminding students that not spending money now means they can spend the money later.

We can construct a point estimate for this difference as (T for treatment and C for control):

$$\hat{p}_T - \hat{p}_C = \frac{34}{75} - \frac{19}{75} = 0.453 - 0.253 = 0.200$$

The proportion of students who chose not to buy the video was 20 percentage points higher in the treatment group than the control group. Is this 20% difference between the two groups so prominent that it is unlikely to have occurred from chance alone, if there is no difference between the spending habits of the two groups?

11.2.2 Variability of the statistic

The primary goal in this data analysis is to understand what sort of differences we might see if the null hypothesis were true, i.e., the treatment had no effect on students. Because this is an experiment, we'll use the same procedure we applied in Section 11.1: randomization.

Let's think about the data in the context of the hypotheses. If the null hypothesis (H_0) was true and the treatment had no impact on student decisions, then the observed difference between the two groups of 20% could be attributed entirely to random chance. If, on the other hand, the alternative hypothesis (H_A) is true, then the difference indicates that reminding students about saving for later purchases actually impacts their buying decisions.

---

[7]Success is often defined in a study as the outcome of interest, and a "success" may or may not actually be a positive outcome. For example, researchers working on a study on COVID prevalence might define a "success" in the statistical sense as a patient who has COVID-19. A more complete discussion of the term **success** will be given in Chapter 16.

11.2.3 Observed statistic vs. null statistics

Just like with the sex discrimination study, we can perform a statistical analysis. Using the same randomization technique from the last section, let's see what happens when we simulate the experiment under the scenario where there is no effect from the treatment.

While we would in reality do this simulation on a computer, it might be useful to think about how we would go about carrying out the simulation without a computer. We start with 150 index cards and label each card to indicate the distribution of our response variable: decision. That is, 53 cards will be labeled "not buy video" to represent the 53 students who opted not to buy, and 97 will be labeled "buy video" for the other 97 students. Then we shuffle these cards thoroughly and divide them into two stacks of size 75, representing the simulated treatment and control groups. Because we have shuffled the cards from both groups together, assuming no difference in their purchasing behavior, any observed difference between the proportions of "not buy video" cards (what we earlier defined as *success*) can be attributed entirely to chance.

EXAMPLE

If we are randomly assigning the cards into the simulated treatment and control groups, how many "not buy video" cards would we expect to end up in each simulated group? What would be the expected difference between the proportions of "not buy video" cards in each group?

Since the simulated groups are of equal size, we would expect $53/2 = 26.5$, i.e., 26 or 27, "not buy video" cards in each simulated group, yielding a simulated point estimate of the difference in proportions of 0%. However, due to random chance, we might also expect to sometimes observe a number a little above or below 26 and 27.

The results of a single randomization is shown in Table 11.5.

Table 11.5: Summary of student choices against their simulated groups. The group assignment had no connection to the student decisions, so any difference between the two groups is due to chance.

	decision		
group	buy video	not buy video	Total
control	46	29	75
treatment	51	24	75
Total	97	53	150

From this table, we can compute a difference that occurred from the first shuffle of the data (i.e., from chance alone):

$$\hat{p}_{T,shfl1} - \hat{p}_{C,shfl1} = \frac{24}{75} - \frac{29}{75} = 0.32 - 0.387 = -0.067$$

Just one simulation will not be enough to get a sense of what sorts of differences would happen from chance alone.

We'll simulate another set of simulated groups and compute the new difference: 0.04.

And again: 0.12.

And again: -0.013.

We'll do this 1,000 times.

The results are summarized in a dot plot in Figure 11.6, where each point represents the difference from one randomization.

11.2. OPPORTUNITY COST CASE STUDY

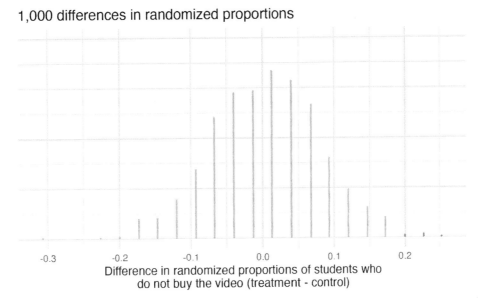

Figure 11.6: A stacked dot plot of 1,000 simulated (null) differences produced under the null hypothesis, H_0. Six of the 1,000 simulations had a difference of at least 20%, which was the difference observed in the study.

Since there are so many points and it is difficult to discern one point from the other, it is more convenient to summarize the results in a histogram such as the one in Figure 11.7, where the height of each histogram bar represents the number of simulations resulting in an outcome of that magnitude.

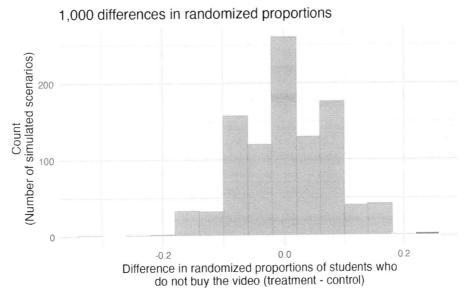

Figure 11.7: A histogram of 1,000 chance differences produced under the null hypothesis. Histograms like this one are a convenient representation of data or results when there are a large number of simulations.

Under the null hypothesis (no treatment effect), we'd observe a difference of at least +20% about 0.6% of the time. That is really rare! Instead, we will conclude the data provide strong evidence there is a treatment effect: reminding students before a purchase that they could instead spend the money later on something else lowers the chance that they will continue with the purchase. Notice that we are able to make a causal statement for this study since the study is an experiment, although we don't know why the reminder induces a lower purchase rate.

11.3 Hypothesis testing

In the last two sections, we utilized a **hypothesis test**, which is a formal technique for evaluating two competing possibilities. In each scenario, we described a **null hypothesis**, which represented either a skeptical perspective or a perspective of no difference. We also laid out an **alternative hypothesis**, which represented a new perspective such as the possibility of a relationship between two variables or a treatment effect in an experiment. The alternative hypothesis is usually the reason the scientists set out to do the research in the first place.

Null and alternative hypotheses.

The **null hypothesis** (H_0) often represents either a skeptical perspective or a claim of "no difference" to be tested.

The **alternative hypothesis** (H_A) represents an alternative claim under consideration and is often represented by a range of possible values for the value of interest.

If a person makes a somewhat unbelievable claim, we are initially skeptical. However, if there is sufficient evidence that supports the claim, we set aside our skepticism. The hallmarks of hypothesis testing are also found in the US court system.

11.3.1 The US court system

In the US course system, jurors evaluate the evidence to see whether it convincingly shows a defendant is guilty. Defendants are considered to be innocent until proven otherwise.

EXAMPLE

The US court considers two possible claims about a defendant: they are either innocent or guilty.

If we set these claims up in a hypothesis framework, which would be the null hypothesis and which the alternative?

The jury considers whether the evidence is so convincing (strong) that there is no reasonable doubt regarding the person's guilt. That is, the skeptical perspective (null hypothesis) is that the person is innocent until evidence is presented that convinces the jury that the person is guilty (alternative hypothesis).

Jurors examine the evidence to see whether it convincingly shows a defendant is guilty. Notice that if a jury finds a defendant *not guilty*, this does not necessarily mean the jury is confident in the person's innocence. They are simply not convinced of the alternative, that the person is guilty. This is also the case with hypothesis testing: *even if we fail to reject the null hypothesis, we do not accept the null hypothesis as truth.*

Failing to find evidence in favor of the alternative hypothesis is not equivalent to finding evidence that the null hypothesis is true. We will see this idea in greater detail in Section 14.

11.3.2 p-value and statistical significance

In Section 11.1 we encountered a study from the 1970's that explored whether there was strong evidence that female candidates were less likely to be promoted than male candidates. The research

11.3. HYPOTHESIS TESTING

question – are female candidates discriminated against in promotion decisions? – was framed in the context of hypotheses:

- H_0 : Sex has no effect on promotion decisions.
- H_A : Female candidates are discriminated against in promotion decisions.

The null hypothesis (H_0) was a perspective of no difference in promotion. The data on sex discrimination provided a point estimate of a 29.2% difference in recommended promotion rates between male and female candidates. We determined that such a difference from chance alone, assuming the null hypothesis was true, would be rare: it would only happen about 2 in 100 times. When results like these are inconsistent with H_0, we reject H_0 in favor of H_A. Here, we concluded there was discrimination against female candidates.

The 2-in-100 chance is what we call a **p-value**, which is a probability quantifying the strength of the evidence against the null hypothesis, given the observed data.

p-value.

The **p-value** is the probability of observing data at least as favorable to the alternative hypothesis as our current dataset, if the null hypothesis were true. We typically use a summary statistic of the data, such as a difference in proportions, to help compute the p-value and evaluate the hypotheses. This summary value that is used to compute the p-value is often called the **test statistic**.

EXAMPLE

In the sex discrimination study, the difference in discrimination rates was our test statistic. What was the test statistic in the opportunity cost study covered in Section 11.2?

The test statistic in the opportunity cost study was the difference in the proportion of students who decided against the video purchase in the treatment and control groups. In each of these examples, the **point estimate** of the difference in proportions was used as the test statistic.

When the p-value is small, i.e., less than a previously set threshold, we say the results are **statistically significant**. This means the data provide such strong evidence against H_0 that we reject the null hypothesis in favor of the alternative hypothesis. The threshold, called the **significance level** and often represented by α (the Greek letter *alpha*). The value of α represents how rare an event needs to be in order for the null hypothesis to be rejected. Historically, many fields have set $\alpha = 0.05$, meaning that the results need to occur less than 5% of the time, if the null hypothesis is to be rejected. The value of α can vary depending on the the field or the application.

Although in everyday language "significant" would indicate that a difference is large or meaningful, that is not necessarily the case here. The term "statistically significant" only indicates that the p-value from a study fell below the chosen significance level. For example, in the sex discrimination study, the p-value was found to be approximately 0.002. Using a significance level of $\alpha = 0.05$, we would say that the data provided statistically significant evidence against the null hypothesis. However, this conclusion gives us no information regarding the size of the difference in promotion rates!

Statistical significance.

We say that the data provide **statistically significant** evidence against the null hypothesis if the p-value is less than some predetermined threshold (e.g., 0.01, 0.05, 0.1).

EXAMPLE

In the opportunity cost study in Section 11.2, we analyzed an experiment where study participants had a 20% drop in likelihood of continuing with a video purchase if they were reminded that the money, if not spent on the video, could be used for other purchases in the future. We determined that such a large difference would only occur 6-in-1,000 times if the reminder actually had no influence on student decision-making. What is the p-value in this study? Would you classify the result as "statistically significant"?

The p-value was 0.006. Since the p-value is less than 0.05, the data provide statistically significant evidence that US college students were actually influenced by the reminder.

What's so special about 0.05?

We often use a threshold of 0.05 to determine whether a result is statistically significant. But why 0.05? Maybe we should use a bigger number, or maybe a smaller number. If you're a little puzzled, that probably means you're reading with a critical eye – good job! We've made a video to help clarify *why 0.05*:

https://www.openintro.org/book/stat/why05/

Sometimes it's also a good idea to deviate from the standard. We'll discuss when to choose a threshold different than 0.05 in Section 14.

11.4 Chapter review

11.4.1 Summary

Figure 11.8 provides a visual summary of the randomization testing procedure.

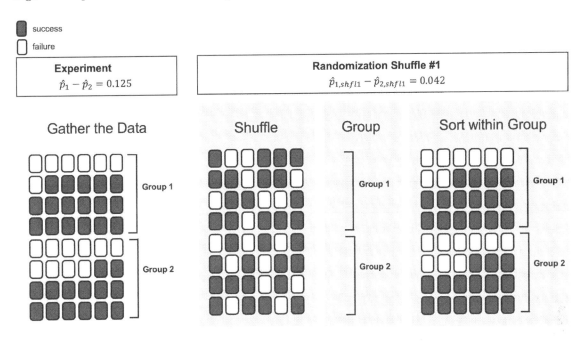

Figure 11.8: An example of one simulation of the full randomization procedure from a hypothetical dataset as visualized in the first panel. We repeat the steps hundreds or thousands of times.

We can summarize the randomization test procedure as follows:

- **Frame the research question in terms of hypotheses.** Hypothesis tests are appropriate for research questions that can be summarized in two competing hypotheses. The null hypothesis (H_0) usually represents a skeptical perspective or a perspective of no relationship between the variables. The alternative hypothesis (H_A) usually represents a new view or the existance of a relationship between the variables.
- **Collect data with an observational study or experiment.** If a research question can be formed into two hypotheses, we can collect data to run a hypothesis test. If the research question focuses on associations between variables but does not concern causation, we would use an observational study. If the research question seeks a causal connection between two or more variables, then an experiment should be used.
- **Model the randomness that would occur if the null hypothesis was true.** In the examples above, the variability has been modeled as if the treatment (e.g., sexual identity, opportunity) allocation was independent of the outcome of the study. The computer generated null distribution is the result of many different randomizations and quantifies the variability that would be expected if the null hypothesis was true.
- **Analyze the data.** Choose an analysis technique appropriate for the data and identify the p-value. So far, we've only seen one analysis technique: randomization. Throughout the rest of this textbook, we'll encounter several new methods suitable for many other contexts.
- **Form a conclusion.** Using the p-value from the analysis, determine whether the data provide evidence against the null hypothesis. Also, be sure to write the conclusion in plain language so casual readers can understand the results.

Table 11.6 is another look at the randomization test summary.

Table 11.6: Summary of randomization as an inferential statistical method.

Question	Answer
What does it do?	Shuffles the explanatory variable to mimic the natural variability found in a randomized experiment
What is the random process described?	Randomized experiment
What other random processes can be approximated?	Can also be used to describe random sampling in an observational model
What is it best for?	Hypothesis testing (can also be used for confidence intervals, but not covered in this text).
What physical object represents the simulation process?	Shuffling cards

11.4.2 Terms

We introduced the following terms in the chapter. If you're not sure what some of these terms mean, we recommend you go back in the text and review their definitions. We are purposefully presenting them in alphabetical order, instead of in order of appearance, so they will be a little more challenging to locate. However you should be able to easily spot them as **bolded text**.

alternative hypothesis	permutation test	statistical inference
confidence interval	point estimate	statistically significant
hypothesis test	randomization test	success
independent	significance level	test statistic
null hypothesis	simulation	
p-value	statistic	

11.5 Exercises

Answers to odd numbered exercises can be found in Appendix A.11.

11.1. **Identify the parameter, I** For each of the following situations, state whether the parameter of interest is a mean or a proportion. It may be helpful to examine whether individual responses are numerical or categorical.

   a. In a survey, one hundred college students are asked how many hours per week they spend on the Internet.

   b. In a survey, one hundred college students are asked: "What percentage of the time you spend on the Internet is part of your course work?"

   c. In a survey, one hundred college students are asked whether or not they cited information from Wikipedia in their papers.

   d. In a survey, one hundred college students are asked what percentage of their total weekly spending is on alcoholic beverages.

   e. In a sample of one hundred recent college graduates, it is found that 85 percent expect to get a job within one year of their graduation date.

11.2. **Identify the parameter, II.** For each of the following situations, state whether the parameter of interest is a mean or a proportion.

   a. A poll shows that 64% of Americans personally worry a great deal about federal spending and the budget deficit.

   b. A survey reports that local TV news has shown a 17% increase in revenue within a two year period while newspaper revenues decreased by 6.4% during this time period.

   c. In a survey, high school and college students are asked whether or not they use geolocation services on their smart phones.

   d. In a survey, smart phone users are asked whether or not they use a web-based taxi service.

   e. In a survey, smart phone users are asked how many times they used a web-based taxi service over the last year.

11.3. **Hypotheses.** For each of the research statements below, note whether it represents a null hypothesis claim or an alternative hypothesis claim.

   a. The number of hours that grade-school children spend doing homework predicts their future success on standardized tests.

   b. King cheetahs on average run the same speed as standard spotted cheetahs.

   c. For a particular student, the probability of correctly answering a 5-option multiple choice test is larger than 0.2 (i.e., better than guessing).

   d. The mean length of African elephant tusks has changed over the last 100 years.

   e. The risk of facial clefts is equal for babies born to mothers who take folic acid supplements compared with those from mothers who do not.

   f. Caffeine intake during pregnancy affects mean birth weight.

   g. The probability of getting in a car accident is the same if using a cell phone than if not using a cell phone.

11.4. **True null hypothesis.** Unbeknownst to you, let's say that the null hypothesis is actually true in the population. You plan to run a study anyway.

   a. If the level of significance you choose (i.e., the cutoff for your p-value) is 0.05, how likely is it that you will mistakenly reject the null hypothesis?

   b. If the level of significance you choose (i.e., the cutoff for your p-value) is 0.01, how likely is it that you will mistakenly reject the null hypothesis?

   c. If the level of significance you choose (i.e., the cutoff for your p-value) is 0.10, how likely is it that you will mistakenly reject the null hypothesis?

11.5. **Identify hypotheses, I.** Write the null and alternative hypotheses in words and then symbols for each of the following situations.

   a. New York is known as "the city that never sleeps". A random sample of 25 New Yorkers were asked how much sleep they get per night. Do these data provide convincing evidence that New Yorkers on average sleep less than 8 hours a night?

   b. Employers at a firm are worried about the effect of March Madness, a basketball championship held each spring in the US, on employee productivity. They estimate that on a regular business day employees spend on average 15 minutes of company time checking personal email, making personal phone calls, etc. They also collect data on how much company time employees spend on such non- business activities during March Madness. They want to determine if these data provide convincing evidence that employee productivity decreases during March Madness.

11.6. **Identify hypotheses, II.** Write the null and alternative hypotheses in words and using symbols for each of the following situations.

   a. Since 2008, chain restaurants in California have been required to display calorie counts of each menu item. Prior to menus displaying calorie counts, the average calorie intake of diners at a restaurant was 1100 calories. After calorie counts started to be displayed on menus, a nutritionist collected data on the number of calories consumed at this restaurant from a random sample of diners. Do these data provide convincing evidence of a difference in the average calorie intake of a diners at this restaurant?

   b. Based on the performance of those who took the GRE exam between July 1, 2004 and June 30, 2007, the average Verbal Reasoning score was calculated to be 462. In 2021 the average verbal score was slightly higher. Do these data provide convincing evidence that the average GRE Verbal Reasoning score has changed since 2021?

11.7. **Side effects of Avandia.** Rosiglitazone is the active ingredient in the controversial type 2 diabetes medicine Avandia and has been linked to an increased risk of serious cardiovascular problems such as stroke, heart failure, and death. A common alternative treatment is Pioglitazone, the active ingredient in a diabetes medicine called Actos. In a nationwide retrospective observational study of 227,571 Medicare beneficiaries aged 65 years or older, it was found that 2,593 of the 67,593 patients using Rosiglitazone and 5,386 of the 159,978 using Pioglitazone had serious cardiovascular problems. These data are summarized in the contingency table below.[8] (Graham et al., 2010)

treatment	No	Yes	Total
Pioglitazone	154,592	5,386	159,978
Rosiglitazone	65,000	2,593	67,593
Total	219,592	7,979	227,571

a. Determine if each of the following statements is true or false. If false, explain why. *Be careful:* The reasoning may be wrong even if the statement's conclusion is correct. In such cases, the statement should be considered false.

  i. Since more patients on Pioglitazone had cardiovascular problems (5,386 vs. 2,593), we can conclude that the rate of cardiovascular problems for those on a Pioglitazone treatment is higher.

  ii. The data suggest that diabetic patients who are taking Rosiglitazone are more likely to have cardiovascular problems since the rate of incidence was (2,593 / 67,593 = 0.038) 3.8% for patients on this treatment, while it was only (5,386 / 159,978 = 0.034) 3.4% for patients on Pioglitazone.

  iii. The fact that the rate of incidence is higher for the Rosiglitazone group proves that Rosiglitazone causes serious cardiovascular problems.

  iv. Based on the information provided so far, we cannot tell if the difference between the rates of incidences is due to a relationship between the two variables or due to chance.

b. What proportion of all patients had cardiovascular problems?

c. If the type of treatment and having cardiovascular problems were independent, about how many patients in the Rosiglitazone group would we expect to have had cardiovascular problems?

See next page for part d.

---

[8] The `avandia` data used in this exercise can be found in the **openintro** R package.

d. We can investigate the relationship between outcome and treatment in this study using a randomization technique. While in reality we would carry out the simulations required for randomization using statistical software, suppose we actually simulate using index cards. In order to simulate from the independence model, which states that the outcomes were independent of the treatment, we write whether or not each patient had a cardiovascular problem on cards, shuffled all the cards together, then deal them into two groups of size 67,593 and 159,978. We repeat this simulation 100 times and each time record the difference between the proportions of cards that say "Yes" in the Rosiglitazone and Pioglitazone groups. Use the histogram of these differences in proportions to answer the following questions.

   i. What are the claims being tested?

   ii. Compared to the number calculated in part (b), which would provide more support for the alternative hypothesis, *higher* or *lower* proportion of patients with cardiovascular problems in the Rosiglitazone group?

   iii. What do the simulation results suggest about the relationship between taking Rosiglitazone and having cardiovascular problems in diabetic patients?

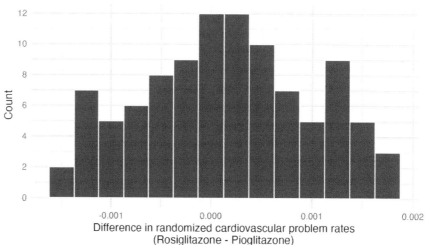

11.8. **Heart transplants.** The Stanford University Heart Transplant Study was conducted to determine whether an experimental heart transplant program increased lifespan. Each patient entering the program was designated an official heart transplant candidate, meaning that they were gravely ill and would most likely benefit from a new heart. Some patients got a transplant and some did not. The variable `transplant` indicates which group the patients were in; patients in the treatment group got a transplant and those in the control group did not. Of the 34 patients in the control group, 30 died. Of the 69 people in the treatment group, 45 died. Another variable called `survived` was used to indicate whether or not the patient was alive at the end of the study.[9] (Turnbull et al., 1974)

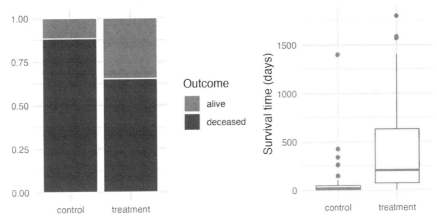

a. Does the stacked bar plot indicate that survival is independent of whether or not the patient got a transplant? Explain your reasoning.

b. What do the box plots above suggest about the efficacy (effectiveness) of the heart transplant treatment.

c. What proportion of patients in the treatment group and what proportion of patients in the control group died?

See next page for part d.

---

[9] The `heart_transplant` data used in this exercise can be found in the **openintro** R package.

d. One approach for investigating whether or not the treatment is effective is to use a randomization technique.

   i. What are the claims being tested?

   ii. The paragraph below describes the set up for such approach, if we were to do it without using statistical software. Fill in the blanks with a number or phrase, whichever is appropriate.

   We write *alive* on _____ cards representing patients who were alive at the end of the study, and *deceased* on _____ cards representing patients who were not. Then, we shuffle these cards and split them into two groups: one group of size _____ representing treatment, and another group of size _____ representing control. We calculate the difference between the proportion of *deceased* cards in the treatment and control groups (treatment - control) and record this value. We repeat this 100 times to build a distribution centered at _____. Lastly, we calculate the fraction of simulations where the simulated differences in proportions are _____. If this fraction is low, we conclude that it is unlikely to have observed such an outcome by chance and that the null hypothesis should be rejected in favor of the alternative.

   iii. What do the simulation results shown below suggest about the effectiveness of the transplant program?

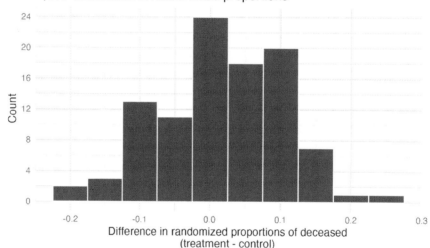

Chapter 12

Confidence intervals with bootstrapping

In this chapter, we expand on the familiar idea of using a sample proportion to estimate a population proportion. That is, we create what is called a **confidence interval**, which is a range of plausible values where we may find the true population value. The process for creating a confidence interval is based on understanding how a statistic (here the sample proportion) *varies* around the parameter (here the population proportion) when many different statistics are calculated from many different samples.

If we could, we would measure the variability of the statistics by repeatedly taking sample data from the population and compute the sample proportion. Then we could do it again. And again. And so on until we have a good sense of the variability of our original estimate.

When the variability across the samples is large, we would assume that the original statistic is possibly far from the true population parameter of interest (and the interval estimate will be wide). When the variability across the samples is small, we expect the sample statistic to be close to the true parameter of interest (and the interval estimate will be narrow).

The ideal world where sampling data is free or extremely cheap is almost never the case, and taking repeated samples from a population is usually impossible.

So, instead of using a "resample from the population" approach, bootstrapping uses a "resample from the sample" approach. In this chapter we provide examples and details about the bootstrapping process.

As seen in Chapter 11, randomization is a statistical technique suitable for evaluating whether a difference in sample proportions is due to chance.

Randomization tests are best suited for modeling experiments where the treatment (explanatory variable) has been randomly assigned to the observational units and there is an attempt to answer a simple yes/no research question.

For example, consider the following research questions that can be well assessed with a randomization test:

- Does this vaccine make it less likely that a person will get malaria?
- Does drinking caffeine affect how quickly a person can tap their finger?
- Can we predict whether candidate A will win the upcoming election?

In this chapter, however, we are instead interested in a different approach to understanding population parameters. Instead of testing a claim, the goal now is to estimate the unknown value of a population parameter.

For example,

- How much less likely am I to get malaria if I get the vaccine?
- How much faster (or slower) can a person tap their finger, on average, if they drink caffeine first?
- What proportion of the vote will go to candidate A?

Here, we explore the situation where the focus is on a single proportion, and we introduce a new simulation method: **bootstrapping**.

Bootstrapping is best suited for modeling studies where the data have been generated through random sampling from a population.

As with randomization tests, our goal with bootstrapping is to understand variability of a statistic.

Unlike randomization tests (which modeled how the statistic would change if the treatment had been allocated differently), the bootstrap will model how a statistic varies from one sample to another taken from the population. This will provide information about how different the statistic is from the parameter of interest.

Quantifying the variability of a statistic from sample to sample is a hard problem.

Fortunately, sometimes the mathematical theory for how a statistic varies (across different samples) is well-known; this is the case for the sample proportion as seen in Chapter 13.

However, some statistics don't have simple theory for how they vary, and bootstrapping provides a computational approach for providing interval estimates for almost any population parameter. In this chapter we will focus on bootstrapping to estimate a single proportion, and we will revisit bootstrapping in Chapters 19 through 21, so you'll get plenty of practice as well as exposure to bootstrapping in many different datasettings.

Our goal with bootstrapping will be to produce an interval estimate (a range of plausible values) for the population parameter.

12.1 Medical consultant case study

People providing an organ for donation sometimes seek the help of a special medical consultant. These consultants assist the patient in all aspects of the surgery, with the goal of reducing the possibility of complications during the medical procedure and recovery. Patients might choose a consultant based in part on the historical complication rate of the consultant's clients.

12.1.1 Observed data

One consultant tried to attract patients by noting the average complication rate for liver donor surgeries in the US is about 10%, but her clients have had only 3 complications in the 62 liver donor surgeries she has facilitated. She claims this is strong evidence that her work meaningfully contributes to reducing complications (and therefore she should be hired!).

12.1. MEDICAL CONSULTANT CASE STUDY

EXAMPLE

We will let p represent the true complication rate for liver donors working with this consultant. (The "true" complication rate will be referred to as the **parameter**.) We estimate p using the data, and label the estimate $\hat{p}$.

The sample proportion for the complication rate is 3 complications divided by the 62 surgeries the consultant has worked on: $\hat{p} = 3/62 = 0.048$.

EXAMPLE

Is it possible to assess the consultant's claim (that the reduction in complications is due to her work) using the data?

No.

The claim is that there is a causal connection, but the data are observational, so we must be on the lookout for confounding variables.

For example, maybe patients who can afford a medical consultant can afford better medical care, which can also lead to a lower complication rate.

While it is not possible to assess the causal claim, it is still possible to understand the consultant's true rate of complications.

Parameter.

A **parameter** is the "true" value of interest.

We typically estimate the parameter using a point estimate from a sample of data. The point estimate is also known as the **statistic**.

For example, we estimate the probability p of a complication for a client of the medical consultant by examining the past complications rates of her clients:

$$\hat{p} = 3/62 = 0.048 \text{ is used to estimate } p$$

12.1.2 Variability of the statistic

In the medical consultant case study, the parameter is p, the true probability of a complication for a client of the medical consultant. There is no reason to believe that p is exactly $\hat{p} = 3/62$, but there is also no reason to believe that p is particularly far from $\hat{p} = 3/62$. By sampling with replacement from the dataset (a process called bootstrapping), the variability of the possible $\hat{p}$ values can be approximated.

Most of the inferential procedures covered in this text are grounded in quantifying how one dataset would differ from another when they are both taken from the same population. It doesn't make sense to take repeated samples from the same population because if you have the means to take more samples, a larger sample size will benefit you more than separately evaluating two sample of the exact same size. Instead, we measure how the samples behave under an estimate of the population.

Figure 12.1 shows how the unknown original population can be estimated by using the sample to approximate the proportion of successes and failures (in our case, the proportion of complications and no complications for the medical consultant).

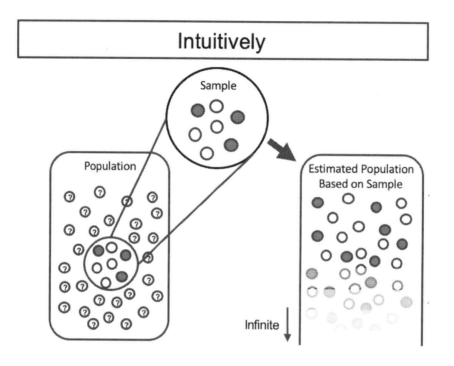

Figure 12.1: The unknown population is estimated using the observed sample data. Note that we can use the sample to create an estimated or bootstrapped population from which to sample. The observed data include three red and four white marbles, so the estimated population contains 3/7 red marbles and 4/7 white marbles.

By taking repeated samples from the estimated population, the variability from sample to sample can be observed. In Figure 12.2 the repeated bootstrap samples are obviously different both from each other and from the original population. Recall that the bootstrap samples were taken from the same (estimated) population, and so the differences are due entirely to natural variability in the sampling procedure.

By summarizing each of the bootstrap samples (here, using the sample proportion), we see, directly, the variability of the sample proportion, $\hat{p}$, from sample to sample. The distribution of $\hat{p}_{boot}$ for the example scenario is shown in Figure 12.3, and the full bootstrap distribution for the medical consultant data is shown in Figure 12.6.

It turns out that in practice, it is very difficult for computers to work with an infinite population (with the same proportional breakdown as in the sample). However, there is a physical and computational method which produces an equivalent bootstrap distribution of the sample proportion in a computationally efficient manner.

Consider the observed data to be a bag of marbles 3 of which are success (red) and 4 of which are failures (white). By drawing the marbles out of the bag with replacement, we depict the exact same sampling **process** as was done with the infinitely large estimated population.

If we apply the bootstrap sampling process to the medical consultant example, we consider each client to be one of the marbles in the bag. There will be 59 white marbles (no complication) and 3 red marbles (complication). If we choose 62 marbles out of the bag (one at a time with replacement) and compute the proportion of simulated patients with complications, $\hat{p}_{boot}$, then this "bootstrap" proportion represents a single simulated proportion from the "resample from the sample" approach.

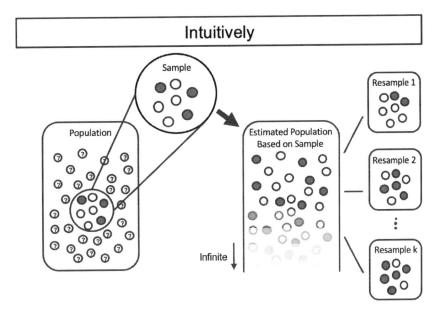

Figure 12.2: Bootstrap sampling provides a measure of the sample to sample variability. Note that we are taking samples from the estimated population that was created from the observed data.

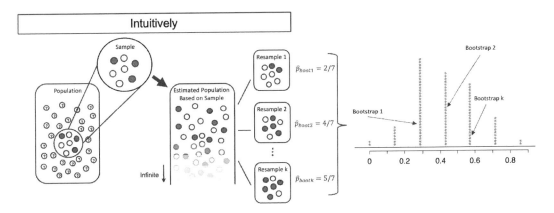

Figure 12.3: The bootstrapped proportion is estimated for each bootstrap sample. The resulting bootstrap distribution (dotplot) provides a measure for how the proportions vary from sample to sample

 GUIDED PRACTICE

In a simulation of 62 patients, about how many would we expect to have had a complication?[1]

One simulation isn't enough to get a sense of the variability from one bootstrap proportion to another bootstrap proportion, so we repeat the simulation 10,000 times using a computer.

Figure 12.6 shows the distribution from the 10,000 bootstrap simulations. The bootstrapped proportions vary from about zero to 11.3%. The variability in the bootstrapped proportions leads us to believe that the true probability of complication (the parameter, p) is likely to fall somewhere between 0 and 11.3%, as these numbers capture 95% of the bootstrap resampled values.

The range of values for the true proportion is called a **bootstrap percentile confidence interval**, and we will see it again throughout the next few sections and chapters.

---

[1] About 4.8% of the patients (3 on average) in the simulation will have a complication, as this is what was seen in the sample. We will, however, see a little variation from one simulation to the next.

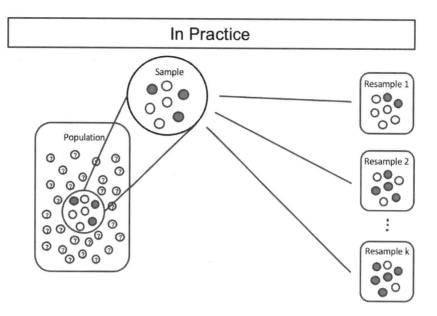

Figure 12.4: Taking repeated resamples from the sample data is the same process as creating an infinitely large estimate of the population. It is computationally more feasible to take resamples directly from the sample. Note that the resampling is now done with replacement (that is, the original sample does not ever change) so that the original sample and estimated hypothetical population are equivalent.

EXAMPLE

The original claim was that the consultant's true rate of complication was under the national rate of 10%. Does the interval estimate of 0 to 11.3% for the true probability of complication indicate that the surgical consultant has a lower rate of complications than the national average? Explain.

No. Because the interval overlaps 10%, it might be that the consultant's work is associated with a lower risk of complications, or it might be that the consultant's work is associated with a higher risk (i.e., greater than 10%) of complications! Additionally, as previously mentioned, because this is an observational study, even if an association can be measured, there is no evidence that the consultant's work is the cause of the complication rate (being higher or lower).

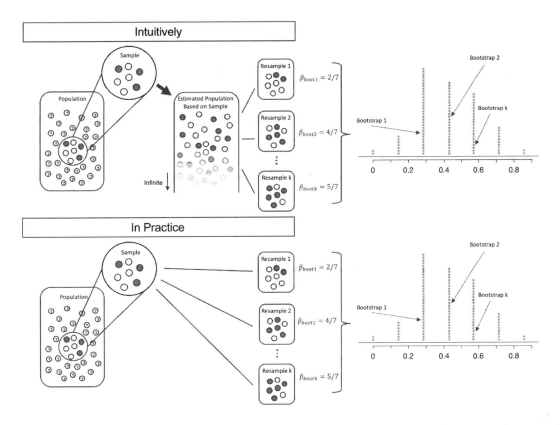

Figure 12.5: A comparison of the process of sampling from the estimate infinite population and resampling with replacement from the original sample. Note that the dotplot of bootstrapped proportions is the same because the process by which the statistics were estimated is equivalent.

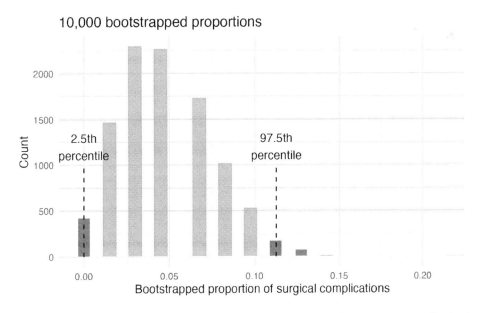

Figure 12.6: The original medical consultant data is bootstrapped 10,000 times. Each simulation creates a sample from the original data where the probability of a complication is $\hat{p} = 3/62$. The bootstrap 2.5 percentile proportion is 0 and the 97.5 percentile is 0.113. The result is: we are confident that, in the population, the true probability of a complication is between 0% and 11.3%.

12.2 Tappers and listeners case study

Here's a game you can try with your friends or family: pick a simple, well-known song, tap that tune on your desk, and see if the other person can guess the song. In this simple game, you are the tapper, and the other person is the listener.

12.2.1 Observed data

A Stanford University graduate student named Elizabeth Newton conducted an experiment using the tapper-listener game.[2] In her study, she recruited 120 tappers and 120 listeners into the study. About 50% of the tappers expected that the listener would be able to guess the song. Newton wondered, is 50% a reasonable expectation?

In Newton's study, only 3 out of 120 listeners ($\hat{p} = 0.025$) were able to guess the tune! That seems like quite a low number which leads the researcher to ask: what is the true proportion of people who can guess the tune?

12.2.2 Variability of the statistic

To answer the question, we will again use a simulation. To simulate 120 games, this time we use a bag of 120 marbles 3 are red (for those who guessed correctly) and 117 are white (for those who could not guess the song). Sampling from the bag 120 times (remembering to replace the marble back into the bag each time to keep constant the population proportion of red) produces one bootstrap sample.

For example, we can start by simulating 5 tapper-listener pairs by sampling 5 marbles from the bag of 3 red and 117 white marbles.

W	W	W	R	W
Wrong	Wrong	Wrong	Correct	Wrong

After selecting 120 marbles, we counted 2 red for $\hat{p}_{boot1} = 0.0167$. As we did with the randomization technique, seeing what would happen with one simulation isn't enough. In order to understand how far the observed proportion of 0.025 might be from the true parameter, we should generate more simulations. Here we've repeated the entire simulation ten times:

$$0.0417 \quad 0.025 \quad 0.025 \quad 0.0083 \quad 0.05 \quad 0.0333 \quad 0.025 \quad 0 \quad 0.0083 \quad 0$$

As before, we'll run a total of 10,000 simulations using a computer. As seen in Figure 12.7, the range of 95% of the resampled values of $\hat{p}_{boot}$ is 0.000 to 0.0583. That is, we expect that between 0% and 5.83% of people are truly able to guess the tapper's tune.

GUIDED PRACTICE

Do the data provide convincing evidence against the claim that 50% of listeners can guess the tapper's tune?[3]

---

[2]This case study is described in Made to Stick by Chip and Dan Heath. Little known fact: the teaching principles behind many OpenIntro resources are based on *Made to Stick*.
[3]Because 50% is not in the interval estimate for the true parameter, we can say that there is convincing evidence against the hypothesis that 50% of listeners can guess the tune. Moreover, 50% is a substantial distance from the largest resample statistic, suggesting that there is **very** convincing evidence against this hypothesis.

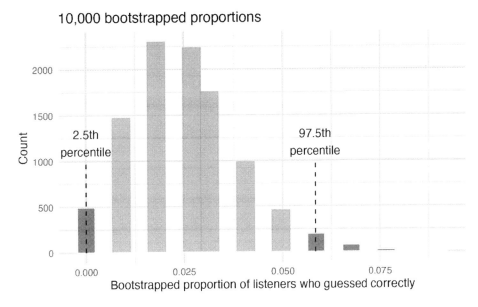

Figure 12.7: The original listener-tapper data is bootstrapped 10,000 times. Each simulation creates a sample where the probability of being correct is $\hat{p} = 3/120$. The 2.5 percentile proportion is 0 and the 97.5 percentile is 0.0583. The result is that we are confident that, in the population, the true percent of people who can guess correctly is between 0% and 5.83%.

12.3 Confidence intervals

A point estimate provides a single plausible value for a parameter. However, a point estimate is rarely perfect; usually there is some error in the estimate. In addition to supplying a point estimate of a parameter, a next logical step would be to provide a plausible *range of values* for the parameter.

12.3.1 Plausible range of values for the population parameter

A plausible range of values for the population parameter is called a **confidence interval**. Using only a single point estimate is like fishing in a murky lake with a spear, and using a confidence interval is like fishing with a net. We can throw a spear where we saw a fish, but we will probably miss. On the other hand, if we toss a net in that area, we have a good chance of catching the fish.

If we report a point estimate, we probably will not hit the exact population parameter. On the other hand, if we report a range of plausible values – a confidence interval – we have a good shot at capturing the parameter.

 GUIDED PRACTICE

If we want to be very certain we capture the population parameter, should we use a wider interval (e.g., 99%) or a smaller interval (e.g., 80%)?[4]

12.3.2 Bootstrap confidence interval

As we saw above, a **bootstrap sample** is a sample of the original sample. In the case of the medical complications data, we proceed as follows:

---

[4] If we want to be more certain we will capture the fish, we might use a wider net. Likewise, we use a wider confidence interval if we want to be more certain that we capture the parameter.

- Randomly sample one observation from the 62 patients (replace the marble back into the bag so as to keep the population constant).
- Randomly sample a second observation from the 62 patients. Because we sample with replacement (i.e., we don't actually remove the marbles from the bag), there is a 1-in-62 chance that the second observation will be the same one sampled in the first step!
- Keep going one sampled observation at a time ...
- Randomly sample the 62nd observation from the 62 patients.

Bootstrap sampling is often called **sampling with replacement**.

A bootstrap sample behaves similarly to how an actual sample from a population would behave, and we compute the point estimate of interest (here, compute $\hat{p}_{boot}$).

Due to theory that is beyond this text, we know that the bootstrap proportions $\hat{p}_{boot}$ vary around $\hat{p}$ in a similar way to how different sample proportions (i.e., values of $\hat{p}$) vary around the true parameter p.

Therefore, an interval estimate for p can be produced using the $\hat{p}_{boot}$ values themselves.

95% Bootstrap percentile confidence interval for a parameter p.

The 95% bootstrap confidence interval for the parameter p can be obtained directly using the ordered $\hat{p}_{boot}$ values.

Consider the sorted $\hat{p}_{boot}$ values. Call the 2.5% bootstrapped proportion value "lower", and call the 97.5% bootstrapped proportion value "upper".

The 95% confidence interval is given by: (lower, upper)

In Section 16.1 we will discuss different percentages for the confidence interval (e.g., 90% confidence interval or 99% confidence interval).

Section 16.1 also provides a longer discussion on what "95% confidence" actually means.

12.4 Chapter review

12.4.1 Summary

Figure 12.8 provides a visual summary of creating bootstrap confidence intervals.

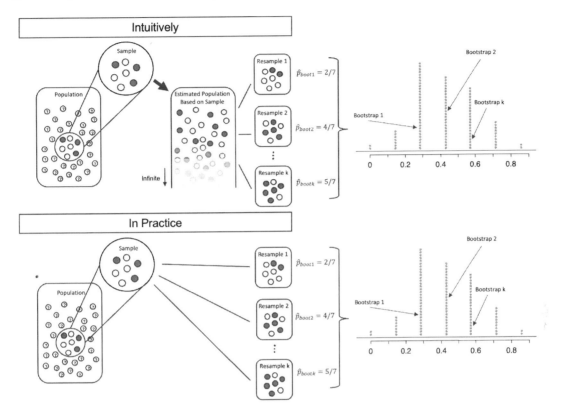

Figure 12.8: We will use sampling with replacement to measure the variability of the statistic of interest (here the proportion). Sampling with replacement is a computational tool which is equivalent to using the sample as a way of estimating an infinitely large population from which to sample.

We can summarize the bootstrap process as follows:

- **Frame the research question in terms of a parameter to estimate.** Confidence Intervals are appropriate for research questions that aim to estimate a number from the population (called a parameter).
- **Collect data with an observational study or experiment.** If a research question can be formed as a query about the parameter, we can collect data to calculate a statistic which is the best guess we have for the value of the parameter. However, we know that the statistic won't be exactly equal to the parameter due to natural variability.
- **Model the randomness by using the data values as a proxy for the population.** In order to assess how far the statistic might be from the parameter, we take repeated resamples from the dataset to measure the variability in bootstrapped statistics. The variability of the bootstrapped statistics around the observed statistic (a quantity which can be measured through computational technique) should be approximately the same as the variability of many observed sample statistics around the parameter (a quantity which is very difficult to measure because in real life we only get exactly one sample).
- **Create the interval.** After choosing a particular confidence level, use the variability of the bootstrapped statistics to create an interval estimate which will hope to capture the true parameter. While the interval estimate associated with the particular sample at hand may or may not capture the parameter, the researcher knows that over their lifetime, the confidence level will determine the percentage of their research confidence intervals that do capture the true parameter.

- **Form a conclusion.** Using the confidence interval from the analysis, report on the interval estimate for the parameter of interest. Also, be sure to write the conclusion in plain language so casual readers can understand the results.

Table 12.2 is another look at the Bootstrap process summary.

Table 12.2: Summary of bootstrapping as an inferential statistical method.

Question	Answer
What does it do?	Resamples (with replacement) from the observed data to mimic the sampling variability found by collecting data from a population
What is the random process described?	Random sampling from a population
What other random processes can be approximated?	Can also be used to describe random allocation in an experiment
What is it best for?	Confidence intervals (can also be used for bootstrap hypothesis testing for one proportion as well).
What physical object represents the simulation process?	Pulling marbles from a bag with replacement

12.4.2 Terms

We introduced the following terms in the chapter. If you're not sure what some of these terms mean, we recommend you go back in the text and review their definitions. We are purposefully presenting them in alphabetical order, instead of in order of appearance, so they will be a little more challenging to locate. However you should be able to easily spot them as **bolded text**.

bootstrap percentile confidence interval	parameter	statistic
bootstrap sample	point estimate	
bootstrapping	sampling with replacement	

12.5 Exercises

Answers to odd numbered exercises can be found in Appendix A.12.

12.1. **Outside YouTube videos.** Let's say that you want to estimate the proportion of YouTube videos which take place outside (define "outside" to be if any part of the video takes place outdoors). You take a random sample of 128 YouTube videos[5] and determine that 37 of them take place outside. You'd like to estimate the proportion of all YouTube videos which take place outside, so you decide to create a bootstrap interval from the original sample of 128 videos.

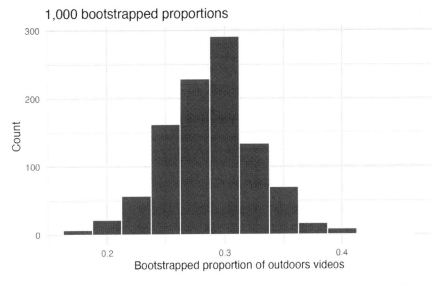

a. Describe in words the relevant statistic and parameter for this problem. If you know the numerical value for either one, provide it. If you don't know the numerical value, explain why the value is unknown.

b. What notation is used to describe, respectively, the statistic and the parameter?

c. If using software to bootstrap the original dataset, what is the statistic calculated on each bootstrap sample?

d. When creating a bootstrap sampling distribution (histogram) of the bootstrapped sample proportions, where should the center of the histogram lie?

e. The histogram provides a bootstrap sampling distribution for the sample proportion (with 1000 bootstrap repetitions). Using the histogram, estimate a 90% confidence interval for the proportion of YouTube videos which take place outdoors.

f. In words of the problem, interpret the confidence interval which was estimated in the previous part.

---

[5]There are many choices for implementing a random selection of YouTube videos, but it isn't clear how "random" they are.

12.2. **Chronic illness.** In 2012 the Pew Research Foundation reported that "45% of US adults report that they live with one or more chronic conditions." However, this value was based on a sample, so it may not be a perfect estimate for the population parameter of interest on its own. The study was based on a sample of 3014 adults. Below is a distribution of 1000 bootstrapped sample proportions from the Pew dataset. (Pew Research Center, 2013a)

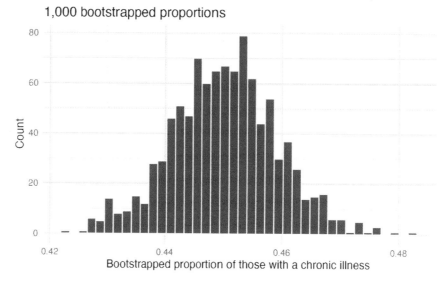

Using the distribution of 1000 bootstrapped proportions, approximate a 92% confidence interval for the true proportion of US adults who live with one or more chronic conditions. Interpret the interval in the context of the problem.

12.3. **Twitter users and news.** A poll conducted in 2013 found that 52% of all US adult Twitter users get at least some news on Twitter. However, this value was based on a sample, so it may not be a perfect estimate for the population parameter of interest on its own. The study was based on a sample of 736 adults. Below is a distribution of 1000 bootstrapped sample proportions from the Pew dataset. (Pew Research Center, 2013b)

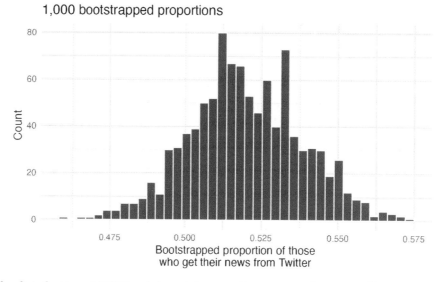

Using the distribution of 1000 bootstrapped proportions, approximate a 98% confidence interval for the true proportion of US adult Twitter users (in 2013) who get at least some of their news from Twitter. Interpret the interval in the context of the problem.

12.4. **Bootstrap distributions of $\hat{p}$, I.** Each of the following four distributions was created using a different dataset. Each dataset was based on $n = 23$ observations. The original datasets had the following proportions of successes:

$$\hat{p} = 0.13 \quad \hat{p} = 0.22 \quad \hat{p} = 0.30 \quad \hat{p} = 0.43.$$

Match each histogram with the original data proportion of success.

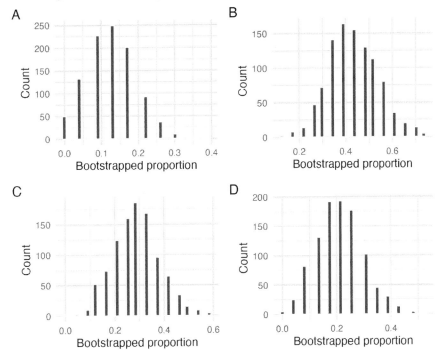

12.5. Bootstrap distributions of $\hat{p}$, II. Each of the following four distributions was created using a different dataset. Each dataset was based on $n = 23$ observations.

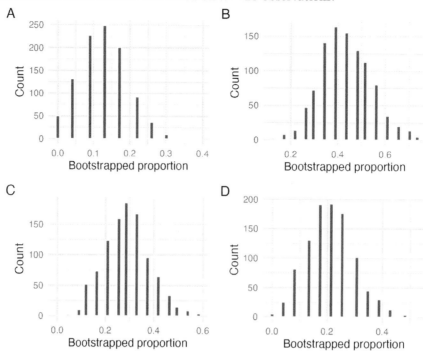

Consider each of the following values for the true popluation p (proportion of success). Datasets A, B, C, D were bootstrapped 1000 times, with bootstrap proportions as given in the histograms provided. For each parameter value, list the datasets which could plausibly have come from that population. (Hint: there may be more than one dataset for each parameter value.)

a. $p = 0.05$

b. $p = 0.25$

c. $p = 0.45$

d. $p = 0.55$

e. $p = 0.75$

12.6. **Bootstrap distributions of $\hat{p}$, III.** Each of the following four distributions was created using a different dataset. Each dataset had the same proportion of successes ($\hat{p} = 0.4$) but a different sample size. The four datasets were given by $n = 10, 100, 500,$ and 1000.

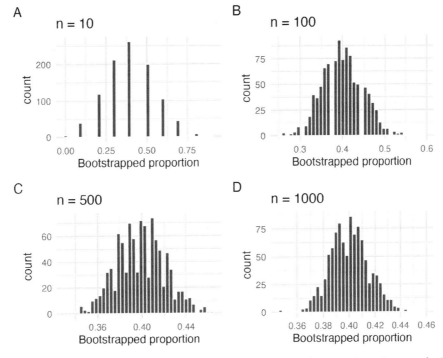

Consider each of the following values for the true popluation p (proportion of success). Datasets A, B, C, D were bootstrapped 1000 times, with bootstrap proportions as given in the histograms provided. For each parameter value, list the datasets which could plausibly have come from that population. (Hint: there may be more than one dataset for each parameter value.)

a. $p = 0.05$

b. $p = 0.25$

c. $p = 0.45$

d. $p = 0.55$

e. $p = 0.75$

12.7. **Cyberbullying rates.** Teens were surveyed about cyberbullying, and 54% to 64% reported experiencing cyberbullying (95% confidence interval). Answer the following questions based on this interval. (Pew Research Center, 2018)

a. A newspaper claims that a majority of teens have experienced cyberbullying. Is this claim supported by the confidence interval? Explain your reasoning.

b. A researcher conjectured that 70% of teens have experienced cyberbullying. Is this claim supported by the confidence interval? Explain your reasoning.

c. Without actually calculating the interval, determine if the claim of the researcher from part (b) would be supported based on a 90% confidence interval?

12.8. **Waiting at an ER.** A 95% confidence interval for the mean waiting time at an emergency room (ER) of (128 minutes, 147 minutes). Answer the following questions based on this interval.

a. A local newspaper claims that the average waiting time at this ER exceeds 3 hours. Is this claim supported by the confidence interval? Explain your reasoning.

b. The Dean of Medicine at this hospital claims the average wait time is 2.2 hours. Is this claim supported by the confidence interval? Explain your reasoning.

c. Without actually calculating the interval, determine if the claim of the Dean from part (b) would be supported based on a 99% confidence interval?

Chapter 13

Inference with mathematical models

 In Chapters 11 and 12 questions about population parameters were addressed using computational techniques. With randomization tests, the data were permuted assuming the null hypothesis. With bootstrapping, the data were resampled in order to measure the variability. In many cases (indeed, with sample proportions), the variability of the statistic can be described by the computational method (as in previous chapters) or by a mathematical formula (as in this chapter).

The normal distribution is presented here to describe the variability associated with sample proportions which are taken from either repeated samples or repeated experiments. The normal distribution is quite powerful in that it describes the variability of many different statistics, and we will encounter the normal distribution throughout the remainder of the book.

For now, however, focus is on the parallels between how data can provide insight about a research question either through computational methods or through mathematical models.

13.1 Central Limit Theorem

In recent chapters, we've encountered four case studies. While they differ in the settings, in their outcomes, and also in the technique we've used to analyze the data, they all have something in common: the general shape of the distribution of the statistics (called the **sampling distribution**). You may have noticed that the distributions were symmetric and bell-shaped.

Sampling distribution.

A sampling distribution is the distribution of all possible values of a *sample statistic* from samples of a given sample size from a given population. We can think about the sample distribution as describing as how sample statistics (e.g. the sample proportion $\hat{p}$ or the sample mean $\bar{x}$) varies from one study to another. A sampling distribution is contrasted with a data distribution which shows the variability of the *observed* data values. The data distribution can be visualized from the observations themselves. However, because a sampling distribution describes sample statistics computed from many studies, it cannot be visualized directly from a single dataset. Instead, we use either computational or mathematical structures to estimate the sampling distribution and hence to describe the expected variability of the sample statistic in repeated studies.

Figure 13.1 shows the null distributions in each of the four case studies where we ran 10,000 simulations. Note that the **null distribution** is the sampling distribution of the statistic created under the setting where the null hypothesis is true. Therefore, the null distribution will always be centered at the value of the parameter given by the null hypothesis. In the case of the opportunity cost study, which originally had just 1,000 simulations, we've included an additional 9,000 simulations.

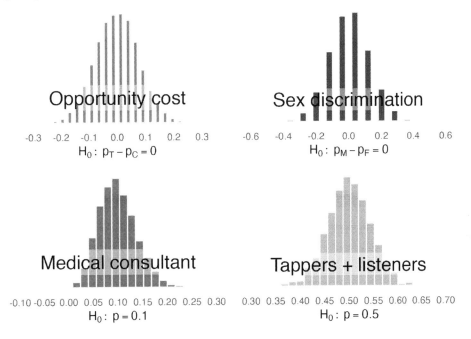

Figure 13.1: The null distribution for each of the four case studies presented previously. Note that the center of each distribution is given by the value of the parameter set in the null hypothesis.

GUIDED PRACTICE

Describe the shape of the distributions and note anything that you find interesting.[1]

The case study for the medical consultant is the only distribution with any evident skew. As we observed in Chapter 1, it's common for distributions to be skewed or contain outliers. However, the null distributions we've so far encountered have all looked somewhat similar and, for the most part, symmetric. They all resemble a bell-shaped curve. The bell-shaped curve similarity is not a

---

[1]In general, the distributions are reasonably symmetric. The case study for the medical consultant is the only distribution with any evident skew (the distribution is skewed right).

coincidence, but rather, is guaranteed by mathematical theory.

 Central Limit Theorem for proportions.

If we look at a proportion (or difference in proportions) and the scenario satisfies certain conditions, then the sample proportion (or difference in proportions) will appear to follow a bell-shaped curve called the *normal distribution*.

An example of a perfect normal distribution is shown in Figure 13.2. Imagine laying a normal curve over each of the four null distributions in Figure 13.1. While the mean (center) and standard deviation (width or spread) may change for each plot, the general shape remains roughly intact.

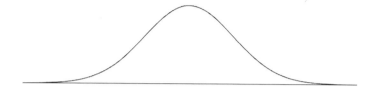

Figure 13.2: A normal curve.

Mathematical theory guarantees that if repeated samples are taken a sample proportion or a difference in sample proportions will follow something that resembles a normal distribution when certain conditions are met. (Note: we typically only take **one** sample, but the mathematical model lets us know what to expect if we *had* taken repeated samples.) These conditions fall into two general categories describing the independence between observations and the need to take a sufficiently large sample size.

1. Observations in the sample are **independent**. Independence is guaranteed when we take a random sample from a population. Independence can also be guaranteed if we randomly divide individuals into treatment and control groups.

2. The sample is **large enough**. The sample size cannot be too small. What qualifies as "small" differs from one context to the next, and we'll provide suitable guidelines for proportions in Chapter 16.

So far we've had no need for the normal distribution. We've been able to answer our questions somewhat easily using simulation techniques. However, soon this will change. Simulating data can be non-trivial. For example, some of the scenarios encountered in Chapter 8 where we introduced regression models with multiple predictors would require complex simulations in order to make inferential conclusions. Instead, the normal distribution and other distributions like it offer a general framework for statistical inference that applies to a very large number of settings.

 Technical Conditions.

In order for the normal approximation to describe the sampling distribution of the sample proportion as it varies from sample to sample, two conditions must hold. If these conditions do not hold, it is unwise to use the normal distribution (and related concepts like Z scores, probabilities from the normal curve, etc.) for inferential analyses.

1. **Independent observations**
2. **Large enough sample:** For proportions, at least 10 expected successes and 10 expected failures in the sample.

13.2 Normal Distribution

Among all the distributions we see in statistics, one is overwhelmingly the most common. The symmetric, unimodal, bell curve is ubiquitous throughout statistics. It is so common that people know it as a variety of names including the **normal curve, normal model, or normal distribution**.[2] Under certain conditions, sample proportions, sample means, and sample differences can be modeled using the normal distribution. Additionally, some variables such as SAT scores and heights of US adult males closely follow the normal distribution.

Normal distribution facts.

Many summary statistics and variables are nearly normal, but none are exactly normal. Thus the normal distribution, while not perfect for any single problem, is very useful for a variety of problems. We will use it in data exploration and to solve important problems in statistics.

In this section, we will discuss the normal distribution in the context of data to become familiar with normal distribution techniques. In the following sections and beyond, we'll move our discussion to focus on applying the normal distribution and other related distributions to model point estimates for hypothesis tests and for constructing confidence intervals.

13.2.1 Normal distribution model

The normal distribution always describes a symmetric, unimodal, bell-shaped curve. However, normal curves can look different depending on the details of the model. Specifically, the normal model can be adjusted using two parameters: mean and standard deviation. As you can probably guess, changing the mean shifts the bell curve to the left or right, while changing the standard deviation stretches or constricts the curve. Figure 13.3 shows the normal distribution with mean 0 and standard deviation 1 (which is commonly referred to as the **standard normal distribution**) on the left. A normal distributions with mean 19 and standard deviation 4 is shown on the right. Figure 13.4 shows the same two normal distributions on the same axis.

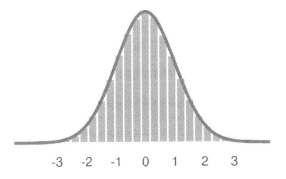
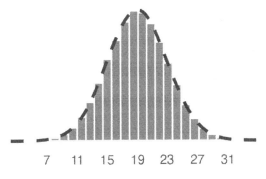

Figure 13.3: Both curves represent the normal distribution, however, they differ in their center and spread. The normal distribution with mean 0 and standard deviation 1 (blue solid line, on the left) is called the **standard normal distribution**. The other distribution (green dashed line, on the right) has mean 19 and standard deviation 4.

If a normal distribution has mean μ and standard deviation σ, we may write the distribution as $N(\mu, \sigma)$. The two distributions in Figure 13.4 can be written as

$$N(\mu = 0, \sigma = 1) \quad \text{and} \quad N(\mu = 19, \sigma = 4)$$

---

[2]It is also introduced as the Gaussian distribution after Frederic Gauss, the first person to formalize its mathematical expression.

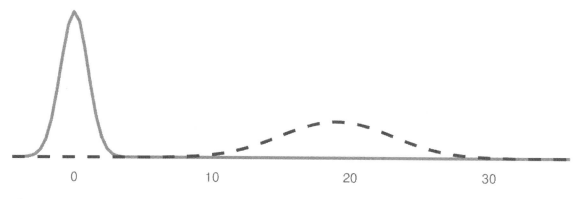

Figure 13.4: The two normal models shown in Figure 13.3 but plotted together on the same scale.

Because the mean and standard deviation describe a normal distribution exactly, they are called the distribution's **parameters**.

 EXAMPLE

Write down the short-hand for a normal distribution with the following parameters.

a. mean 5 and standard deviation 3
b. mean -100 and standard deviation 10
c. mean 2 and standard deviation 9

---

a. $N(\mu = 5, \sigma = 3)$
b. $N(\mu = -100, \sigma = 10)$
c. $N(\mu = 2, \sigma = 9)$

13.2.2 Standardizing with Z scores

 GUIDED PRACTICE

SAT scores follow a nearly normal distribution with a mean of 1500 points and a standard deviation of 300 points. ACT scores also follow a nearly normal distribution with mean of 21 points and a standard deviation of 5 points. Suppose Nel scored 1800 points on their SAT and Sian scored 24 points on their ACT. Who performed better?[3]

The solution to the previous example relies on a standardization technique called a Z score, a method most commonly employed for nearly normal observations (but that may be used with any distribution). The **Z score** of an observation is defined as the number of standard deviations it falls above or below the mean. If the observation is one standard deviation above the mean, its Z score is 1. If it is 1.5 standard deviations *below* the mean, then its Z score is -1.5. If x is an observation from a distribution $N(\mu, \sigma)$, we define the Z score mathematically as

---

[3]We use the standard deviation as a guide. Nel is 1 standard deviation above average on the SAT: $1500 + 300 = 1800$. Sian is 0.6 standard deviations above the mean on the ACT: $21 + 0.6 \times 5 = 24$. In Figure 13.5, we can see that Nel did better compared to other test takers than Sian did, so their score was better.

13.2. NORMAL DISTRIBUTION

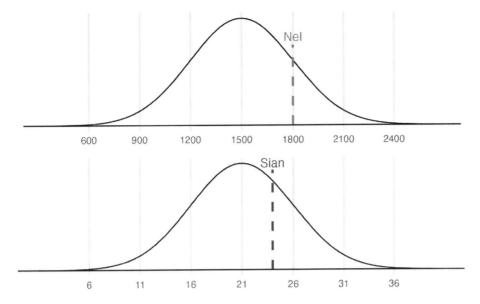

Figure 13.5: Nel's and Sian's scores shown with the distributions of SAT and ACT scores.

$$Z = \frac{x - \mu}{\sigma}$$

Using $\mu_{SAT} = 1500$, $\sigma_{SAT} = 300$, and $x_{Nel} = 1800$, we find Nel's Z score:

$$Z_{Nel} = \frac{x_{Nel} - \mu_{SAT}}{\sigma_{SAT}} = \frac{1800 - 1500}{300} = 1$$

The Z score.

The Z score of an observation is the number of standard deviations it falls above or below the mean. We compute the Z score for an observation x that follows a distribution with mean μ and standard deviation σ using

$$Z = \frac{x - \mu}{\sigma}$$

If the observation x comes from a *normal* distribution centered at μ with standard deviation of σ, then the Z score will be distributed according to a *normal* distribution with a center of 0 and a standard deviation of 1. That is, the normality remains when transforming from x to Z with a shift in both the center as well as the spread.

GUIDED PRACTICE

Use Sian's ACT score, 24, along with the ACT mean and standard deviation to compute their Z score.[4]

Observations above the mean always have positive Z scores while those below the mean have negative Z scores. If an observation is equal to the mean (e.g., SAT score of 1500), then the Z score is 0.

---

[4]$Z_{Sian} = \frac{x_{Sian} - \mu_{ACT}}{\sigma_{ACT}} = \frac{24 - 21}{5} = 0.6$

EXAMPLE

Let X represent a random variable from $N(\mu = 3, \sigma = 2)$, and suppose we observe $x = 5.19$. Find the Z score of x. Then, use the Z score to determine how many standard deviations above or below the mean x falls.

Its Z score is given by $Z = \frac{x-\mu}{\sigma} = \frac{5.19-3}{2} = 2.19/2 = 1.095$. The observation x is 1.095 standard deviations *above* the mean. We know it must be above the mean since Z is positive.

GUIDED PRACTICE

Head lengths of brushtail possums follow a nearly normal distribution with mean 92.6 mm and standard deviation 3.6 mm. Compute the Z scores for possums with head lengths of 95.4 mm and 85.8 mm.[5]

We can use Z scores to roughly identify which observations are more unusual than others. One observation x_1 is said to be more unusual than another observation x_2 if the absolute value of its Z score is larger than the absolute value of the other observation's Z score: $|Z_1| > |Z_2|$. This technique is especially insightful when a distribution is symmetric.

GUIDED PRACTICE

Which of the two brushtail possum observations in the previous guided practice is more *unusual*?[6]

13.2.3 Normal probability calculations

EXAMPLE

Nel from the SAT Guided Practice earned a score of 1800 on their SAT with a corresponding $Z = 1$. They would like to know what percentile they fall in among all SAT test-takers.

Nel's **percentile** is the percentage of people who earned a lower SAT score than Nel. We shade the area representing those individuals in Figure 13.6. The total area under the normal curve is always equal to 1, and the proportion of people who scored below Nel on the SAT is equal to the *area* shaded in Figure 13.6: 0.8413. In other words, Nel is in the 84th percentile of SAT takers.

We can use the normal model to find percentiles or probabilities. A **normal probability table**, which lists Z scores and corresponding percentiles, can be used to identify a percentile based on the Z score (and vice versa). Statistical software can also be used.

Normal probabilities are most commonly found using statistical software which we will show here using R. We use the software to identify the percentile corresponding to any particular Z score. For

---

[5]For $x_1 = 95.4$ mm: $Z_1 = \frac{x_1-\mu}{\sigma} = \frac{95.4-92.6}{3.6} = 0.78$. For $x_2 = 85.8$ mm: $Z_2 = \frac{85.8-92.6}{3.6} = -1.89$.
[6]Because the *absolute value* of Z score for the second observation is larger than that of the first, the second observation has a more unusual head length.

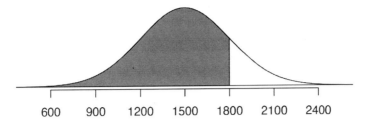

Figure 13.6: The normal model for SAT scores, shading the area of those individuals who scored below Nel.

instance, the percentile of $Z = 0.43$ is 0.6664, or the 66.64^{th} percentile. The `pnorm()` function is available in default R and will provide the percentile associated with any cutoff on a normal curve. The `normTail()` function is available in the **openintro** R package and will draw the associated normal distribution curve.

```
pnorm(0.43, mean = 0, sd = 1)
#> [1] 0.666
normTail(m = 0, s = 1, L = 0.43)
```

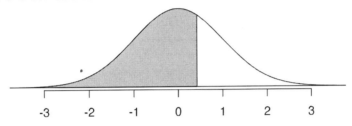

We can also find the Z score associated with a percentile. For example, to identify Z for the 80th percentile, we use `qnorm()` which identifies the **quantile** for a given percentage. The quantile represents the cutoff value. (To remember the function `qnorm()` as providing a cutoff, notice that both `qnorm()` and "cutoff" start with the sound "kuh". To remember the `pnorm()` function as providing a probability from a given cutoff, notice that both `pnorm()` and probability start with the sound "puh".) We determine the Z score for the 80th percentile using `qnorm()`: 0.84.

```
qnorm(0.80, mean = 0, sd = 1)
#> [1] 0.842
openintro::normTail(m = 0, s = 1, L = 0.842)
```

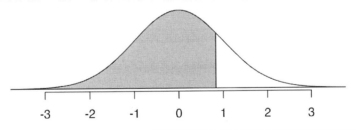

 GUIDED PRACTICE

Determine the proportion of SAT test takers who scored better than Nel on the SAT.[7]

---

[7] If 84% had lower scores than Nel, the number of people who had better scores must be 16%. (Generally ties are ignored when the normal model, or any other continuous distribution, is used.)

13.2.4 Normal probability examples

Cumulative SAT scores are approximated well by a normal model, $N(\mu = 1500, \sigma = 300)$.

> **EXAMPLE**
>
> Shannon is a randomly selected SAT taker, and nothing is known about Shannon's SAT aptitude. What is the probability that Shannon scores at least 1630 on their SATs?
>
> ---
>
> First, always draw and label a picture of the normal distribution. (Drawings need not be exact to be useful.) We are interested in the chance they score above 1630, so we shade the upper tail. See the normal curve below.
>
> The x-axis identifies the mean and the values at 2 standard deviations above and below the mean. The simplest way to find the shaded area under the curve makes use of the Z score of the cutoff value. With $\mu = 1500$, $\sigma = 300$, and the cutoff value $x = 1630$, the Z score is computed as
>
> $$Z = \frac{x - \mu}{\sigma} = \frac{1630 - 1500}{300} = \frac{130}{300} = 0.43$$
>
> We use software to find the percentile of $Z = 0.43$, which yields 0.6664. However, the percentile describes those who had a Z score *lower* than 0.43. To find the area *above* $Z = 0.43$, we compute one minus the area of the lower tail, as seen below.
>
> The probability Shannon scores at least 1630 on the SAT is 0.3336. This calculation is visualized in Figure 13.7.

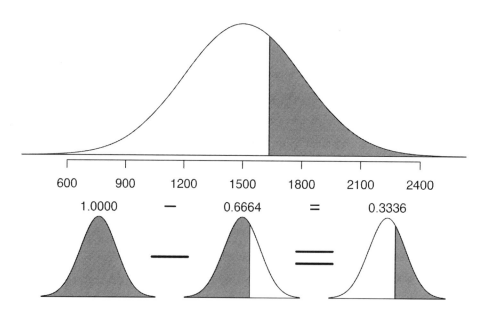

Figure 13.7: Visual calculation of the probability that Shannon scores at least 1630 on the SAT.

13.2. NORMAL DISTRIBUTION

Always draw a picture first, and find the Z score second.

For any normal probability situation, *always always always* draw and label the normal curve and shade the area of interest first. The picture will provide an estimate of the probability.

After drawing a figure to represent the situation, identify the Z score for the observation of interest.

GUIDED PRACTICE

If the probability of Shannon scoring at least 1630 is 0.3336, then what is the probability they score less than 1630? Draw the normal curve representing this exercise, shading the lower region instead of the upper one.[8]

EXAMPLE

Edward earned a 1400 on their SAT. What is their percentile?

First, a picture is needed. Edward's percentile is the proportion of people who do not get as high as a 1400. These are the scores to the left of 1400.

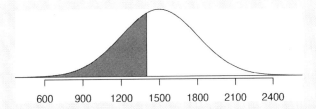

The mean $\mu = 1500$, the standard deviation $\sigma = 300$, and the cutoff for the tail area $x = 1400$ are used to compute the Z score:

$$Z = \frac{x - \mu}{\sigma} = \frac{1400 - 1500}{300} = -0.33$$

Statistical software can be used to find the proportion of the $N(0, 1)$ curve to the left of -0.33 which is 0.3707. Edward is at the 37th percentile.

---

[8] We found the probability to be 0.6664. A picture for this exercise is represented by the shaded area below "0.6664".

EXAMPLE

Use the results of the previous example to compute the proportion of SAT takers who did better than Edward. Also draw a new picture.

If Edward did better than 37% of SAT takers, then about 63% must have done better than them.

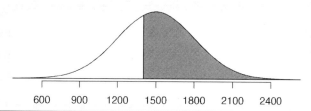

Areas to the right.

Most statistical software, as well as normal probability tables in most books, give the area to the left. If you would like the area to the right, first find the area to the left and then subtract the amount from one.

GUIDED PRACTICE

Stuart earned an SAT score of 2100. Draw a picture for each part. (a) What is their percentile? (b) What percent of SAT takers did better than Stuart?[9]

Based on a sample of 100 men,[10] the heights of adults who identify as male, between the ages 20 and 62 in the US is nearly normal with mean 70.0'' and standard deviation 3.3''.

---

[9] Numerical answers: (a) 0.9772. (b) 0.0228.
[10] This sample was taken from the USDA Food Commodity Intake Database.

13.2. NORMAL DISTRIBUTION

EXAMPLE

Kamron is 5'7'' (67 inches) and Adrian is 6'4'' (76 inches). (a) What is Kamron's height percentile? (b) What is Adrian's height percentile? Also draw one picture for each part.

Numerical answers, calculated using statistical software (e.g., `pnorm()` in R): (a) 18.17th percentile. (b) 96.55th percentile.

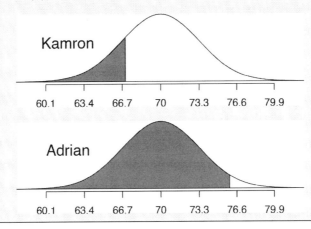

The last several problems have focused on finding the probability or percentile for a particular observation. What if you would like to know the observation corresponding to a particular percentile?

EXAMPLE

Yousef's height is at the 40th percentile. How tall are they?

As always, first draw the picture.

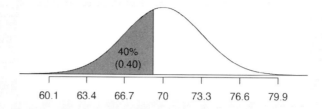

In this case, the lower tail probability is known (0.40), which can be shaded on the diagram. We want to find the observation that corresponds to the known probability of 0.4. As a first step in this direction, we determine the Z score associated with the 40th percentile.

Because the percentile is below 50%, we know Z will be negative. Statistical software provides the Z value to be -0.25.

```
qnorm(0.4, mean = 0, sd = 1)
#> [1] -0.253
```

Knowing $Z_{Yousef} = -0.25$ and the population parameters $\mu = 70$ and $\sigma = 3.3$ inches, the Z score formula can be set up to determine Yousef's unknown height, labeled x_{Yousef}:

$$-0.25 = Z_{Yousef} = \frac{x_{Yousef} - \mu}{\sigma} = \frac{x_{Yousef} - 70}{3.3}$$

Solving for x_{Yousef} yields the height 69.18 inches. That is, Yousef is about 5'9'' (this is notation for 5-feet, 9-inches).

EXAMPLE

What is the adult male height at the 82nd percentile?

Again, we draw the figure first.

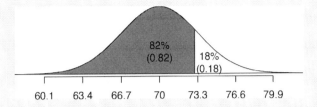

And calculate the Z value associated with the 82nd percentile:

```
qnorm(0.82, m = 0, s = 1)
#> [1] 0.915
```

Next, we want to find the Z score at the 82nd percentile, which will be a positive value (because the percentile is bigger than 50%). Using `qnorm()`, the 82nd percentile corresponds to $Z = 0.92$. Finally, the height x is found using the Z score formula with the known mean μ, standard deviation σ, and Z score $Z = 0.92$:

$$0.92 = Z = \frac{x - \mu}{\sigma} = \frac{x - 70}{3.3}$$

This yields 73.04 inches or about 6'1'' as the height at the 82nd percentile.

GUIDED PRACTICE

(a) What is the 95th percentile for SAT scores?

(b) What is the 97.5^{th} percentile of the male heights? As always with normal probability problems, first draw a picture.[11]

GUIDED PRACTICE

(a) What is the probability that a randomly selected male adult is at least 6'2'' (74 inches)?

(b) What is the probability that a male adult is shorter than 5'9'' (69 inches)?[12]

---

[11] Remember: draw a picture first, then find the Z score. (We leave the pictures to you.) The Z score can be found by using the percentiles and the normal probability table. (a) We look for 0.95 in the probability portion (middle part) of the normal probability table, which leads us to row 1.6 and (about) column 0.05, i.e., $Z_{95} = 1.65$. Knowing $Z_{95} = 1.65$, $\mu = 1500$, and $\sigma = 300$, we setup the Z score formula: $1.65 = \frac{x_{95} - 1500}{300}$. We solve for x_{95}: $x_{95} = 1995$. (b) Similarly, we find $Z_{97.5} = 1.96$, again setup the Z score formula for the heights, and calculate $x_{97.5} = 76.5$.

[12] Numerical answers: (a) 0.1131. (b) 0.3821.

EXAMPLE

What is the probability that a randomly selected adult male is between 5'9'' and 6'2''?

These heights correspond to 69 inches and 74 inches. First, draw the figure. The area of interest is no longer an upper or lower tail.

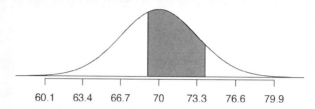

The total area under the curve is 1. If we find the area of the two tails that are not shaded (from the previous Guided Practice, these areas are 0.3821 and 0.1131), then we can find the middle area:

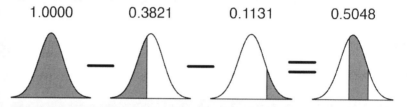

That is, the probability of being between 5'9'' and 6'2'' is 0.5048.

GUIDED PRACTICE

What percent of SAT takers get between 1500 and 2000?[13]

GUIDED PRACTICE

What percent of adult males are between 5'5'' and 5'7''?[14]

13.3 Quantifying the variability of a statistic

As seen in later chapters, it turns out that many of the statistics used to summarize data (e.g., the sample proportion, the sample mean, differences in two sample proportions, differences in two sample means, the sample slope from a linear model, etc.) vary according to the normal distribution seen above. The mathematical models are derived from the normal theory, but even the computational methods (and the intuitive thinking behind both approaches) use the general bell-shaped variability seen in most of the distributions constructed so far.

---

[13] This is an abbreviated solution. (Be sure to draw a figure!) First find the percent who get below 1500 and the percent that get above 2000: $Z_{1500} = 0.00 \rightarrow 0.5000$ (area below), $Z_{2000} = 1.67 \rightarrow 0.0475$ (area above). Final answer: $1.0000 - 0.5000 - 0.0475 = 0.4525$.

[14] 5'5'' is 65 inches. 5'7'' is 67 inches. Numerical solution: $1.000 - 0.0649 - 0.8183 = 0.1168$, i.e., 11.68%.

13.3.1 68-95-99.7 rule

Here, we present a useful general rule for the probability of falling within 1, 2, and 3 standard deviations of the mean in the normal distribution. The rule will be useful in a wide range of practical settings, especially when trying to make a quick estimate without a calculator or Z table.

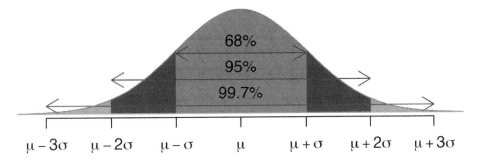

Figure 13.8: Probabilities for falling within 1, 2, and 3 standard deviations of the mean in a normal distribution.

GUIDED PRACTICE

Use `pnorm()` (or a Z table) to confirm that about 68%, 95%, and 99.7% of observations fall within 1, 2, and 3, standard deviations of the mean in the normal distribution, respectively. For instance, first find the area that falls between $Z = -1$ and $Z = 1$, which should have an area of about 0.68. Similarly there should be an area of about 0.95 between $Z = -2$ and $Z = 2$.[15]

It is possible for a normal random variable to fall 4, 5, or even more standard deviations from the mean. However, these occurrences are very rare if the data are nearly normal. The probability of being further than 4 standard deviations from the mean is about 1-in-30,000. For 5 and 6 standard deviations, it is about 1-in-3.5 million and 1-in-1 billion, respectively.

GUIDED PRACTICE

SAT scores closely follow the normal model with mean $\mu = 1500$ and standard deviation $\sigma = 300$. About what percent of test takers score 900 to 2100? What percent score between 1500 and 2100 ?[16]

13.3.2 Standard error

Point estimates vary from sample to sample, and we quantify this variability with what is called the **standard error (SE)**. The standard error is equal to the standard deviation associated with the statistic. So, for example, to quantify the variability of a point estimate from one sample to the next, the variability is called the standard error of the point estimate. Almost always, the standard error is itself an estimate, calculated from the sample of data.

The way we determine the standard error varies from one situation to the next. However, typically it is determined using a formula based on the Central Limit Theorem.

---

[15] First draw the pictures. To find the area between $Z = -1$ and $Z = 1$, use `pnorm()` or the normal probability table to determine the areas below $Z = -1$ and above $Z = 1$. Next verify the area between $Z = -1$ and $Z = 1$ is about 0.68. Repeat this for $Z = -2$ to $Z = 2$ and also for $Z = -3$ to $Z = 3$.

[16] 900 and 2100 represent two standard deviations above and below the mean, which means about 95% of test takers will score between 900 and 2100. Since the normal model is symmetric, then half of the test takers from part (a) ($\frac{95\%}{2} = 47.5\%$ of all test takers) will score 900 to 1500 while 47.5% score between 1500 and 2100.

13.3.3 Margin of error

Very related to the standard error is the **margin of error**. The margin of error describes how far away observations are from their mean.

For example, to describe where most (i.e., 95%) observations lie, we say that the margin of error is approximately $2 \times SE$. That is, 95% of the observations are within one margin of error of the mean.

Margin of error for sample proportions.

The distance given by $z^\star \times SE$ is called the **margin of error**.

$z^\star$ is the cutoff value found on the normal distribution. The most common value of $z^\star$ is 1.96 (often approximated to be 2) indicating that the margin of error describes the variability associated with 95% of the sampled statistics.

Notice that if the spread of the observations goes from some lower bound to some upper bound, a rough approximation of the SE is to divide the range by 4. That is, if you notice the sample proportions go from 0.1 to 0.4, the SE can be approximated to be 0.075.

13.4 Case Study (test): Opportunity cost

The approach for using the normal model in the context of inference is very similar to the practice of applying the model to individual observations that are nearly normal. We will replace null distributions we previously obtained using the randomization or simulation techniques and verify the results once again using the normal model. When the sample size is sufficiently large, the normal approximation generally provides us with the same conclusions as the simulation model.

13.4.1 Observed data

In Section 11.2 we were introduced to the opportunity cost study, which found that students became thriftier when they were reminded that not spending money now means the money can be spent on other things in the future. Let's re-analyze the data in the context of the normal distribution and compare the results.

The `opportunity_cost` data can be found in the **openintro** R package.

13.4.2 Variability of the statistic

Figure 13.9 summarizes the null distribution as determined using the randomization method. The best fitting normal distribution for the null distribution has a mean of 0. We can calculate the standard error of this distribution by borrowing a formula that we will become familiar with in Chapter 17, but for now let's just take the value $SE = 0.078$ as a given. Recall that the point estimate of the difference was 0.20, as shown in Figure 13.9. Next, we'll use the normal distribution approach to compute the p-value.

13.4.3 Observed statistic vs. null statistics

As we learned in Section 13.2, it is helpful to draw and shade a picture of the normal distribution so we know precisely what we want to calculate. Here we want to find the area of the tail beyond 0.2, representing the p-value.

13.4. CASE STUDY (TEST): OPPORTUNITY COST

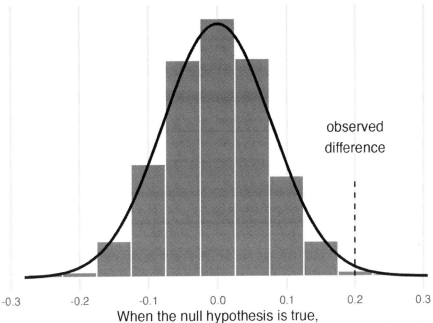

Figure 13.9: Null distribution of differences with an overlaid normal curve for the opportunity cost study. 10,000 simulations were run for this figure.

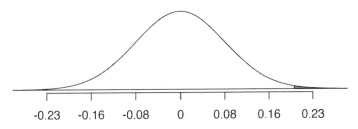

Next, we can calculate the Z score using the observed difference, 0.20, and the two model parameters. The standard error, $SE = 0.078$, is the equivalent of the model's standard deviation.

$$Z = \frac{\text{observed difference} - 0}{SE} = \frac{0.20 - 0}{0.078} = 2.56$$

We can either use statistical software or look up $Z = 2.56$ in the normal probability table to determine the right tail area: 0.0052, which is about the same as what we got for the right tail using the randomization approach (0.006). Using this area as the p-value, we see that the p-value is less than 0.05, we conclude that the treatment did indeed impact students' spending.

Z score in a hypothesis test.

In the context of a hypothesis test, the Z score for a point estimate is

$$Z = \frac{\text{point estimate} - \text{null value}}{SE}$$

The standard error in this case is the equivalent of the standard deviation of the point estimate, and the null value comes from the claim made in the null hypothesis.

We have confirmed that the randomization approach we used earlier and the normal distribution approach provide almost identical p-values and conclusions in the opportunity cost case study. Next,

13.5 Case study (test): Medical consultant

13.5.1 Observed data

In Section 12.1 we learned about a medical consultant who reported that only 3 of their 62 clients who underwent a liver transplant had complications, which is less than the more common complication rate of 0.10. In that work, we did not model a null scenario, but we will discuss a simulation method for a one proportion null distribution in Section 16.1, such a distribution is provided in Figure 13.10. We have added the best-fitting normal curve to the figure, which has a mean of 0.10. Borrowing a formula that we'll encounter in Chapter 16, the standard error of this distribution was also computed: $SE = 0.038$.

13.5.2 Variability of the statistic

Before we begin, we want to point out a simple detail that is easy to overlook: the null distribution we generated from the simulation is slightly skewed, and the histogram is not particularly smooth. In fact, the normal distribution only sort-of fits this model.

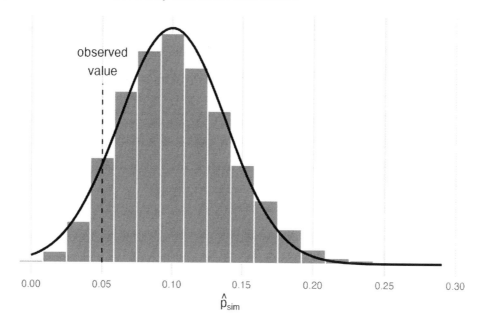

Figure 13.10: The null distribution for the sample proportion, created from 10,000 simulated studies, along with the best-fitting normal model.

13.5.3 Observed statistic vs. null statistics

As always, we'll draw a picture before finding the normal probabilities. Below is a normal distribution centered at 0.10 with a standard error of 0.038.

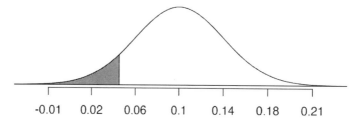

Next, we can calculate the Z score using the observed complication rate, $\hat{p} = 0.048$ along with the mean and standard deviation of the normal model. Here again, we use the standard error for the standard deviation.

$$Z = \frac{\hat{p} - p_0}{SE_{\hat{p}}} = \frac{0.048 - 0.10}{0.038} = -1.37$$

Identifying $Z = -1.37$ using statistical software or in the normal probability table, we can determine that the left tail area is 0.0853 which is the estimated p-value for the hypothesis test. There is a small problem: the p-value of 0.0853 is almost 30% smaller than the simulation p-value of 0.1222 which will be calculated in Section 16.1.

The discrepancy is explained by the normal model's poor representation of the null distribution in Figure 13.10. As noted earlier, the null distribution from the simulations is not very smooth, and the distribution itself is slightly skewed. That's the bad news. The good news is that we can foresee these problems using some simple checks. We'll learn more about these checks in the following chapters.

In Section 13.1 we noted that the two common requirements to apply the Central Limit Theorem are (1) the observations in the sample must be independent, and (2) the sample must be sufficiently large. The guidelines for this particular situation – which we will learn in Chapter 16 – would have alerted us that the normal model was a poor approximation.

13.5.4 Conditions for applying the normal model

The success story in this section was the application of the normal model in the context of the opportunity cost data. However, the biggest lesson comes from the less successful attempt to use the normal approximation in the medical consultant case study.

Statistical techniques are like a carpenter's tools. When used responsibly, they can produce amazing and precise results. However, if the tools are applied irresponsibly or under inappropriate conditions, they will produce unreliable results. For this reason, with every statistical method that we introduce in future chapters, we will carefully outline conditions when the method can reasonably be used. These conditions should be checked in each application of the technique.

After covering the introductory topics in this course, advanced study may lead to working with complex models which, for example, bring together many variables with different variability structure. Working with data that come from normal populations makes higher-order models easier to estimate and interpret. There are times when simulation, randomization, or bootstrapping are unwieldy in either structure or computational demand. Normality can often lead to excellent approximations of the data using straightforward modeling techniques.

13.6 Case study (interval): Stents

A point estimate is our best guess for the value of the parameter, so it makes sense to build the confidence interval around that value. The standard error, which is a measure of the uncertainty associated with the point estimate, provides a guide for how large we should make the confidence interval. The 68-95-99.7 rule tells us that, in general, 95% of observations are within 2 standard errors of the mean. Here, we use the value 1.96 to be slightly more precise.

Constructing a 95% confidence interval.

When the sampling distribution of a point estimate can reasonably be modeled as normal, the point estimate we observe will be within 1.96 standard errors of the true value of interest about 95% of the time. Thus, a **95% confidence interval** for such a point estimate can be constructed:

$$\text{point estimate} \pm 1.96 \times SE$$

We can be **95% confident** this interval captures the true value.

GUIDED PRACTICE

Compute the area between -1.96 and 1.96 for a normal distribution with mean 0 and standard deviation 1.[17]

EXAMPLE

The point estimate from the opportunity cost study was that 20% fewer students would buy a video if they were reminded that money not spent now could be spent later on something else. The point estimate from this study can reasonably be modeled with a normal distribution, and a proper standard error for this point estimate is $SE = 0.078$. Construct a 95% confidence interval.

Since the conditions for the normal approximation have already been verified, we can move forward with the construction of the 95% confidence interval:

$$\text{point estimate} \pm 1.96 \times SE = 0.20 \pm 1.96 \times 0.078 = (0.047, 0.353)$$

We are 95% confident that the video purchase rate resulting from the treatment is between 4.7% and 35.3% lower than in the control group. Since this confidence interval does not contain 0, it is consistent with our earlier result where we rejected the notion of "no difference" using a hypothesis test.

Note that we've used SE = 0.078 from the last section. However, it would more generally be appropriate to recompute the SE slightly differently for this confidence interval using sample proportions. Don't worry about this detail for now since the two resulting standard errors are, in this case, almost identical.

13.6.1 Observed data

Consider an experiment that examined whether implanting a stent in the brain of a patient at risk for a stroke helps reduce the risk of a stroke. The results from the first 30 days of this study, which included 451 patients, are summarized in Table 13.1. These results are surprising! The point estimate suggests that patients who received stents may have a *higher* risk of stroke: $p_{trmt} - p_{ctrl} = 0.090$.

---

[17]We will leave it to you to draw a picture. The Z scores are $Z_{left} = -1.96$ and $Z_{right} = 1.96$. The area between these two Z scores is $0.9750 - 0.0250 = 0.9500$. This is where "1.96" comes from in the 95% confidence interval formula.

13.6. CASE STUDY (INTERVAL): STENTS

Table 13.1: Descriptive statistics for 30-day results for the stent study.

Group	Stroke	No event	Total
control	214	13	227
treatment	191	33	224
Total	405	46	451

The stent30 data can be found in the **openintro** R package.

13.6.2 Variability of the statistic

EXAMPLE

Consider the stent study and results. The conditions necessary to ensure the point estimate $p_{trmt} - p_{ctrl} = 0.090$ is nearly normal have been verified for you, and the estimate's standard error is $SE = 0.028$. Construct a 95% confidence interval for the change in 30-day stroke rates from usage of the stent.

The conditions for applying the normal model have already been verified, so we can proceed to the construction of the confidence interval:

$$\text{point estimate} \pm 1.96 \times SE = 0.090 \pm 1.96 \times 0.028 = (0.035, 0.145)$$

We are 95% confident that implanting a stent in a stroke patient's brain increased the risk of stroke within 30 days by a rate of 0.035 to 0.145. This confidence interval can also be used in a way analogous to a hypothesis test: since the interval does not contain 0 (is completely above 0), it means the data provide convincing evidence that the stent used in the study changed the risk of stroke within 30 days.

As with hypothesis tests, confidence intervals are imperfect. About 1-in-20 properly constructed 95% confidence intervals will fail to capture the parameter of interest, simply due to natural variability in the observed data. Figure 13.11 shows 25 confidence intervals for a proportion that were constructed from 25 different datasets that all came from the same population where the true proportion was $p = 0.3$. However, 1 of these 25 confidence intervals happened not to include the true value. The interval which does not capture $p = 0.3$ is not due to bad science. Instead, it is due to natural variability, and we should *expect* some of our intervals to miss the parameter of interest. Indeed, over a lifetime of creating 95% intervals, you should expect 5% of your reported intervals to miss the parameter of interest (unfortunately, you will not ever know which of your reported intervals captured the parameter and which missed the parameter).

GUIDED PRACTICE

In Figure 13.11, one interval does not contain the true proportion, $p = 0.3$. Does this imply that there was a problem with the datasets that were selected?[18]

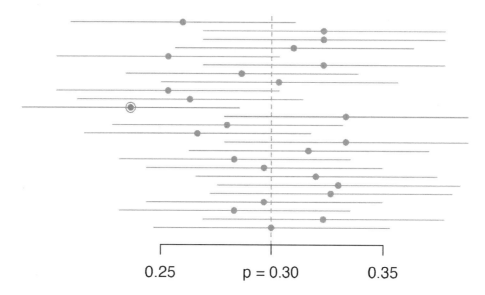

Figure 13.11: Twenty-five samples of size $n = 300$ were collected from a population with $p = 0.30$. For each sample, a confidence interval was created to try to capture the true proportion p. However, 1 of these 25 intervals did not capture $p = 0.30$.

13.6.3 Interpreting confidence intervals

A careful eye might have observed the somewhat awkward language used to describe confidence intervals.

 Correct confidence interval interpretation.

We are XX% confident that the population parameter is between *lower* and *upper* (where *lower* and *upper* are both numerical values).

Incorrect language might try to describe the confidence interval as capturing the population parameter with a certain probability.

This is one of the most common errors: while it might be useful to think of it as a probability, the confidence level only quantifies how plausible it is that the parameter is in the interval.

Another especially important consideration of confidence intervals is that they *only try to capture the population parameter*. Our intervals say nothing about the confidence of capturing individual observations, a proportion of the observations, or about capturing point estimates. Confidence intervals provide an interval estimate for and attempt to capture **population parameters**.

---

[18]No. Just as some observations occur more than 1.96 standard deviations from the mean, some point estimates will be more than 1.96 standard errors from the parameter. A confidence interval only provides a plausible range of values for a parameter. While we might say other values are implausible based on the data, this does not mean they are impossible.

13.7 Chapter review

13.7.1 Summary

We can summarise the process of using the normal model as follows:

- **Frame the research question.** The mathematical model can be applied to both the hypothesis testing and the confidence interval framework. Make sure that your research question is being addressed by the most appropriate inference procedure.
- **Collect data with an observational study or experiment.** To address the research question, collect data on the variables of interest. Note that your data may be a random sample from a population or may be part of a randomized experiment.
- **Model the randomness of the statistic.** In many cases, the normal distribution will be an excellent model for the randomness associated with the statistic of interest. The Central Limit Theorem tells us that if the sample size is large enough, sample averages (which can be calculated as either a proportion or a sample mean) will be approximately normally distributed when describing how the statistics change from sample to sample.
- **Calculate the variability of the statistic.** Using formulas, come up with the standard deviation (or more typically, an estimate of the standard deviation called the standard error) of the statistic. The SE of the statistic will give information on how far the observed statistic is from the null hypothesized value (if performing a hypothesis test) or from the unknown population parameter (if creating a confidence interval).
- **Use the normal distribution to quantify the variability.** The normal distribution will provide a probability which measures how likely it is for your observed and hypothesized (or observed and unknown) parameter to differ by the amount measured. The unusualness (or not) of the discrepancy will form the conclusion to the research question.
- **Form a conclusion.** Using the p-value or the confidence interval from the analysis, report on the research question of interest. Also, be sure to write the conclusion in plain language so casual readers can understand the results.

Table 13.2 is another look at the mathematical model approach to inference.

Table 13.2: Summary of mathematical models as an inferential statistical method.

Question	Answer
What does it do?	Uses theory (primarily the Central Limit Theorem) to describe the hypothetical variability resulting from either repeated randomized experiments or random samples
What is the random process described?	Randomized experiment or random sampling
What other random processes can be approximated?	Randomized experiment or random sampling
What is it best for?	Quick analyses through, for example, calculating a Z score.
What physical object represents the simulation process?	Not applicable

13.7.2 Terms

We introduced the following terms in the chapter. If you're not sure what some of these terms mean, we recommend you go back in the text and review their definitions. We are purposefully presenting them in alphabetical order, instead of in order of appearance, so they will be a little more challenging to locate. However you should be able to easily spot them as **bolded text**.

95% confidence interval	normal distribution	percentile
95% confident	normal model	sampling distribution
Central Limit Theorem	normal probability table	standard error
margin of error	null distribution	standard normal distribution
normal curve	parameter	Z score

13.8 Exercises

Answers to odd numbered exercises can be found in Appendix A.13.

13.1. **Chronic illness.** In 2013, the Pew Research Foundation reported that "45% of U.S. adults report that they live with one or more chronic conditions". However, this value was based on a sample, so it may not be a perfect estimate for the population parameter of interest on its own. The study reported a standard error of about 1.2%, and a normal model may reasonably be used in this setting.

    a. Create a 95% confidence interval for the proportion of U.S. adults who live with one or more chronic conditions. Also interpret the confidence interval in the context of the study. (Pew Research Center, 2013a)

    b. Identify each of the following statements as true or false. Provide an explanation to justify each of your answers.

        i. We can say with certainty that the confidence interval from part (a) contains the true percentage of U.S. adults who suffer from a chronic illness.

        ii. If we repeated this study 1,000 times and constructed a 95% confidence interval for each study, then approximately 950 of those confidence intervals would contain the true fraction of U.S. adults who suffer from chronic illnesses.

        iii. The poll provides statistically significant evidence (at the $\alpha = 0.05$ level) that the percentage of U.S. adults who suffer from chronic illnesses is below 50%.

        iv. Since the standard error is 1.2%, only 1.2% of people in the study communicated uncertainty about their answer.

13.2. **Twitter users and news.** A poll conducted in 2013 found that 52% of U.S. adult Twitter users get at least some news on Twitter. The standard error for this estimate was 2.4%, and a normal distribution may be used to model the sample proportion. (Pew Research Center, 2013b)

    a. Construct a 99% confidence interval for the fraction of U.S. adult Twitter users who get some news on Twitter, and interpret the confidence interval in context.

    b. Identify each of the following statements as true or false. Provide an explanation to justify each of your answers.

        i. The data provide statistically significant evidence that more than half of U.S. adult Twitter users get some news through Twitter. Use a significance level of $\alpha = 0.01$.

        ii. Since the standard error is 2.4%, we can conclude that 97.6% of all U.S. adult Twitter users were included in the study.

        iii. If we want to reduce the standard error of the estimate, we should collect less data.

        iv. If we construct a 90% confidence interval for the percentage of U.S. adults Twitter users who get some news through Twitter, this confidence interval will be wider than a corresponding 99% confidence interval.

13.3. **Interpreting a Z score from a sample proportion.** Suppose that you conduct a hypothesis test about a population proportion and calculate the Z score to be 0.47. Which of the following is the best interpretation of this value? For the problems which are not a good interpretation, indicate the statistical idea being described.[19]

    a. The probability is 0.47 that the null hypothesis is true.

    b. If the null hypothesis were true, the probability would be 0.47 of obtaining a sample proportion as far as observed from the hypothesized value of the population proportion.

    c. The sample proportion is 0.47 standard errors greater than the hypothesized value of the population proportion.

    d. The sample proportion is equal to 0.47 times the standard error.

    e. The sample proportion is 0.47 away from the hypothesized value of the population.

    f. The sample proportion is 0.47.

13.4. **Mental health.** The General Social Survey asked the question: "For how many days during the past 30 days was your mental health, which includes stress, depression, and problems with emotions, not good?" Based on responses from 1,151 US residents, the survey reported a 95% confidence interval of 3.40 to 4.24 days in 2010.

    a. Interpret this interval in context of the data.

    b. What does "95% confident" mean? Explain in the context of the application.

    c. Suppose the researchers think a 99% confidence level would be more appropriate for this interval. Will this new interval be smaller or wider than the 95% confidence interval?

    d. If a new survey were to be done with 500 Americans, do you think the standard error of the estimate be larger, smaller, or about the same.

13.5. **Repeated water samples.** A nonprofit wants to understand the fraction of households that have elevated levels of lead in their drinking water. They expect at least 5% of homes will have elevated levels of lead, but not more than about 30%. They randomly sample 800 homes and work with the owners to retrieve water samples, and they compute the fraction of these homes with elevated lead levels. They repeat this 1,000 times and build a distribution of sample proportions.

    a. What is this distribution called?

    b. Would you expect the shape of this distribution to be symmetric, right skewed, or left skewed? Explain your reasoning.

    c. What is the name of the variability of this distribution.

    d. Suppose the researchers' budget is reduced, and they are only able to collect 250 observations per sample, but they can still collect 1,000 samples. They build a new distribution of sample proportions. How will the variability of this new distribution compare to the variability of the distribution when each sample contained 800 observations?

13.6. **Repeated student samples.** Of all freshman at a large college, 16% made the dean's list in the current year. As part of a class project, students randomly sample 40 students and check if those students made the list. They repeat this 1,000 times and build a distribution of sample proportions.

    a. What is this distribution called?

    b. Would you expect the shape of this distribution to be symmetric, right skewed, or left skewed? Explain your reasoning.

    c. What is the name of the variability of this distribution?

    d. Suppose the students decide to sample again, this time collecting 90 students per sample, and they again collect 1,000 samples. They build a new distribution of sample proportions. How will the variability of this new distribution compare to the variability of the distribution when each sample contained 40 observations?

---

[19]This exercise was inspired by discussion on Dr. Allan Rossman's blog Ask Good Questions.

Chapter 14

Decision Errors

 Using data to make inferential decisions about larger populations is not a perfect process. As seen in Chapter 11, a small p-value typically leads the researcher to a decision to reject the null claim or hypothesis. Sometimes, however, data can produce a small p-value when the null hypothesis is actually true and the data are just inherently variable. Here we describe the errors which can arise in hypothesis testing, how to define and quantify the different errors, and suggestions for mitigating errors if possible.

Hypothesis tests are not flawless. Just think of the court system: innocent people are sometimes wrongly convicted and the guilty sometimes walk free. Similarly, data can point to the wrong conclusion. However, what distinguishes statistical hypothesis tests from a court system is that our framework allows us to quantify and control how often the data lead us to the incorrect conclusion.

In a hypothesis test, there are two competing hypotheses: the null and the alternative. We make a statement about which one might be true, but we might choose incorrectly. There are four possible scenarios in a hypothesis test, which are summarized in Table 14.1.

Table 14.1: Four different scenarios for hypothesis tests.

	Test conclusion	
Truth	Reject null hypothesis	Fail to reject null hypothesis
Null hypothesis is true	Type 1 Error	Good decision
Alternative hypothesis is true	Good decision	Type 2 Error

A **Type 1 Error** is rejecting the null hypothesis when H_0 is actually true. Since we rejected the null hypothesis in the sex discrimination and opportunity cost studies, it is possible that we made a Type 1 Error in one or both of those studies. A **Type 2 Error** is failing to reject the null hypothesis when the alternative is actually true.

EXAMPLE

In a US court, the defendant is either innocent (H_0) or guilty (H_A). What does a Type 1 Error represent in this context? What does a Type 2 Error represent? Table 14.1 may be useful.

If the court makes a Type 1 Error, this means the defendant is innocent (H_0 true) but wrongly convicted. A Type 2 Error means the court failed to reject H_0 (i.e., failed to convict the person) when they were in fact guilty (H_A true).

GUIDED PRACTICE

Consider the opportunity cost study where we concluded students were less likely to make a DVD purchase if they were reminded that money not spent now could be spent later. What would a Type 1 Error represent in this context?[1]

EXAMPLE

How could we reduce the Type 1 Error rate in US courts? What influence would this have on the Type 2 Error rate?

To lower the Type 1 Error rate, we might raise our standard for conviction from "beyond a reasonable doubt" to "beyond a conceivable doubt" so fewer people would be wrongly convicted. However, this would also make it more difficult to convict the people who are actually guilty, so we would make more Type 2 Errors.

GUIDED PRACTICE

How could we reduce the Type 2 Error rate in US courts? What influence would this have on the Type 1 Error rate?[2]

The example and guided practice above provide an important lesson: if we reduce how often we make one type of error, we generally make more of the other type.

---

[1] Making a Type 1 Error in this context would mean that reminding students that money not spent now can be spent later does not affect their buying habits, despite the strong evidence (the data suggesting otherwise) found in the experiment. Notice that this does *not* necessarily mean something was wrong with the data or that we made a computational mistake. Sometimes data simply point us to the wrong conclusion, which is why scientific studies are often repeated to check initial findings.

[2] To lower the Type 2 Error rate, we want to convict more guilty people. We could lower the standards for conviction from "beyond a reasonable doubt" to "beyond a little doubt". Lowering the bar for guilt will also result in more wrongful convictions, raising the Type 1 Error rate.

14.1 Significance level

The **significance level** provides the cutoff for the p-value which will lead to a decision of "reject the null hypothesis." Choosing a significance level for a test is important in many contexts, and the traditional level is 0.05. However, it is sometimes helpful to adjust the significance level based on the application. We may select a level that is smaller or larger than 0.05 depending on the consequences of any conclusions reached from the test.

If making a Type 1 Error is dangerous or especially costly, we should choose a small significance level (e.g., 0.01 or 0.001). If we want to be very cautious about rejecting the null hypothesis, we demand very strong evidence favoring the alternative H_A before we would reject H_0.

If a Type 2 Error is relatively more dangerous or much more costly than a Type 1 Error, then we should choose a higher significance level (e.g., 0.10). Here we want to be cautious about failing to reject H_0 when the null is actually false.

Significance levels should reflect consequences of errors.

The significance level selected for a test should reflect the real-world consequences associated with making a Type 1 or Type 2 Error.

14.2 Two-sided hypotheses

In Chapter 11 we explored whether women were discriminated against and whether a simple trick could make students a little thriftier. In these two case studies, we've actually ignored some possibilities:

- What if *men* are actually discriminated against?
- What if the money trick actually makes students *spend more*?

These possibilities weren't considered in our original hypotheses or analyses. The disregard of the extra alternatives may have seemed natural since the data pointed in the directions in which we framed the problems. However, there are two dangers if we ignore possibilities that disagree with our data or that conflict with our world view:

1. Framing an alternative hypothesis simply to match the direction that the data point will generally inflate the Type 1 Error rate. After all the work we've done (and will continue to do) to rigorously control the error rates in hypothesis tests, careless construction of the alternative hypotheses can disrupt that hard work.

2. If we only use alternative hypotheses that agree with our worldview, then we're going to be subjecting ourselves to **confirmation bias**, which means we are looking for data that supports our ideas. That's not very scientific, and we can do better!

The original hypotheses we've seen are called **one-sided hypothesis tests** because they only explored one direction of possibilities. Such hypotheses are appropriate when we are exclusively interested in the single direction, but usually we want to consider all possibilities. To do so, let's learn about **two-sided hypothesis tests** in the context of a new study that examines the impact of using blood thinners on patients who have undergone CPR.

Cardiopulmonary resuscitation (CPR) is a procedure used on individuals suffering a heart attack when other emergency resources are unavailable. This procedure is helpful in providing some blood circulation to keep a person alive, but CPR chest compression can also cause internal injuries. Internal bleeding and other injuries that can result from CPR complicate additional treatment efforts. For instance, blood thinners may be used to help release a clot that is causing the heart attack once a patient arrives in the hospital. However, blood thinners negatively affect internal injuries.

Here we consider an experiment with patients who underwent CPR for a heart attack and were subsequently admitted to a hospital. Each patient was randomly assigned to either receive a blood thinner (treatment group) or not receive a blood thinner (control group). The outcome variable of interest was whether the patient survived for at least 24 hours. (Böttiger et al., 2001)

The cpr data can be found in the **openintro** R package.

EXAMPLE

Form hypotheses for this study in plain and statistical language. Let p_C represent the true survival rate of people who do not receive a blood thinner (corresponding to the control group) and p_T represent the survival rate for people receiving a blood thinner (corresponding to the treatment group).

We want to understand whether blood thinners are helpful or harmful. We'll consider both of these possibilities using a two-sided hypothesis test.

- H_0 : Blood thinners do not have an overall survival effect, i.e., the survival proportions are the same in each group. $p_T - p_C = 0$.

- H_A : Blood thinners have an impact on survival, either positive or negative, but not zero. $p_T - p_C \neq 0$.

Note that if we had done a one-sided hypothesis test, the resulting hypotheses would have been:

- H_0 : Blood thinners do not have a positive overall survival effect, i.e., the survival proportions for the blood thinner group is the same or lower than the control group. $p_T - p_C \leq 0$.

- H_A : Blood thinners have a positive impact on survival. $p_T - p_C > 0$.

There were 50 patients in the experiment who did not receive a blood thinner and 40 patients who did. The study results are shown in Table 14.2.

Table 14.2: Results for the CPR study. Patients in the treatment group were given a blood thinner, and patients in the control group were not.

Group	Died	Survived	Total
Control	39	11	50
Treatment	26	14	40
Total	65	25	90

GUIDED PRACTICE

What is the observed survival rate in the control group? And in the treatment group? Also, provide a point estimate $(\hat{p}_T - \hat{p}_C)$ for the true difference in population survival proportions across the two groups: $p_T - p_C$.[3]

---

[3] Observed control survival rate: $\hat{p}_C = \frac{11}{50} = 0.22$. Treatment survival rate: $\hat{p}_T = \frac{14}{40} = 0.35$. Observed difference: $\hat{p}_T - \hat{p}_C = 0.35 - 0.22 = 0.13$.

14.2. TWO-SIDED HYPOTHESES

According to the point estimate, for patients who have undergone CPR outside of the hospital, an additional 13% of these patients survive when they are treated with blood thinners. However, we wonder if this difference could be easily explainable by chance, if the treatment has no effect on survival.

As we did in past studies, we will simulate what type of differences we might see from chance alone under the null hypothesis. By randomly assigning each of the patient's files to a "simulated treatment" or "simulated control" allocation, we get a new grouping. If we repeat this simulation 1,000 times, we can build a **null distribution** of the differences shown in Figure 14.1.

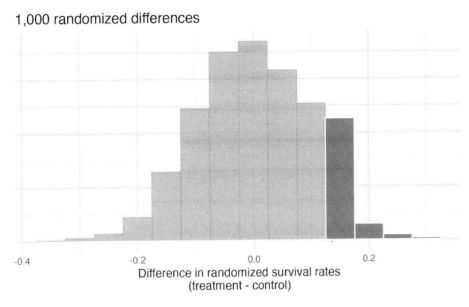

Figure 14.1: Null distribution of the point estimate for the difference in proportions, $\hat{p}_T - \hat{p}_C$. The shaded right tail shows observations that are at least as large as the observed difference, 0.13.

The right tail area is 0.135. (Note: it is only a coincidence that we also have $\hat{p}_T - \hat{p}_C = 0.13$.) However, contrary to how we calculated the p-value in previous studies, the p-value of this test is not actually the tail area we calculated, i.e., it's not 0.135!

The p-value is defined as the probability we observe a result at least as favorable to the alternative hypothesis as the result (i.e., the difference) we observe. In this case, any differences less than or equal to -0.13 would also provide equally strong evidence favoring the alternative hypothesis as a difference of +0.13 did. A difference of -0.13 would correspond to 13% higher survival rate in the control group than the treatment group. In Figure 14.2 we've also shaded these differences in the left tail of the distribution. These two shaded tails provide a visual representation of the p-value for a two-sided test.

For a two-sided test, take the single tail (in this case, 0.131) and double it to get the p-value: 0.262. Since this p-value is larger than 0.05, we do not reject the null hypothesis. That is, we do not find convincing evidence that the blood thinner has any influence on survival of patients who undergo CPR prior to arriving at the hospital.

 Default to a two-sided test.

We want to be rigorous and keep an open mind when we analyze data and evidence. Use a one-sided hypothesis test only if you truly have interest in only one direction.

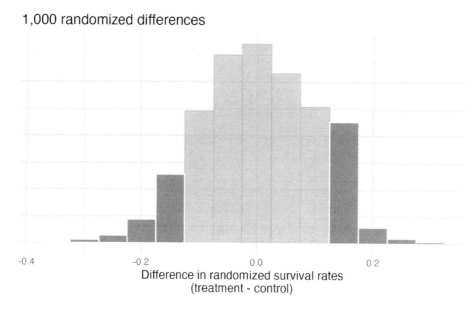

Figure 14.2: Null distribution of the point estimate for the difference in proportions, $\hat{p}_T - \hat{p}_C$. All values that are at least as extreme as +0.13 but in either direction away from 0 are shaded.

 Computing a p-value for a two-sided test.

First compute the p-value for one tail of the distribution, then double that value to get the two-sided p-value. That's it!

 EXAMPLE

Consider the situation of the medical consultant. Now that you know about one-sided and two-sided tests, which type of test do you think is more appropriate?

The setting has been framed in the context of the consultant being helpful (which is what led us to a one-sided test originally), but what if the consultant actually performed *worse* than the average? Would we care? More than ever! Since it turns out that we care about a finding in either direction, we should run a two-sided test. The p-value for the two-sided test is double that of the one-sided test, here the simulated p-value would be 0.2444.

Generally, to find a two-sided p-value we double the single tail area, which remains a reasonable approach even when the distribution is asymmetric. However, the approach can result in p-values larger than 1 when the point estimate is very near the mean in the null distribution; in such cases, we write that the p-value is 1. Also, very large p-values computed in this way (e.g., 0.85), may also be slightly inflated. Typically, we do not worry too much about the precision of very large p-values because they lead to the same analysis conclusion, even if the value is slightly off.

14.3 Controlling the Type 1 Error rate

Now that we understand the difference between one-sided and two-sided tests, we must recognize when to use each type of test. Because of the result of increased error rates, it is never okay to change two-sided tests to one-sided tests after observing the data. We explore the consequences of ignoring this advice in the next example.

EXAMPLE

Using $\alpha = 0.05$, we show that freely switching from two-sided tests to one-sided tests will lead us to make twice as many Type 1 Errors as intended.

Suppose we are interested in finding any difference from 0. We've created a smooth-looking **null distribution** representing differences due to chance in Figure 14.3.

Suppose the sample difference was larger than 0. Then if we can flip to a one-sided test, we would use H_A : difference > 0. Now if we obtain any observation in the upper 5% of the distribution, we would reject H_0 since the p-value would just be a the single tail. Thus, if the null hypothesis is true, we incorrectly reject the null hypothesis about 5% of the time when the sample mean is above the null value, as shown in Figure 14.3.

Suppose the sample difference was smaller than 0. Then if we change to a one-sided test, we would use H_A : difference < 0. If the observed difference falls in the lower 5% of the figure, we would reject H_0. That is, if the null hypothesis is true, then we would observe this situation about 5% of the time.

By examining these two scenarios, we can determine that we will make a Type 1 Error $5\% + 5\% = 10\%$ of the time if we are allowed to swap to the "best" one-sided test for the data. This is twice the error rate we prescribed with our significance level: $\alpha = 0.05$ (!).

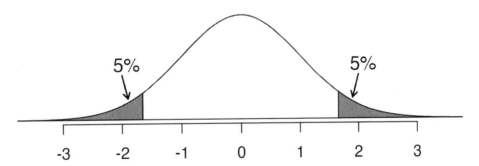

Figure 14.3: The shaded regions represent areas where we would reject H_0 under the bad practices considered in when $\alpha = 0.05$.

Hypothesis tests should be set up *before* seeing the data.

After observing data, it is tempting to turn a two-sided test into a one-sided test. Avoid this temptation. Hypotheses should be set up *before* observing the data.

14.4 Power

Although we won't go into extensive detail here, power is an important topic for follow-up consideration after understanding the basics of hypothesis testing. A good power analysis is a vital preliminary step to any study as it will inform whether the data you collect are sufficient for being able to conclude your research broadly.

Often times in experiment planning, there are two competing considerations:

- We want to collect enough data that we can detect important effects.
- Collecting data can be expensive, and, in experiments involving people, there may be some risk to patients.

When planning a study, we want to know how likely we are to detect an effect we care about. In other words, if there is a real effect, and that effect is large enough that it has practical value, then what is the probability that we detect that effect? This probability is called the **power**, and we can compute it for different sample sizes or different effect sizes.

Power.

The power of the test is the probability of rejecting the null claim when the alternative claim is true.

How easy it is to detect the effect depends on both how big the effect is (e.g., how good the medical treatment is) as well as the sample size.

We think of power as the probability that you will become rich and famous from your science. In order for your science to make a splash, you need to have good ideas! That is, you won't become famous if you happen to find a single Type 1 error which rejects the null hypothesis. Instead, you'll become famous if your science is very good and important (that is, if the alternative hypothesis is true). The better your science is (i.e., the better the medical treatment), the larger the *effect size* and the easier it will be for you to convince people of your work.

Not only does your science need to be solid, but you also need to have evidence (i.e., data) that shows the effect. A few observations (e.g., $n = 2$) is unlikely to be convincing because of well known ideas of natural variability. Indeed, the larger the dataset which provides evidence for your scientific claim, the more likely you are to convince the community that your idea is correct.

14.5 Chapter review

14.5.1 Summary

Although hypothesis testing provides a strong framework for making decisions based on data, as the analyst, you need to understand how and when the process can go wrong. That is, always keep in mind that the conclusion to a hypothesis test may not be right! Sometimes when the null hypothesis is true, we will accidentally reject it and commit a type 1 error; sometimes when the alternative hypothesis is true, we will fail to reject the null hypothesis and commit a type 2 error. The power of the test quantifies how likely it is to obtain data which will reject the null hypothesis when indeed the alternative is true; the power of the test is increased when larger sample sizes are taken.

14.5.2 Terms

We introduced the following terms in the chapter. If you're not sure what some of these terms mean, we recommend you go back in the text and review their definitions. We are purposefully presenting them in alphabetical order, instead of in order of appearance, so they will be a little more challenging to locate. However you should be able to easily spot them as **bolded text**.

confirmation bias	power	type 1 error
null distribution	significance level	type 2 error
one-sided hypothesis test	two-sided hypothesis test	

14.6 Exercises

Answers to odd numbered exercises can be found in Appendix A.14.

14.1. **Testing for Fibromyalgia.** A patient named Diana was diagnosed with Fibromyalgia, a long-term syndrome of body pain, and was prescribed anti-depressants. Being the skeptic that she is, Diana didn't initially believe that anti-depressants would help her symptoms. However after a couple months of being on the medication she decides that the anti-depressants are working, because she feels like her symptoms are in fact getting better.

  a. Write the hypotheses in words for Diana's skeptical position when she started taking the anti-depressants.

  b. What is a Type 1 Error in this context?

  c. What is a Type 2 Error in this context?

14.2. **Which is higher?** In each part below, there is a value of interest and two scenarios: (i) and (ii). For each part, report if the value of interest is larger under scenario (i), scenario (ii), or whether the value is equal under the scenarios.

  a. The standard error of $\hat{p}$ when (i) $n = 125$ or (ii) $n = 500$.

  b. The margin of error of a confidence interval when the confidence level is (i) 90% or (ii) 80%.

  c. The p-value for a Z-statistic of 2.5 calculated based on a (i) sample with $n = 500$ or based on a (ii) sample with $n = 1000$.

  d. The probability of making a Type 2 Error when the alternative hypothesis is true and the significance level is (i) 0.05 or (ii) 0.10.

14.3. **Testing for food safety.** A food safety inspector is called upon to investigate a restaurant with a few customer reports of poor sanitation practices. The food safety inspector uses a hypothesis testing framework to evaluate whether regulations are not being met. If he decides the restaurant is in gross violation, its license to serve food will be revoked.

  a. Write the hypotheses in words.

  b. What is a Type 1 Error in this context?

  c. What is a Type 2 Error in this context?

  d. Which error is more problematic for the restaurant owner? Why?

  e. Which error is more problematic for the diners? Why?

  f. As a diner, would you prefer that the food safety inspector requires strong evidence or very strong evidence of health concerns before revoking a restaurant's license? Explain your reasoning.

14.4. **True or false.** Determine if the following statements are true or false, and explain your reasoning. If false, state how it could be corrected.

  a. If a given value (for example, the null hypothesized value of a parameter) is within a 95% confidence interval, it will also be within a 99% confidence interval.

  b. Decreasing the significance level (α) will increase the probability of making a Type 1 Error.

  c. Suppose the null hypothesis is $p = 0.5$ and we fail to reject H_0. Under this scenario, the true population proportion is 0.5.

  d. With large sample sizes, even small differences between the null value and the observed point estimate, a difference often called the effect size, will be identified as statistically significant.

14.5. **Online communication.** A study suggests that 60% of college student spend 10 or more hours per week communicating with others online. You believe that this is incorrect and decide to collect your own sample for a hypothesis test. You randomly sample 160 students from your dorm and find that 70% spent 10 or more hours a week communicating with others online. A friend of yours, who offers to help you with the hypothesis test, comes up with the following set of hypotheses. Indicate any errors you see.

$$H_0 : \hat{p} < 0.6 \qquad H_A : \hat{p} > 0.7$$

14.6. **Same observation, different sample size.** Suppose you conduct a hypothesis test based on a sample where the sample size is $n = 50$, and arrive at a p-value of 0.08. You then refer back to your notes and discover that you made a careless mistake, the sample size should have been $n = 500$. Will your p-value increase, decrease, or stay the same? Explain.

Chapter 15

Applications: Foundations

15.1 Recap: Foundations

In the Foundations of inference chapters, we have provided three different methods for statistical inference. We will continue to build on all three of the methods throughout the text, and by the end, you should have an understanding of the similarities and differences between them. Meanwhile, it is important to note that the methods are designed to mimic variability with data, and we know that variability can come from different sources (e.g., random sampling vs. random allocation, see Figure 2.8). In Table 15.1, we have summarized some of the ways the inferential procedures feature specific sources of variability. We hope that you refer back to the table often as you dive more deeply into inferential ideas in future chapters.

You might have noticed that the word *distribution* is used throughout this part (and will continue to be used in future chapters). A distribution always describes variability, but sometimes it is worth reflecting on *what* is varying. Typically the distribution either describes how the observations vary or how a statistic varies. But even when describing how a statistic varies, there is a further consideration with respect to the study design, e.g., does the statistic vary from random sample to random sample or does it vary from random allocation to random allocation? The methods presented in this text (and used in science generally) are typically used interchangeably across ideas of random samples or random allocations of the treatment. Often, the two different analysis methods will give equivalent conclusions. The most important thing to consider is how to contextualize the conclusion in terms of the problem. See Figure 2.8 to confirm that your conclusions are appropriate.

Below, we synthesize the different types of distributions discussed throughout the text. Reading through the different definitions and solidifying your understanding will help as you come across these distributions in future chapters and you can always return back here to refresh your understanding of the differences between the various distributions.

15.1. RECAP: FOUNDATIONS

Table 15.1: Summary and comparison of randomization, bootstrapping, and mathematical models as inferential statistical methods.

Question	Answer		
	Randomization	Bootstrapping	Mathematical models
What does it do?	Shuffles the explanatory variable to mimic the natural variability found in a randomized experiment	Resamples (with replacement) from the observed data to mimic the sampling variability found by collecting data from a population	Uses theory (primarily the Central Limit Theorem) to describe the hypothetical variability resulting from either repeated randomized experiments or random samples
What is the random process described?	Randomized experiment	Random sampling from a population	Randomized experiment or random sampling
What other random processes can be approximated?	Can also be used to describe random sampling in an observational model	Can also be used to describe random allocation in an experiment	Randomized experiment or random sampling
What is it best for?	Hypothesis testing (can also be used for confidence intervals, but not covered in this text).	Confidence intervals (can also be used for bootstrap hypothesis testing for one proportion as well).	Quick analyses through, for example, calculating a Z score.
What physical object represents the simulation process?	Shuffling cards	Pulling marbles from a bag with replacement	Not applicable

Distributions.

- A **data distribution** describes the shape, center, and variability of the **observed data**.

  This can also be referred to as the **sample distribution** but we'll avoid that phrase as it sounds too much like sampling distribution, which is different.

- A **population distribution** describes the shape, center, and variability of the entire **population of data**.

  Except in very rare circumstances of very small, very well-defined populations, this is never observed.

- A **sampling distribution** describes the shape, center, and variability of all possible values of a **sample statistic** from samples of a given sample size from a given population.

  Since the population is never observed, it's never possible to observe the true sampling distribution either. However, when certain conditions hold, the Central Limit Theorem tells us what the sampling distribution is.

- A **randomization distribution** describes the shape, center, and variability of all possible values of a **sample statistic** from random allocations of the treatment variable.

  We computationally generate the randomization distribution, though usually, it's not feasible to generate the full distribution of all possible values of the sample statistic, so we instead generate a large number of them. Almost always, by randomly allocating the treatment variable, the randomization distribution describes the null hypothesis, i.e., it is centered at the null hypothesized value of the parameter.

- A **bootstrap distribution** describes the shape, center, and variability of all possible values of a **sample statistic** from resamples of the observed data.

  We computationally generate the bootstrap distribution, though usually, it's not feasible to generate all possible resamples of the observed data, so we instead generate a large number of them. Since bootstrap distributions are generated by randomly resampling from the observed data, they are centered at the sample statistic. Bootstrap distributions are most often used for estimation, i.e., we base confidence intervals off of them.

15.2 Case study: Malaria vaccine

In this case study, we consider a new malaria vaccine called PfSPZ. In the malaria study, volunteer patients were randomized into one of two experiment groups: 14 patients received an experimental vaccine and 6 patients received a placebo vaccine. Nineteen weeks later, all 20 patients were exposed to a drug-sensitive malaria virus strain; the motivation of using a drug-sensitive strain of virus here is for ethical considerations, allowing any infections to be treated effectively.

The malaria data can be found in the **openintro** R package.

The results are summarized in Table 15.2, where 9 of the 14 treatment patients remained free of signs of infection while all of the 6 patients in the control group showed some baseline signs of infection.

Table 15.2: Summary results for the malaria vaccine experiment.

treatment	infection	no infection	Total
placebo	6	0	6
vaccine	5	9	14
Total	11	9	20

GUIDED PRACTICE

Is this an observational study or an experiment? What implications does the study type have on what can be inferred from the results?[1]

15.2.1 Variability within data

In this study, a smaller proportion of patients who received the vaccine showed signs of an infection (35.7% versus 100%). However, the sample is very small, and it is unclear whether the difference provides *convincing evidence* that the vaccine is effective.

EXAMPLE

Statisticians and data scientists are sometimes called upon to evaluate the strength of evidence. When looking at the rates of infection for patients in the two groups in this study, what comes to mind as we try to determine whether the data show convincing evidence of a real difference?

The observed infection rates (35.7% for the treatment group versus 100% for the control group) suggest the vaccine may be effective. However, we cannot be sure if the observed difference represents the vaccine's efficacy or if there is no treatment effect and the observed difference is just from random chance. Generally there is a little bit of fluctuation in sample data, and we wouldn't expect the sample proportions to be *exactly* equal, even if the truth was that the infection rates were independent of getting the vaccine. Additionally, with such small samples, perhaps it's common to observe such large differences when we randomly split a group due to chance alone!

---

[1] The study is an experiment, as patients were randomly assigned an experiment group. Since this is an experiment, the results can be used to evaluate a causal relationship between the malaria vaccine and whether patients showed signs of an infection.

This example is a reminder that the observed outcomes in the data sample may not perfectly reflect the true relationships between variables since there is **random noise**. While the observed difference in rates of infection is large, the sample size for the study is small, making it unclear if this observed difference represents efficacy of the vaccine or whether it is simply due to chance. We label these two competing claims, H_0 and H_A:

- H_0: **Independence model.** The variables and are independent. They have no relationship, and the observed difference between the proportion of patients who developed an infection in the two groups, 64.3%, was due to chance.

- H_A: **Alternative model.** The variables are *not* independent. The difference in infection rates of 64.3% was not due to chance. Here (because an experiment was done), if the difference in infection rate is not due to chance, it was the vaccine that affected the rate of infection.

What would it mean if the independence model, which says the vaccine had no influence on the rate of infection, is true? It would mean 11 patients were going to develop an infection *no matter which group they were randomized into*, and 9 patients would not develop an infection *no matter which group they were randomized into*. That is, if the vaccine did not affect the rate of infection, the difference in the infection rates was due to chance alone in how the patients were randomized.

Now consider the alternative model: infection rates were influenced by whether a patient received the vaccine or not. If this was true, and especially if this influence was substantial, we would expect to see some difference in the infection rates of patients in the groups.

We choose between these two competing claims by assessing if the data conflict so much with H_0 that the independence model cannot be deemed reasonable. If this is the case, and the data support H_A, then we will reject the notion of independence and conclude the vaccine was effective.

15.2.2 Simulating the study

We're going to implement **simulation** under the setting where we will pretend we know that the malaria vaccine being tested does *not* work. Ultimately, we want to understand if the large difference we observed in the data is common in these simulations that represent independence. If it is common, then maybe the difference we observed was purely due to chance. If it is very uncommon, then the possibility that the vaccine was helpful seems more plausible.

Table 15.2 shows that 11 patients developed infections and 9 did not. For our simulation, we will suppose the infections were independent of the vaccine and we were able to *rewind* back to when the researchers randomized the patients in the study. If we happened to randomize the patients differently, we may get a different result in this hypothetical world where the vaccine doesn't influence the infection. Let's complete another **randomization** using a simulation.

In this **simulation**, we take 20 notecards to represent the 20 patients, where we write down "infection" on 11 cards and "no infection" on 9 cards. In this hypothetical world, we believe each patient that got an infection was going to get it regardless of which group they were in, so let's see what happens if we randomly assign the patients to the treatment and control groups again. We thoroughly shuffle the notecards and deal 14 into a pile and 6 into a pile. Finally, we tabulate the results, which are shown in Table 15.3.

Table 15.3: Simulation results, where any difference in infection ratio is purely due to chance.

treatment	placebo	vaccine	Total
infection	4	7	11
no infection	2	7	9
Total	6	14	20

15.2. CASE STUDY: MALARIA VACCINE

GUIDED PRACTICE

How does this compare to the observed 64.3% difference in the actual data?[2]

15.2.3 Independence between treatment and outcome

We computed one possible difference under the independence model in the previous Guided Practice, which represents one difference due to chance, assuming there is no vaccine effect. While in this first simulation, we physically dealt out notecards to represent the patients, it is more efficient to perform the simulation using a computer.

Repeating the simulation on a computer, we get another difference due to chance:

$$\frac{2}{6} - \frac{9}{14} = -0.310$$

And another:

$$\frac{3}{6} - \frac{8}{14} = -0.071$$

And so on until we repeat the simulation enough times to create a *distribution of differences that could have occurred if the null hypothesis was true*.

Figure 15.1 shows a stacked plot of the differences found from 100 simulations, where each dot represents a simulated difference between the infection rates (control rate minus treatment rate).

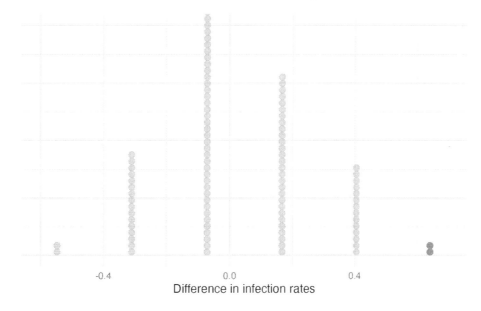

Figure 15.1: A stacked dot plot of differences from 100 simulations produced under the independence mode, H_0, where in these simulations infections are unaffected by the vaccine. Two of the 100 simulations had a difference of at least 64.3%, the difference observed in the study.

Note that the distribution of these simulated differences is centered around 0. We simulated these differences assuming that the independence model was true, and under this condition, we expect the difference to be near zero with some random fluctuation, where *near* is pretty generous in this case since the sample sizes are so small in this study.

---

[2]$4/6 - 7/14 = 0.167$ or about 16.7% in favor of the vaccine. This difference due to chance is much smaller than the difference observed in the actual groups.

EXAMPLE

How often would you observe a difference of at least 64.3% (0.643) according to Figure 15.1? Often, sometimes, rarely, or never?

It appears that a difference of at least 64.3% due to chance alone would only happen about 2% of the time according to Figure 15.1. Such a low probability indicates a rare event.

The difference of 64.3% being a rare event suggests two possible interpretations of the results of the study:

- H_0: **Independence model.** The vaccine has no effect on infection rate, and we just happened to observe a difference that would only occur on a rare occasion.

- H_A: **Alternative model.** The vaccine has an effect on infection rate, and the difference we observed was actually due to the vaccine being effective at combating malaria, which explains the large difference of 64.3%.

Based on the simulations, we have two options. (1) We conclude that the study results do not provide strong evidence against the independence model. That is, we do not have sufficiently strong evidence to conclude the vaccine had an effect in this clinical setting. (2) We conclude the evidence is sufficiently strong to reject H_0 and assert that the vaccine was useful. When we conduct formal studies, usually we reject the notion that we just happened to observe a rare event. So in the vaccine case, we reject the independence model in favor of the alternative. That is, we are concluding the data provide strong evidence that the vaccine provides some protection against malaria in this clinical setting.

One field of statistics, statistical inference, is built on evaluating whether such differences are due to chance. In statistical inference, data scientists evaluate which model is most reasonable given the data. Errors do occur, just like rare events, and we might choose the wrong model. While we do not always choose correctly, statistical inference gives us tools to control and evaluate how often decision errors occur.

15.3 Interactive R tutorials

Navigate the concepts you've learned in this chapter in R using the following self-paced tutorials. All you need is your browser to get started!

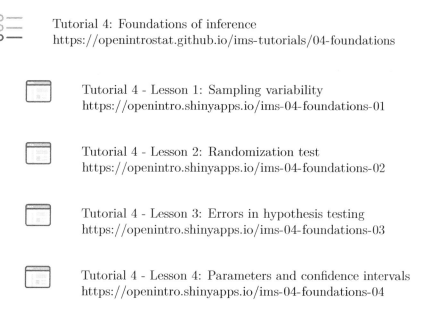

Tutorial 4: Foundations of inference
https://openintrostat.github.io/ims-tutorials/04-foundations

Tutorial 4 - Lesson 1: Sampling variability
https://openintro.shinyapps.io/ims-04-foundations-01

Tutorial 4 - Lesson 2: Randomization test
https://openintro.shinyapps.io/ims-04-foundations-02

Tutorial 4 - Lesson 3: Errors in hypothesis testing
https://openintro.shinyapps.io/ims-04-foundations-03

Tutorial 4 - Lesson 4: Parameters and confidence intervals
https://openintro.shinyapps.io/ims-04-foundations-04

You can also access the full list of tutorials supporting this book at
https://openintrostat.github.io/ims-tutorials.

15.4 R labs

Further apply the concepts you've learned in this part in R with computational labs that walk you through a data analysis case study.

Sampling distributions - Does science benefit you?
https://www.openintro.org/go?id=ims-r-lab-foundations-1

Confidence intervals - Climate change
https://www.openintro.org/go?id=ims-r-lab-foundations-2

You can also access the full list of labs supporting this book at
https://www.openintro.org/go?id=ims-r-labs.

PART V

STATISTICAL INFERENCE

Chapter 16

Inference for a single proportion

Focusing now on statistical inference for categorical data, we will revisit many of the foundational aspects of hypothesis testing from Chapter 11.

The three data structures we detail are one binary variable, summarized using a single proportion; two binary variables, summarized using a difference of two proportions; and two categorical variables, summarized using a two-way table. When appropriate, each of the data structures will be analyzed using the three methods from Chapters 11, 12, and 13: randomization test, bootstrapping, and mathematical models, respectively.

As we build on the inferential ideas, we will visit new foundational concepts in statistical inference. For example, we will cover the conditions for when a normal model is appropriate; the two different error rates in hypothesis testing; and choosing the confidence level for a confidence interval.

We encountered inference methods for a single proportion in Chapter 12, exploring point estimates, confidence intervals, and hypothesis tests. In this section, we'll do a review of these topics and also how to choose an appropriate sample size when collecting data for single proportion contexts.

Note that there is only one variable being measured in a study which focuses on one proportion. For each observational unit, the single variable is measured as either a success or failure (e.g., "surgical complication" vs. "no surgical complication"). Because the nature of the research question at hand focuses on only a single variable, there is not a way to randomize the variable across a different (explanatory) variable. For this reason, we will not use randomization as an analysis tool when focusing on a single proportion. Instead, we will apply bootstrapping techniques to test a given hypothesis, and we will also revisit the associated mathematical models.

16.1 Bootstrap test for a proportion

The bootstrap simulation concept when H_0 is true is similar to the ideas used in the case studies presented in Chapter 12 where we bootstrapped without an assumption about H_0. Because we will be testing a hypothesized value of p (referred to as p_0), the bootstrap simulation for hypothesis testing has a fantastic advantage that it can be used for any sample size (a huge benefit for small samples, a nice alternative for large samples).

We expand on the medical consultant example, see Section 12.1, where instead of finding an interval estimate for the true complication rate, we work to test a specific research claim.

16.1.1 Observed data

Recall the set-up for the example:

People providing an organ for donation sometimes seek the help of a special "medical consultant". These consultants assist the patient in all aspects of the surgery, with the goal of reducing the possibility of complications during the medical procedure and recovery. Patients might choose a consultant based in part on the historical complication rate of the consultant's clients. One consultant tried to attract patients by noting the average complication rate for liver donor surgeries in the US is about 10%, but her clients have only had 3 complications in the 62 liver donor surgeries she has facilitated. She claims this is strong evidence that her work meaningfully contributes to reducing complications (and therefore she should be hired!).

EXAMPLE

Using the data, is it possible to assess the consultant's claim that her complication rate is less than 10%?

No. The claim is that there is a causal connection, but the data are observational. Patients who hire this medical consultant may have lower complication rates for other reasons.

While it is not possible to assess this causal claim, it is still possible to test for an association using these data. For this question we ask, could the low complication rate of $\hat{p} = 0.048$ have simply occurred by chance, if her complication rate does not differ from the US standard rate?

GUIDED PRACTICE

Write out hypotheses in both plain and statistical language to test for the association between the consultant's work and the true complication rate, p, for the consultant's clients.[1]

Because, as it turns out, the conditions of working with the normal distribution are not met (see Section 16.2), the uncertainty associated with the sample proportion should not be modeled using the normal distribution, as doing so would underestimate the uncertainty associated with the sample statistic. However, we would still like to assess the hypotheses from the previous Guided Practice in absence of the normal framework. To do so, we need to evaluate the possibility of a sample value ($\hat{p}$) as far below the null value, $p_0 = 0.10$ as what was observed. The deviation of the sample value from the hypothesized parameter is usually quantified with a p-value.

The p-value is computed based on the null distribution, which is the distribution of the test statistic if the null hypothesis is true. Supposing the null hypothesis is true, we can compute the p-value by identifying the probability of observing a test statistic that favors the alternative hypothesis at least as strongly as the observed test statistic. Here we will use a bootstrap simulation to calculate the p-value.

---

[1] H_0: There is no association between the consultant's contributions and the clients' complication rate. In statistical language, $p = 0.10$. H_A: Patients who work with the consultant tend to have a complication rate lower than 10%, i.e., $p < 0.10$.

16.1.2 Variability of the statistic

We want to identify the sampling distribution of the test statistic ($\hat{p}$) if the null hypothesis was true. In other words, we want to see the variability we can expect from sample proportions if the null hypothesis was true. Then we plan to use this information to decide whether there is enough evidence to reject the null hypothesis.

Under the null hypothesis, 10% of liver donors have complications during or after surgery. Suppose this rate was really no different for the consultant's clients (for *all* the consultant's clients, not just the 62 previously measured). If this was the case, we could *simulate* 62 clients to get a sample proportion for the complication rate from the null distribution. Simulating observations using a hypothesized null parameter value is often called a **parametric bootstrap simulation**.

Similar to the process described in Chapter 12, each client can be simulated using a bag of marbles with 10% red marbles and 90% white marbles. Sampling a marble from the bag (with 10% red marbles) is one way of simulating whether a patient has a complication *if the true complication rate is 10%*. If we select 62 marbles and then compute the proportion of patients with complications in the simulation, $\hat{p}_{sim1}$, then the resulting sample proportion is a sample from the null distribution.

There were 5 simulated cases with a complication and 57 simulated cases without a complication, i.e., $\hat{p}_{sim1} = 5/62 = 0.081$.

EXAMPLE

Is this one simulation enough to determine whether or not we should reject the null hypothesis?

No. To assess the hypotheses, we need to see a distribution of many values of $\hat{p}_{sim}$, not just a *single* draw from this sampling distribution.

16.1.3 Observed statistic vs. null statistics

One simulation isn't enough to get a sense of the null distribution; many simulation studies are needed. Roughly 10,000 seems sufficient. However, paying someone to simulate 10,000 studies by hand is a waste of time and money. Instead, simulations are typically programmed into a computer, which is much more efficient.

Figure 16.1 shows the results of 10,000 simulated studies. The proportions that are equal to or less than $\hat{p} = 0.048$ are shaded. The shaded areas represent sample proportions under the null distribution that provide at least as much evidence as $\hat{p}$ favoring the alternative hypothesis. There were 420 simulated sample proportions with $\hat{p}_{sim} \leq 0.048$. We use these to construct the null distribution's left-tail area and find the p-value:

$$\text{left tail area} = \frac{\text{Number of observed simulations with } \hat{p}_{sim} \leq 0.048}{10000}$$

Of the 10,000 simulated $\hat{p}_{sim}$, 420 were equal to or smaller than $\hat{p}$. Since the hypothesis test is one-sided, the estimated p-value is equal to this tail area: 0.042.

GUIDED PRACTICE

Because the estimated p-value is 0.042, which is smaller than the significance level 0.05, we reject the null hypothesis. Explain what this means in plain language in the context of the problem.[2]

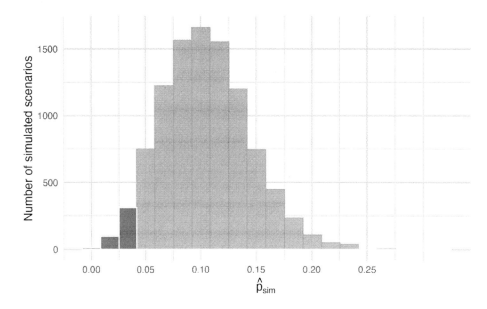

Figure 16.1: The null distribution for $\hat{p}$, created from 10,000 simulated studies. The left tail, representing the p-value for the hypothesis test, contains 4.2% of the simulations.

 GUIDED PRACTICE

Does the conclusion in the previous Guided Practice imply the consultant is good at their job? Explain.[3]

 Null distribution of $\hat{p}$ with bootstrap simulation.

Regardless of the statistical method chosen, the p-value is always derived by analyzing the null distribution of the test statistic. The normal model poorly approximates the null distribution for $\hat{p}$ when the success-failure condition is not satisfied. As a substitute, we can generate the null distribution using simulated sample proportions and use this distribution to compute the tail area, i.e., the p-value.

In the previous Guided Practice, the p-value is *estimated*. It is not exact because the simulated null distribution itself is only a close approximation of the sampling distribution of the sample statistic. An exact p-value can be generated using the binomial distribution, but that method will not be covered in this text.

---

[2]There is sufficiently strong evidence to reject the null hypothesis in favor of the alternative hypothesis. We would conclude that there is evidence that the consultant's surgery complication rate is lower than the US standard rate of 10%.

[3]No. Not necessarily. The evidence supports the alternative hypothesis that the consultant's complication rate is lower, but it's not a measurement of their performance.

16.2 Mathematical model for a proportion

16.2.1 Conditions

In Section 13.2, we introduced the normal distribution and showed how it can be used as a mathematical model to describe the variability of a statistic. There are conditions under which a sample proportion $\hat{p}$ is well modeled using a normal distribution. When the sample observations are independent and the sample size is sufficiently large, the normal model will describe the sampling distribution of the sample proportion quite well; when the observations violate the conditions, the normal model can be inaccurate. Particularly, it can underestimate the variability of the sample proportion.

Sampling distribution of $\hat{p}$.

The sampling distribution for $\hat{p}$ based on a sample of size n from a population with a true proportion p is nearly normal when:

1. The sample's observations are independent, e.g., are from a simple random sample.
2. We expected to see at least 10 successes and 10 failures in the sample, i.e., $np \geq 10$ and $n(1-p) \geq 10$. This is called the **success-failure condition**.

When these conditions are met, then the sampling distribution of $\hat{p}$ is nearly normal with mean p and standard error of $\hat{p}$ as $SE = \sqrt{\frac{\hat{p}(1-\hat{p})}{n}}$.

Recall that the margin of error is defined by the standard error. The margin of error for $\hat{p}$ can be directly obtained from $SE(\hat{p})$.

Margin of error for $\hat{p}$.

The margin of error is $z^{\star} \times \sqrt{\frac{\hat{p}(1-\hat{p})}{n}}$ where $z^{\star}$ is calculated from a specified percentile on the normal distribution.

Typically we don't know the true proportion p, so we substitute some value to check conditions and estimate the standard error. For confidence intervals, the sample proportion $\hat{p}$ is used to check the success-failure condition and compute the standard error. For hypothesis tests, typically the null value – that is, the proportion claimed in the null hypothesis – is used in place of p.

The independence condition is a more nuanced requirement. When it isn't met, it is important to understand how and why it is violated. For example, there exist no statistical methods available to truly correct the inherent biases of data from a convenience sample. On the other hand, if we took a cluster sample (see Section 2.1.5), the observations wouldn't be independent, but suitable statistical methods are available for analyzing the data (but they are beyond the scope of even most second or third courses in statistics).

EXAMPLE

In the examples based on large sample theory, we modeled $\hat{p}$ using the normal distribution. Why is this not appropriate for the case study on the medical consultant?

The independence assumption may be reasonable if each of the surgeries is from a different surgical team. However, the success-failure condition is not satisfied. Under the null hypothesis, we would anticipate seeing $62 \times 0.10 = 6.2$ complications, not the 10 required for the normal approximation.

While this book is scoped to well-constrained statistical problems, do remember that this is just the first book in what is a large library of statistical methods that are suitable for a very wide range of data and contexts.

16.2.2 Confidence interval for a proportion

A confidence interval provides a range of plausible values for the parameter p, and when $\hat{p}$ can be modeled using a normal distribution, the confidence interval for p takes the form $\hat{p} \pm z^\star \times SE$. We have seen $\hat{p}$ to be the sample proportion. The value $z^\star$ determines the confidence level (previously set to be 1.96) and will be discussed in detail in the examples following. The value of the standard error, SE, depends heavily on the sample size.

Standard error of one proportion, $\hat{p}$.

When the conditions are met so that the distribution of $\hat{p}$ is nearly normal, the **variability** of a single proportion, $\hat{p}$ is well described by:

$$SE(\hat{p}) = \sqrt{\frac{p(1-p)}{n}}$$

Note that we almost never know the true value of p. A more helpful formula to use is:

$$SE(\hat{p}) \approx \sqrt{\frac{(\text{best guess of } p)(1 - \text{best guess of } p)}{n}}$$

For hypothesis testing, we often use p_0 as the best guess of p. For confidence intervals, we typically use $\hat{p}$ as the best guess of p.

GUIDED PRACTICE

Consider taking many polls of registered voters (i.e., random samples) of size 300 asking them if they support legalized marijuana. It is suspected that about 2/3 of all voters support legalized marijuana. To understand how the sample proportion ($\hat{p}$) would vary across the samples, calculate the standard error of $\hat{p}$.[4]

---

[4]Because the p is unknown but expected to be around 2/3, we will use 2/3 in place of p in the formula for the standard error.
$SE = \sqrt{\frac{p(1-p)}{n}} \approx \sqrt{\frac{2/3(1-2/3)}{300}} = 0.027.$

16.2.3 Variability of the sample proportion

EXAMPLE

A simple random sample of 826 payday loan borrowers was surveyed to better understand their interests around regulation and costs. 70% of the responses supported new regulations on payday lenders.

1. Is it reasonable to model the variability of $\hat{p}$ from sample to sample using a normal distribution?

2. Estimate the standard error of $\hat{p}$.

3. Construct a 95% confidence interval for p, the proportion of payday borrowers who support increased regulation for payday lenders.

---

1. The data are a random sample, so it is reasonable to assume that the observations are independent and representative of the population of interest.

We also must check the success-failure condition, which we do using $\hat{p}$ in place of p when computing a confidence interval:

$$\text{Support: } np \approx 826 \times 0.70 = 578$$
$$\text{Not: } n(1-p) \approx 826 \times (1-0.70) = 248$$

Since both values are at least 10, we can use the normal distribution to model $\hat{p}$.

2. Because p is unknown and the standard error is for a confidence interval, use $\hat{p}$ in place of p in the formula.

$$SE = \sqrt{\frac{p(1-p)}{n}} \approx \sqrt{\frac{0.70(1-0.70)}{826}} = 0.016.$$

3. Using the point estimate 0.70, $z^\star = 1.96$ for a 95% confidence interval, and the standard error $SE = 0.016$ from the previous Guided Practice, the confidence interval is

$$\text{point estimate } \pm \ z^\star \times \ SE$$
$$0.70 \ \pm \ 1.96 \ \times \ 0.016$$
$$(0.669 \ , \ 0.731)$$

We are 95% confident that the true proportion of payday borrowers who supported regulation at the time of the poll was between 0.669 and 0.731.

Constructing a confidence interval for a single proportion.

There are three steps to constructing a confidence interval for p.

1. Check if it seems reasonable to assume the observations are independent and check the success-failure condition using $\hat{p}$. If the conditions are met, the sampling distribution of $\hat{p}$ may be well-approximated by the normal model.
2. Construct the standard error using $\hat{p}$ in place of p in the standard error formula.
3. Apply the general confidence interval formula.

For additional one-proportion confidence interval examples, see Section 12.3.

16.2.4 Changing the confidence level

Suppose we want to consider confidence intervals where the confidence level is somewhat higher than 95%: perhaps we would like a confidence level of 99%. Think back to the analogy about trying to catch a fish: if we want to be more sure that we will catch the fish, we should use a wider net. To create a 99% confidence level, we must also widen our 95% interval. On the other hand, if we want an interval with lower confidence, such as 90%, we could make our original 95% interval slightly slimmer.

The 95% confidence interval structure provides guidance in how to make intervals with new confidence levels. Below is a general 95% confidence interval for a point estimate that comes from a nearly normal distribution:

$$\text{point estimate} \pm 1.96 \times SE$$

There are three components to this interval: the point estimate, "1.96", and the standard error. The choice of $1.96 \times SE$ was based on capturing 95% of the data since the estimate is within 1.96 standard errors of the true value about 95% of the time. The choice of 1.96 corresponds to a 95% confidence level.

GUIDED PRACTICE

If X is a normally distributed random variable, how often will X be within 2.58 standard deviations of the mean?[5]

To create a 99% confidence interval, change 1.96 in the 95% confidence interval formula to be 2.58. The previous Guided Practice highlights that 99% of the time a normal random variable will be within 2.58 standard deviations of its mean. This approach – using the Z scores in the normal model to compute confidence levels – is appropriate when the point estimate is associated with a normal distribution and we can properly compute the standard error. Thus, the formula for a 99% confidence interval is:

$$\text{point estimate} \pm 2.58 \times SE$$

The normal approximation is crucial to the precision of the $z^\star$ confidence intervals (in contrast to the bootstrap percentile confidence intervals). When the normal model is not a good fit, we will use alternative distributions that better characterize the sampling distribution or we will use bootstrapping procedures.

---

[5]This is equivalent to asking how often the Z score will be larger than -2.58 but less than 2.58. (For a picture, see Figure 16.2.) To determine this probability, look up -2.58 and 2.58 in the normal probability table (0.0049 and 0.9951). Thus, there is a $0.9951 - 0.0049 \approx 0.99$ probability that the unobserved random variable X will be within 2.58 standard deviations of the mean.

16.2. MATHEMATICAL MODEL FOR A PROPORTION

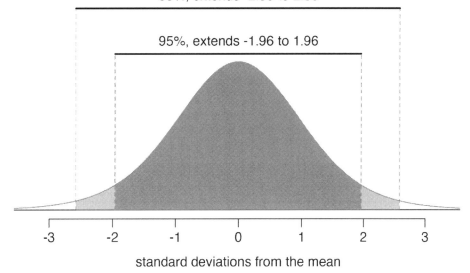

Figure 16.2: The area between $-z^\star$ and $z^\star$ increases as $|z^\star|$ becomes larger. If the confidence level is 99%, we choose $z^\star$ such that 99% of the normal curve is between $-z^\star$ and $z^\star$, which corresponds to 0.5% in the lower tail and 0.5% in the upper tail: $z^\star = 2.58$.

 GUIDED PRACTICE

Create a 99% confidence interval for the impact of the stent on the risk of stroke using the data from Section 1.1. The point estimate is 0.090, and the standard error is $SE = 0.028$. It has been verified for you that the point estimate can reasonably be modeled by a normal distribution.[6]

 Mathematical model confidence interval for any confidence level.

If the point estimate follows the normal model with standard error SE, then a confidence interval for the population parameter is

$$\text{point estimate} \pm z^\star \times SE$$

where $z^\star$ corresponds to the confidence level selected.

Figure 16.2 provides a picture of how to identify $z^\star$ based on a confidence level. We select $z^\star$ so that the area between $-z^\star$ and $z^\star$ in the normal model corresponds to the confidence level.

 GUIDED PRACTICE

Previously, we found that implanting a stent in the brain of a patient at risk for a stroke *increased* the risk of a stroke. The study estimated a 9% increase in the number of patients who had a stroke, and the standard error of this estimate was about $SE = 2.8$ Compute a 90% confidence interval for the effect.[7]

---

[6]Since the necessary conditions for applying the normal model have already been checked for us, we can go straight to the construction of the confidence interval: point estimate $\pm 2.58 \times SE$ Which gives an interval of $(0.018, 0.162)$.$ We are 99% confident that implanting a stent in the brain of a patient who is at risk of stroke increases the risk of stroke within 30 days by a rate of 0.018 to 0.162 (assuming the patients are representative of the population).

16.2.5 Hypothesis test for a proportion

The test statistic for assessing a single proportion is a Z.

The **Z score** is a ratio of how the sample proportion differs from the hypothesized proportion as compared to the expected variability of the $\hat{p}$ values.

$$Z = \frac{\hat{p} - p_0}{\sqrt{p_0(1-p_0)/n}}$$

When the null hypothesis is true and the conditions are met, Z has a standard normal distribution.

Conditions:

- independent observations
- large samples ($np_0 \geq 10$ and $n(1-p_0) \geq 10$)

One possible regulation for payday lenders is that they would be required to do a credit check and evaluate debt payments against the borrower's finances. We would like to know: would borrowers support this form of regulation?

GUIDED PRACTICE

Set up hypotheses to evaluate whether borrowers have a majority support for this type of regulation.[8]

To apply the normal distribution framework in the context of a hypothesis test for a proportion, the independence and success-failure conditions must be satisfied. In a hypothesis test, the success-failure condition is checked using the null proportion: we verify np_0 and $n(1-p_0)$ are at least 10, where p_0 is the null value.

GUIDED PRACTICE

Do payday loan borrowers support a regulation that would require lenders to pull their credit report and evaluate their debt payments? From a random sample of 826 borrowers, 51% said they would support such a regulation. Is it reasonable to use a normal distribution to model $\hat{p}$ for a hypothesis test here?[9]

---

[7]We must find $z^\star$ such that 90% of the distribution falls between $-z^\star$ and $z^\star$ in the standard normal model, $N(\mu = 0, \sigma = 1)$. We can look up $-z^\star$ in the normal probability table by looking for a lower tail of 5% (the other 5% is in the upper tail), thus $z^\star = 1.65$. The 90% confidence interval can then be computed as point estimate $\pm 1.65 \times SE \rightarrow$ (4.4%, 13.6%). (Note: the conditions for normality had earlier been confirmed for us.) That is, we are 90% confident that implanting a stent in a stroke patient's brain increased the risk of stroke within 30 days by 4.4% to 13.6%. Note, the problem was set up as 90% to indicate that there was not a need for a high level of confidence (such as 95% or 99%). A lower degree of confidence increases potential for error, but it also produces a more narrow interval.

[8]H_0 : there is not support for the regulation; $H_0 : p \leq 0.50$. H_A : the majority of borrowers support the regulation; $H_A : p > 0.50$.

[9]Independence holds since the poll is based on a random sample. The success-failure condition also holds, which is checked using the null value ($p_0 = 0.5$) from $H_0: np_0 = 826 \times 0.5 = 413, n(1-p_0) = 826 \times 0.5 = 413$. Recall that here, the best guess for p is p_0 which comes from the null hypothesis (because we assume the null hypothesis is true when performing the testing procedure steps). H_0 : there is not support for the regulation; $H_0 : p \leq 0.50$. H_A : the majority of borrowers support the regulation; $H_A : p > 0.50$.

EXAMPLE

Using the hypotheses and data from the previous Guided Practices, evaluate whether the poll on lending regulations provides convincing evidence that a majority of payday loan borrowers support a new regulation that would require lenders to pull credit reports and evaluate debt payments.

With hypotheses already set up and conditions checked, we can move onto calculations. The standard error in the context of a one-proportion hypothesis test is computed using the null value, p_0 :

$$SE = \sqrt{\frac{p_0(1-p_0)}{n}} = \sqrt{\frac{0.5(1-0.5)}{826}} = 0.017$$

A picture of the normal model is shown with the p-value represented by the shaded region.

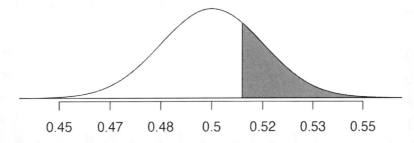

Based on the normal model, the test statistic can be computed as the Z score of the point estimate:

$$\begin{aligned} Z &= \frac{\text{point estimate} - \text{null value}}{SE} \\ &= \frac{0.51 - 0.50}{0.017} \\ &= 0.59 \end{aligned}$$

The single tail area which represents the p-value is 0.2776. Because the p-value is larger than 0.05, we do not reject H_0. The poll does not provide convincing evidence that a majority of payday loan borrowers support regulations around credit checks and evaluation of debt payments.

In Section 17.1 we discuss two-sided hypothesis tests of which the payday example may have been better structured. That is, we might have wanted to ask whether the borrows **support or oppose** the regulations (to study opinion in either direction away from the 50% benchmark). In that case, the p-value would have been doubled to 0.5552 (again, we would not reject H_0). In the two-sided hypothesis setting, the appropriate conclusion would be to claim that the poll does not provide convincing evidence that a majority of payday loan borrowers support or oppose regulations around credit checks and evaluation of debt payments.

In both the one-sided or two-sided setting, the conclusion is somewhat unsatisfactory because there is no conclusion. That is, there is no resolution one way or the other about public opinion. We cannot claim that exactly 50% of people support the regulation, but we cannot claim a majority in either direction.

Mathematical model hypothesis test for a proportion.

Set up hypotheses and verify the conditions using the null value, p_0, to ensure $\hat{p}$ is nearly normal under H_0. If the conditions hold, construct the standard error, again using p_0, and show the p-value in a drawing. Lastly, compute the p-value and evaluate the hypotheses.

For additional one-proportion hypothesis test examples, see Section 11.3.

16.2.6 Violating conditions

We've spent a lot of time discussing conditions for when $\hat{p}$ can be reasonably modeled by a normal distribution. What happens when the success-failure condition fails? What about when the independence condition fails? In either case, the general ideas of confidence intervals and hypothesis tests remain the same, but the strategy or technique used to generate the interval or p-value change.

When the success-failure condition isn't met for a hypothesis test, we can simulate the null distribution of $\hat{p}$ using the null value, p_0, as seen in Section 16.1. Unfortunately, methods for dealing with observations which are not independent (e.g., repeated measurements on subject, such as in studies where measurements from the same subjects are taken pre and post study) are outside the scope of this book.

16.3 Chapter review

16.3.1 Summary

Building on the foundational ideas from the previous few ideas, this chapter focused exclusively on the single population proportion as the parameter of interest. Note that it is not possible to do a randomization test with only one variable, so to do computational hypothesis testing, we applied a bootstrapping framework. The bootstrap confidence interval and the mathematical framework for both hypothesis testing and confidence intervals are similar to those applied to other data structures and parameters. When using the mathematical model, keep in mind the success-failure conditions. Additionally, know that bootstrapping is always more accurate with larger samples.

16.3.2 Terms

We introduced the following terms in the chapter. If you're not sure what some of these terms mean, we recommend you go back in the text and review their definitions. We are purposefully presenting them in alphabetical order, instead of in order of appearance, so they will be a little more challenging to locate. However you should be able to easily spot them as **bolded text**.

parametric bootstrap	success-failure condition
SE single proportion	Z score

16.4 Exercises

Answers to odd numbered exercises can be found in Appendix A.16.

16.1. **Do aliens exist?** In May 2021, YouGov asked 4,839 adult Great Britain residents whether they think aliens exist, and if so, if they have or have not visited Earth. You want to evaluate if more than a quarter (25%) of Great Britain adults think aliens don't exist. In the survey 22% responded "I think they exist, and have visited Earth", 28% responded "I think they exist, but have not visited Earth", 29% responded "I don't think they exist", and 22% responded "Don't know". A friend of yours offers to help you with setting up the hypothesis test and comes up with the following hypotheses. Indicate any errors you see.

$H_0 : \hat{p} = 0.29 \qquad H_A : \hat{p} > 0.29$

16.2. **Married at 25.** A study suggests that the 25% of 25 year olds have gotten married. You believe that this is incorrect and decide to collect your own sample for a hypothesis test. From a random sample of 25 year olds in census data with size 776, you find that 24% of them are married. A friend of yours offers to help you with setting up the hypothesis test and comes up with the following hypotheses. Indicate any errors you see.

$H_0 : \hat{p} = 0.24 \qquad H_A : \hat{p} \neq 0.24$

16.3. **Defund the police.** A Survey USA poll conducted in Seattle, WA in May 2021 reports that of the 650 respondents (adults living in this area), 159 support proposals to defund police departments. (Survey USA, 2021)

a. A journalist writing a news story on the poll results wants to use the headline "More than 1 in 5 adults living in Seattle support proposals to defund police departments." You caution the journalist that they should first conduct a hypothesis test to see if the poll data provide concincing evidence for this claim Write the hypotheses for this test.

b. Calculate the proportion of Seattle adults in thsi sample who support proposals to defund police departments.

c. Describe a setup for a simulation that would be appropriate in this situation and how the p-value can be calculated using the simulation results.

d. Below is a histogram showing the distribution of $\hat{p}_{sim}$ in 1,000 simulations under the null hypothesis. Estimate the p-value using the plot and use it to evaluate the hypotheses.

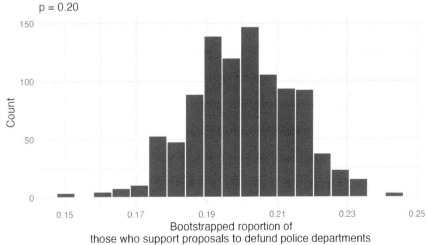

16.4. **Assisted reproduction.** Assisted Reproductive Technology (ART) is a collection of techniques that help facilitate pregnancy (e.g., in vitro fertilization). The 2018 ART Fertility Clinic Success Rates Report published by the Centers for Disease Control and Prevention reports that ART has been successful in leading to a live birth in 48.8% of cases where the patient is under 35 years old. (CDC, 2018) A new fertility clinic claims that their success rate is higher than average for this age group. A random sample of 30 of their patients yielded a success rate of 60%. A consumer watchdog group would like to determine if this provides strong evidence to support the company's claim.

a. Write the hypotheses to test if the success rate for ART at this clinic is significantly higher than the success rate reported by the CDC.

b. Describe a setup for a simulation that would be appropriate in this situation and how the p-value can be calculated using the simulation results.

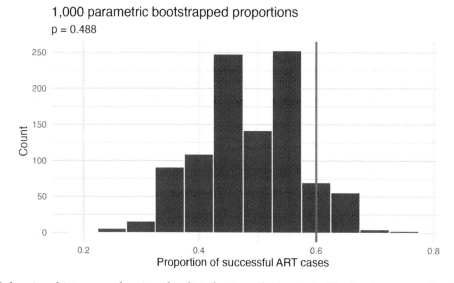

c. Below is a histogram showing the distribution of $\hat{p}_{sim}$ in 1,000 simulations under the null hypothesis. Estimate the p-value using the plot and use it to evaluate the hypotheses.

d. After performing this analysis, the consumer group releases the following news headline: "Infertility clinic falsely advertises better success rates". Comment on the appropriateness of this statement.

16.5. **If I fits, I sits, bootstrap test.** A citizen science project on which type of enclosed spaces cats are most likely to sit in compared (among other options) two different spaces taped to the ground. The first was a square, and the second was a shape known as Kanizsa square illusion. When comparing the two options given to 7 cats, 5 chose the square, and 2 chose the Kanizsa square illusion. We are interested to know whether these data provide convincing evidence that cats prefer one of the shapes over the other. (Smith et al., 2021)

   a. What are the null and alternative hypotheses for evaluating whether these data provide convincing evidence that cats have preference for one of the shapes

   b. A parametric bootstrap simulation (with 1,000 bootstrap samples) was run and the resulting null distribution is displayed in the histogram below. Find the p-value using this distribution and conclude the hypothesis test in the context of the problem.

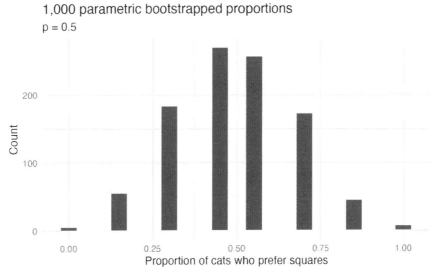

16.6. **Legalization of marijuana, bootstrap test.** The 2018 General Social Survey asked a random sample of 1,563 US adults: "Do you think the use of marijuana should be made legal, or not?" 60% of the respondents said it should be made legal. (NORC, 2018) Consider a scenario where, in order to become legal, 55% (or more) of voters must approve.

   a. What are the null and alternative hypotheses for evaluating whether these data provide convincing evidence that, if voted on, marijuana would be legalized in the US.

   b. A parametric bootstrap simulation (with 1,000 bootstrap samples) was run and the resulting null distribution is displayed in the histogram below. Find the p-value using this distribution and conclude the hypothesis test in the context of the problem.

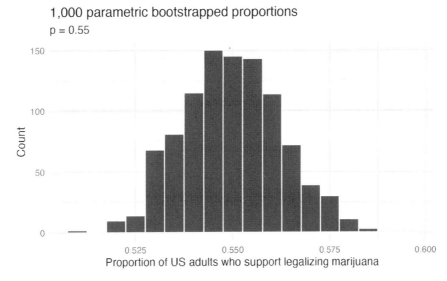

16.7. **If I fits, I sits, standard errors.** The results of a study on the type of enclosed spaces cats are most likely to sit in show that 5 out of 7 cats chose a square taped to the ground over a shape known as Kanizsa square illusion, which was preferred by the remaining 2 cats. To evaluate whether these data provide convincing evidence that cats prefer one of the shapes over the other, we set $H_0 : p = 0.5$, where p is the population proportion of cats who prefer square over the Kanizsa square illusion and $H_A : p \neq 0.5$, which suggests some preference, without specifying which shape is more preferred. (Smith et al., 2021)

   a. Using the mathematical model, calculate the standard error of the sample proportion in repeated samples of size 7.

   b. A parametric bootstrap simulation (with 1,000 bootstrap samples) was run and the resulting null distribution is displayed in the histogram below. This distribution shows the variability of the sample proportion in samples of size 7 when 50% of cats prefer the square shape over the Kanizsa square illusion. What is the approximate standard error of the sample proportion based on this distribution?

   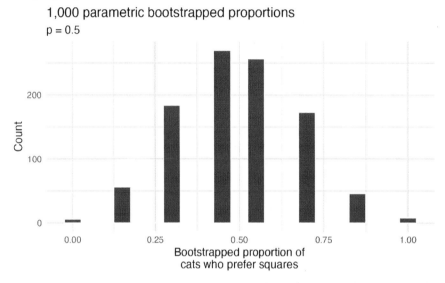

   c. Do the mathematical model and parametric bootstrap give similar standard errors?

   d. In order to approach the problem using the mathematical model, is the success-failure condition met for this study? Explain.

   e. What about the null distribution shown above (generated using the parametric bootstrap) tells us that the mathematical model should probably not be used?

16.8. **Legalization of marijuana, standard errors.** According to the 2018 General Social Survey, in a random sample of 1,563 US adults, 60% think marijuana should be made legal. (NORC, 2018) Consider a scenario where, in order to become legal, 55% (or more) of voters must approve.

a. Calculate the standard error of the sample proportion using the mathematical model.

b. A parametric bootstrap simulation (with 1,000 bootstrap samples) was run and the resulting null distribution is displayed in the histogram below. This distribution shows the variability of the sample proportion in samples of size 1,563 when 55% of voters approve legalizing marijuana. What is the approximate standard error of the sample proportion based on this distribution?

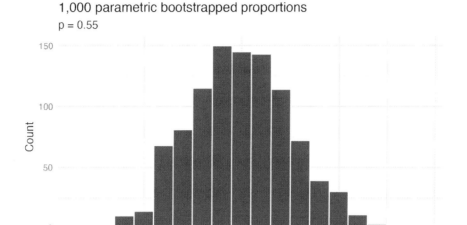

c. Do the mathematical model and parametric bootstrap give similar standard errors?

d. In this setting (to test whether the true underlying population proportion is greater than 0.55), would there be a strong reason to choose the mathematical model over the parametric bootstrap (or vice versa)?

16.9. **Statistics and employment, describe the bootstrap.** A large university knows that about 70% of the full-time students are employed at least 5 hours per week. The members of the Statistics Department wonder if the same proportion of their students work at least 5 hours per week. They randomly sample 25 majors and find that 15 of the students work 5 or more hours each week.

Two bootstrap sampling distributions are created to describe the variability in the proportion of statistics majors who work at least 5 hours per week. The parametric bootstrap imposes a true population proportion of $p = 0.7$ while the data bootstrap resamples from the actual data (which has 60% of the observations who work at least 5 hours per week).

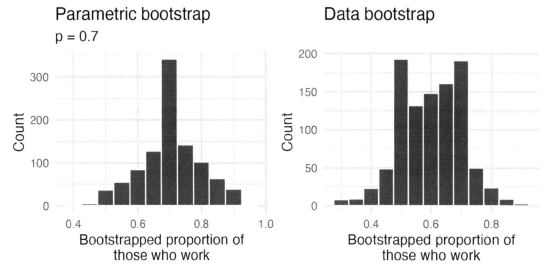

a. The bootstrap sampling was done under two different settings to generate each of the distributions shown above. Describe the two different settings.

b. Where are each of the two distributions centered? Are they centered at roughly the same place?

c. Estimate the standard error of the simulated proportions based on each distribution. Are the two standard errors you estimate roughly equal?

d. Describe the shapes of the two distributions. Are they roughly the same?

16.10. **National Health Plan, parametric bootstrap.** A Kaiser Family Foundation poll for a random sample of US adults in 2019 found that 79% of Democrats, 55% of Independents, and 24% of Republicans supported a generic "National Health Plan". There were 347 Democrats, 298 Republicans, and 617 Independents surveyed. (Foundation, 2019)

A political pundit on TV claims that a majority of Independents support a National Health Plan. Do these data provide strong evidence to support this type of statement? One approach to assessing the question of whether a majority of Independents support a National Health Plan is to simulate 1,000 parametric bootstrap samples with $p = 0.5$ as the proportion of Independents in support.

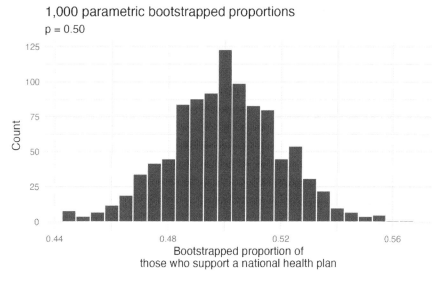

a. The histogram above displays 1000 values of what?

b. Is the observed proportion of Independents consistent with the parametric bootstrap proportions under the setting where $p = 0.5$?

c. In order to test the claim that "a majority of Independents support a National Health Plan" what are the null and alternative hypotheses?

d. Using the parametric bootstrap distribution, find the p-value and conclude the hypothesis test in the context of the problem.

16.11. **Statistics and employment, use the bootstrap.** In a large university where 70% of the full-time students are employed at least 5 hours per week, the members of the Statistics Department wonder if the same proportion of their students work at least 5 hours per week. They randomly sample 25 majors and find that 15 of the students work 5 or more hours each week.

Two bootstrap sampling distributions are created to describe the variability in the proportion of statistics majors who work at least 5 hours per week. The parametric bootstrap imposes a true population proportion of $p = 0.7$ while the data bootstrap resamples from the actual data (which has 60% of the observations who work at least 5 hours per week).

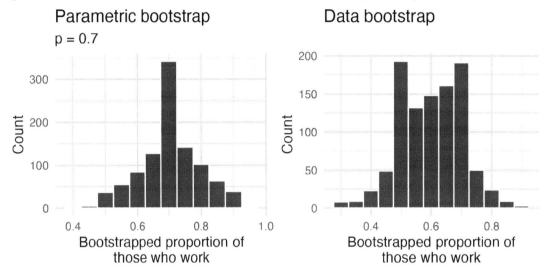

a. Which bootstrap distribution should be used to test whether the proportion of all statistics majors who work at least 5 hours per week is 70%? And which bootstrap distribution should be used to find a confidence interval for the true poportion of statistics majors who work at least 5 hours per week?

b. Using the appropriate histogram, test the claim that 70% of statistics majors, like their peers, work at least 5 hours per week. State the null and alternative hypotheses, find the p-value, and conclude the test in the context of the problem.

c. Using the appropriate histogram, find a 98% bootstrap percentile confidence interval for the true proportion of statistics majors who work at least 5 hours per week. Interpret the confidence interval in the context of the problem.

d. Using the appropriate historgram, find a 98% bootstrap SE confidence interval for the true proportion of statistics majors who work at least 5 hours per week. Interpret the confidence interval in the context of the problem.

16.12. **CLT for proportions.** Define the term "sampling distribution" of the sample proportion, and describe how the shape, center, and spread of the sampling distribution change as the sample size increases when $p = 0.1$.

16.4. EXERCISES

16.13. **Vegetarian college students.** Suppose that 8% of college students are vegetarians. Determine if the following statements are true or false, and explain your reasoning.

   a. The distribution of the sample proportions of vegetarians in random samples of size 60 is approximately normal since $n \geq 30$.

   b. The distribution of the sample proportions of vegetarian college students in random samples of size 50 is right skewed.

   c. A random sample of 125 college students where 12% are vegetarians would be considered unusual.

   d. A random sample of 250 college students where 12% are vegetarians would be considered unusual.

   e. The standard error would be reduced by one-half if we increased the sample size from 125 to 250.

16.14. **Young Americans, American dream.** About 77% of young adults think they can achieve the American dream. Determine if the following statements are true or false, and explain your reasoning. (Vaughn, 2011)

   a. The distribution of sample proportions of young Americans who think they can achieve the American dream in random samples of size 20 is left skewed.

   b. The distribution of sample proportions of young Americans who think they can achieve the American dream in random samples of size 40 is approximately normal since $n \geq 30$.

   c. A random sample of 60 young Americans where 85% think they can achieve the American dream would be considered unusual.

   d. A random sample of 120 young Americans where 85% think they can achieve the American dream would be considered unusual.

16.15. **Orange tabbies.** Suppose that 90% of orange tabby cats are male. Determine if the following statements are true or false, and explain your reasoning.

   a. The distribution of sample proportions of random samples of size 30 is left skewed.

   b. Using a sample size that is 4 times as large will reduce the standard error of the sample proportion by one-half.

   c. The distribution of sample proportions of random samples of size 140 is approximately normal.

   d. The distribution of sample proportions of random samples of size 280 is approximately normal.

16.16. **Young Americans, starting a family.** About 25% of young Americans have delayed starting a family due to the continued economic slump. Determine if the following statements are true or false, and explain your reasoning. (Demos, 2011)

   a. The distribution of sample proportions of young Americans who have delayed starting a family due to the continued economic slump in random samples of size 12 is right skewed.

   b. In order for the distribution of sample proportions of young Americans who have delayed starting a family due to the continued economic slump to be approximately normal, we need random samples where the sample size is at least 40.

   c. A random sample of 50 young Americans where 20% have delayed starting a family due to the continued economic slump would be considered unusual.

   d. A random sample of 150 young Americans where 20% have delayed starting a family due to the continued economic slump would be considered unusual.

   e. Tripling the sample size will reduce the standard error of the sample proportion by one-third.

16.17. **Sex equality.** The General Social Survey asked a random sample of 1,390 Americans the following question: "On the whole, do you think it should or should not be the government's responsibility to promote equality between men and women?" 82% of the respondents said it "should be". At a 95% confidence level, this sample has 2% margin of error. Based on this information, determine if the following statements are true or false, and explain your reasoning. (NORC, 2016)

   a. We are 95% confident that between 80% and 84% of Americans in this sample think it's the government's responsibility to promote equality between men and women.

   b. We are 95% confident that between 80% and 84% of all Americans think it's the government's responsibility to promote equality between men and women.

   c. If we considered many random samples of 1,390 Americans, and we calculated 95% confidence intervals for each, 95% of these intervals would include the true population proportion of Americans who think it's the government's responsibility to promote equality between men and women.

   d. In order to decrease the margin of error to 1%, we would need to quadruple (multiply by 4) the sample size.

   e. Based on this confidence interval, there is sufficient evidence to conclude that a majority of Americans think it's the government's responsibility to promote equality between men and women.

16.18. **Elderly drivers.** The Marist Poll published a report stating that 66% of adults nationally think licensed drivers should be required to retake their road test once they reach 65 years of age. It was also reported that interviews were conducted on a random sample of 1,018 American adults, and that the margin of error was 3% using a 95% confidence level. (Poll, 2011)

   a. Verify the margin of error reported by The Marist Poll using a mathematical model.

   b. Based on a 95% confidence interval, does the poll provide convincing evidence that *more than* two thirds of the population think that licensed drivers should be required to retake their road test once they turn 65?

16.19. **Fireworks on July 4th.** A local news outlet reported that 56% of 600 randomly sampled Kansas residents planned to set off fireworks on July 4th. Determine the margin of error for the 56% point estimate using a 95% confidence level using a mathematical model. (Survey USA, 2012)

16.20. **Proof of COVID-19 vaccination.** A Gallup poll surveyed 3,731 randomly sampled US in April 2021, asking how they felt about requiring proof of COVID-19 vaccination for travel by airplane. The poll found that 57% said they would favor it. (Gallup, 2021b)

   a. Describe the population parameter of interest. What is the value of the point estimate of this parameter?

   b. Check if the conditions required for constructing a confidence interval using a mathematical model based on these data are met.

   c. Construct a 95% confidence interval for the proportion of US adults who favor requiring proof of COVID-19 vaccination for travel by airplane.

   d. Without doing any calculations, describe what would happen to the confidence interval if we decided to use a higher confidence level.

   e. Without doing any calculations, describe what would happen to the confidence interval if we used a larger sample.

16.4. EXERCISES

16.21. **Study abroad.** A survey on 1,509 high school seniors who took the SAT and who completed an optional web survey shows that 55% of high school seniors are fairly certain that they will participate in a study abroad program in college. (AEC, 2008)

  a. Is this sample a representative sample from the population of all high school seniors in the US? Explain your reasoning.

  b. Let's suppose the conditions for inference are met. Even if your answer to part (a) indicated that this approach would not be reliable, this analysis may still be interesting to carry out (though not report). Using a mathematical model, construct a 90% confidence interval for the proportion of high school seniors (of those who took the SAT) who are fairly certain they will participate in a study abroad program in college, and interpret this interval in context.

  c. What does "90% confidence" mean?

  d. Based on this interval, would it be appropriate to claim that the majority of high school seniors are fairly certain that they will participate in a study abroad program in college?

16.22. **Legalization of marijuana, mathematical interval.** The General Social Survey asked a random sample of 1,563 US adults: "Do you think the use of marijuana should be made legal, or not?" 60% of the respondents said it should be made legal. (NORC, 2018)

  a. Is 60% a sample statistic or a population parameter? Explain.

  b. Using a mathematical model, construct a 95% confidence interval for the proportion of US adults who think marijuana should be made legal, and interpret it.

  c. A critic points out that this 95% confidence interval is only accurate if the statistic follows a normal distribution, or if the normal model is a good approximation. Is this true for these data? Explain.

  d. A news piece on this survey's findings states, "Majority of US adults think marijuana should be legalized." Based on your confidence interval, is this statement justified?

16.23. **National Health Plan, mathematical inference.** A Kaiser Family Foundation poll for a random sample of US adults in 2019 found that 79% of Democrats, 55% of Independents, and 24% of Republicans supported a generic "National Health Plan". There were 347 Democrats, 298 Republicans, and 617 Independents surveyed. (Foundation, 2019)

  a. A political pundit on TV claims that a majority of Independents support a National Health Plan. Do these data provide strong evidence to support this type of statement? Your response should use a mathematical model.

  b. Would you expect a confidence interval for the proportion of Independents who oppose the public option plan to include 0.5? Explain.

16.24. **Is college worth it?** Among a simple random sample of 331 American adults who do not have a four-year college degree and are not currently enrolled in school, 48% said they decided not to go to college because they could not afford school. (Pew Research Center, 2011)

  a. A newspaper article states that only a minority of the Americans who decide not to go to college do so because they cannot afford it and uses the point estimate from this survey as evidence. Conduct a hypothesis test to determine if these data provide strong evidence supporting this statement.

  b. Would you expect a confidence interval for the proportion of American adults who decide not to go to college because they cannot afford it to include 0.5? Explain.

16.25. **Taste test.** Some people claim that they can tell the difference between a diet soda and a regular soda in the first sip. A researcher wanting to test this claim randomly sampled 80 such people. He then filled 80 plain white cups with soda, half diet and half regular through random assignment, and asked each person to take one sip from their cup and identify the soda as diet or regular. 53 participants correctly identified the soda.

   a. Do these data provide strong evidence that these people are able to detect the difference between diet and regular soda, in other words, are the results significantly better than just random guessing? Your response should use a mathematical model.

   b. Interpret the p-value in this context.

16.26. **Will the coronavirus bring the world closer together?** An April 2021 YouGov poll asked 4,265 UK adults whether they think the coronavirus bring the world closer together or leave us further apart. 12% of the respondents said it will bring the world closer together. 37% said it would leave us further apart, 39% said it won't make a difference and the remainder didn't have an opinion on the matter. (YouGov, 2021)

   a. Calculate, using a mathematical model, a 90% confidence interval for the proportion of UK adults who think the coronavirus will bring the world closer together, and interpret the interval in context.

   b. Suppose we wanted the margin of error for the 90% confidence level to be about 0.5%. How large of a sample size would you recommend for the poll?

16.27. **Quality control.** As part of a quality control process for computer chips, an engineer at a factory randomly samples 212 chips during a week of production to test the current rate of chips with severe defects. She finds that 27 of the chips are defective.

   a. What population is under consideration in the data set?

   b. What parameter is being estimated?

   c. What is the point estimate for the parameter?

   d. What is the name of the statistic that can be used to measure the uncertainty of the point estimate?

   e. Compute the value of the statistic from part (d) using a mathematical model.

   f. The historical rate of defects is 10%. Should the engineer be surprised by the observed rate of defects during the current week?

   g. Suppose the true population value was found to be 10%. If we use this proportion to recompute the value in part (d) using $p = 0.1$ instead of $\hat{p}$, how much does the resulting value of the statistic change?

16.28. **Nearsighted children.** Nearsightedness (myopia) is a common vision condition in which you can see objects near to you clearly, but objects farther away are blurry. It is believed that nearsightedness affects about 8% of all children. In a random sample of 194 children, 21 are nearsighted. Using a mathematical model, conduct a hypothesis test for the following question: do these data provide evidence that the 8% value is inaccurate?

16.29. **Website registration.** A website is trying to increase registration for first-time visitors, exposing 1% of these visitors to a new site design. Of 752 randomly sampled visitors over a month who saw the new design, 64 registered.

   a. Check the conditions for constructing a confidence interval using a mathematical model.

   b. Compute the standard error which would describe the variability associated with repeated samples of size 752.

   c. Construct and interpret a 90% confidence interval for the fraction of first-time visitors of the site who would register under the new design (assuming stable behaviors by new visitors over time).

16.30. **Coupons driving visits.** A store randomly samples 603 shoppers over the course of a year and finds that 142 of them made their visit because of a coupon they'd received in the mail. Using a mathematical model, construct a 95% confidence interval for the fraction of all shoppers during the year whose visit was because of a coupon they'd received in the mail.

Chapter 17

Inference for comparing two proportions

> We now extend the methods from Chapter 16 to apply confidence intervals and hypothesis tests to differences in population proportions that come from two groups, Group 1 and Group 2: $p_1 - p_2$.
>
> In our investigations, we'll identify a reasonable point estimate of $p_1 - p_2$ based on the sample, and you may have already guessed its form: $\hat{p}_1 - \hat{p}_2$. Then we'll look at the inferential analysis in three different ways: using a randomization test, applying bootstrapping for interval estimates, and, if we verify that the point estimate can be modeled using a normal distribution, we compute the estimate's standard error, and we apply the mathematical framework.

17.1 Randomization test for the difference in proportions

17.1.1 Observed data

Let's take another look at the cardiopulmonary resuscitation (CPR) study we introduced in Chapter 14.2. The experiment consisted of two treatments on patients who underwent CPR for a heart attack and were subsequently admitted to a hospital. Each patient was randomly assigned to either receive a blood thinner (treatment group) or not receive a blood thinner (control group). The outcome variable of interest was whether the patient survived for at least 24 hours. (Böttiger et al., 2001)

The cpr data can be found in the **openintro** R package.

The results are summarized in Table 17.1 (which is a replica of Table 14.2). 11 out of the 50 patients in the control group and 14 out of the 40 patients in the treatment group survived.

17.1. RANDOMIZATION TEST FOR THE DIFFERENCE IN PROPORTIONS

Table 17.1: Results for the CPR study. Patients in the treatment group were given a blood thinner, and patients in the control group were not.

Group	Died	Survived	Total
Control	39	11	50
Treatment	26	14	40
Total	65	25	90

GUIDED PRACTICE

Is this an observational study or an experiment? What implications does the study type have on what can be inferred from the results?[1]

In this study, a larger proportion of patients who received blood thinner after CPR, $\hat{p}_T = \frac{14}{40} = 0.35$, survived compared to those who did not receive blood thinner, $\hat{p}_C = \frac{11}{50} = 0.22$. However, based on these observed proportions alone, we cannot determine whether the difference ($\hat{p}_T - \hat{p}_C = 0.35 - 0.22 = 0.13$) provides *convincing evidence* that blood thinner usage after CPR is effective.

As we saw in Chapter 11, we can re-randomize the responses (survived or died) to the treatment conditions assuming the null hypothesis is true and compute possible differences in proportions. The process by which we randomize observations to two groups is summarized and visualized in Figure 11.8.

17.1.2 Variability of the statistic

Figure 17.1 shows a stacked plot of the differences found from 100 randomization simulations (i.e., repeated iterations as described in Figure 11.8), where each dot represents a simulated difference between the infection rates (control rate minus treatment rate).

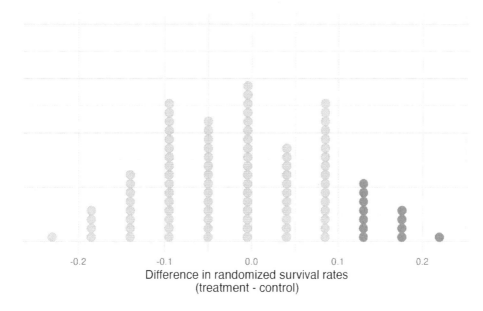

Figure 17.1: A stacked dot plot of differences from 100 simulations produced under the independence model H_0, where in these simulations survival is unaffected by the treatment. Twelve of the 100 simulations had a difference of at least 13%, the difference observed in the study.

---

[1]The study is an experiment, as patients were randomly assigned an experiment group. Since this is an experiment, the results can be used to evaluate a causal relationship between blood thinner use after CPR and whether patients survived.

17.1.3 Observed statistic vs null statistics

Note that the distribution of these simulated differences is centered around 0. We simulated the differences assuming that the independence model was true, that blood thinners after CPR have no effect on survival. Under the null hypothesis, we expect the difference to be near zero with some random fluctuation, where *near* is pretty generous in this case since the sample sizes are so small in this study.

EXAMPLE

How often would you observe a difference of at least 13% (0.13) according to Figure 17.1? Is this a rare event?

It appears that a difference of at least 13% due to chance alone, if the null hypothesis was true would happen about 12% of the time according to Figure 17.1. This is not a very rare event.

The difference of 13% not being a rare event suggests two possible interpretations of the results of the study:

- H_0 Independence model. Blood thinners after CPR have no effect on survival, and we just happened to observe a difference that would only occur on a rare occasion.
- H_A Alternative model. Blood thinners after CPR increase chance of survival, and the difference we observed was actually due to the blood thinners after CPR being effective at increasing the chance of survival, which explains the difference of 13%.

Since we determined that the outcome is not that rare (12% chance of observing a difference of 13% or more under the assumption that blood thinners after CPR have no effect on survival), we fail to reject H_0, and conclude that the study results do not provide strong evidence against the independence model. This doesn't mean that we have proved that blood thinners are not effective, it just means that this study does not provide convincing evidence that they are effective in this setting.

Statistical inference, is built on evaluating how likely such differences are to occur due to chance if in fact the null hypothesis is true. In statistical inference, data scientists evaluate which model is most reasonable given the data. Errors do occur, just like rare events, and we might choose the wrong model. While we do not always choose correctly, statistical inference gives us tools to control and evaluate how often these errors occur.

17.2 Bootstrap confidence interval for the difference in proportions

In Section 17.1, we worked with the randomization distribution to understand the distribution of $\hat{p}_1 - \hat{p}_2$ when the null hypothesis $H_0 : p_1 - p_2 = 0$ is true. Now, through bootstrapping, we study the variability of $\hat{p}_1 - \hat{p}_2$ without assuming the null hypothesis is true.

17.2.1 Observed data

Reconsider the CPR data from Section 17.1 which is provided in Table 14.2. Again, we use the difference in sample proportions as the observed statistic of interest. Here, the value of the statistic is: $\hat{p}_T - \hat{p}_C = 0.35 - 0.22 = 0.13$.

17.2.2 Variability of the difference in sample proportions

The bootstrap method applied to two samples is an extension of the method described in Chapter 12. Now, we have two samples, so each sample estimates the population from which they came. In the CPR setting, the `treatment` sample estimates the population of all individuals who have gotten (or will get) the treatment; the `control` sample estimate the population of all individuals who do not get the treatment and are controls. Figure 17.2 extends Figure 12.1 to show the bootstrapping process from two samples simultaneously.

Figure 17.2: Creating two populations from which to take each of the bootstrap samples.

As before, once the population is estimated, we can randomly resample observations to create bootstrap samples, as seen in Figure 17.3.

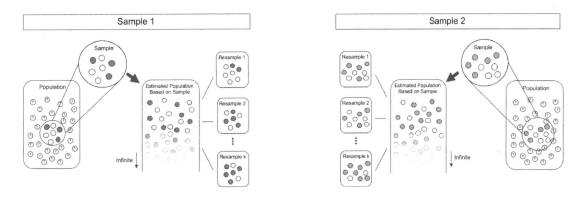

Figure 17.3: Taking each bootstrap sample from the estimated population.

The variability of the statistic (the difference in sample proportions) can be calculated by taking one bootstrap resample from Sample 1 and one bootstrap resample from Sample 2 and calculating the difference in the bootstrap proportions.

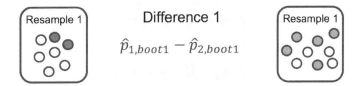

Figure 17.4: For example, the first bootstrap resamples from Sample 1 and Sample 2 provide resample proportions of 2/7 and 5/9, respectively.

As always, the variability of the difference in proportions can only be estimated by repeated simulations, in this case, repeated bootstrap resamples. Figure 17.4 shows multiple bootstrap differences calculated for each of the repeated bootstrap samples.

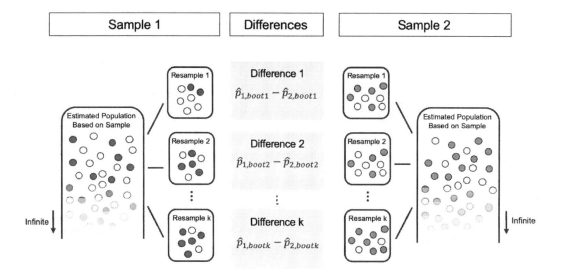

Figure 17.5: For each pair of bootstrap samples, we calculate the difference in sample proportions

Repeated bootstrap simulations lead to a bootstrap sampling distribution of the statistic of interest, here the difference in sample proportions. Figure 17.6 visualizes the process and Figure 17.7 shows 1,000 bootstrap differences in proportions for the CPR data. Note that the CPR data includes 40 and 50 people in the respective groups, and the illustrated example includes 7 and 9 people in the two groups. Accordingly, the variability in the distribution of sample proportions is higher for the illustrated example. As you will see in the mathematical models discussed in Section 17.3, large sample sizes lead to smaller standard errors for a difference in proportions.

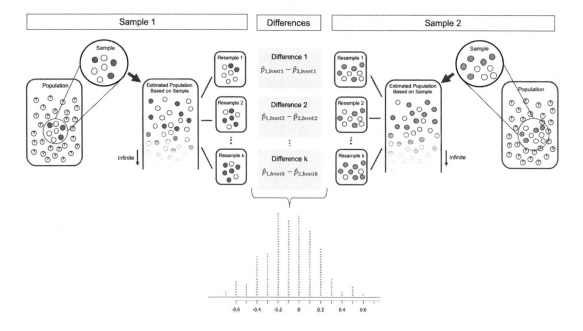

Figure 17.6: The differences in each bootstrapped pair of proportions are combined to create the sampling distribution of the differences in proportions.

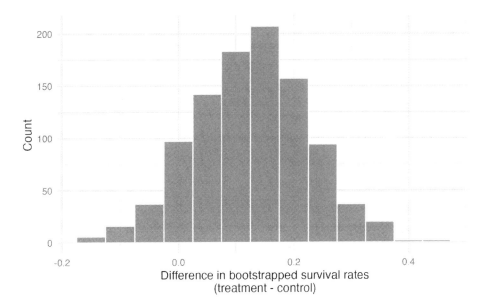

Figure 17.7: A histogram of differences in proportions from 1000 bootstrap simulations of the CPR data. Note that because the CPR data has a larger sample size than the illustrated example, the variability of the difference in proportions is much smaller with the CPR histogram.

17.2.3 Bootstrap percentile vs. SE confidence intervals

Figure 17.7 provides an estimate for the variability of the difference in survival proportions from sample to sample. The values in the histogram can be used in two different ways to create a confidence interval for the parameter of interest: $p_1 - p_2$.

Bootstrap percentile confidence interval

As in Chapter 12, the bootstrap confidence interval can be calculated directly from the bootstrapped differences in Figure 17.7. The interval created from the percentiles of the distribution is called the **percentile interval**. Note that here we calculate the 90% confidence interval by finding the 5th and 95th percentile values from the bootstrapped differences. The bootstrap 5 percentile proportion is -0.032 and the 95 percentile is 0.284. The result is: we are 90% confident that, in the population, the true difference in probability of survival for individuals receiving blood thinners after CPR is between -0.032 lower and 0.284 higher than those who did not receive blood thinners. The interval shows that we do not have much definitive evidence of the affect of blood thinners, one way or another.

Bootstrap SE confidence interval

Alternatively, we can use the variability in the bootstrapped differences to calculate a standard error of the difference. The resulting interval is called the **SE interval**. Section 17.3 details the mathematical model for the standard error of the difference in sample proportions, but the bootstrap distribution typically does an excellent job of estimating the variability of the sampling distribution of the sample statistic.

$$SE(\hat{p}_T - \hat{p}_C) \approx SE(\hat{p}_{T,boot} - \hat{p}_{C,boot}) = 0.098$$

The variability of the difference in proportions was calculated in R using the sd() function, but any statistical software will calculate the standard deviation of the differences, here, the exact quantity we hope to approximate.

Note that we do not know know the true distribution of $\hat{p}_T - \hat{p}_C$, so we will use a rough approximation to find a confidence interval for $p_T - p_C$. As seen in the bootstrap histograms, the shape of the distribution is roughly symmetric and bell-shaped. So for a rough approximation, we will apply the 67-95-99.7 rule which tells us that 95% of observed differences should be roughly no farther than 2 SE from the true parameter (difference in proportions). A 95% confidence interval for $p_T - p_C$ is given by:

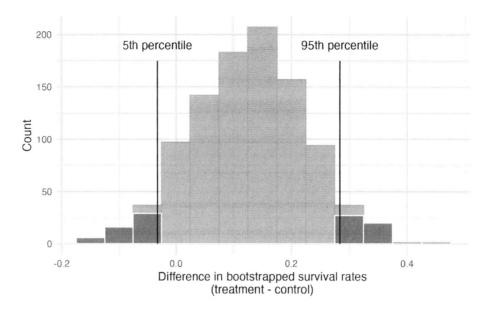

Figure 17.8: The CPR data is bootstrapped 1,000 times. Each simulation creates a sample from the original data where the probability of survival in the treatment group is $\hat{p}_T = 14/40$ and the probability of survival in the control group is $\hat{p}_C = 11/50$.

$$\hat{p}_T - \hat{p}_C \pm 2 \cdot SE \quad \rightarrow \quad 14/40 - 11/50 \pm 2 \cdot 0.098 \quad \rightarrow \quad (-0.067, 0.327)$$

We are 95% confident that the true value of $p_T - p_C$ is between -0.067 and 0.327. Again, the wide confidence interval that contains zero indicates that the study provides very little evidence about the effectiveness of blood thinners. For other percentages, e.g., a 90% bootstrap SE confidence interval, we will use quantiles given by the standard normal distribution, as seen in Section 13.2 and Figure 13.8.

17.2.4 What does 95% mean?

Recall that the goal of a confidence interval is to find a plausible range of values for a *parameter* of interest. The estimated statistic is not the value of interest, but it is typically the best guess for the unknown parameter. The confidence level (often 95%) is a number that takes a while to get used to. Surprisingly, the percentage doesn't describe the dataset at hand, it describes many possible datasets. One way to understand a confidence interval is to think about all the confidence intervals that you have ever made or that you will ever make as a scientist, the confidence level describes **those** intervals.

Figure 17.9 demonstrates a hypothetical situation in which 25 different studies are performed on the exact same population (with the same goal of estimating the true parameter value of $p_1 - p_2 = 0.47$). The study at hand represents one point estimate (a dot) and a corresponding interval. It is not possible to know whether the interval at hand is to the right of the unknown true parameter value (the black line) or to the left of that line. It is also impossible to know whether the interval captures the true parameter (is blue) or doesn't (is red). If we are making 95% intervals, then about 5% of the intervals we create over our lifetime will *not* capture the parameter of interest (e.g., will be red as in Figure 17.9). What we know is that over our lifetimes as scientists, about 95% of the intervals created and reported on will capture the parameter value of interest: thus the language "95% confident."

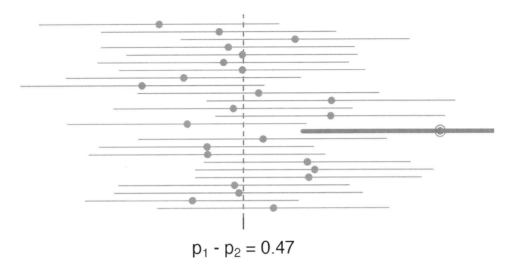

Figure 17.9: One hypothetical population, parameter value of: $p_1 - p_2 = 0.47$. Twenty-five different studies all which led to a different point estimate, SE, and confidence interval. The study at hand is one of the horizontal lines (hopefully a blue line!).

The choice of 95% or 90% or even 99% as a confidence level is admittedly somewhat arbitrary; however, it is related to the logic we used when deciding that a p-value should be declared as "significant" if it is lower than 0.05 (or 0.10 or 0.01, respectively). Indeed, one can show mathematically, that a 95% confidence interval and a two-sided hypothesis test at a cutoff of 0.05 will provide the same conclusion when the same data and mathematical tools are applied for the analysis. A full derivation of the explicit connection between confidence intervals and hypothesis tests is beyond the scope of this text.

17.3 Mathematical model for the difference in proportions

17.3.1 Variability of the difference between two proportions

Like with $\hat{p}$, the difference of two sample proportions $\hat{p}_1 - \hat{p}_2$ can be modeled using a normal distribution when certain conditions are met. First, we require a broader independence condition, and secondly, the success-failure condition must be met by both groups.

Conditions for the sampling distribution of $\hat{p}_1 - \hat{p}_2$ to be normal.

The difference $\hat{p}_1 - \hat{p}_2$ can be modeled using a normal distribution when

1. *Independence* (extended). The data are independent within and between the two groups. Generally this is satisfied if the data come from two independent random samples or if the data come from a randomized experiment.
2. *Success-failure condition.* The success-failure condition holds for both groups, where we check successes and failures in each group separately. That is, we should have at least 10 successes and 10 failures in each of the two groups.

When these conditions are satisfied, the standard error of $\hat{p}_1 - \hat{p}_2$ is:

$$SE(\hat{p}_1 - \hat{p}_2) = \sqrt{\frac{p_1(1-p_1)}{n_1} + \frac{p_2(1-p_2)}{n_2}}$$

where p_1 and p_2 represent the population proportions, and n_1 and n_2 represent the sample sizes.

Note that in most cases, the standard error is approximated using the observed data:

$$SE(\hat{p}_1 - \hat{p}_2) = \sqrt{\frac{\hat{p}_1(1-\hat{p}_1)}{n_1} + \frac{\hat{p}_2(1-\hat{p}_2)}{n_2}}$$

where $\hat{p}_1$ and $\hat{p}_2$ represent the observed sample proportions, and n_1 and n_2 represent the sample sizes.

Recall that the margin of error is defined by the standard error. The margin of error for $\hat{p}_1 - \hat{p}_2$ can be directly obtained from $SE(\hat{p}_1 - \hat{p}_2)$.

Margin of error for $\hat{p}_1 - \hat{p}_2$.

The margin of error is $z^\star \times \sqrt{\frac{\hat{p}_1(1-\hat{p}_1)}{n_1} + \frac{\hat{p}_2(1-\hat{p}_2)}{n_2}}$ where $z^\star$ is calculated from a specified percentile on the normal distribution.

17.3.2 Confidence interval for the difference between two proportions

We can apply the generic confidence interval formula for a difference of two proportions, where we use $\hat{p}_1 - \hat{p}_2$ as the point estimate and substitute the SE formula:

$$\text{point estimate} \pm z^\star \times SE$$

$$(\hat{p}_1 - \hat{p}_2) \pm z^\star \times \sqrt{\frac{\hat{p}_1(1-\hat{p}_1)}{n_1} + \frac{\hat{p}_2(1-\hat{p}_2)}{n_2}}$$

Standard error of the difference in two proportions, $\hat{p}_1 - \hat{p}_2$.

When the conditions for the normal model are are met, the **variability** of the difference in proportions, $\hat{p}_1 - \hat{p}_2$, is well described by:

$$SE(\hat{p}_1 - \hat{p}_2) = \sqrt{\frac{\hat{p}_1(1-\hat{p}_1)}{n_1} + \frac{\hat{p}_2(1-\hat{p}_2)}{n_2}}$$

EXAMPLE

We reconsider the experiment for patients who underwent cardiopulmonary resuscitation (CPR) for a heart attack and were subsequently admitted to a hospital. These patients were randomly divided into a treatment group where they received a blood thinner or the control group where they did not receive a blood thinner. The outcome variable of interest was whether the patients survived for at least 24 hours. The results are shown in Table 14.2. Check whether we can model the difference in sample proportions using the normal distribution.

We first check for independence: since this is a randomized experiment, it seems reasonable to assume that the observations are idependent.

Next, we check the success-failure condition for each group. We have at least 10 successes and 10 failures in each experiment arm (11, 14, 39, 26), so this condition is also satisfied.

With both conditions satisfied, the difference in sample proportions can be reasonably modeled using a normal distribution for these data.

EXAMPLE

Create and interpret a 90% confidence interval of the difference for the survival rates in the CPR study.

We'll use p_T for the survival rate in the treatment group and p_C for the control group:

$$\hat{p}_T - \hat{p}_C = \frac{14}{40} - \frac{11}{50} = 0.35 - 0.22 = 0.13$$

We use the standard error formula previously provided. As with the one-sample proportion case, we use the sample estimates of each proportion in the formula in the confidence interval context:

$$SE \approx \sqrt{\frac{0.35(1-0.35)}{40} + \frac{0.22(1-0.22)}{50}} = 0.095$$

For a 90% confidence interval, we use $z^\star = 1.65$:

$$\text{point estimate} \pm z^\star \times SE$$
$$0.13 \pm 1.65 \times 0.095$$
$$(-0.027, 0.287)$$

We are 90% confident that individuals receiving blood thinners have between a 2.7% less chance of survival to a 28.7% greater chance of survival than those in the control group. Because 0% is contained in the interval, we do not have enough information to say whether blood thinners help or harm heart attack patients who have been admitted after they have undergone CPR.

Note, the problem was set up as 90% to indicate that there was not a need for a high level of confidence (such a 95% or 99%). A lower degree of confidence increases potential for error, but it also produces a more narrow interval.

GUIDED PRACTICE

A 5-year experiment was conducted to evaluate the effectiveness of fish oils on reducing cardiovascular events, where each subject was randomized into one of two treatment groups (Manson et al., 2019). We'll consider heart attack outcomes in the patients listed in Table 17.2.

Create a 95% confidence interval for the effect of fish oils on heart attacks for patients who are well-represented by those in the study. Also interpret the interval in the context of the study.[2]

---

[2]Because the patients were randomized, the subjects are independent, both within and between the two groups. The success-failure condition is also met for both groups as all counts are at least 10. This satisfies the conditions necessary to model the difference in proportions using a normal distribution. Compute the sample proportions ($\hat{p}_{\text{fish oil}} = 0.0112$, $\hat{p}_{\text{placebo}} = 0.0155$), point estimate of the difference ($0.0112 - 0.0155 = -0.0043$), and standard error $SE = \sqrt{\frac{0.0112 \times 0.9888}{12933} + \frac{0.0155 \times 0.9845}{12938}}$, $SE = 0.00145$. Next, plug the values into the general formula for a confidence interval, where we'll use a 95% confidence level with $z^\star = 1.96$: $-0.0043 \pm 1.96 \times 0.00145 = (-0.0071, -0.0015)$. We are 95% confident that fish oils decreases heart attacks by 0.15 to 0.71 percentage points (off of a baseline of about 1.55%) over a 5-year period for subjects who are similar to those in the study. Because the interval is entirely below 0, and the treatment was randomly assigned the data provide strong evidence that fish oil supplements reduce heart attacks in patients like those in the study.

17.3. MATHEMATICAL MODEL FOR THE DIFFERENCE IN PROPORTIONS

Table 17.2: Results for the study on n-3 fatty acid supplement and related health benefits.

	heart attack	no event	Total
fish oil	145	12788	12933
placebo	200	12738	12938

The `fish_oil_18` data can be found in the **openintro** R package.

17.3.3 Hypothesis test for the difference between two proportions

The details for calculating a SE and for checking technical conditions are very similar to that of confidence intervals. However, when the null hypothesis is that $p_1 - p_2 = 0$, we use a special proportion called the **pooled proportion** to estimate the SE and to check the success-failure condition.

Use the pooled proportion when H_0 is $p_1 - p_2 = 0$.

When the null hypothesis is that the proportions are equal, use the pooled proportion ($\hat{p}_{pool}$) of successes to verify the success-failure condition and estimate the standard error:

$$\hat{p}_{pool} = \frac{\text{number of successes}}{\text{number of cases}} = \frac{\hat{p}_1 n_1 + \hat{p}_2 n_2}{n_1 + n_2}$$

Here $\hat{p}_1 n_1$ represents the number of successes in sample 1 because $\hat{p}_1 = \frac{\text{number of successes in sample 1}}{n_1}$.

Similarly, $\hat{p}_2 n_2$ represents the number of successes in sample 2.

The test statistic for assessing two proportions is a Z.

The Z score is a ratio of how the two sample proportions differ as compared to the expected variability of difference between the proportions.

$$Z = \frac{(\hat{p}_1 - \hat{p}_2) - 0}{\sqrt{\hat{p}_{pool}(1 - \hat{p}_{pool})\left(\frac{1}{n_1} + \frac{1}{n_2}\right)}}$$

When the null hypothesis is true and the conditions are met, Z has a standard normal distribution. See the box below for calculation of the pooled proportion of successes.

Conditions:

- Independent observations
- Large samples: ($n_1 p_1 \geq 10$ and $n_1(1 - p_1) \geq 10$ and $n_2 p_2 \geq 10$ and $n_2(1 - p_2) \geq 10$)
- Check conditions using: ($n_1 \hat{p}_{pool} \geq 10$ and $n_1(1 - \hat{p}_{pool}) \geq 10$ and $n_2 \hat{p}_{pool} \geq 10$ and $n_2(1 - \hat{p}_{pool}) \geq 10$)

A mammogram is an X-ray procedure used to check for breast cancer. Whether mammograms should be used is part of a controversial discussion, and it's the topic of our next example where we learn about 2-proportion hypothesis tests when H_0 is $p_1 - p_2 = 0$ (or equivalently, $p_1 = p_2$).

A 30-year study was conducted with nearly 90,000 participants who identified as female. During a 5-year screening period, each participant was randomized to one of two groups: in the first group, participants received regular mammograms to screen for breast cancer, and in the second group, participants received regular non-mammogram breast cancer exams. No intervention was made during the following 25 years of the study, and we'll consider death resulting from breast cancer over the full 30-year period. Results from the study are summarized in Figure 17.3.

The mammogram data can be found in the **openintro** R package.

If mammograms are much more effective than non-mammogram breast cancer exams, then we would expect to see additional deaths from breast cancer in the control group. On the other hand, if mammograms are not as effective as regular breast cancer exams, we would expect to see an increase in breast cancer deaths in the mammogram group.

Table 17.3: Summary results for breast cancer study.

	Death from breast cancer?	
Treatment	Yes	No
control	505	44,405
mammogram	500	44,425

GUIDED PRACTICE

Is this study an experiment or an observational study?[3]

GUIDED PRACTICE

Set up hypotheses to test whether there was a difference in breast cancer deaths in the mammogram and control groups.[4]

The research question describing mammograms is set up to address specific hypotheses (in contrast to a confidence interval for a parameter). In order to fully take advantage of the hypothesis testing structure, we asses the randomness under the condition that the null hypothesis is true (as we always do for hypothesis testing). Using the data from Table 17.3, we will check the conditions for using a normal distribution to analyze the results of the study using a hypothesis test.

$$\hat{p}_{pool} = \frac{\text{number of patients who died from breast cancer in the entire study}}{\text{number of patients in the entire study}}$$
$$= \frac{500 + 505}{500 + 44{,}425 + 505 + 44{,}405}$$
$$= 0.0112$$

---

[3]This is an experiment. Patients were randomized to receive mammograms or a standard breast cancer exam. We will be able to make causal conclusions based on this study.

[4]H_0 : the breast cancer death rate for patients screened using mammograms is the same as the breast cancer death rate for patients in the control, $p_{MGM} - p_C = 0$. H_A : the breast cancer death rate for patients screened using mammograms is different than the breast cancer death rate for patients in the control, $p_{MGM} - p_C \neq 0$.

17.3. MATHEMATICAL MODEL FOR THE DIFFERENCE IN PROPORTIONS

This proportion is an estimate of the breast cancer death rate across the entire study, and it's our best estimate of the proportions p_{MGM} and p_C if the null hypothesis is true that $p_{MGM} = p_C$. We will also use this pooled proportion when computing the standard error.

EXAMPLE

Is it reasonable to model the difference in proportions using a normal distribution in this study?

Because the patients were randomized, observations can be assumed to be independent, both within each group and between treatment groups. We also must check the success-failure condition for each group. Under the null hypothesis, the proportions p_{MGM} and p_C are equal, so we check the success-failure condition with our best estimate of these values under H_0, the pooled proportion from the two samples, $\hat{p}_{pool} = 0.0112$:

$$\hat{p}_{pool} \times n_{MGM} = 0.0112 \times 44{,}925 = 503$$
$$(1 - \hat{p}_{pool}) \times n_{MGM} = 0.9888 \times 44{,}925 = 44{,}422$$
$$\hat{p}_{pool} \times n_C = 0.0112 \times 44{,}910 = 503$$
$$(1 - \hat{p}_{pool}) \times n_C = 0.9888 \times 44{,}910 = 44{,}407$$

The success-failure condition is satisfied since all values are at least 10. With both conditions satisfied, we can safely model the difference in proportions using a normal distribution.

In the previous example, the pooled proportion was used to check the success-failure condition[5]. In the next example, we see an additional place where the pooled proportion comes into play: the standard error calculation.

EXAMPLE

Compute the point estimate of the difference in breast cancer death rates in the two groups, and use the pooled proportion $\hat{p}_{pool} = 0.0112$ to calculate the standard error.

The point estimate of the difference in breast cancer death rates is

$$\hat{p}_{MGM} - \hat{p}_C = \frac{500}{500 + 44{,}425} - \frac{505}{505 + 44{,}405}$$
$$= 0.01113 - 0.01125$$
$$= -0.00012$$

The breast cancer death rate in the mammogram group was 0.012% less than in the control group. Next, the standard error is calculated *using the pooled proportion, $\hat{p}_{pool}$*:

$$SE = \sqrt{\frac{\hat{p}_{pool}(1 - \hat{p}_{pool})}{n_{MGM}} + \frac{\hat{p}_{pool}(1 - \hat{p}_{pool})}{n_C}} = 0.00070$$

---

[5] For an example of a two-proportion hypothesis test that does not require the success-failure condition to be met, see Section 17.1.

> **EXAMPLE**
>
> Using the point estimate $\hat{p}_{MGM} - \hat{p}_C = -0.00012$ and standard error $SE = 0.00070$, calculate a p-value for the hypothesis test and write a conclusion.
>
> ---
>
> Just like in past tests, we first compute a test statistic and draw a picture:
>
> $$Z = \frac{\text{point estimate} - \text{null value}}{SE} = \frac{-0.00012 - 0}{0.00070} = -0.17$$
>
>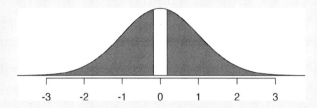
>
> The lower tail area is 0.4325, which we double to get the p-value: 0.8650. Because this p-value is larger than 0.05, we do not reject the null hypothesis. That is, the difference in breast cancer death rates is likely to have occurred just by chance, if the null hypothesis is true. Thus, we do not observe benefits or harm from mammograms relative to a regular breast exam.

Can we conclude that mammograms have no benefits or harm? Here are a few considerations to keep in mind when reviewing the mammogram study as well as any other medical study:

- We do not accept the null hypothesis, which means we don't have sufficient evidence to conclude that mammograms reduce or increase breast cancer deaths.
- If mammograms are helpful or harmful, the data suggest the effect isn't very large.
- Are mammograms more or less expensive than a non-mammogram breast exam? If one option is much more expensive than the other and doesn't offer clear benefits, then we should lean towards the less expensive option.
- The study's authors also found that mammograms led to over-diagnosis of breast cancer, which means some breast cancers were found (or thought to be found) but that these cancers would not cause symptoms during patients' lifetimes. That is, something else would kill the patient before breast cancer symptoms appeared. This means some patients may have been treated for breast cancer unnecessarily, and this treatment is another cost to consider. It is also important to recognize that over-diagnosis can cause unnecessary physical or emotional harm to patients.

These considerations highlight the complexity around medical care and treatment recommendations. Experts and medical boards who study medical treatments use considerations like those above to provide their best recommendation based on the current evidence.

17.4 Chapter review

17.4.1 Summary

When the parameter of interest is the difference in population proportions across two groups, randomization tests, bootstrapping, and mathematical modeling can be applied. For confidence intervals, bootstrapping from each group separately will provide a sampling distribution for the difference in sample proportions; the mathematical model shows a similar distributional shape as long as the sample size is large enough to fulfill the success-failure conditions and so that the data are representative of the entire population. Keep in mind that some datasets will produce a confidence interval which does not capture the true parameter, this is the nature of variability! Over your lifetime, about 95% of the confidence intervals you create will capture the parameter of interest, and about 5% won't. For hypothesis testing, repeated randomization of the explanatory variable creates a null distribution of differences in sample proportions that could have occurred under the null hypothesis. Randomization and the mathematical model will have similar null distributions, as long as the sample size is large enough to fulfill the success-failure conditions.

17.4.2 Terms

We introduced the following terms in the chapter. If you're not sure what some of these terms mean, we recommend you go back in the text and review their definitions. We are purposefully presenting them in alphabetical order, instead of in order of appearance, so they will be a little more challenging to locate. However you should be able to easily spot them as **bolded text**.

percentile interval	pooled proportion	SE interval
point estimate	SE difference in proportions	Z score two proportions

17.5 Exercises

Answers to odd numbered exercises can be found in Appendix A.17.

17.1. **Disaggregating Asian American tobacco use, hypothesis testing.** Understanding cultural differences in tobacco use across different demographic groups can lead to improved health care education and treatment. A recent study disaggregated tobacco use across Asian American ethnic groups including Asian-Indian (n = 4,373), Chinese (n = 4,736), and Filipino (n = 4,912), in comparison to non-Hispanic Whites (n = 275,025). The number of current smokers in each group was reported as Asian-Indian (n = 223), Chinese (n = 279), Filipino (n = 609), and non-Hispanic Whites (n = 50,880). (Rao et al., 2021)

To determine whether or not the proportion of Asian-Indian Americans who are current smokers is different from the proportion of Chinese Americans who are smokers, a randomization simulation was performed.

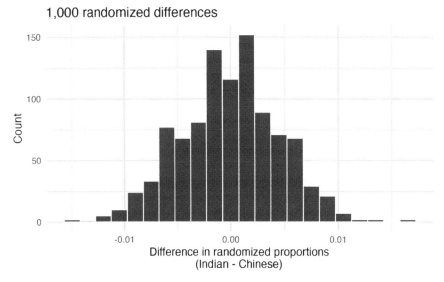

a. In both words and symbols provide the parameter and statistic of interest for this study. Do you know the numerical value of either the parameter or statisic of interest? If so, provide the numerical value.

b. The histogram above provides the sampling distribution (under randomization) for $\hat{p}_{Asian-Indian} - \hat{p}_{Chinese}$ under repeated null randomizations ($\hat{p}$ is the proportion in the sample who are current smokers). Estimate the standard error of $\hat{p}_{Asian-Indian} - \hat{p}_{Chinese}$ based on the randomization histogram.

c. Consider the hypothesis test to determine if there is a difference in proportion of Asian-Indian Americans as compared to Chinese Americans who are current smokers. Write out the null and alternative hypotheses, estimate a p-value using the randomization histogram, and conclude the test in the context of the problem.

17.2. **Malaria vaccine effectiveness, hypothesis test.** With no currently licensed vaccines to inhibit malaria, good news was welcomed with a recent study reporting long-awaited vaccine success for children in Burkina Faso. With 450 children randomized to either one of two different doses of the malaria vaccine or a control vaccine, 89 of 292 malaria vaccine and 106 out of 147 control vaccine children contracted malaria within 12 months after the treatment. (Datoo et al., 2021)

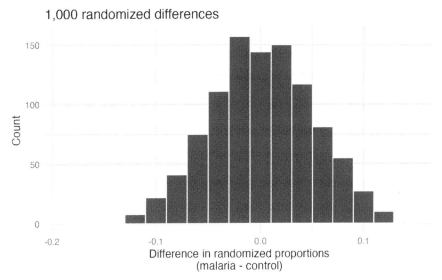

a. In both words and symbols provide the parameter and statistic of interest for this study. Do you know the numerical value of either the parameter or statisic of interest? If so, provide the numerical value.

b. The histogram above provides the sampling distribution (under randomization) for $\hat{p}_{malaria} - \hat{p}_{control}$ under repeated null randomizations ($\hat{p}$ is the proportion of children in the sample who contracted malaria). Estimate the standard error of $\hat{p}_{malaria} - \hat{p}_{control}$ based on the randomization histogram.

c. Consider the hypothesis test constructed to show a lower proportion of children contracting malaria on the malaria vaccine as compared to the control vaccine. Write out the null and alternative hypotheses, estimate a p-value using the randomization histogram, and conclude the test in the context of the problem.

17.3. **Disaggregating Asian American tobacco use, confidence interval.** Based on a study on the degree to which smoking practices differ across ethnic groups. a confidence interval for the difference in current smoking status for Filipino versus Chinese Americans is desired. (Rao et al., 2021)

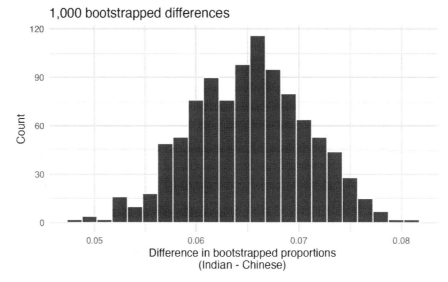

a. Consider the bootstrap distribution of difference in sample proportions of current smokers (Filipino Americans minus Chinese Americans) in 1,000 bootstrap repetitions as above. Estimate the standard error of the difference in sample proportions, as seen in the histogram.

b. Using the standard error from the bootstrap distribution, find a 95% bootstrap SE confidence interval for the true difference in proportion of current smokers (Filipino Americans minus Chinese Americans) in the population. Interpret the interval in the context of the problem.

c. Using the entire bootstrap distribution, find a 95% bootstrap percentile confidence interval for the true difference in proportion of current smokers (Filipino Americans minus Chinese Americans) in the population. Interpret the interval in the context of the problem.

17.4. **Malaria vaccine effectiveness, confidence interval.** With no currently licensed vaccines to inhibit malaria, good news was welcomed with a recent study reporting long-awaited vaccine success for children in Burkina Faso. With 450 children randomized to either one of two different doses of the malaria vaccine or a control vaccine, 89 of 292 malaria vaccine and 106 out of 147 control vaccine children contracted malaria within 12 months after the treatment. (Datoo et al., 2021)

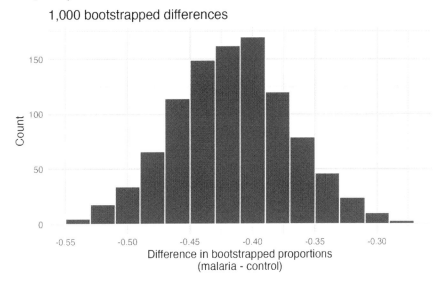

a. Consider the bootstrap distribution of difference in sample proportions of children who contracted malaria (malaria vaccine minus control vaccine) in 1000 bootstrap repetitions as above. Estimate the standard error of the difference in sample proportions, as seen in the histogram.

b. Using the standard error from the bootstrap distribution, find a 95% bootstrap SE confidence interval for the true difference in proportion of children who contract malaria (malaria vaccine minus control vaccine) in the population. Interpret the interval in the context of the problem.

c. Using the entire bootstrap distribution, find a 95% bootstrap percentile confidence interval for the true difference in proportion of children who contract malaria (malaria vaccine minus control vaccine) in the population. Interpret the interval in the context of the problem.

17.5. **COVID-19 and degree completion.** A 2021 Gallup poll surveyed 3,941 students pursuing a bachelor's degree and 2,064 students pursuing an associate degree (students were not randomly selected but were weighted so as to represent a random selection of currently enrolled US college students). The poll found that 51% of the bachelor's degree students and 44% of associate degree students said that the COVID-19 pandemic will negatively impact their ability to complete the degree. (Gallup, 2021a)

Below are two histograms which represent different computational approaches (both use 1,000 repetitions) to research questions which could be asked from the Gallup data which was provided. One of the histograms can be used to do a randomization test on whether the proportions of bachelor's and associate students who think the COVID-19 pandemic will negatively impact their ability to complete the degree. The other histogram is a bootstrap distribution used to quantify the difference in the proportions of bachelor's and associate's students who feel this way.

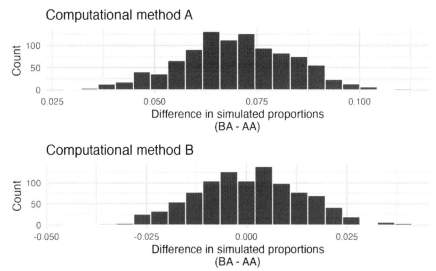

a. Are the center and standard error of the two graphs approximately the same? Explain.

b. Write a research question which could be addressed using this hitogram with computational method A.

c. Write a research question which could be addressed using this hitogram with computational method B.

17.6. **Renewable energy.** A 2021 Gallup poll surveyed 5,447 randomly sampled US adults who are Republican (or Republican leaning) and 7,962 who are Democrats (or Democrat leaning). 31% of Republicans and 81% of Democrats said "government regulations are necessary to encourage businesses and consumers to rely more on renewable energy sources". (Gallup, 2021a)

Below are two histograms which represent different computational approaches (both use 1,000 repetitions) to research questions which could be asked from the Gallup data which was provided. One of the histograms can be used to do a randomization test on whether the proportions of Republicans and Democrats who think government regulations are necessary to encourage businesses and consumers to rely more on renewable energy sources are different. The other histogram is a bootstrap distribution used to quantify the difference in the proportions of Republicans and Democrats who agree with this statement.

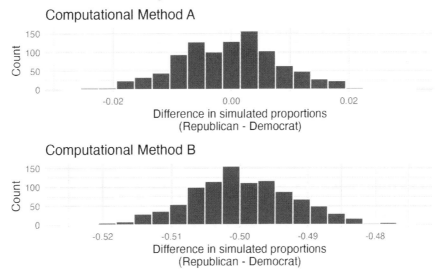

a. Are the center and standard error of the two graphs approximately the same? Explain.

b. Write a research question which could be addressed using this hitogram with computational method A.

c. Write a research question which could be addressed using this hitogram with computational method B.

17.7. **HIV in sub-Saharan Africa.** In July 2008 the US National Institutes of Health announced that it was stopping a clinical study early because of unexpected results. The study population consisted of HIV-infected women in sub-Saharan Africa who had been given single dose Nevaripine (a treatment for HIV) while giving birth, to prevent transmission of HIV to the infant. The study was a randomized comparison of continued treatment of a woman (after successful childbirth) with Nevaripine vs Lopinavir, a second drug used to treat HIV. 240 women participated in the study; 120 were randomized to each of the two treatments. Twenty-four weeks after starting the study treatment, each woman was tested to determine if the HIV infection was becoming worse (an outcome called *virologic failure*). Twenty-six of the 120 women treated with Nevaripine experienced virologic failure, while 10 of the 120 women treated with the other drug experienced virologic failure. (Lockman et al., 2007)

a. Create a two-way table presenting the results of this study.

b. State appropriate hypotheses to test for difference in virologic failure rates between treatment groups.

c. Complete the hypothesis test and state an appropriate conclusion. (Reminder: Verify any necessary conditions for the test.)

17.8. **Supercommuters.** The fraction of workers who are considered "supercommuters", because they commute more than 90 minutes to get to work, varies by state. Suppose the 1% of Nebraska residents and 6% of New York residents are supercommuters. Now suppose that we plan a study to survey 1000 people from each state, and we will compute the sample proportions $\hat{p}_{NE}$ for Nebraska and $\hat{p}_{NY}$ for New York.

  a. What is the associated mean and standard deviation of $\hat{p}_{NE}$ in repeated samples of size 1000?

  b. What is the associated mean and standard deviation of $\hat{p}_{NY}$ in repeated samples of size 1000?

  c. Calculate and interpret the mean and standard deviation associated with the difference in sample proportions for the two groups, $\hat{p}_{NY} - \hat{p}_{NE}$ in repeated samples of 1000 in each group.

  d. How are the standard deviations from parts (a), (b), and (c) related?

17.9. **National Health Plan.** A Kaiser Family Foundation poll for US adults in 2019 found that 79% of Democrats, 55% of Independents, and 24% of Republicans supported a generic "National Health Plan". There were 347 Democrats, 298 Republicans, and 617 Independents surveyed. 79% of 347 Democrats and 55% of 617 Independents support a National Health Plan. (Foundation, 2019)

  a. Calculate a 95% confidence interval for the difference between the proportion of Democrats and Independents who support a National Health Plan ($p_D - p_I$), and interpret it in this context. We have already checked conditions for you.

  b. True or false: If we had picked a random Democrat and a random Independent at the time of this poll, it is more likely that the Democrat would support the National Health Plan than the Independent.

17.10. **Sleep deprivation, CA vs. OR, confidence interval.** According to a report on sleep deprivation by the Centers for Disease Control and Prevention, the proportion of California residents who reported insufficient rest or sleep during each of the preceding 30 days is 8.0%, while this proportion is 8.8% for Oregon residents. These data are based on simple random samples of 11,545 California and 4,691 Oregon residents. Calculate a 95% confidence interval for the difference between the proportions of Californians and Oregonians who are sleep deprived and interpret it in context of the data. (CDC, 2008)

17.11. **Gender pay gap in medicine.** A study examined the average pay for men and women entering the workforce as doctors for 21 different positions. (AT et al., 2011)

  a. If each gender was equally paid, then we would expect about half of those positions to have men paid more than women and women would be paid more than men in the other half of positions. Write appropriate hypotheses to test this scenario.

  b. Men were, on average, paid more in 19 of those 21 positions. Complete a hypothesis test using your hypotheses from part (a).

17.12. **Sleep deprivation, CA vs. OR, hypothesis test.** A CDC report on sleep deprivation rates shows that the proportion of California residents who reported insufficient rest or sleep during each of the preceding 30 days is 8.0%, while this proportion is 8.8% for Oregon residents. These data are based on simple random samples of 11,545 California and 4,691 Oregon residents.

  a. Conduct a hypothesis test to determine if these data provide strong evidence that the rate of sleep deprivation is different for the two states. (Reminder: Check conditions)

  b. It is possible the conclusion of the test in part (a) is incorrect. If this is the case, what type of error was made?

17.5. EXERCISES

17.13. **Is yawning contagious?** An experiment conducted by the MythBusters, a science entertainment TV program on the Discovery Channel, tested if a person can be subconsciously influenced into yawning if another person near them yawns. 50 people were randomly assigned to two groups: 34 to a group where a person near them yawned (treatment) and 16 to a group where there wasn't a person yawning near them (control). The visualization below displays how many participants yawned in each group.[6]

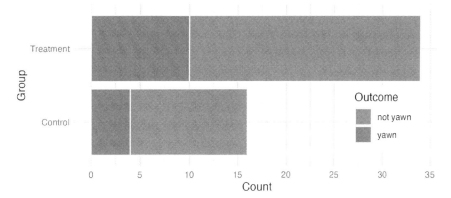

Suppose we are interested in estimating the difference in yawning rates between the control and treatment groups using a confidence interval. Explain why we cannot construct such an interval using the normal approximation. What might go wrong if we constructed the confidence interval despite this problem?

17.14. **Heart transplant success.** The Stanford University Heart Transplant Study was conducted to determine whether an experimental heart transplant program increased lifespan. Each patient entering the program was officially designated a heart transplant candidate, meaning that he was gravely ill and might benefit from a new heart. Patients were randomly assigned into treatment and control groups. Patients in the treatment group received a transplant, and those in the control group did not. The visualization below displays how many patients survived and died in each group.[7] (Turnbull et al., 1974)

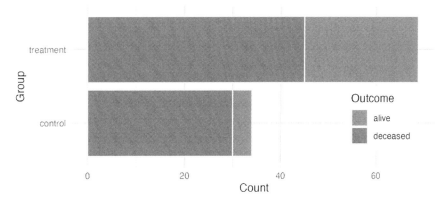

Suppose we are interested in estimating the difference in survival rate between the control and treatment groups using a confidence interval. Explain why we cannot construct such an interval using the normal approximation. What might go wrong if we constructed the confidence interval despite this problem?

---

[6]The yawn data used in this exercise can be found in the **openintro** R package.
[7]The heart_transplant data used in this exercise can be found in the **openintro** R package.

17.15. **Government shutdown.** The United States federal government shutdown of 2018–2019 occurred from December 22, 2018 until January 25, 2019, a span of 35 days. A Survey USA poll of 614 randomly sampled Americans during this time period reported that 48% of those who make less than $40,000 per year and 55% of those who make $40,000 or more per year said the government shutdown has not at all affected them personally. A 95% confidence interval for $(p_{<40K} - p_{\geq 40K})$, where p is the proportion of those who said the government shutdown has not at all affected them personally, is (-0.16, 0.02). Based on this information, determine if the following statements are true or false, and explain your reasoning if you identify the statement as false. (Survey USA, 2019)

   a. At the 5% significance level, the data provide convincing evidence of a real difference in the proportion who are not affected personally between Americans who make less than $40,000 annually and Americans who make $40,000 annually.

   b. We are 95% confident that 16% more to 2% fewer Americans who make less than $40,000 per year are not at all personally affected by the government shutdown compared to those who make $40,000 or more per year.

   c. A 90% confidence interval for $(p_{<40K} - p_{\geq 40K})$ would be wider than the $(-0.16, 0.02)$ interval.

   d. A 95% confidence interval for $(p_{\geq 40K} - p_{<40K})$ is (-0.02, 0.16).

17.16. **Online harassment.** A Pew Research poll asked US adults aged 18-29 and 30-49 whether they have personally experienced harassment online. A 95% confidence interval for the difference between the proportions of 18-29 year olds and 30-49 year olds who have personally experienced harassment online $(p_{18-29} - p_{30-49})$ was calculated to be (0.115, 0.185). Based on this information, determine if the following statements are true or false, and explain your reasoning for each statement you identify as false. (Pew Research Center, 2021a)

   a. We are 95% confident that the true proportion of 18-29 year olds who have personally experienced harassment online is 11.5% to 18.5% lower than the true proportion of 30-49 year olds who have personally experienced harassment online.

   b. We are 95% confident that the true proportion of 18-29 year olds who have personally experienced harassment online is 11.5% to 18.5% higher than the true proportion of 30-49 year olds who have personally experienced harassment online.

   c. 95% of random samples will produce 95% confidence intervals that include the true difference between the population proportions of 18-29 year olds and 30-49 year olds who have personally experienced harassment online.

   d. We can conclude that there is a significant difference between the proportions of 18-29 year olds and 30-49 year olds who have personally experienced harassment online is too large to plausibly be due to chance, if in fact there is no difference between the two proportions.

   e. The 90% confidence interval for $(p_{18-29} - p_{30-49})$ cannot be calculated with only the information given in this exercise.

17.17. **Decision errors and comparing proportions, I.** In the following research studies, conclusions were made based on the data provided. It is always possible that the analysis conclusion could be wrong, although we will almost never actually know if an error has been made or not. For each study conclusion, specify which of a Type 1 or Type 2 error could have been made, and state the error in the context of the problem.

   a. The malaria vaccine was seen to be effective at lowering the rate of contracting malaria (when compared to the control vaccine).

   b. In the US population, Asian-Indian Americans and Chinese Americans are not observed to have different proportions of current smokers.

   c. There is no evidence to claim a difference in the proportion of Americans who are not affected personally by a government shutdown when comparing Americans who make less than $40,000 annually and Americans who make $40,000 annually.

17.5. EXERCISES

17.18. **Decision errors and comparing proportions, II.** In the following research studies, conclusions were made based on the data provided. It is always possible that the analysis conclusion could be wrong, although we will almost never actually know if an error has been made or not. For each study conclusion, specify which of a Type 1 or Type 2 error could have been made, and state the error in the context of the problem.

  a. Of registered voters in California, the proportion who report not knowing enough to voice an opinion on whether or not they support off shore drilling is different across those who have a college degree and those who do not.

  b. In comparing Californians and Oregonians, there is no evidence to support a difference in the proportion of each who are sleep deprived.

17.19. **Active learning.** A teacher wanting to increase the active learning component of her course is concerned about student reactions to changes she is planning to make. She conducts a survey in her class, asking students whether they believe more active learning in the classroom (hands on exercises) instead of traditional lecture will helps improve their learning. She does this at the beginning and end of the semester and wants to evaluate whether students' opinions have changed over the semester. Can she used the methods we learned in this chapter for this analysis? Explain your reasoning.

17.20. **An apple a day keeps the doctor away.** A physical education teacher at a high school wanting to increase awareness on issues of nutrition and health asked her students at the beginning of the semester whether they believed the expression "an apple a day keeps the doctor away", and 40% of the students responded yes. Throughout the semester she started each class with a brief discussion of a study highlighting positive effects of eating more fruits and vegetables. She conducted the same apple-a-day survey at the end of the semester, and this time 60% of the students responded yes. Can she used a two-proportion method from this section for this analysis? Explain your reasoning.

17.21. **Malaria vaccine effectiveness, effect size.** A randomized controlled trial on malaria vaccine effectiveness randomly assigned 450 children intro either one of two different doses of the malaria vaccine or a control vaccine. 89 of 292 malaria vaccine and 106 out of 147 control vaccine children contracted malaria within 12 months after the treatment. (Datoo et al., 2021)

Recall that in order to reject the null hypothesis that the two vaccines (malaria and control) are equivalent, we'd need the sample proportion to be about 2 standard errors below the hypothesized value of zero.

Say that the true difference (in the population) is given as δ, the sample sizes are the same in both groups ($n_{malaria} = n_{control}$), and the true proportion who contract malaria on the control vaccine is $p_{control} = 0.7$. If you ran your own study (in the future), how likely is it that you would get a difference in sample proportions that was sufficiently far from zero that you could reject under each of the conditions below. (*Hint:* Use the mathematical model.)

  a. $\delta = -0.1$ and $n_{malaria} = n_{control} = 20$

  b. $\delta = -0.4$ and $n_{malaria} = n_{control} = 20$

  c. $\delta = -0.1$ and $n_{malaria} = n_{control} = 100$

  d. $\delta = -0.4$ and $n_{malaria} = n_{control} = 100$

  e. What can you conclude about values of δ and the sample size?

17.22. **Diabetes and unemployment.** A Gallup poll surveyed Americans about their employment status and whether or not they have diabetes. The survey results indicate that 1.5% of the 47,774 employed (full or part time) and 2.5% of the 5,855 unemployed 18-29 year olds have diabetes. (Gallup, 2012)

  a. Create a two-way table presenting the results of this study.

  b. State appropriate hypotheses to test for difference in proportions of diabetes between employed and unemployed Americans.

  c. The sample difference is about 1%. If we completed the hypothesis test, we would find that the p-value is very small (about 0), meaning the difference is statistically significant. Use this result to explain the difference between statistically significant and practically significant findings.

Chapter 18

Inference for two-way tables

 In Section 17 our focus was on the difference in proportions, a statistic calculated from finding the success proportions (from the binary response variable) measured across two groups (the binary explanatory variable). As we will see in the examples below, sometimes the explanatory or response variables have more than two possible options. In that setting, a difference across two groups is not sufficient, and the proportion of "success" is not well defined if there are 3 or 4 or more possible response levels. The primary way to summarize categorical data where the explanatory and response variables both have 2 or more levels is through a two-way table as in Table 18.1.

Note that with two-way tables, there is not an obvious single parameter of interest. Instead, research questions usually focus on how the proportions of the response variable changes (or not) across the different levels of the explanatory variable. Because there is not a population parameter to estimate, bootstrapping to find the standard error of the estimate is not meaningful. As such, for two-way tables, we will focus on the randomization test and corresponding mathematical approximation (and not bootstrapping).

18.1 Randomization test of independence

We all buy used products – cars, computers, textbooks, and so on – and we sometimes assume the sellers of those products will be forthright about any underlying problems with what they're selling. This is not something we should take for granted. Researchers recruited 219 participants in a study where they would sell a used iPod[1] that was known to have frozen twice in the past. The participants were incentivized to get as much money as they could for the iPod since they would receive a 5% cut of the sale on top of $10 for participating. The researchers wanted to understand what types of questions would elicit the seller to disclose the freezing issue.

Unbeknownst to the participants who were the sellers in the study, the buyers were collaborating with the researchers to evaluate the influence of different questions on the likelihood of getting the

---

[1] For readers not as old as the authors, an iPod is basically an iPhone without any cellular service, assuming it was one of the later generations. Earlier generations were more basic.

sellers to disclose the past issues with the iPod. The scripted buyers started with "Okay, I guess I'm supposed to go first. So you've had the iPod for 2 years ..." and ended with one of three questions:

- General: What can you tell me about it?
- Positive Assumption: It doesn't have any problems, does it?
- Negative Assumption: What problems does it have?

The question is the treatment given to the sellers, and the response is whether the question prompted them to disclose the freezing issue with the iPod. The results are shown in Table 18.1, and the data suggest that asking the, *What problems does it have?*, was the most effective at getting the seller to disclose the past freezing issues. However, you should also be asking yourself: could we see these results due to chance alone if there really is no difference in the question asked, or is this in fact evidence that some questions are more effective for getting at the truth?

Table 18.1: Summary of the iPod study, where a question was posed to the study participant who acted.

Question	Disclose problem	Hide problem	Total
General	2	71	73
Positive assumption	23	50	73
Negative assumption	36	37	73
Total	61	158	219

The ask data can be found in the **openintro** R package.

The hypothesis test for the iPod experiment is really about assessing whether there is convincing evidence that there was a difference in the success rates that each question had on getting the participant to disclose the problem with the iPod. In other words, the goal is to check whether the buyer's question was independent of whether the seller disclosed a problem.

18.1.1 Expected counts in two-way tables

While we would not expect the number of disclosures to be exactly the same across the three question classes, the rate of disclosure seems substantially different across the three groups. In order to investigate whether the differences in rates is due to natural variability in people's honesty or due to a treatment effect (i.e., the question causing the differences), we need to compute estimated counts for each cell in a two-way table.

EXAMPLE

From the experiment, we can compute the proportion of all sellers who disclosed the freezing problem as $61/219 = 0.2785$. If there really is no difference among the questions and 27.85% of sellers were going to disclose the freezing problem no matter the question they were asked, how many of the 73 people in the `General` group would we have expected to disclose the freezing problem?

We would predict that $0.2785 \times 73 = 20.33$ sellers would disclose the problem. Obviously we observed fewer than this, though it is not yet clear if that is due to chance variation or whether that is because the questions vary in how effective they are at getting to the truth.

GUIDED PRACTICE

If the questions were actually equally effective, meaning about 27.85% of respondents would disclose the freezing issue regardless of what question they were asked, about how many sellers would we expect to *hide* the freezing problem from the Positive Assumption group?[2]

We can compute the expected number of sellers who we would expect to disclose or hide the freezing issue for all groups, if the questions had no impact on what they disclosed, using the same strategies employed in the previous Example and Guided Practice to compute expected counts. These expected counts were used to construct Table 18.2, which is the same as Table 18.1, except now the expected counts have been added in parentheses.

Table 18.2: The observed counts and the expected counts for the iPod experiment.

	Disclose problem		Hide problem		Total
General	2	*(20.33)*	71	*(52.67)*	73
Positive assumption	23	*(20.33)*	50	*(52.67)*	73
Negative assumption	36	*(20.33)*	37	*(52.67)*	73
Total	61		158		219

The examples and exercises above provided some help in computing expected counts. In general, expected counts for a two-way table may be computed using the row totals, column totals, and the table total. For instance, if there was no difference between the groups, then about 27.85% of each row should be in the first column:

$$0.2785 \times (\text{row 1 total}) = 20.33$$
$$0.2785 \times (\text{row 2 total}) = 20.33$$
$$0.2785 \times (\text{row 3 total}) = 20.33$$

Looking back to how 0.2785 was computed – as the fraction of sellers who disclosed the freezing issue (158/219) – these three expected counts could have been computed as

$$\left(\frac{\text{row 1 total}}{\text{table total}}\right)(\text{column 1 total}) = 20.33$$
$$\left(\frac{\text{row 1 total}}{\text{table total}}\right)(\text{column 2 total}) = 20.33$$
$$\left(\frac{\text{row 1 total}}{\text{table total}}\right)(\text{column 3 total}) = 20.33$$

This leads us to a general formula for computing expected counts in a two-way table when we would like to test whether there is strong evidence of an association between the column variable and row variable.

---

[2] We would expect $(1 - 0.2785) \times 73 = 52.67$. It is okay that this result, like the result from Example ??, is a fraction.

Computing expected counts in a two-way table.

To calculate the expected count for the ith row and j^{th} column, compute

$$\text{Expected Count}_{\text{row } i, \text{ col } j} = \frac{(\text{row } i \text{ total}) \times (\text{column } j \text{ total})}{\text{table total}}$$

18.1.2 The observed chi-squared statistic

The chi-squared test statistic for a two-way table is found by finding the ratio of how far the observed counts are from the expected counts, as compared to the expected counts, for every cell in the table. For each table count, compute:

General formula $\quad \dfrac{(\text{observed count } - \text{ expected count})^2}{\text{expected count}}$

Row 1, Col 1 $\quad \dfrac{(2 - 20.33)^2}{20.33} = 16.53$

Row 2, Col 1 $\quad \dfrac{(23 - 20.33)^2}{20.33} = 0.35$

$\vdots \qquad\qquad \vdots$

Row 2, Col 3 $\quad \dfrac{(37 - 52.67)^2}{52.67} = 4.66$

Adding the computed value for each cell gives the chi-squared test statistic X^2:

$$X^2 = 16.53 + 0.35 + \cdots + 4.66 = 40.13$$

Is 40.13 a big number? That is, does it indicate that the observed and expected values are really different? Or is 40.13 a value of the statistic that we'd expect to see just due to natural variability? Previously, we applied the randomization test to the setting where the research question investigated a difference in proportions. The same idea of shuffling the data under the null hypothesis can be used in the setting of the two-way table.

18.1.3 Variability of the statistic

Assuming that the individuals would disclose or hide the problems **regardless** of the question they are given (i.e., that the null hypothesis is true), we can randomize the data by reassigning the 61 disclosed problems and 158 hidden problems to the three groups at random. Table 18.3 shows a possible randomization of the observed data under the condition that the null hypothesis is true (in contrast to the original observed data in Table 18.1).

Table 18.3: Summary of the iPod study, where a question was posed to the study participant who acted.

Question	Disclose problem	Hide problem	Total
General	29	44	73
Positive assumption	15	58	73
Negative assumption	17	56	73
Total	61	158	219

As before, the randomized data is used to find a single value for the test statistic (here a chi-squared statistic). The chi-squared statistic for the randomized two-way table is found by comparing the observed and expected counts for each cell in the *randomized* table. For each cell, compute:

$$\text{General formula} \quad \frac{(\text{observed count} - \text{expected count})^2}{\text{expected count}}$$

$$\text{Row 1, Col 1} \quad \frac{(29 - 20.33)^2}{20.33} = 3.7$$

$$\text{Row 2, Col 1} \quad \frac{(15 - 20.33)^2}{20.33} = 1.4$$

$$\vdots \quad \vdots$$

$$\text{Row 3, Col 2} \quad \frac{(56 - 52.67)^2}{52.67} = 0.211$$

Adding the computed value for each cell gives the chi-squared test statistic X^2:

$$X^2 = 3.7 + 1.4 + \cdots + 0.211 = 8$$

18.1.4 Observed statistic vs. null chi-squared statistics

As before, one randomization will not be sufficient for understanding if the observed data are particularly different from the expected chi-squared statistics when H_0 is true. To investigate whether 40.13 is large enough to indicate the observed and expected counts are substantially different, we need to understand the variability in the values of the chi-squared statistic we would expect to see if the null hypothesis was true. Figure 18.1 plots 1,000 chi-squared statistics generated under the null hypothesis. We can see that the observed value is so far from the null statistics that the simulated p-value is zero. That is, the probability of seeing the observed statistic when the null hypothesis is true is virtually zero. In this case we can conclude that the decision of whether or not to disclose the iPod's problem is changed by the question asked. We use the causal language of "changed" because the study was an experiment. Note that with a chi-squared test, we only know that the two variables (`question_class` and `response`) are related (i.e., not independent). We are not able to claim which type of question causes which type of response.

18.2 Mathematical model for test of independence

18.2.1 The chi-squared test of independence

Previously, in Section 17.3, we applied the Central Limit Theorem to the sampling variability of $\hat{p}_1 - \hat{p}_2$. The result was that we could use the normal distribution (e.g., z^* values (see Figure 16.2) and p-values from Z scores) to complete the mathematical inferential procedure. The chi-squared test statistic has a different mathematical distribution called the Chi-squared distribution. The important specification to make in describing the chi-squared distribution is something called degrees of freedom. The degrees of freedom change the shape of the chi-squared distribution to fit the problem at hand. Figure 18.2 visualizes different chi-squared distributions corresponding to different degrees of freedom.

18.2.2 Variability of the chi-squared statistic

As it turns out, the chi-squared test statistic follows a Chi-squared distribution when the null hypothesis is true. For two way tables, the degrees of freedom is equal to: $df =$ (number of rows minus 1) $\times$ (number of columns minus 1). In our example, the degrees of freedom parameter is $df = (2-1) \times (3-1) = 2$.

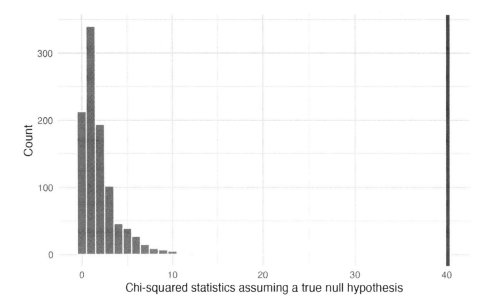

Figure 18.1: A histogram of chi-squared statisics from 1,000 simulations produced under the null hypothesis, H_0, where the question is independent of the response. The observed statistic of 40.13 is marked by the red line. None of the 1,000 simulations had a chi-squared value of at least 40.13. In fact, none of the simulated chi-squared statistics came anywhere close to the observed statistic!

18.2.3 Observed statistic vs. null chi-squared statistics

The test statistic for assessing the independence between two categorical variables is a X^2.

The X^2 statistic is a ratio of how the observed counts vary from the expected counts as compared to the expected counts (which are a measure of how large the sample size is).

$$X^2 = \sum_{i,j} \frac{(\text{observed count} - \text{expected count})^2}{\text{expected count}}$$

When the null hypothesis is true and the conditions are met, X^2 has a Chi-squared distribution with $df = (r-1) \times (c-1)$.

Conditions:

- Independent observations
- Large samples: 5 expected counts in each cell

To bring it back to the example, we can safely assume that the observations are independent, as the question groups were randomly assigned. Additionally, there are over 5 expected counts in each cell, so the conditions for using the Chi-square distribution are met. If the null hypothesis is true (i.e., the questions had no impact on the sellers in the experiment), then the test statistic $X^2 = 40.13$ is expected to follow a Chi-squared distribution with 2 degrees of freedom. Using this information, we can compute the p-value for the test, which is depicted in Figure 18.3.

18.2. MATHEMATICAL MODEL FOR TEST OF INDEPENDENCE

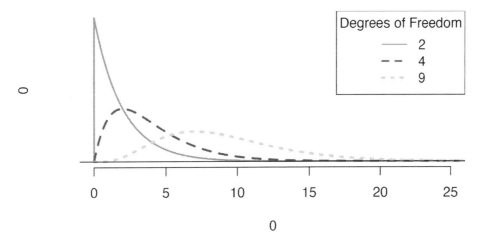

Figure 18.2: The chi-squared distribution for differing degrees of freedom. The larger the degrees of freedom, the longer the right tail extends. The smaller the degrees of freedom, the more peaked the mode on the left becomes.

 Computing degrees of freedom for a two-way table.

When applying the chi-squared test to a two-way table, we use $df = (R-1) \times (C-1)$ where R is the number of rows in the table and C is the number of columns.

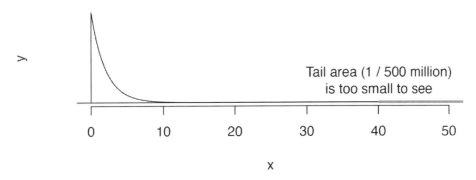

Figure 18.3: Visualization of the p-value for $X^2 = 40.13$ when $df = 2$.

The software R can be used to find the p-value with the function pchisq(). Just like pnorm(), pchisq() always gives the area to the left of the cutoff value. Because, in this example, the p-value is represented by the area to the right of 40.13, we subtract the output of pchisq() from 1.

```
1 - pchisq(40.13, df = 2)
#> [1] 1.93e-09
```

 EXAMPLE

Find the p-value and draw a conclusion about whether the question affects the sellers likelihood of reporting the freezing problem.

Using a computer, we can compute a very precise value for the tail area above $X^2 = 40.13$ for a chi-squared distribution with 2 degrees of freedom: 0.000000002.

Using a significance level of $\alpha = 0.05$, the null hypothesis is rejected since the p-value is smaller. That is, the data provide convincing evidence that the question asked did affect a seller's likelihood to tell the truth about problems with the iPod.

EXAMPLE

Table 18.4 summarizes the results of an experiment evaluating three treatments for Type 2 Diabetes in patients aged 10-17 who were being treated with metformin. The three treatments considered were continued treatment with metformin (met), treatment with metformin combined with rosiglitazone (rosi), or a lifestyle intervention program. Each patient had a primary outcome, which was either lacked glycemic control (failure) or did not lack that control (success). What are appropriate hypotheses for this test?

- H_0 : There is no difference in the effectiveness of the three treatments.
- H_A : There is some difference in effectiveness between the three treatments, e.g., perhaps the rosi treatment performed better than lifestyle.

Table 18.4: Results for the Type 2 Diabetes study.

Treatment	Failure	Success	Total
lifestyle	109	125	234
met	120	112	232
rosi	90	143	233
Total	319	380	699

The diabetes2 data can be found in the **openintro** R package.

Typically we will use a computer to do the computational work of finding the chi-squared statistic. However, it is always good to have a sense for what the computer is doing, and in particular, calculating the values which would be expected if the null hypothesis is true can help to understand the null hypothesis claim. Additionally, comparing the expected and observed values by eye often gives the researcher some insight into why or why not the null hypothesis for a given test is rejected or not.

GUIDED PRACTICE

A chi-squared test for a two-way table may be used to test the hypotheses in the diabetes Example above. To get a sense for the statistic used in the chi-squared test, first compute the expected values for each of the six table cells.[3]

Note, when analyzing 2-by-2 contingency tables (that is, when both variables only have two possible options), one guideline is to use the two-proportion methods introduced in Chapter 17.

---

[3]The expected count for row one / column one is found by multiplying the row one total (234) and column one total (319), then dividing by the table total (699): $\frac{234 \times 319}{699} = 106.8$. Similarly for the second column and the first row: $\frac{234 \times 380}{699} = 127.2$. Row 2: 105.9 and 126.1. Row 3: 106.3 and 126.7.

18.3 Chapter review

18.3.1 Summary

In this chapter we extended the randomization / bootstrap / mathematical model paradigm to research questions involving categorical variables. We continued working with one population proportion as well as the difference in populations proportions, but the test of independence allowed for hypothesis testing on categorical variables with more than two levels. We note that the normal model was an excellent mathematical approximation to the sampling distribution of sample proportions (or differences in sample proportions), but that the questions with categorical variables with more than 2 levels required a new mathematical model, the chi-squared distribution. As seen in Chapters 11, 12 and 13, almost all the research questions can be approached using computational methods (e.g., randomization tests or bootstrapping) or using mathematical models. We continue to emphasize the importance of experimental design in making conclusions about research claims. In particular, recall that variability can come from different sources (e.g., random sampling vs. random allocation, see Figure 2.8).

18.3.2 Terms

We introduced the following terms in the chapter. If you're not sure what some of these terms mean, we recommend you go back in the text and review their definitions. We are purposefully presenting them in alphabetical order, instead of in order of appearance, so they will be a little more challenging to locate. However you should be able to easily spot them as **bolded text**.

Chi-squared distribution	expected counts
chi-squared statistic	independence

18.4 Exercises

Answers to odd numbered exercises can be found in Appendix A.18.

18.1. **Quitters.** Does being part of a support group affect the ability of people to quit smoking? A county health department enrolled 300 smokers in a randomized experiment. 150 participants were randomly assigned to a group that used a nicotine patch and met weekly with a support group; the other 150 received the patch and did not meet with a support group. At the end of the study, 40 of the participants in the patch plus support group had quit smoking while only 30 smokers had quit in the other group.

  a. Create a two-way table presenting the results of this study.

  b. Answer each of the following questions under the null hypothesis that being part of a support group does not affect the ability of people to quit smoking, and indicate whether the expected values are higher or lower than the observed values.

18.2. **Act on climate change.** The table below summarizes results from a Pew Research poll which asked respondents whether they have personally taken action to help address climate change within the last year and their generation. The differences in each generational group may be due to chance. Complete the following computations under the null hypothesis of independence between an individual's generation and whether or not they have personally taken action to help address climate change within the last year. (Pew Research Center, 2021b)

| | Response | | |
Generation	Took action	Didn't take action	Total
Gen Z	292	620	912
Millenial	885	2,275	3,160
Gen X	809	2,709	3,518
Boomer & older	1,276	4,798	6,074
Total	3,262	10,402	13,664

  a. If there is no relationship between age and action, how many Gen Z'ers would you expect to have personally taken action to help address climate change within the last year?

  b. If there is no relationship between age and action, how many Millenials would you expect to have personally taken action to help address climate change within the last year?

  c. If there is no relationship between age and action, how many Gen X'ers would you expect to have personally taken action to help address climate change within the last year?

  d. If there is no relationship between age and action, how many Boomers and older would you expect to have personally taken action to help address climate change within the last year?

18.4. EXERCISES

18.3. **Lizard habitats, data.** In order to assess whether habitat conditions are related to the sunlight choices a lizard makes for resting, Western fence lizard (*Sceloporus occidentalis*) were observed across three different microhabitats.[4] (Adolph, 1990; Asbury and Adolph, 2007)

site	sunlight			Total
	sun	partial	shade	
desert	16	32	71	119
mountain	56	36	15	107
valley	42	40	24	106
Total	114	108	110	332

a. If the variables describing the habitat and the amount of sunlight are independent, what proporiton of lizards (total) would be expected in each of the three sunlight categories?

b. Given the proportions of each sunlight condition, how many lizards of each type would you expect to see in the sun? in the partial sun? in the shade?

c. Compare the observed (original data) and expected (part b.) tables. From a first glance, does it seem as though the habitat and choice of sunlight may be associated?

d. Regardless of your answer to part (c), is it possible to tell from looking only at the expected and observed counts whether the two variables are associated?

18.4. **Disaggregating Asian American tobacco use, data.** Understanding cultural differences in tobacco use across different demographic groups can lead to improved health care education and treatment. A recent study disaggregated tobacco use across Asian American ethnic groups including Asian-Indian (n = 4,373), Chinese (n = 4,736), and Filipino (n = 4,912), in comparison to non-Hispanic Whites (n = 275,025). The number of current smokers in each group was reported as Asian-Indian (n = 223), Chinese (n = 279), Filipino (n = 609), and non-Hispanic Whites (n = 50,880). (Rao et al., 2021)

In order to assess whether there is a difference in current smoking rates across three Asian American ethnic groups, the observed data is compared to the data that would be expected if there were no association between the variables.

ethnicity	Smoking		Total
	don't smoke	smoke	
Asian-Indian	4,150	223	4,373
Chinese	4,457	279	4,736
Filipino	4,303	609	4,912
Total	12,910	1,111	14,021

a. If the variables on ethnicity and smoking status are independent, estimate the proporiton of individuals (total) who smoke?

b. Given the overall proportion who smoke, how many of each Asian American ethnicity would you expect to smoke?

c. Compare the observed (original data) and expected (part b.) tables. From a first glance, does it seem as though the Asian American ethnicity and choice of smoking may be associated?

d. Regardless of your answer to part (c), is it possible to tell from looking only at the expected and observed counts whether the two variables are associated?

---

[4]The `lizard_habitat` data used in this exercise can be found in the **openintro** R package.

18.5. **Lizard habitats, randomize once.** In order to assess whether habitat conditions are related to the sunlight choices a lizard makes for resting, Western fence lizard (*Sceloporus occidentalis*) were observed across three different microhabitats. The original data is shown below. (Adolph, 1990; Asbury and Adolph, 2007)

	Original data			
	sunlight			
site	sun	partial	shade	Total
desert	16	32	71	119
mountain	56	36	15	107
valley	42	40	24	106
Total	114	108	110	332

Then, the data were randomized once, where sunlight preference was randomly assigned to the lizards across different sites. The results of the randomization is shown below.

	Randomized data			
	sunlight			
site	sun	partial	shade	Total
desert	44	42	33	119
mountain	39	31	37	107
valley	31	35	40	106
Total	114	108	110	332

Recall that the Chi-squared statistic (X^2) measures the difference between the expected and observed counts. Without calculating the actual statistic, report on whether the original data or the randomized data will have a larger Chi-squared statistic. Explain your choice.

18.4. EXERCISES

18.6. **Disaggregating Asian American tobacco use, randomize once.** In a study that aims to disaggregate tobacco use across Asian American ethnic groups (Asian-Indian, Chinese, and Filipino, in comparison to non-Hispanic Whites), respondents were asked whether they smoke tobacco or not. The original data is shown below. (Rao et al., 2021)

	Original data		
	Smoking		
ethnicity	don't smoke	smoke	Total
Asian-Indian	4,150	223	4,373
Chinese	4,457	279	4,736
Filipino	4,303	609	4,912
Total	12,910	1,111	14,021

Then, the data were randomized once, where smoking status was randomly assigned to the participants across different ethnicities. The results of the randomization is shown below.

	Randomized data		
	Smoking		
ethnicity	don't smoke	smoke	Total
Asian-Indian	4,015	358	4,373
Chinese	4,385	351	4,736
Filipino	4,510	402	4,912
Total	12,910	1,111	14,021

Recall that the Chi-squared statistic (X^2) measures the difference between the expected and observed counts. Without calculating the actual statistic, report on whether the original data or the randomized data will have a larger Chi-squared statistic. Explain your choice.

18.7. **Lizard habitats, randomization test.** In order to assess whether habitat conditions are related to the sunlight choices a lizard makes for resting, Western fence lizard (*Sceloporus occidentalis*) were observed across three different microhabitats. (Adolph, 1990; Asbury and Adolph, 2007)

The original data were randomized 1,000 times (sunlight variable randomly assigned to the observations across different habitats), and the histogram of the Chi-squared statistic on each randomization is displayed.

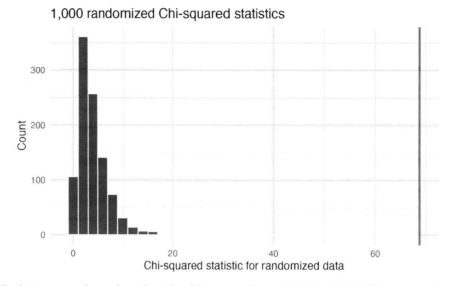

a. The histogram above describes the Chi-squared statistics for 1,000 different randomization datasets. When randomizing the data, is the imposed structure that the variables are independent or that the variables are associated? Explain.

b. What is the range of plausible values for the randomized Chi-squared statistic?

c. The observed Chi-squared statistic is 68.8 (and seen in red on the graph). Does the observed value provide evidence against the null hypothesis? To answer the question, state the null and alternative hypotheses, approximate the p-value, and conclude the test in the context of the problem.

18.8. **Disaggregating Asian American tobacco use, randomization test.** Understanding cultural differences in tobacco use across different demographic groups can lead to improved health care education and treatment. A recent study disaggregated tobacco use across Asian American ethnic groups including Asian-Indian (n = 4373), Chinese (n = 4736), and Filipino (n = 4912), in comparison to non-Hispanic Whites (n = 275,025). The number of current smokers in each group was reported as Asian-Indian (n = 223), Chinese (n = 279), Filipino (n = 609), and non-Hispanic Whites (n = 50,880). (Rao et al., 2021)

The original data were randomized 1000 times (smoking status randomly assigned to the observations across ethnicities), and the histogram of the Chi-squared statistic on each randomization is displayed.

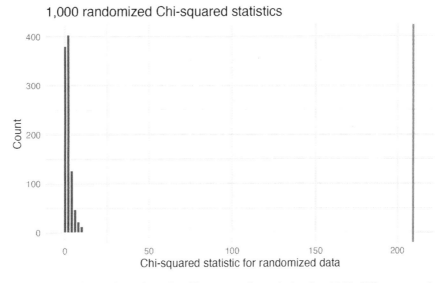

a. The histogram above describes the Chi-squared statistics for 1000 different randomization datasets. When randomizing the data, is the imposed structure that the variables are independent or that the variables are associated? Explain.

b. What is the (approximate) range of plausible values for the randomized Chi-squared statistic?

c. The observed Chi-squared statistic is 209.42 (and seen in red on the graph). Does the observed value provide evidence against the null hypothesis? To answer the question, state the null and alternative hypotheses, approximate the p-value, and conclude the test in the context of the problem.

18.9. **Lizard habitats, larger data.** In order to assess whether habitat conditions are related to the sunlight choices a lizard makes for resting, Western fence lizard (*Sceloporus occidentalis*) were observed across three different microhabitats. (Adolph, 1990; Asbury and Adolph, 2007)

Consider the situation where the data set is 5 times *larger* than the original data (but have the same proportional representation in each category). The distribution of lizards in each of the sites resting in the sun, partial sun, and shade are as follows.

	Larger data			
	sunlight			
site	sun	partial	shade	Total
desert	80	160	355	595
mountain	280	180	75	535
valley	210	200	120	530
Total	570	540	550	1,660

The larger dataset was randomized 1,000 times (sunlight preference randomly assigned to the observations across sites), and the histogram of the Chi-squared statistic on each randomization is displayed.

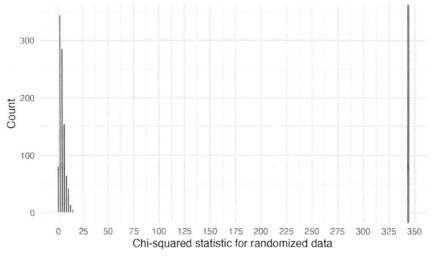

a. The histogram above describes the Chi-squared statistics for 1,000 different randomization of the larger dataset. When randomizing the data, is the imposed structure that the variables are independent or that the variables are associated? Explain.

b. What is the (approximate) range of plausible values for the randomized Chi-squared statistic?

c. The observed Chi-squared statistic is 343.865 (and seen in red on the graph). Does the observed value provide evidence against the null hypothesis? To answer the question, state the null and alternative hypotheses, approximate the p-value, and conclude the test in the context of the problem.

d. If the alternative hypothesis is true, how does the sample size effect the ability to reject the null hypothesis? (*Hint:* Consider the original data as compared with the larger dataset that have the same proportional values.)

18.10. **Disaggregating Asian American tobacco use, smaller data.** Understanding cultural differences in tobacco use across different demographic groups can lead to improved health care education and treatment. A recent study disaggregated tobacco use across Asian American ethnic groups (Rao et al., 2021).

Consider the situation where the data set is 50 times *smaller* than the original data (but have the same proportional representation in each category). The distribution of smokers in each of the ethnicity groups in the smaller data are as follows.

	Smaller data		
	Smoking		
ethnicity	don't smoke	smoke	Total
Asian-Indian	83	4	87
Chinese	89	6	95
Filipino	86	12	98
Total	258	22	280

The smaller dataset was randomized 1,000 times (smoking status randomly assigned to the observations across ethnicities), and the histogram of the Chi-squared statistic on each randomization is displayed.

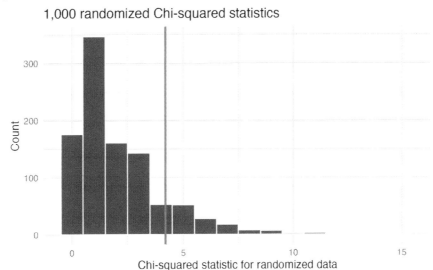

1,000 randomized Chi-squared statistics

a. The histogram above describes the Chi-squared statistics for 1,000 different randomization of the smaller dataset. When randomizing the data, is the imposed structure that the variables are independent or that the variables are associated? Explain.

b. What is the (approximate) range of plausible values for the randomized Chi-squared statistic?

c. The observed Chi-squared statistic is 4.19 (and seen in red on the graph). Does the observed value provide evidence against the null hypothesis? To answer the question, state the null and alternative hypotheses, approximate the p-value, and conclude the test in the context of the problem.

d. If the alternative hypothesis is true, how does the sample size effect the ability to reject the null hypothesis? (*Hint:* Consider the original data as compared with the smaller dataset that have the same proportional values.)

18.11. True or false, I. Determine if the statements below are true or false. For each false statement, suggest an alternative wording to make it a true statement.

a. The Chi-square distribution, just like the normal distribution, has two parameters, mean and standard deviation.

b. The Chi-square distribution is always right skewed, regardless of the value of the degrees of freedom parameter.

c. The Chi-square statistic is always greater than or equal to 0.

d. As the degrees of freedom increases, the shape of the Chi-square distribution becomes more skewed.

18.12. True or false, II. Determine if the statements below are true or false. For each false statement, suggest an alternative wording to make it a true statement.

a. As the degrees of freedom increases, the mean of the Chi-square distribution increases.

b. If you found $\chi^2 = 10$ with $df = 5$ you would fail to reject H_0 at the 5% significance level.

c. When finding the p-value of a Chi-square test, we always shade the tail areas in both tails.

d. As the degrees of freedom increases, the variability of the Chi-square distribution decreases.

18.13. Sleep deprived transportation workers. The National Sleep Foundation conducted a survey on the sleep habits of randomly sampled transportation workers and randomly sampled non-transportation workers that serve as a "control" for comparison. The results of the survey are shown below. (Foundation, 2012)

Profession	Less than 6 hours	6 to 8 hours	More than 8 hours	Total
Non-transportation workers	35	193	64	292
Transportation workers	104	499	192	795
Total	139	692	256	1,087

Conduct a hypothesis test to evaluate if these data provide evidence of an association between sleep levels and profession.

18.14. Parasitic worm. Lymphatic filariasis is a disease caused by a parasitic worm. Complications of the disease can lead to extreme swelling and other complications. Here we consider results from a randomized experiment that compared three different drug treatment options to clear people of the this parasite, which people are working to eliminate entirely. The results for the second year of the study are given below: (King et al., 2018)

	Outcome		
group	Clear at Year 2	Not Clear at Year 2	Total
Three drugs	52	2	54
Two drugs	31	24	55
Two drugs annually	42	14	56
Total	125	40	165

a. Set up hypotheses for evaluating whether there is any difference in the performance of the treatments, and also check conditions.

b. Statistical software was used to run a Chi-square test, which output:

$$X^2 = 23.7 \quad df = 2 \quad \text{p-value} < 0.0001$$

Use these results to evaluate the hypotheses from part (a), and provide a conclusion in the context of the problem.

18.15. **Shipping holiday gifts.** A local news survey asked 500 randomly sampled Los Angeles residents which shipping carrier they prefer to use for shipping holiday gifts. The table below shows the distribution of responses by age group as well as the expected counts for each cell (shown in italics).

Shipping method	Age 18-34		Age 35-54		Age 55+		Total
USPS	72	*81*	97	*102*	76	*62*	245
UPS	52	*53*	76	*68*	34	*41*	162
FedEx	31	*21*	24	*27*	9	*16*	64
Something else	7	*5*	6	*7*	3	*4*	16
Not sure	3	*5*	6	*5*	4	*3*	13
Total	165		209		126		500

a. State the null and alternative hypotheses for testing for independence of age and preferred shipping method for holiday gifts among Los Angeles residents.

b. Are the conditions for inference using a Chi-square test satisfied?

18.16. **Coffee and depression.** Researchers conducted a study investigating the relationship between caffeinated coffee consumption and risk of depression in women. They collected data on 50,739 women free of depression symptoms at the start of the study in the year 1996, and these women were followed through 2006. The researchers used questionnaires to collect data on caffeinated coffee consumption, asked each individual about physician- diagnosed depression, and also asked about the use of antidepressants. The table below shows the distribution of incidences of depression by amount of caffeinated coffee consumption. (Lucas et al., 2011)

Clinical depression	Caffeinated coffee consumption					Total
	1 cup / week or fewer	2-6 cups / week	1 cups / day	2-3 cups / day	4 cups / day or more	
Yes	670		905	564	95	2,607
No	11,545	6,244	16,329	11,726	2,288	48,132
Total	12,215	6,617	17,234	12,290	2,383	50,739

a. What type of test is appropriate for evaluating if there is an association between coffee intake and depression?

b. Write the hypotheses for the test you identified in part (a).

c. Calculate the overall proportion of women who do and do not suffer from depression.

d. Identify the expected count for the empty cell, and calculate the contribution of this cell to the test statistic.

e. The test statistic is $\chi^2 = 20.93$. What is the p-value?

f. What is the conclusion of the hypothesis test?

g. One of the authors of this study was quoted on the NYTimes as saying it was "too early to recommend that women load up on extra coffee" based on just this study. (O'Connor, 2011) Do you agree with this statement? Explain your reasoning.

Chapter 19

Inference for a single mean

Focusing now on Statistical Inference for **numerical data**, again, we will revisit and expand upon the foundational aspects of hypothesis testing from Chapter 11.

The important data structure for this chapter is a numeric response variable (that is, the outcome is quantitative). The four data structures we detail are one numeric response variable, one numeric response variable which is a difference across a pair of observations, a numeric response variable broken down by a binary explanatory variable, and a numeric response variable broken down by an explanatory variable that has two or more levels. When appropriate, each of the data structures will be analyzed using the three methods from Chapters 11, 12, and 13: randomization test, bootstrapping, and mathematical models, respectively.

As we build on the inferential ideas, we will visit new foundational concepts in statistical inference. One key new idea rests in estimating how the sample mean (as opposed to the sample proportion) varies from sample to sample; the resulting value is referred to as the standard error of the mean. We will also introduce a new important mathematical model, the t-distribution (as the foundation for the t-test).

In this chapter, we focus on the sample mean (instead of, for example, the sample median or the range of the observations) because of the well-studied mathematical model which describes the behavior of the sample mean. We will not cover mathematical models which describe other statistics, but the bootstrap and randomization techniques described below are immediately extendable to any function of the observed data. The sample mean will be calculated in one group, two paired groups, two independent groups, and many groups settings. The techniques described for each setting will vary slightly, but you will be well served to find the structural similarities across the different settings.

Similar to how we can model the behavior of the sample proportion $\hat{p}$ using a normal distribution, the sample mean $\bar{x}$ can also be modeled using a normal distribution when certain conditions are met. However, we'll soon learn that a new distribution, called the t-distribution, tends to be more useful when working with the sample mean. We'll first learn about this new distribution, then we'll use it to construct confidence intervals and conduct hypothesis tests for the mean.

19.1 Bootstrap confidence interval for a mean

Consider a situation where you want to know whether you should buy a franchise of the used car store Awesome Autos. As part of your planning, you'd like to know for how much an average car from Awesome Autos sells. In order to go through the example more clearly, let's say that you are only able to randomly sample five cars from Awesome Auto. (If this were a real example, you would surely be able to take a much larger sample size, possibly even being able to measure the entire population!)

19.1.1 Observed data

Figure 19.1 shows a (small) random sample of observations from Awesome Auto. The actual cars as well as their selling price is shown.

Figure 19.1: A sample of five cars from Awesome Auto.

The sample average car price of $17140.00 is a first guess at the price of the average car price at Awesome Auto. However, as a student of statistics, you understand that one sample mean based on a sample of five observations will not necessarily equal the true population average car price for all the cars at Awesome Auto. Indeed, you can see that the observed car prices vary with a standard deviation of $7170.29, and surely the average car price would be different if a different sample of size five had been taken from the population. Fortunately, as it did in previous chapters for the sample proportion, bootstrapping will approximate the variability of the sample mean from sample to sample.

19.1.2 Variability of the statistic

As with the inferential ideas covered in Chapters 11, 12, and 13, the inferential analysis methods in this chapter are grounded in quantifying how one dataset differs from another when they are both taken from the same population. To repeat, the idea is that we want to know how datasets differ from one another, but we aren't ever going to take more than one sample of observations. It doesn't make sense to take repeated samples from the same population because if you have the ability to take more samples, a larger sample size will benefit you more than taking two samples from the population. Instead of taking repeated samples from the actual population, we use bootstrapping to measure how the samples behave under an estimate of the population.

As mentioned previously, to get a sense of the cars at Awesome Auto, you take a sample of five cars from the Awesome Auto branch near you as a way to gauge the price of the cars being sold. Figure 19.2 shows how the unknown original population can be estimated by using the sample to approximate the distribution of car prices from the population of cars at Awesome Auto.

By taking repeated samples from the estimated population, the variability from sample to sample can be observed. In Figure 12.2 the repeated bootstrap samples are seen to be different both from each other and from the original population. Recall that the bootstrap samples were taken from

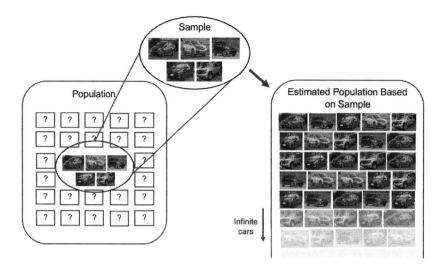

Figure 19.2: As seen previously, the idea behind bootstrapping is to consider the sample at hand as an estimate of the population. Sampling from the sample (of 5 cars) is identical to sampling from an infinite population which is made up of only the cars in the original sample.

the same (estimated) population, and so the differences in bootstrap samples are due entirely to natural variability in the sampling procedure. For the situation at hand where the sample mean is the statistic of interest, the variability from sample to sample can be seen in Figure 19.3.

By summarizing each of the bootstrap samples (here, using the sample mean), we see, directly, the variability of the sample mean, $\bar{x}$, from sample to sample. The distribution of $\bar{x}_{bs}$ for the Awesome Auto cars is shown in Figure 19.4.

Figure 19.5 summarizes one thousand bootstrap samples in a histogram of the bootstrap sample means. The bootstrapped average car prices vary from about \$10,000 to \$25,000. The bootstrap percentile confidence interval is found by locating the middle 90% (for a 90% confidence interval) or a 95% (for a 95% confidence interval) of the bootstrapped statistics.

EXAMPLE

Using Figure 19.5, find the 90% and 95% bootstrap percentile confidence intervals for the true average price of a car from Awesome Auto.

A 90% confidence interval is given by \$12,140 and \$22,007. The conclusion is that we are 90% confident that the true average car price at Awesome Auto lies somewhere between \$12,140 and \$22,007.

A 95% confidence interval is given by \$11,778 to \$22,500. The conclusion is that we are 95% confident that the true average car price at Awesome Auto lies somewhere between \$11,778 to \$22,500.

19.1.3 Bootstrap SE confidence interval

As seen in Section 17.2, another method for creating bootstrap confidence intervals directly uses a calculation of the variability of the bootstrap statistics (here, the bootstrap means). If the bootstrap distribution is relatively symmetric and bell-shaped, then the 95% bootstrap SE confidence interval can be constructed with the formula familiar from the mathematical models in previous chapters:

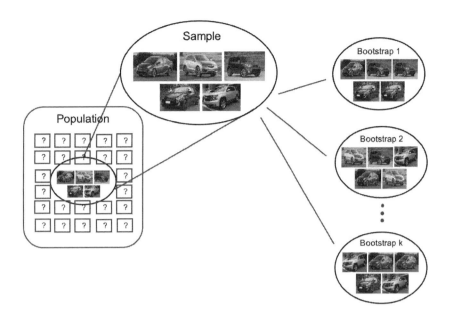

Figure 19.3: To estimate the natural variability in the sample mean, different bootstrap samples are taken from the original sample. Notice that each bootstrap resample is different from each other as well as from the original sample

$$\text{point estimate} \pm 2 \cdot SE_{BS}$$

The number 2 is an approximation connected to the "95%" part of the confidence interval (remember the 68-95-99.7 rule). As will be seen in Section 19.2, a new distribution (the t-distribution) will be applied to most mathematical inference on numerical variables. However, because bootstrapping is not grounded in the same theory as the mathematical approach given in this text, we stick with the standard normal quantiles (in R use the function `qnorm()` to find normal percentiles other than 95%) for different confidence percentages.[1]

 EXAMPLE

Explain how the standard error (SE) of the bootstrapped means is calculated and what it is measuring.

The SE of the bootstrapped means measures how variable the means are from resample to resample. The bootstrap SE is a good approximation to the SE of means as if we had taken repeated samples from the original population (which we agreed isn't something we would do because of wasted resources).

Logistically, we can find the standard deviation of the bootstrapped means using the same calculations from Chapter 5. That is, the bootstrapped means are the individual observations about which we measure the variability.

Although we won't spend a lot of energy on this concept, you may be wondering some of the differences between a standard error and a standard deviation. The **standard error** describes how a statistic (e.g., sample mean or sample proportion) varies from sample to sample. The **standard deviation** can be thought of as a function applied to any list of numbers which measures how far

---

[1] There is a large literature on understanding and improving bootstrap intervals, see Hesterberg (2015) titled "What Teachers Should Know About the Bootstrap" and Hayden (2019) titled "Questionable Claims for Simple Versions of the Bootstrap" for more information.

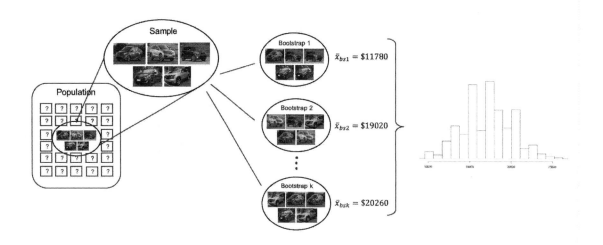

Figure 19.4: Because each of the bootstrap resamples respresents a different set of cars, the mean of the each bootstrap resample will be a different value. Each of the bootstrapped means is calculated, and a histogram of the values describes the inherent natural variability of the sample mean which is due to the sampling process.

those numbers vary from their own average. So, you can have a standard deviation calculated on a column of dog heights or a standard deviation calculated on a column of bootstrapped means from the resampled data. Note that the standard deviation calculated on the bootstrapped means is referred to as the bootstrap standard error of the mean.

 GUIDED PRACTICE

It turns out that the standard deviation of the bootstrapped means from Figure 19.5 is $2,891.87 (a value which is an excellent approximation for the standard error of sample means if we were to take repeated samples from the population). [Note: in R the calculation was done using the function `sd()`.] The average of the observed prices is $17,140, ad we will consider the sample average to be the best guess point estimate for μ. .

Find and interpret the confidence interval for μ (the true average cost of a car at Awesome Auto) using the bootstrap SE confidence interval formula.[2]

---

[2]Using the formula for the bootstrap SE interval, we find the 95% confidence interval for μ is: $17,140 \pm 2 \cdot 2,891.87 \rightarrow$ ($11,356.26, $22,923.74). We are 95% confident that the true average car price at Awesome Auto is somewhere between $11,356.26 and $22,923.74.

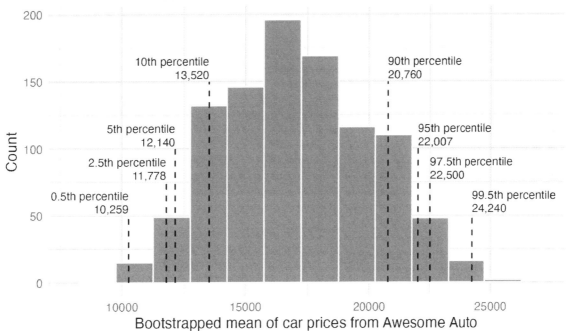

Figure 19.5: The original Awesome Auto data is bootstrapped 1,000 times. The histogram provides a sense for the variability of the average car price from sample to sample.

EXAMPLE

Compare and contrast the two different 95% confidence intervals for μ created by finding the percentiles of the bootstrapped means and created by finding the SE of the bootstrapped means. Do you think the intervals *should* be identical?

- Percentile interval: ($11,778, $22,500)
- SE interval: ($11,356.26, $22,923.74)

The intervals were created using different methods, so it is not surprising that they are not identical. However, we are pleased to see that the two methods provide very similar interval approximations.

The technical details surrounding which data structures are best for percentile intervals and which are best for SE intervals is beyond the scope of this text. However, the larger the samples are, the better (and closer) the interval estimates will be.

19.1.4 Bootstrap percentile confidence interval for a standard deviation

Suppose that the research question at hand seeks to understand how variable the prices of the cars are at Awesome Auto. That is, your interest is no longer in the average car price but in the *standard deviation* of the prices of all cars at Awesome Auto, σ. You may have already realized that the sample standard deviation, s, will work as a good **point estimate** for the parameter of interest: the population standard deviation, σ. The point estimate of the five observations is calculated to be $s = \$7,170.286$. While $s = \$7,170.286$ might be a good guess for σ, we prefer to have an interval estimate for the parameter of interest. Although there is a mathematical model which describes how s varies from sample to sample, the mathematical model will not be presented in this text. Even without the mathematical model, bootstrapping can be used to find a confidence interval for the parameter σ. Using the same technique as presented for a confidence interval for μ, here we find the bootstrap percentile confidence interval for σ.

EXAMPLE

Describe the bootstrap distribution for the standard deviation shown in Figure 19.6.

The distribution is skewed left and centered near $7,170.286, which is the point estimate from the original data. Most observations in this distribution lie between $0 and $10,000.

GUIDED PRACTICE

Using Figure 19.6, find *and interpret* a 90% bootstrap percentile confidence interval for the population standard deviation for car prices at Awesome Auto.[3]

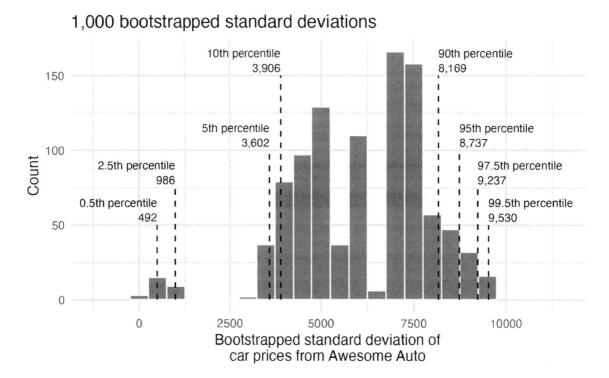

Figure 19.6: The original Awesome Auto data is bootstrapped 1,000 times. The histogram provides a sense for the variability of the standard deviation of the car prices from sample to sample.

19.1.5 Bootstrapping is not a solution to small sample sizes!

The example presented above is done for a sample with only five observations. As with analysis techniques that build on mathematical models, bootstrapping works best when a large random sample has been taken from the population. Bootstrapping is a method for capturing the variability

---

[3]By looking at the percentile values in Figure 19.6, the middle 90% of the bootstrap standard deviations are given by the 5 percentile ($3,602.5) and 95 percentile ($8,737.2). That is, we are 90% confident that the true standard deviation of car prices is between $3,602.5 and $8,737.2. Note, the problem was set up as 90% to indicate that there was not a need for a high level of confidence (such a 95% or 99%). A lower degree of confidence increases potential for error, but it also produces a more narrow interval.

19.2 Mathematical model for a mean

As with the sample proportion, the variability of the sample mean is well described by the mathematical theory given by the Central Limit Theorem. However, because of missing information about the inherent variability in the population (σ), a t-distribution is used in place of the standard normal when performing hypothesis test or confidence interval analyses.

19.2.1 Mathematical distribution of the sample mean

The sample mean tends to follow a normal distribution centered at the population mean, μ, when certain conditions are met. Additionally, we can compute a standard error for the sample mean using the population standard deviation σ and the sample size n.

Central Limit Theorem for the sample mean.

When we collect a sufficiently large sample of n independent observations from a population with mean μ and standard deviation σ, the sampling distribution of $\bar{x}$ will be nearly normal with

$$\text{Mean} = \mu \qquad \text{Standard Error } (SE) = \frac{\sigma}{\sqrt{n}}$$

Before diving into confidence intervals and hypothesis tests using $\bar{x}$, we first need to cover two topics:

- When we modeled $\hat{p}$ using the normal distribution, certain conditions had to be satisfied. The conditions for working with $\bar{x}$ are a little more complex, and below, we will discuss how to check conditions for inference using a mathematical model.
- The standard error is dependent on the population standard deviation, σ. However, we rarely know σ, and instead we must estimate it. Because this estimation is itself imperfect, we use a new distribution called the t-distribution to fix this problem, which we discuss below.

19.2.2 Evaluating the two conditions required for modeling $\bar{x}$

Two conditions are required to apply the Central Limit Theorem for a sample mean $\bar{x}$:

- **Independence.** The sample observations must be independent, The most common way to satisfy this condition is when the sample is a simple random sample from the population. If the data come from a random process, analogous to rolling a die, this would also satisfy the independence condition.
- **Normality.** When a sample is small, we also require that the sample observations come from a normally distributed population. We can relax this condition more and more for larger and larger sample sizes. This condition is obviously vague, making it difficult to evaluate, so next we introduce a couple rules of thumb to make checking this condition easier.

General rule for performing the normality check.

There is no perfect way to check the normality condition, so instead we use two general rules based on the number and magnitude of extreme observations. Note, it often takes practice to get a sense for whether or not a normal approximation is appropriate.

- **n < 30**: If the sample size n is less than 30 and there are **no clear outliers** in the data, then we typically assume the data come from a nearly normal distribution to satisfy the condition.
- **n ≥ 30**: If the sample size n is at least 30 and there are no **particularly extreme** outliers, then we typically assume the sampling distribution of $\bar{x}$ is nearly normal, even if the underlying distribution of individual observations is not.

In this first course in statistics, you aren't expected to develop perfect judgment on the normality condition. However, you are expected to be able to handle clear cut cases based on the rules of thumb.[4]

EXAMPLE

Consider the four plots provided in Figure 19.7 that come from simple random samples from different populations. Their sample sizes are $n_1 = 15$ and $n_2 = 50$.

Are the independence and normality conditions met in each case?

Each samples is from a simple random sample of its respective population, so the independence condition is satisfied. Let's next check the normality condition for each using the rule of thumb.

The first sample has fewer than 30 observations, so we are watching for any clear outliers. None are present; while there is a small gap in the histogram on the right, this gap is small and over 20% of the observations in this small sample are represented to the left of the gap, so we can hardly call these clear outliers. With no clear outliers, the normality condition can be reasonably assumed to be met.

The second sample has a sample size greater than 30 and includes an outlier that appears to be roughly 5 times further from the center of the distribution than the next furthest observation. This is an example of a particularly extreme outlier, so the normality condition would not be satisfied.

It's often helpful to also visualize the data using a box plot to assess skewness and existence of outliers. The box plots provided underneath each histogram confirms our conclusions that the first sample does not have any outliers and the second sample does, with one outlier being particularly more extreme than the others.

In practice, it's typical to also do a mental check to evaluate whether we have reason to believe the underlying population would have moderate skew (if $n < 30$) or have particularly extreme outliers ($n \geq 30$) beyond what we observe in the data. For example, consider the number of followers for each individual account on Twitter, and then imagine this distribution. The large majority of accounts have built up a couple thousand followers or fewer, while a relatively tiny fraction have amassed tens of millions of followers, meaning the distribution is extremely skewed. When we know the data come from such an extremely skewed distribution, it takes some effort to understand what sample size is

---

[4]More nuanced guidelines would consider further relaxing the *particularly extreme outlier* check when the sample size is very large. However, we'll leave further discussion here to a future course.

19.2. MATHEMATICAL MODEL FOR A MEAN

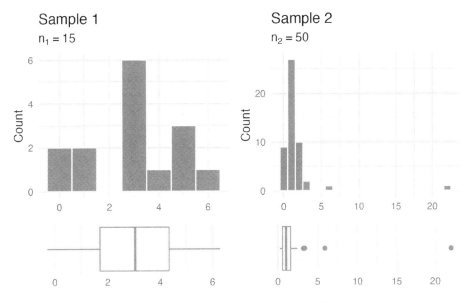

Figure 19.7: Histograms of samples from two different populations.

large enough for the normality condition to be satisfied.

19.2.3 Introducing the t-distribution

In practice, we cannot directly calculate the standard error for $\bar{x}$ since we do not know the population standard deviation, σ. We encountered a similar issue when computing the standard error for a sample proportion, which relied on the population proportion, p. Our solution in the proportion context was to use the sample value in place of the population value when computing the standard error. We'll employ a similar strategy for computing the standard error of $\bar{x}$, using the sample standard deviation s in place of σ:

$$SE = \frac{\sigma}{\sqrt{n}} \approx \frac{s}{\sqrt{n}}$$

This strategy tends to work well when we have a lot of data and can estimate σ using s accurately. However, the estimate is less precise with smaller samples, and this leads to problems when using the normal distribution to model $\bar{x}$.

We'll find it useful to use a new distribution for inference calculations called the t-distribution. A t-distribution, shown as a solid line in Figure 19.8, has a bell shape. However, its tails are thicker than the normal distribution's, meaning observations are more likely to fall beyond two standard deviations from the mean than under the normal distribution.

The extra thick tails of the t-distribution are exactly the correction needed to resolve the problem (due to extra variability of the T score) of using s in place of σ in the SE calculation.

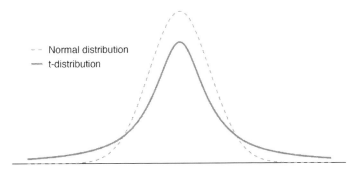

Figure 19.8: Comparison of a t-distribution and a normal distribution.

The t-distribution is always centered at zero and has a single parameter: degrees of freedom. The **degrees of freedom** describes the precise form of the bell-shaped t-distribution. Several t-distributions are shown in Figure 19.9 in comparison to the normal distribution. Similar to the Chi-square distribution, the shape of the t-distribution also depends on the degrees of freedom.

In general, we'll use a t-distribution with $df = n - 1$ to model the sample mean when the sample size is n. That is, when we have more observations, the degrees of freedom will be larger and the t-distribution will look more like the standard normal distribution; when the degrees of freedom is about 30 or more, the t-distribution is nearly indistinguishable from the normal distribution.

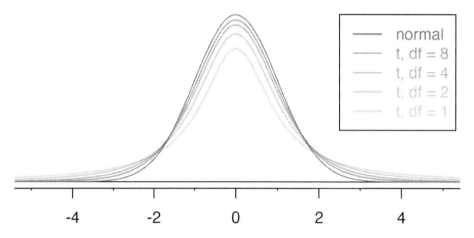

Figure 19.9: The larger the degrees of freedom, the more closely the t-distribution resembles the standard normal distribution.

 Degrees of freedom: df.

The degrees of freedom describes the shape of the t-distribution. The larger the degrees of freedom, the more closely the distribution approximates the normal distribution.

When modeling $\bar{x}$ using the t-distribution, use $df = n - 1$.

The t-distribution allows us greater flexibility than the normal distribution when analyzing numerical data. In practice, it's common to use statistical software, such as R, Python, or SAS for these analyses. In R, the function used for calculating probabilities under a t-distribution is pt() (which should seem similar to previous R functions, pnorm() and pchisq()). Don't forget that with the t-distribution, the degrees of freedom must always be specified!

For the examples and guided practices below, you may have to use a table or statistical software to find the answers. We recommend trying the problems so as to get a sense for how the t-distribution can vary in width depending on the degrees of freedom. No matter the approach you choose, apply your method using the examples below to confirm your working understanding of the t-distribution.

 EXAMPLE

What proportion of the t-distribution with 18 degrees of freedom falls below -2.10?

Just like a normal probability problem, we first draw the picture in Figure 19.10 and shade the area below -2.10.

Using statistical software, we can obtain a precise value: 0.0250.

```
# use pt() to find probability under the $t$-distribution
pt(-2.10, df = 18)
#> [1] 0.025
```

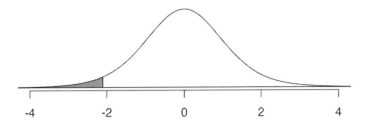

Figure 19.10: The t-distribution with 18 degrees of freedom. The area below -2.10 has been shaded.

 EXAMPLE

A t-distribution with 20 degrees of freedom is shown in Figure 19.11. Estimate the proportion of the distribution falling above 1.65.

Note that with 20 degrees of freedom, the t-distribution is relatively close to the normal distribution. With a normal distribution, this would correspond to about 0.05, so we should expect the t-distribution to give us a value in this neighborhood. Using statistical software: 0.0573.

```
# use pt() to find probability under the $t$-distribution
1 - pt(1.65, df = 20)
#> [1] 0.0573
```

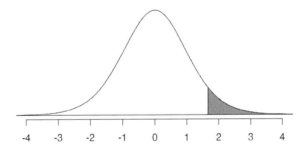

Figure 19.11: Top: The t-distribution with 20 degrees of freedom, with the area above 1.65 shaded.

 EXAMPLE

A t-distribution with 2 degrees of freedom is shown in Figure 19.12. Estimate the proportion of the distribution falling more than 3 units from the mean (above or below).

With so few degrees of freedom, the t-distribution will give a more notably different value than the normal distribution. Under a normal distribution, the area would be about 0.003 using the 68-95-99.7 rule. For a t-distribution with $df = 2$, the area in both tails beyond 3 units totals 0.0955. This area is dramatically different than what we obtain from the normal distribution.

```
# use pt() to find probability under the $t$-distribution
pt(-3, df = 2) + (1 - pt(3, df = 2))
#> [1] 0.0955
```

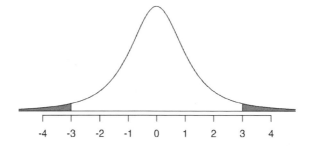

Figure 19.12: The t-distribution with 2 degrees of freedom, with the area further than 3 units from 0 shaded.

 GUIDED PRACTICE

What proportion of the t-distribution with 19 degrees of freedom falls above -1.79 units? Use your preferred method for finding tail areas.[5]

19.2.4 One sample t-intervals

Let's get our first taste of applying the t-distribution in the context of an example about the mercury content of dolphin muscle. Elevated mercury concentrations are an important problem for both dolphins and other animals, like humans, who occasionally eat them.

Figure 19.13: A Risso's dolphin. Photo by Mike Baird, www.bairdphotos.com

We will identify a confidence interval for the average mercury content in dolphin muscle using a sample of 19 Risso's dolphins from the Taiji area in Japan. The data are summarized in Table 19.1. The minimum and maximum observed values can be used to evaluate whether or not there are clear outliers.

---

[5]We want to find the shaded area *above* -1.79 (we leave the picture to you). The lower tail area has an area of 0.0447, so the upper area would have an area of $1 - 0.0447 = 0.9553$.

19.2. MATHEMATICAL MODEL FOR A MEAN

Table 19.1: Summary of mercury content in the muscle of 19 Risso's dolphins from the Taiji area. Measurements are in micrograms of mercury per wet gram of muscle (μg/wet g).

n	Mean	SD	Min	Max
19	4.4	2.3	1.7	9.2

EXAMPLE

Are the independence and normality conditions satisfied for this dataset?

The observations are a simple random sample, therefore it is reasonable to assume that the dolphins are independent. The summary statistics in Table 19.1 do not suggest any clear outliers, with all observations within 2.5 standard deviations of the mean. Based on this evidence, the normality condition seems reasonable.

In the normal model, we used $z^\star$ and the standard error to determine the width of a confidence interval. We revise the confidence interval formula slightly when using the t-distribution:

$$\text{point estimate} \ \pm \ t^\star_{df} \times SE$$

$$\bar{x} \ \pm \ t^\star_{df} \times \frac{s}{\sqrt{n}}$$

EXAMPLE

Using the summary statistics in Table 19.1, compute the standard error for the average mercury content in the $n = 19$ dolphins.

We plug in s and n into the formula: $SE = \frac{s}{\sqrt{n}} = \frac{2.3}{\sqrt{19}} = 0.528$.

The value $t^\star_{df}$ is a cutoff we obtain based on the confidence level and the t-distribution with df degrees of freedom. That cutoff is found in the same way as with a normal distribution: we find $t^\star_{df}$ such that the fraction of the t-distribution with df degrees of freedom within a distance $t^\star_{df}$ of 0 matches the confidence level of interest.

EXAMPLE

When $n = 19$, what is the appropriate degrees of freedom? Find $t^\star_{df}$ for this degrees of freedom and the confidence level of 95%

The degrees of freedom is easy to calculate: $df = n - 1 = 18$.

Using statistical software, we find the cutoff where the upper tail is equal to 2.5%: $t^\star_{18} = 2.10$. The area below -2.10 will also be equal to 2.5%. That is, 95% of the t-distribution with $df = 18$ lies within 2.10 units of 0.

```
# use qt() to find the t-cutoff (with 95% in the middle)
qt(0.025, df = 18)
#> [1] -2.1
qt(0.975, df = 18)
#> [1] 2.1
```

Degrees of freedom for a single sample.

If the sample has n observations and we are examining a single mean, then we use the t-distribution with $df = n - 1$ degrees of freedom.

EXAMPLE

Compute and interpret the 95% confidence interval for the average mercury content in Risso's dolphins.

We can construct the confidence interval as

$$\bar{x} \pm t^\star_{18} \times SE$$
$$4.4 \pm 2.10 \times 0.528$$
$$(3.29, 5.51)$$

We are 95% confident the average mercury content of muscles in Risso's dolphins is between 3.29 and 5.51 μg/wet gram, which is considered extremely high.

Calculating a t-confidence interval for the mean, μ.

Based on a sample of n independent and nearly normal observations, a confidence interval for the population mean is

$$\text{point estimate} \pm t^\star_{df} \times SE$$
$$\bar{x} \pm t^\star_{df} \times \frac{s}{\sqrt{n}}$$

where $\bar{x}$ is the sample mean, $t^\star_{df}$ corresponds to the confidence level and degrees of freedom df, and SE is the standard error as estimated by the sample.

GUIDED PRACTICE

The FDA's webpage provides some data on mercury content of fish. Based on a sample of 15 croaker white fish (Pacific), a sample mean and standard deviation were computed as 0.287 and 0.069 ppm (parts per million), respectively. The 15 observations ranged from 0.18 to 0.41 ppm. We will assume these observations are independent. Based on the summary statistics of the data, do you have any objections to the normality condition of the individual observations?[6]

---

[6]The sample size is under 30, so we check for obvious outliers: since all observations are within 2 standard deviations of the mean, there are no such clear outliers.

19.2. MATHEMATICAL MODEL FOR A MEAN

EXAMPLE

Estimate the standard error of $\bar{x} = 0.287$ ppm using the data summaries in the previous Guided Practice. If we are to use the t-distribution to create a 90% confidence interval for the actual mean of the mercury content, identify the degrees of freedom and $t_{df}^\star$.

The standard error: $SE = \frac{0.069}{\sqrt{15}} = 0.0178$.

Degrees of freedom: $df = n - 1 = 14$.

Since the goal is a 90% confidence interval, we choose $t_{14}^\star$ so that the two-tail area is 0.1: $t_{14}^\star = 1.76$.

```
# use qt() to find the t-cutoff (with 90% in the middle)
qt(0.05, df = 14)
#> [1] -1.76
qt(0.95, df = 14)
#> [1] 1.76
```

GUIDED PRACTICE

Using the information and results of the previous Guided Practice and Example, compute a 90% confidence interval for the average mercury content of croaker white fish (Pacific).[7]

GUIDED PRACTICE

The 90% confidence interval from the previous Guided Practice is 0.256 ppm to 0.318 ppm. Can we say that 90% of croaker white fish (Pacific) have mercury levels between 0.256 and 0.318 ppm?[8]

Recall that the margin of error is defined by the standard error. The margin of error for $\bar{x}$ can be directly obtained from $SE(\bar{x})$.

Margin of error for $\bar{x}$.

The margin of error is $t_{df}^\star \times s/\sqrt{n}$ where $t_{df}^\star$ is calculated from a specified percentile on the t-distribution with df degrees of freedom.

19.2.5 One sample t-tests

Now that we've used the t-distribution for making a confidence interval for a mean, let's speed on through to hypothesis tests for the mean.

---

[7]$\bar{x} \pm t_{14}^\star \times SE \rightarrow 0.287 \pm 1.76 \times 0.0178 \rightarrow (0.256, 0.318)$. We are 90% confident that the average mercury content of croaker white fish (Pacific) is between 0.256 and 0.318 ppm.

[8]No, a confidence interval only provides a range of plausible values for a population parameter, in this case the population mean. It does not describe what we might observe for individual observations.

The test statistic for assessing a single mean is a T.

The T score is a ratio of how the sample mean differs from the hypothesized mean as compared to how the observations vary.

$$T = \frac{\bar{x} - \text{null value}}{s/\sqrt{n}}$$

When the null hypothesis is true and the conditions are met, T has a t-distribution with $df = n - 1$.

Conditions:

- Independent observations.
- Large samples and no extreme outliers.

Is the typical US runner getting faster or slower over time? We consider this question in the context of the Cherry Blossom Race, which is a 10-mile race in Washington, DC each spring. The average time for all runners who finished the Cherry Blossom Race in 2006 was 93.29 minutes (93 minutes and about 17 seconds). We want to determine using data from 100 participants in the 2017 Cherry Blossom Race whether runners in this race are getting faster or slower, versus the other possibility that there has been no change.

 The run17 data can be found in the **cherryblossom** R package.

 GUIDED PRACTICE

What are appropriate hypotheses for this context?[9]

 GUIDED PRACTICE

The data come from a simple random sample of all participants, so the observations are independent.

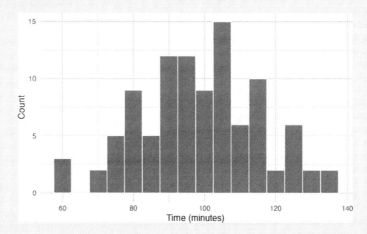

A histogram of the race times is given to evaluate if we can move forward with a t-test. Is the normality condition met?[10]

---

[9]H_0 : The average 10-mile run time was the same for 2006 and 2017. $\mu = 93.29$ minutes. H_A : The average 10-mile run time for 2017 was *different* than that of 2006. $\mu \neq 93.29$ minutes.

19.2. MATHEMATICAL MODEL FOR A MEAN

When completing a hypothesis test for the one-sample mean, the process is nearly identical to completing a hypothesis test for a single proportion. First, we find the Z score using the observed value, null value, and standard error; however, we call it a **T score** since we use a *t*-distribution for calculating the tail area. Then we find the p-value using the same ideas we used previously: find the one-tail area under the sampling distribution, and double it.

EXAMPLE

With both the independence and normality conditions satisfied, we can proceed with a hypothesis test using the *t*-distribution. The sample mean and sample standard deviation of the sample of 100 runners from the 2017 Cherry Blossom Race are 98.78 and 16.59 minutes, respectively. Recall that the average run time in 2006 was 93.29 minutes. Find the test statistic and p-value. What is your conclusion?

To find the test statistic (T score), we first must determine the standard error:

$$SE = 16.6/\sqrt{100} = 1.66$$

Now we can compute the **T score** using the sample mean (98.78), null value (93.29), and SE:

$$T = \frac{98.8 - 93.29}{1.66} = 3.32$$

For $df = 100 - 1 = 99$, we can determine using statistical software (or a *t*-table) that the one-tail area is 0.000631, which we double to get the p-value: 0.00126.

Because the p-value is smaller than 0.05, we reject the null hypothesis. That is, the data provide convincing evidence that the average run time for the Cherry Blossom Run in 2017 is different than the 2006 average.

```
# using pt() to find the left tail and multiply by 2 to get both tails
(1 - pt(3.32, df = 99)) * 2
#> [1] 0.00126
```

When using a *t*-distribution, we use a T score (similar to a Z score).

To help us remember to use the *t*-distribution, we use a T to represent the test statistic, and we often call this a **T score**. The Z score and T score are computed in the exact same way and are conceptually identical: each represents how many standard errors the observed value is from the null value.

---

[10] With a sample of 100, we should only be concerned if there is are particularly extreme outliers. The histogram of the data doesn't show any outliers of concern (and arguably, no outliers at all).

19.3 Chapter review

19.3.1 Summary

In this chapter we extended the randomization / bootstrap / mathematical model paradigm to questions involving quantitative variables of interest. When there is only one variable of interest, we are often hypothesizing or finding confidence intervals about the population mean. Note, however, the bootstrap method can be used for other statistics like the population median or the population IQR. When comparing a quantitative variable across two groups, the question often focuses on the difference in population means (or sometimes a paired difference in means). The questions revolving around one, two, and paired samples of means are addressed using the t-distribution; they are therefore called "t-tests" and "t-intervals." When considering a quantitative variable across 3 or more groups, a method called ANOVA is applied. Again, almost all the research questions can be approached using computational methods (e.g., randomization tests or bootstrapping) or using mathematical models. We continue to emphasize the importance of experimental design in making conclusions about research claims. In particular, recall that variability can come from different sources (e.g., random sampling vs. random allocation, see Figure 2.8).

19.3.2 Terms

We introduced the following terms in the chapter. If you're not sure what some of these terms mean, we recommend you go back in the text and review their definitions. We are purposefully presenting them in alphabetical order, instead of in order of appearance, so they will be a little more challenging to locate. However you should be able to easily spot them as **bolded text**.

Central Limit Theorem	point estimate	T score single mean
degrees of freedom	SD single mean	t-distribution
numerical data	SE single mean	

19.4 Exercises

Answers to odd numbered exercises can be found in Appendix A.19.

19.1. **Statistics vs. parameters: one mean.** Each of the following scenarios were set up to assess an average value. For each one, identify, in words: the statistic and the parameter.

   a. A sample of 25 New Yorkers were asked how much sleep they get per night.

   b. Researchers at two different universities in California collected information on undergraduates' heights.

19.2. **Statistics vs. parameters: one mean.** Each of the following scenarios were set up to assess an average value. For each one, identify, in words: the statistic and the parameter.

   a. Georgianna samples 20 children from a particular city and measures how many years they have each been playing piano.

   b. Traffic police officers (who are regularly exposed to lead from automobile exhaust) had their lead levels measured in their blood.

19.3. **Heights of adults.** Researchers studying anthropometry collected body measurements, as well as age, weight, height and gender, for 507 physically active individuals. Summary statistics for the distribution of heights (measured in centimeters), along with a histogram, are provided below.[11] (Heinz et al., 2003)

Min	Q1	Median	Mean	Q3	Max	SD	IQR
147	164	170	171	178	198	9.4	14

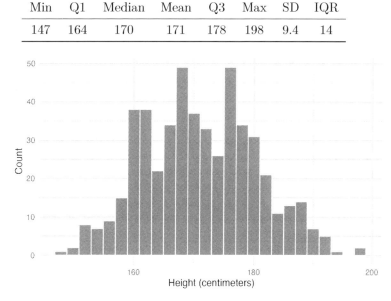

   a. What is the point estimate for the average height of active individuals? What about the median?

   b. What is the point estimate for the standard deviation of the heights of active individuals? What about the IQR?

   c. Is a person who is 1m 80cm (180 cm) tall considered unusually tall? And is a person who is 1m 55cm (155cm) considered unusually short? Explain your reasoning.

   d. The researchers take another random sample of physically active individuals. Would you expect the mean and the standard deviation of this new sample to be the ones given above? Explain your reasoning.

   e. The sample means obtained are point estimates for the mean height of all active individuals, if the sample of individuals is equivalent to a simple random sample. What measure do we use to quantify the variability of such an estimate? Compute this quantity using the data from the original sample under the condition that the data are a simple random sample.

---

[11] The bdims data used in this exercise can be found in the **openintro** R package.

19.4. **Heights of adults, standard error.** Heights of 507 physically active individuals have a mean of 171 centimeters and a standard deviation of 9.4 centimeters. Provide an estimate for the standard error of the mean for samples of following sizes.[12] (Heinz et al., 2003)

   a. n = 10

   b. n = 50

   c. n = 100

   d. n = 1000

   e. The standard error of the mean is a number which describes what?

19.5. **Heights of adults vs. kindergartners.** Heights of 507 physically active individuals have a mean of 171 centimeters and a standard deviation of 9.4 centimeters.[13] (Heinz et al., 2003)

   a. Would the standard deviation of the heights of a few hundred kindergartners be bigger or smaller than 9.4cm? Explain your reasoning.

   b. Suppose many samples of size 100 adults is taken and, separately, many samples of size 100 kindergarteners are taken. For each of the many samples, the average height is computed. Which set of sample averages would have a larger standard error of the mean, the adult sample averages or the kindergartner sample averages?

19.6. **Heights of adults, bootstrap interval.** Researchers studying anthropometry collected body measurements, as well as age, weight, height and gender, for 507 physically active individuals. The histogram below shows the sample distribution of bootstrapped means from 1,000 different bootstrap samples.[14] (Heinz et al., 2003)

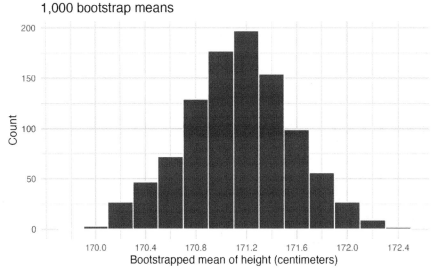

   a. Given the bootstrap sampling distribution for the sample mean, find an approximate value for the standard error of the mean.

   b. By looking at the bootstrap sampling distribution (1,000 bootstrap samples were taken), find an approximate 90% bootstrap percentile confidence interval for the true average adult height in the population from which the data were randomly sampled. Provide the interval as well as a one-sentence interpretation of the interval.

   c. By looking at the bootstrap sampling distribution (1,000 bootstrap samples were taken), find an approximate 90% bootstrap SE confidence interval for the true average adult height in the population from which the data were randomly sampled. Provide the interval as well as a one-sentence interpretation of the interval.

---

[12]The bdims data used in this exercise can be found in the **openintro** R package.
[13]The bdims data used in this exercise can be found in the **openintro** R package.
[14]The bdims data used in this exercise can be found in the **openintro** R package.

19.4. EXERCISES

19.7. **Identify the critical t.** A random sample is selected from an approximately normal population with unknown standard deviation. Find the degrees of freedom and the critical t-value (t^*) for the given sample size and confidence level.

   a. $n = 6$, CL $= 90\%$

   b. $n = 21$, CL $= 98\%$

   c. $n = 29$, CL $= 95\%$

   d. $n = 12$, CL $= 99\%$

19.8. **t-distribution.** The figure below shows three unimodal and symmetric curves: the standard normal (z) distribution, the t-distribution with 5 degrees of freedom, and the t-distribution with 1 degree of freedom. Determine which is which, and explain your reasoning.

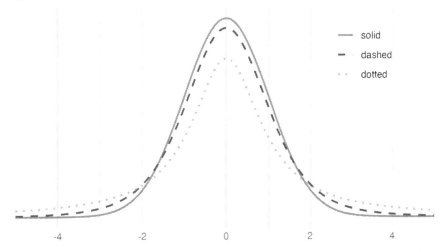

19.9. **Find the p-value, I.** A random sample is selected from an approximately normal population with an unknown standard deviation. Find the p-value for the given sample size and test statistic. Also determine if the null hypothesis would be rejected at $\alpha = 0.05$.

   a. $n = 11$, $T = 1.91$

   b. $n = 17$, $T = -3.45$

   c. $n = 7$, $T = 0.83$

   d. $n = 28$, $T = 2.13$

19.10. **Find the p-value, II.** A random sample is selected from an approximately normal population with an unknown standard deviation. Find the p-value for the given sample size and test statistic. Also determine if the null hypothesis would be rejected at $\alpha = 0.01$.

   a. $n = 26$, $T = 2.485$

   b. $n = 18$, $T = 0.5$

19.11. **Length of gestation, confidence interval.** Every year, the United States Department of Health and Human Services releases to the public a large dataset containing information on births recorded in the country. This dataset has been of interest to medical researchers who are studying the relation between habits and practices of expectant mothers and the birth of their children. In this exercise we work with a random sample of 1,000 cases from the dataset released in 2014. The length of pregnancy, measured in weeks, is commonly referred to as gestation. The histograms below show the distribution of lengths of gestation from the random sample of 1,000 births (on the left) and the distribution of bootstrapped means of gestation from 1,500 different bootstrap samples (on the right).[15]

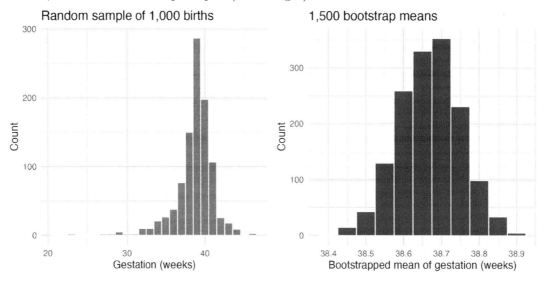

a. Given the bootstrap sampling distribution for the sample mean, find an approximate value for the standard error of the mean.

b. By looking at the bootstrap sampling distribution (1,500 bootstrap samples were taken), find an approximate 99% bootstrap percentile confidence interval for the true average gestation length in the population from which the data were randomly sampled. Provide the interval as well as a one-sentence interpretation of the interval.

c. By looking at the bootstrap sampling distribution (1,500 bootstrap samples were taken), find an approximate 99% bootstrap SE confidence interval for the true average gestation length in the population from which the data were randomly sampled. Provide the interval as well as a one-sentence interpretation of the interval.

---

[15] The `births14` data used in this exercise can be found in the **openintro** R package.

19.12. **Length of gestation, hypothesis test.** In this exercise we work with a random sample of 1,000 cases from the dataset released by the United States Department of Health and Human Services in 2014. Provided below are sample statistics for gestation (length of pregnancy, measured in weeks) of births in this sample.[16]

Min	Q1	Median	Mean	Q3	Max	SD	IQR
21	38	39	38.7	40	46	2.6	2

  a. What is the point estimate for the average length of pregnancy for all women? What about the median?

  b. You might have heard that human gestation is typically 40 weeks. Using the data, perform a complete hypothesis test, using mathematical models, to assess the 40 week claim. State the null and alternative hypotheses, find the T score, find the p-value, and provide a conclusion in context of the data.

  c. A quick internet search validates the claim of "40 weeks gestation" for humans. A friend of yours claims that there are different ways to measure gestation (starting at first day of last period, ovulation, or conception) which will result in estimates that are a week or two different. Another friend mentions that recent increases in cesarean births is likely to have decreased length of gestation. Do the data provide a mechanism to distinguish between your two friends' claims?

19.13. **Interpreting confidence intervals for population mean.** For each of the following statements, indicate if they are a true or false interpretation of the confidence interval. If false, provide a reason or correction to the misinterpretation. You collect a large sample and calculate a 95% confidence interval for the average number of cans of sodas consumed annually per adult in the US to be (440 cans, 520 cans), i.e., on average, adults in the US consume just under two cans of soda per day.

  a. 95% of adults in the US consume between 440 and 520 cans of soda per year.

  b. There is a 95% probability that the true population average per adult yearly soda consumption is between 440 and 520 cans.

  c. The true population average per adult yearly soda consumption is between 440 and 520 cans, with 95% confidence.

  d. The average soda consumption of the people who were sampled is between 440 and 520 cans of soda per year, with 95% confidence.

19.14. **Interpreting p-values for population mean.** For each of the following statements, indicate if they are a true or false interpretation of the p-value. If false, provide a reason or correction to the misinterpretation. You are wondering if the average amount of cereal in a 10oz cereal box is greater than 10oz. You collect 50 boxes of cereal, weigh them carefully, find a T score, and a p-value of 0.23.

  a. The probability that the average weight of all cereal boxes is 10 oz is 0.23.

  b. The probability that the average weight of all cereal boxes is greater than 10 oz is 0.23.

  c. Because the p-value is 0.23, the average weight of all cereal boxes is 10 oz.

  d. Because the p-value is small, the population average must be just barely above 10 oz (small effect).

  e. If H_0 is true, the probability of observing another sample with an average as or more extreme as the data is 0.23.

---

[16] The births14 data used in this exercise can be found in the **openintro** R package.

19.15. **Working backwards, I.** A 95% confidence interval for a population mean, μ, is given as (18.985, 21.015). The population distribution is approximately normal and the population standard deviation is unknown. This confidence interval is based on a simple random sample of 36 observations. Calculate the sample mean, the margin of error, and the sample standard deviation. Assume that all conditions necessary for inference are satisfied. Use the t-distribution in any calculations.

19.16. **Working backwards, II.** A 90% confidence interval for a population mean is (65, 77). The population distribution is approximately normal and the population standard deviation is unknown. This confidence interval is based on a simple random sample of 25 observations. Calculate the sample mean, the margin of error, and the sample standard deviation. Assume that all conditions necessary for inference are satisfied. Use the t-distribution in any calculations.

19.17. **Sleep habits of New Yorkers.** New York is known as "the city that never sleeps". A random sample of 25 New Yorkers were asked how much sleep they get per night. Statistical summaries of these data are shown below. The point estimate suggests New Yorkers sleep less than 8 hours a night on average. Evaluate the claim that New York is the city that never sleeps keeping in mind that, despite this claim, the true average number of hours New Yorkers sleep could be less than 8 hours or more than 8 hours.

n	Mean	SD	Min	Max
25	7.73	0.77	6.17	9.78

a. Write the hypotheses in symbols and in words.

b. Check conditions, then calculate the test statistic, T, and the associated degrees of freedom.

c. Find and interpret the p-value in this context. Drawing a picture may be helpful.

d. What is the conclusion of the hypothesis test?

e. If you were to construct a 90% confidence interval that corresponded to this hypothesis test, would you expect 8 hours to be in the interval?

19.18. **Find the mean.** You are given the hypotheses shown below. We know that the sample standard deviation is 8 and the sample size is 20. For what sample mean would the p-value be equal to 0.05? Assume that all conditions necessary for inference are satisfied.

$$H_0 : \mu = 60 \qquad H_A : \mu \neq 60$$

19.19. **$t^\star$ for the correct confidence level.** As you've seen, the tails of a t-distribution are longer than the standard normal which results in $t^\star_{df}$ being larger than $z^\star$ for any given confidence level. When finding a CI for a population mean, explain how mistakenly using $z^\star$ (instead of the correct $t^\star_{df}$) would affect the confidence level.

19.20. **Possible bootstrap samples.** Consider a simple random sample of the following observations: 47, 4, 92, 47, 12, 8. Which of the following could be a possible bootstrap samples from the observed data above? If the set of values could not be a bootstrap sample, indicate why not.

a. 47, 47, 47, 47, 47, 47

b. 92, 4, 13, 8, 47, 4

c. 92, 47, 12

d. 8, 47, 12, 12, 8, 4, 92

e. 12, 4, 8, 8, 92, 12

19.4. EXERCISES

19.21. **Play the piano.** Georgianna claims that in a small city renowned for its music school, the average child takes less than 5 years of piano lessons. We have a random sample of 20 children from the city, with a mean of 4.6 years of piano lessons and a standard deviation of 2.2 years.

   a. Evaluate Georgianna's claim (or that the opposite might be true) using a hypothesis test.

   b. Construct a 95% confidence interval for the number of years students in this city take piano lessons, and interpret it in context of the data.

   c. Do your results from the hypothesis test and the confidence interval agree? Explain your reasoning.

19.22. **Auto exhaust and lead exposure.** Researchers interested in lead exposure due to car exhaust sampled the blood of 52 police officers subjected to constant inhalation of automobile exhaust fumes while working traffic enforcement in a primarily urban environment. The blood samples of these officers had an average lead concentration of 124.32 μg/l and a SD of 37.74 μg/l; a previous study of individuals from a nearby suburb, with no history of exposure, found an average blood level concentration of 35 μg/l. (Mortada et al., 2000)

   a. Write down the hypotheses that would be appropriate for testing if the police officers appear to have been exposed to a different concentration of lead.

   b. Explicitly state and check all conditions necessary for inference on these data.

   c. Test the hypothesis that the downtown police officers have a higher lead exposure than the group in the previous study. Interpret your results in context.

Chapter 20

Inference for comparing two independent means

We now extend the methods from Chapter 19 to apply confidence intervals and hypothesis tests to differences in population means that come from two groups, Group 1 and Group 2: $\mu_1 - \mu_2$.

In our investigations, we'll identify a reasonable point estimate of $\mu_1 - \mu_2$ based on the sample, and you may have already guessed its form: $\bar{x}_1 - \bar{x}_2$. Then we'll look at the inferential analysis in three different ways: using a randomization test, applying bootstrapping for interval estimates, and, if we verify that the point estimate can be modeled using a normal distribution, we compute the estimate's standard error and apply the mathematical framework.

In this section we consider a difference in two population means, $\mu_1 - \mu_2$, under the condition that the data are not paired. Just as with a single sample, we identify conditions to ensure we can use the t-distribution with a point estimate of the difference, $\bar{x}_1 - \bar{x}_2$, and a new standard error formula.

The details for working through inferential problems in the two independent means setting are strikingly similar to those applied to the two independent proportions setting. We first cover a randomization test where the observations are shuffled under the assumption that the null hypothesis is true. Then we bootstrap the data (with no imposed null hypothesis) to create a confidence interval for the true difference in population means, $\mu_1 - \mu_2$. The mathematical model, here the t-distribution, is able to describe both the randomization test and the bootstrapping as long as the conditions are met.

The inferential tools are applied to three different data contexts: determining whether stem cells can improve heart function, exploring the relationship between pregnant women's smoking habits and birth weights of newborns, and exploring whether there is convincing evidence that one variation of an exam is harder than another variation. This section is motivated by questions like "Is there convincing evidence that newborns from mothers who smoke have a different average birth weight than newborns from mothers who don't smoke?"

20.1 Randomization test for the difference in means

An instructor decided to run two slight variations of the same exam. Prior to passing out the exams, they shuffled the exams together to ensure each student received a random version. Anticipating complaints from students who took Version B, they would like to evaluate whether the difference observed in the groups is so large that it provides convincing evidence that Version B was more difficult (on average) than Version A.

The `classdata` data can be found in the **openintro** R package.

20.1.1 Observed data

Summary statistics for how students performed on these two exams are shown in Table 20.1 and plotted in Figure 20.1.

Table 20.1: Summary statistics of scores for each exam version.

Group	n	Mean	SD	Min	Max
A	58	75.1	13.9	44	100
B	55	72.0	13.8	38	100

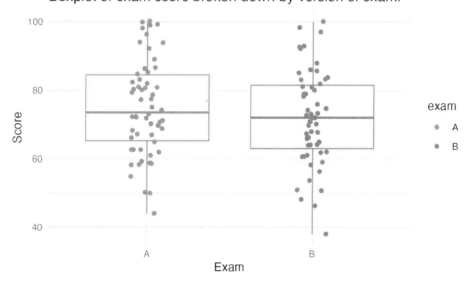

Figure 20.1: Exam scores for students given one of three different exams.

GUIDED PRACTICE

Construct hypotheses to evaluate whether the observed difference in sample means, $\bar{x}_A - \bar{x}_B = 3.1$, is likely to have happened due to chance, if the null hypothesis is true. We will later evaluate these hypotheses using $\alpha = 0.01$.[1]

---

[1] H_0: the exams are equally difficult, on average. $\mu_A - \mu_B = 0$. H_A: one exam was more difficult than the other, on average. $\mu_A - \mu_B \neq 0$.

> **GUIDED PRACTICE**
>
> Before moving on to evaluate the hypotheses in the previous Guided Practice, let's think carefully about the dataset. Are the observations across the two groups independent? Are there any concerns about outliers?[2]

20.1.2 Variability of the statistic

In Section 11, the variability of the statistic (previously: $\hat{p}_1 - \hat{p}_2$) was visualized after shuffling the observations across the two treatment groups many times. The shuffling process implements the null hypothesis model (that there is no effect of the treatment). In the exam example, the null hypothesis is that exam A and exam B are equally difficult, so the average scores across the two tests should be the same. If the exams were equally difficult, *due to natural variability*, we would sometimes expect students to do slightly better on exam A ($\bar{x}_A > \bar{x}_B$) and sometimes expect students to do slightly better on exam B ($\bar{x}_B > \bar{x}_A$). The question at hand is: does $\bar{x}_A - \bar{x}_B = 3.1$ indicate that exam A is easier than exam B.

Figure 20.2 shows the process of randomizing the exam to the observed exam scores. If the null hypothesis is true, then the score on each exam should represent the true student ability on that material. It shouldn't matter whether they were given exam A or exam B. By reallocating which student got which exam, we are able to understand how the difference in average exam scores changes due only to natural variability. There is only one iteration of the randomization process in Figure 20.2, leading to one simulated difference in average scores.

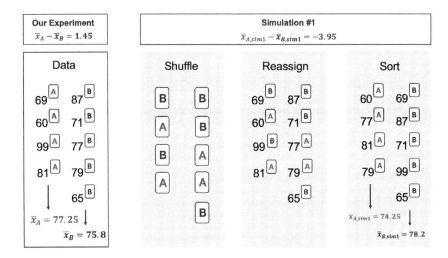

Figure 20.2: The version of the test (A or B) is randomly allocated to the test scores, under the null assumption that the tests are equally difficult.

Building on Figure 20.2, Figure 20.3 shows the values of the simulated statistics $\bar{x}_{1,sim} - \bar{x}_{2,sim}$ over 1,000 random simulations. We see that, just by chance, the difference in scores can range anywhere from -10 points to +10 points.

20.1.3 Observed statistic vs. null statistics

The goal of the randomization test is to assess the observed data, here the statistic of interest is $\bar{x}_A - \bar{x}_B = 3.1$. The randomization distribution allows us to identify whether a difference of 3.1 points is more than one would expect by natural variability of the scores if the two tests were equally

---
[2]Since the exams were shuffled, the "treatment" in this case was randomly assigned, so independence within and between groups is satisfied. The summary statistics suggest the data are roughly symmetric about the mean, and the min/max values don't suggest any outliers of concern.

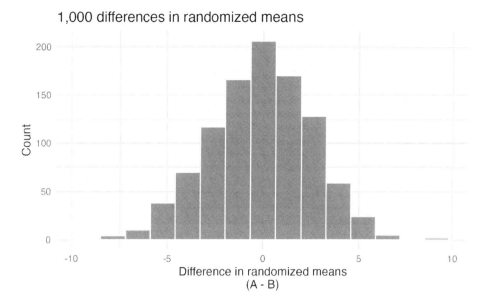

Figure 20.3: Histogram of differences in means, calculated from 1,000 different randomizations of the exam types.

difficult. By plotting the value of 3.1 on Figure 20.4, we can measure how different or similar 3.1 is to the randomized differences which were generated under the null hypothesis.

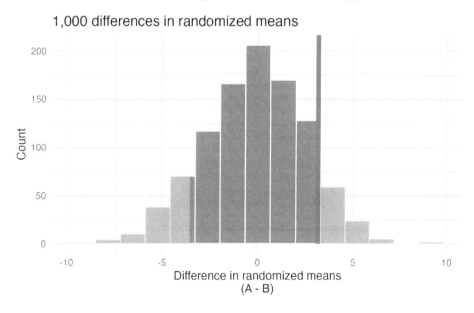

Figure 20.4: Histogram of differences in means, calculated from 1,000 different randomizations of the exam types. The observed difference of 3.1 points is plotted as a vertical line, and the area more extreme than 3.1 is shaded to represent the p-value.

EXAMPLE

Approximate the p-value depicted in Figure 20.4, and provide a conclusion in the context of the case study.

Using software, we can find the number of shuffled differences in means that are less than the observed difference (of 3.14) is 19 (out of 1,000 randomizations). So 10% of the simulations are larger than the observed difference. To get the p-value, we double the proportion of randomized differences which are larger than the observed difference, p-value = 0.2.

Previously, we specified that we would use $\alpha = 0.01$. Since the p-value is larger than α, we do not reject the null hypothesis. That is, the data do not convincingly show that one exam version is more difficult than the other, and the teacher should not be convinced that they should add points to the Version B exam scores.

The large p-value and consistency of $\bar{x}_A - \bar{x}_B = 3.1$ with the randomized differences leads us to *not reject the null hypothesis*. Said differently, there is no evidence to think that one of the tests is easier than the other. One might be inclined to conclude that the tests have the same level of difficulty, but that conclusion would be wrong. The hypothesis testing framework is set up only to reject a null claim, it is not set up to validate a null claim. As we concluded, the data are consistent with exams A and B being equally difficult, but the data are also consistent with exam A being 3.1 points "easier" than exam B. The data are not able to adjudicate on whether the exams are equally hard or whether one of them is slightly easier. Indeed, conclusions where the null hypothesis is not rejected often seem unsatisfactory. However, in this case, the teacher and class are probably all relieved that there is no evidence to demonstrate that one of the exams is more difficult than the other.

20.2 Bootstrap confidence interval for the difference in means

Before providing a full example working through a bootstrap analysis on actual data, we return to the fictional Awesome Auto example as a way to visualize the two sample bootstrap setting. Consider an expanded scenario where the research question centers on comparing the average price of a car at one Awesome Auto franchise (Group 1) to the average price of a car at a different Awesome Auto franchise (Group 2). The process of bootstrapping can be applied to *each* Group separately, and the differences of means recalculated each time. Figure 20.5 visually describes the bootstrap process when interest is in a statistic computed on two separate samples. The analysis proceeds as in the one sample case, but now the (single) statistic of interest is the *difference in sample means*. That is, a bootstrap resample is done on each of the groups separately, but the results are combined to have a single bootstrapped difference in means. Repetition will produce k bootstrapped differences in means, and the histogram will describe the natural sampling variability associated with the difference in means.

20.2.1 Observed data

Does treatment using embryonic stem cells (ESCs) help improve heart function following a heart attack? Table 20.2 contains summary statistics for an experiment to test ESCs in sheep that had a heart attack. Each of these sheep was randomly assigned to the ESC or control group, and the change in their hearts' pumping capacity was measured in the study. (Ménard et al., 2005) Figure 20.8 provides histograms of the two datasets. A positive value corresponds to increased pumping capacity, which generally suggests a stronger recovery. Our goal will be to identify a 95% confidence interval for the effect of ESCs on the change in heart pumping capacity relative to the control group.

20.2. BOOTSTRAP CONFIDENCE INTERVAL FOR THE DIFFERENCE IN MEANS

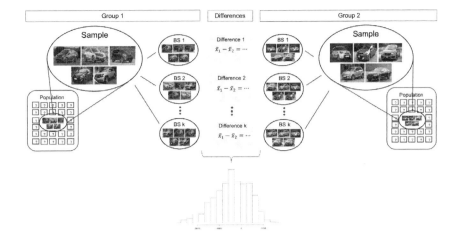

Figure 20.5: For the two group comparison, the bootstrap resampling is done separately on each group, but the statistic is calculated as a difference. The set of k differences is then analyzed as the statistic of interest with conclusions drawn on the parameter of interest.

 The stem_cell data can be found in the **openintro** R package.

Table 20.2: Summary statistics of the embryonic stem cell study.

Group	n	Mean	SD
ESC	9	3.50	5.17
Control	9	-4.33	2.76

The point estimate of the difference in the heart pumping variable is straightforward to find: it is the difference in the sample means.

$$\bar{x}_{esc} - \bar{x}_{control} = 3.50 - (-4.33) = 7.83$$

20.2.2 Variability of the statistic

As we saw in Section 17.2, we will use bootstrapping to estimate the variability associated with the difference in sample means when taking repeated samples. In a method akin to two proportions, a *separate* sample is taken with replacement from each group (here ESCs and control), the sample means are calculated, and their difference is taken. The entire process is repeated multiple times to produce a bootstrap distribution of the difference in sample means (*without* the null hypothesis assumption).

Figure 20.6 displays the variability of the differences in means with the 90% percentile and SE CIs super imposed.

::: {.guidedpractice data-latex=""} Using the histogram of bootstrapped difference in means, estimate the standard error of the differences in sample means, $\bar{x}_{ESC} - \bar{x}_{Control}$.[3] :::

---

[3] The point estimate of the population difference ($\bar{x}_{ESC} - \bar{x}_{Control}$) is 7.83.

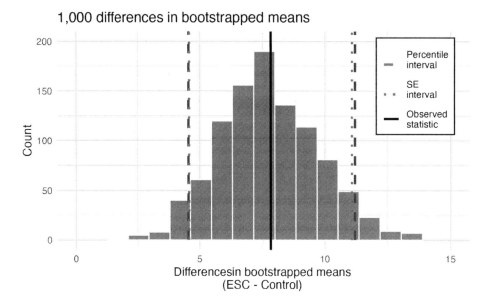

Figure 20.6: Histogram of differences in means after 1,000 bootstrap samples from each of the two groups. The observed difference is plotted as a black vertical line at 7.83. The blue dashed and red dotted lines provide the bootstrap percentile and boostrap SE confidence intervals, respectively, for the difference in true population means.

EXAMPLE

Choose one of the bootstrap confidence intervals for the true difference in average pumping capacity, $\mu_{ESC} - \mu_{Control}$. Does the interval show that there is a difference across the two treatments?

Because neither of the 90% intervals (either percentile or SE) above overlap zero (note that zero is never one of the bootstrapped differences so 95% and 99% intervals would have given the same conclusion!), we conclude that the ESC treatment is substantially better with respect to heart pumping capacity than the treatment.

Because the study is a randomized controlled experiment, we can conclude that it is the treatment (ESC) which is causing the change in pumping capacity.

20.3 Mathematical model for testing the difference in means

Every year, the US releases to the public a large data set containing information on births recorded in the country. This data set has been of interest to medical researchers who are studying the relation between habits and practices of expectant mothers and the birth of their children. We will work with a random sample of 1,000 cases from the data set released in 2014.

The **births14** data can be found in the **openintro** R package.

20.3.1 Observed data

Four cases from this dataset are represented in Table 20.3. We are particularly interested in two variables: weight and smoke. The weight variable represents the weights of the newborns and the smoke variable describes which mothers smoked during pregnancy.

Table 20.3: Four cases from the births14 dataset. The emoty cells indicate missing data.

fage	mage	weeks	visits	gained	weight	sex	habit
34	34	37	14	28	6.96	male	nonsmoker
36	31	41	12	41	8.86	female	nonsmoker
37	36	37	10	28	7.51	female	nonsmoker
	16	38		29	6.19	male	nonsmoker

We would like to know, is there convincing evidence that newborns from mothers who smoke have a different average birth weight than newborns from mothers who don't smoke? We will use data from this sample to try to answer this question.

EXAMPLE

Set up appropriate hypotheses to evaluate whether there is a relationship between a mother smoking and average birth weight.

The null hypothesis represents the case of no difference between the groups.

- H_0 : There is no difference in average birth weight for newborns from mothers who did and did not smoke. In statistical notation: $\mu_n - \mu_s = 0$, where μ_n represents non-smoking mothers and μ_s represents mothers who smoked.
- H_A : There is some difference in average newborn weights from mothers who did and did not smoke $(\mu_n - \mu_s \neq 0)$.

Table 20.4 displays sample statistics from the data. We can see that the average birth weight of babies born to smoker moms is lower than those born to nonsmoker moms.

Table 20.4: Summary statistics for the births14 dataset.

Habit	n	Mean	SD
nonsmoker	867	7.27	1.23
smoker	114	6.68	1.60

20.3.2 Variability of the statistic

We check the two conditions necessary to model the difference in sample means using the t-distribution.

- Because the data come from a simple random sample, the observations are independent, both within and between samples.
- With both groups over 30 observations, we inspect the data in Figure 20.7 for any particularly extreme outliers and find none.

Since both conditions are satisfied, the difference in sample means may be modeled using a t-distribution.

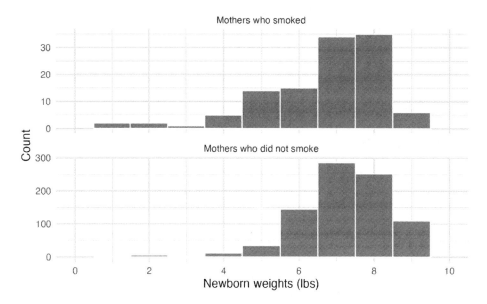

Figure 20.7: The top panel represents birth weights for infants whose mothers smoked during pregnancy. The bottom panel represents the birth weights for infants whose mothers who did not smoke during pregnancy.

GUIDED PRACTICE

The summary statistics in Table 20.4 may be useful for this Guided Practice. What is the point estimate of the population difference, $\mu_n - \mu_s$?[4]

20.3.3 Observed statistic vs. null statistics

The test statistic for comparing two means is a T.

The T score is a ratio of how the groups differ as compared to how the observations within a group vary.

$$T = \frac{(\bar{x}_1 - \bar{x}_2) - 0}{\sqrt{s_1^2/n_1 + s_2^2/n_2}}$$

When the null hypothesis is true and the conditions are met, T has a t-distribution with $df = min(n_1 - 1, n_2 - 1)$.

Conditions:

- Independent observations within and between groups.
- Large samples and no extreme outliers.

---

[4]The point estimate of the population difference $(\bar{x}_n - \bar{x}_s)$ is 0.59.

20.4 MATHEMATICAL MODEL FOR ESTIMATING THE DIFFERENCE IN MEANS

GUIDED PRACTICE

Compute the standard error of the point estimate for the average difference between the weights of babies born to nonsmoker and smoker mothers.[5]

EXAMPLE

Complete the hypothesis test started in the previous Example and Guided Practice on births14 dataset and research question. Use a significance level of $\alpha = 0.05$. For reference, $\bar{x}_n - \bar{x}_s = 0.59$, $SE = 0.16$, and the sample sizes were $n_n = 100$ and $n_s = 50$.

We can find the test statistic for this test using the previous information:

$$T = \frac{0.59 - 0}{0.16} = 3.69$$

We find the single tail area using software. We'll use the smaller of $n_n - 1 = 866$ and $n_s - 1 = 113$ as the degrees of freedom: $df = 113$. The one tail area is roughly 0.00017; doubling this value gives the two-tail area and p-value, 0.00034.

The p-value is much smaller than the significance value, 0.05, so we reject the null hypothesis. The data provide is convincing evidence of a difference in the average weights of babies born to mothers who smoked during pregnancy and those who did not.

This result is likely not surprising. We all know that smoking is bad for you and you've probably also heard that smoking during pregnancy is not just bad for the mother but also for the baby as well. In fact, some in the tobacco industry actually had the audacity to tout that as a *benefit* of smoking:

> It's true. The babies born from women who smoke are smaller, but they're just as healthy as the babies born from women who do not smoke. And some women would prefer having smaller babies. - Joseph Cullman, Philip Morris' Chairman of the Board on CBS' *Face the Nation*, Jan 3, 1971

Furthermore, health differences between babies born to mothers who smoke and those who do not are not limited to weight differences.[6]

20.4 Mathematical model for estimating the difference in means

20.4.1 Observed data

As with hypothesis testing, for the question of whether we can model the difference using a t-distribution, we'll need to check new conditions. Like the 2-proportion cases, we will require a more robust version of independence so we are confident the two groups are also independent. Secondly, we also check for normality in each group separately, which in practice is a check for outliers.

---

[5] $SE(\bar{x}_n - \bar{x}_s) = \sqrt{s_n^2/n_n + s_s^2/n_s}$
$= \sqrt{1.23^2/867 + 1.60^2/114} = 0.16$

[6] You can watch an episode of John Oliver on *Last Week Tonight* to explore the present day offenses of the tobacco industry. Please be aware that there is some adult language.

Using the t-distribution for a difference in means.

The t-distribution can be used for inference when working with the standardized difference of two means if

- *Independence* (extended). The data are independent within and between the two groups, e.g., the data come from independent random samples or from a randomized experiment.
- *Normality.* We check the outliers for each group separately.

The standard error may be computed as

$$SE = \sqrt{\frac{\sigma_1^2}{n_1} + \frac{\sigma_2^2}{n_2}}$$

The official formula for the degrees of freedom is quite complex and is generally computed using software, so instead you may use the smaller of $n_1 - 1$ and $n_2 - 1$ for the degrees of freedom if software isn't readily available.

Recall that the margin of error is defined by the standard error. The margin of error for $\bar{x}_1 - \bar{x}_2$ can be directly obtained from $SE(\bar{x}_1 - \bar{x}_2)$.

Margin of error for $\bar{x}_1 - \bar{x}_2$.

The margin of error is $t^\star_{df} \times \sqrt{\frac{s_1^2}{n_1} + \frac{s_2^2}{n_2}}$ where $t^\star_{df}$ is calculated from a specified percentile on the t-distribution with *df* degrees of freedom.

20.4.2 Variability of the statistic

EXAMPLE

Can the t-distribution be used to make inference using the point estimate, $\bar{x}_{esc} - \bar{x}_{control} = 7.83$?

First, we check for independence. Because the sheep were randomized into the groups, independence within and between groups is satisfied.

Figure 20.8 does not reveal any clear outliers in either group. (The ESC group does look a bit more variable, but this is not the same as having clear outliers.)

With both conditions met, we can use the t-distribution to model the difference of sample means.

Generally, we use statistical software to find the appropriate degrees of freedom, or if software isn't available, we can use the smaller of $n_1 - 1$ and $n_2 - 1$ for the degrees of freedom, e.g., if using a t-table to find tail areas. For transparency in the Examples and Guided Practice, we'll use the latter approach for finding *df*; in the case of the ESC example, this means we'll use $df = 8$.

20.4. MATHEMATICAL MODEL FOR ESTIMATING THE DIFFERENCE IN MEANS

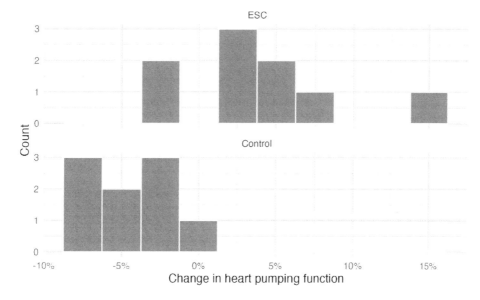

Figure 20.8: Histograms for the difference in heart pumping function after a heart attack for both the treatment group (ESC, which received an the embryonic stem cell treatment) and the control group (which did not receive the treatment).

 EXAMPLE

Calculate a 95% confidence interval for the effect of ESCs on the change in heart pumping capacity of sheep after they've suffered a heart attack.

We will use the sample difference and the standard error that we computed earlier:

$$\bar{x}_{esc} - \bar{x}_{control} = 7.83$$

$$SE = \sqrt{\frac{5.17^2}{9} + \frac{2.76^2}{9}} = 1.95$$

Using $df = 8$, we can identify the critical value of $t_8^\star = 2.31$ for a 95% confidence interval. Finally, we can enter the values into the confidence interval formula:

$$\text{point estimate} \pm t^\star \times SE$$
$$7.83 \pm 2.31 \times 1.95$$
$$(3.32 \, , \, 12.34)$$

We are 95% confident that the heart pumping function in sheep that received embryonic stem cells is between 3.32% and 12.34% higher than for sheep that did not receive the stem cell treatment.

20.5 Chapter review

20.5.1 Summary

In this chapter we extended the single mean inferential methods to questions of differences in means. You may have seen parallels from the chapters that extended a single proportion (Chapter 16) to differences in proportions (Chapter 17). When considering differences in sample means (indeed, when considering many quantitative statistics), we use the t-distribution to describe the sampling distribution of the T score (the standardized difference in sample means). Ideas of confidence level and type of error which might occur from a hypothesis test conclusion are similar to those seen in other chapters (see Section 14).

20.5.2 Terms

We introduced the following terms in the chapter. If you're not sure what some of these terms mean, we recommend you go back in the text and review their definitions. We are purposefully presenting them in alphabetical order, instead of in order of appearance, so they will be a little more challenging to locate. However you should be able to easily spot them as **bolded text**.

difference in means	SE difference in means	t-CI
point estimate	T score	t-test

20.6 Exercises

Answers to odd numbered exercises can be found in Appendix A.20.

20.1. **Experimental baker.** A baker working on perfecting their bagel recipe is experimenting with active dry (AD) and instant (I) yeast. They bake a dozen bagels with each type of yeast and score each bagel on a scale of 1 to 10 on how well the bagels rise. They come up with the following set of hypotheses for evaluating whether there is a difference in the average rise of bagels baked with active dry and instant yeast. What is wrong with the hypotheses as stated?

$$H_0 : \bar{x}_{AD} \leq \bar{x}_I \qquad H_A : \bar{x}_{AD} > \bar{x}_I$$

20.2. **Fill in the blanks.** We use a _____ to evaluate if data provide convincing evidence of a difference between two population means and we use a _____ to estimate this difference.

20.3. **Diamonds, randomization test.** The prices of diamonds go up as the carat weight increases, but the increase is not smooth. For example, the difference between the size of a 0.99 carat diamond and a 1 carat diamond is undetectable to the naked human eye, but the price of a 1 carat diamond tends to be much higher than the price of a 0.99 diamond. In this question we use two random samples of diamonds, 0.99 carats and 1 carat, each sample of size 23, and randomize the carat weight to the price values in order compare the average prices of the diamonds to a null distribution. In order to be able to compare equivalent units, we first divide the price for each diamond by 100 times its weight in carats. That is, for a 0.99 carat diamond, we divide the price by 99. or a 1 carat diamond, we divide the price by 100. The randomization distribution (with 1,000 repetitions) below describes the null distribution of the difference in sample means (of price per carat) if there really was no difference in the population from which these diamonds came.[7] (Wickham, 2016)

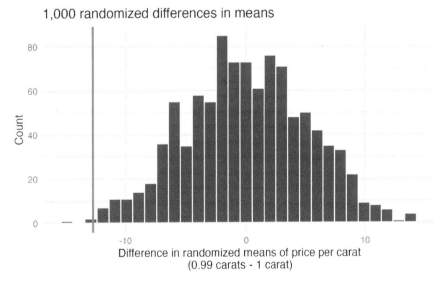

Using the randomization distribution of the difference in average price per carat (1,000 randomizations were run), conduct a hypothesis test to evaluate if there is a difference between the prices per carat of diamonds that weigh 0.99 carats and diamonds that weigh 1 carat. Make sure to state your hypotheses clearly and interpret your results in context of the data. (Wickham, 2016)

---

[7]The `diamonds` data used in this exercise can be found in the **ggplot2** R package.

20.4. **Lizards running, randomization test.** In order to assess physiological characteristics of common lizards, data on top speeds (in m/sec) measured on a laboratory race track for two species of lizards: Western fence lizard (Sceloporus occidentalis) and Sagebrush lizard (Sceloporus graciosus). The original observed difference in lizard speeds is $\bar{x}_{Western fence} - \bar{x}_{Sagebrush} = 0.7$ m/sec. The histogram below shows the distribution of average differences when speed has been randomly allocated across lizard species 1,000 times.[8] (Adolph, 1987)

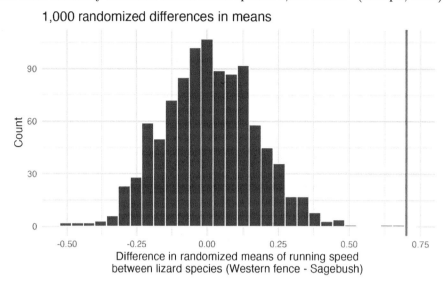

Using the randomization distribution, conduct a hypothesis test to evaluate if there is a difference between the average speed of the Western fence lizard as compared to the Sagebrush lizard. Make sure to state your hypotheses clearly and interpret your results in context of the data.

20.5. **Diamonds, bootstrap interval.** We have data on two random samples of diamonds: one with diamonds that weigh 0.99 carats and one with diamonds that weigh 1 carat. Each sample has 23 diamonds. Provided below is a histogram of bootstrap differences in means of price per carat of diamonds that weight 0.99 carats and diamonds that weigh 1 carat. (Wickham, 2016)

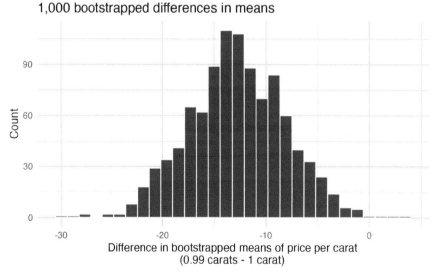

a. Using the bootstrap distribution, create a (rough) 95% bootstrap percentile confidence interval for the true population difference in prices per carat of diamonds that weigh 0.99 carats and 1 carat.

b. Using the bootstrap distribution, create a (rough) 95% bootstrap SE confidence interval for the true population difference in prices per carat of diamonds that weigh 0.99 carats and 1 carat.

---

[8]The `lizard_run` data used in this exercise can be found in the **openintro** R package.

20.6. **Lizards running, bootstrap interval.** We have data on top speeds (in m/sec) measured on a laboratory race track for two species of lizards: Western fence lizard (Sceloporus occidentalis) and Sagebrush lizard (Sceloporus graciosus). The bootstrap distribution below describes the variability of difference in means captured from 1,000 bootstrap samples of the lizard data. (Adolph, 1987)

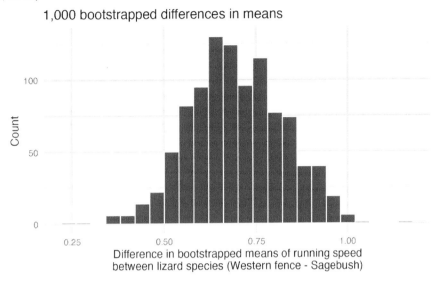

a. Using the bootstrap distribution, create a (rough) 90% percentile bootrap confidence interval for the true population difference in price per carat for Western fence lizard as compared with Sagebrush lizard.

b. Using the bootstrap distribution, create a (rough) 90% bootstrap SE confidence interval for the true population difference in price per carat for Western fence lizard as compared with Sagebrush lizard.

20.7. **Weight loss.** You are reading an article in which the researchers have created a 95% confidence interval for the difference in average weight loss for two diets. They are 95% confident that the true difference in average weight loss over 6 months for the two diets is somewhere between (1 lb, 25 lbs). The authors claim that, "therefore diet A ($\bar{x}_A = 20$ lbs average loss) results in a much larger average weight loss as compared to diet B ($\bar{x}_B = 7$ lbs average loss)." Comment on the authors' claim.

20.8. **Possible randomized means.** Data were collected on data from two groups (A and B). There were 3 measurements taken on Group A and two measurements

Group	Measurement 1	Measurement 2	Measurement 3
A	1	15	5
B	7	3	

If the data are (repeatedly) randomly allocated across the two conditions, provide the following: (1) the values which are assigned to group A, (2) the values which are assigned to group B, and (3) the difference in averages ($\bar{x}_A - \bar{x}_B$) for each of the following:

a. When the randomized difference in averages is as big as possible.

b. When the randomized difference in averages is as small as possible (a big in magnitude negative number).

c. When the randomized difference in averages is as close to zero as possible.

d. When the observed values are randomly assigned to the two groups, to which of the previous parts would you expect the difference in means to fall closest? Explain your reasoning.

20.9. **Diamonds, mathematical test.** We have data on two random samples of diamonds: one with diamonds that weigh 0.99 carats and one with diamonds that weigh 1 carat. Each sample has 23 diamonds. Sample statistics for the price per carat of diamonds in each sample are provided below. Conduct a hypothesis test using a mathematical model to evaluate if there is a difference between the prices per carat of diamonds that weigh 0.99 carats and diamonds that weigh 1 carat Make sure to state your hypotheses clearly, check relevant conditions, and interpret your results in context of the data. (Wickham, 2016)

	Mean	SD	n
0.99 carats	$44.51	$13.32	23
1 carat	$57.20	$18.19	23

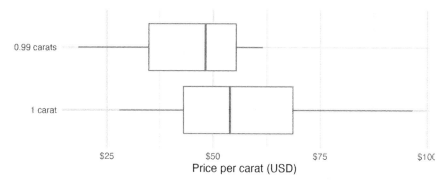

20.10. **A/B testing.** A/B testing is a user experience research methodology where two variants of a page are shown to users at random. A company wants to evaluate whether users will spend more time, on average, on Page A or Page B using an A/B test. Two user experience designers at the company, Lucie and Müge, are tasked with conducting the analysis of the data collected. They agree on how the null hypothesis should be set: on average, users spend the same amount of time on Page A and Page B. Lucie believes that Page B will provide a better experience for users and hence wants to use a one-tailed test, Müge believes that a two-tailed test would be a better choice. Which designer do you agree with, and why?

20.11. **Diamonds, mathematical interval.** We have data on two random samples of diamonds: one with diamonds that weigh 0.99 carats and one with diamonds that weigh 1 carat. Each sample has 23 diamonds. Sample statistics for the price per carat of diamonds in each sample are provided below. Assuming that the conditions for conducting inference using a mathematical model are satisfied, construct a 95% confidence interval for the true population difference in prices per carat of diamonds that weigh 0.99 carats and 1 carat. (Wickham, 2016)

	Mean	SD	n
0.99 carats	$44.51	$13.32	23
1 carat	$57.20	$18.19	23

20.12. **True / False: comparing means.** Determine if the following statements are true or false, and explain your reasoning for statements you identify as false.

   a. As the degrees of freedom increases, the t-distribution approaches normality.

   b. If a 95% confidence interval for the difference between two population means contains 0, a 99% confidence interval calculated based on the same two samples will also contain 0.

   c. If a 95% confidence interval for the difference between two population means contains 0, a 90% confidence interval calculated based on the same two samples will also contain 0.

20.6. EXERCISES

20.13. **Difference of means.** Suppose we will collect two random samples from the following distributions. In each of the parts below, consider the sample means $\bar{x}_1$ and $\bar{x}_2$ that we might observe from these two samples.

	Mean	Standard deviation	Sample size
Sample 1	15	20	50
Sample 2	20	10	30

a. What is the associated mean and standard deviation of $\bar{x}_1$?

b. What is the associated mean and standard deviation of $\bar{x}_2$?

c. Calculate and interpret the mean and standard deviation associated with the difference in sample means for the two groups, $\bar{x}_2 - \bar{x}_1$.

d. How are the standard deviations from parts (a), (b), and (c) related?

20.14. **Gaming, distracted eating, and intake.** A group of researchers who are interested in the possible effects of distracting stimuli during eating, such as an increase or decrease in the amount of food consumption, monitored food intake for a group of 44 patients who were randomized into two equal groups. The treatment group ate lunch while playing solitaire, and the control group ate lunch without any added distractions. Patients in the treatment group ate 52.1 grams of biscuits, with a standard deviation of 45.1 grams, and patients in the control group ate 27.1 grams of biscuits, with a standard deviation of 26.4 grams. Do these data provide convincing evidence that the average food intake (measured in amount of biscuits consumed) is different for the patients in the treatment group compared to the control group? Assume that conditions for conducting inference using mathematical models are satisfied. (Oldham-Cooper et al., 2011)

20.15. **Chicken diet: horsebean vs. linseed.** Chicken farming is a multi-billion dollar industry, and any methods that increase the growth rate of young chicks can reduce consumer costs while increasing company profits, possibly by millions of dollars. An experiment was conducted to measure and compare the effectiveness of various feed supplements on the growth rate of chickens. Newly hatched chicks were randomly allocated into six groups, and each group was given a different feed supplement. In this exercise we consider chicks that were fed horsebean and linseed. Below are some summary statistics from this dataset along with box plots showing the distribution of weights by feed type.[9] (McNeil, 1977)

Feed type	Mean	SD	n
horsebean	160.20	38.63	10
linseed	218.75	52.24	12

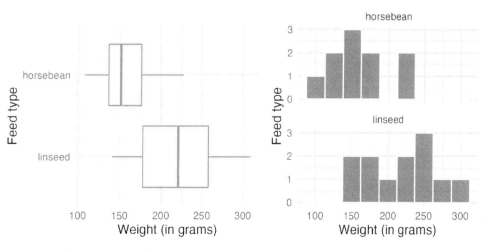

See next page for parts a to d.

---

[9] The `chickwts` data used in this exercise can be found in the **datasets** R package.

a. Describe the distributions of weights of chickens that were fed horsebean and linseed.

b. Do these data provide strong evidence that the average weights of chickens that were fed linseed and horsebean are different? Use a 5% significance level.

c. What type of error might we have committed? Explain.

d. Would your conclusion change if we used $\alpha = 0.01$?

20.16. **Fuel efficiency in the city.** Each year the US Environmental Protection Agency (EPA) releases fuel economy data on cars manufactured in that year. Below are summary statistics on fuel efficiency (in miles/gallon) from random samples of cars with manual and automatic transmissions manufactured in 2021. Do these data provide strong evidence of a difference between the average fuel efficiency of cars with manual and automatic transmissions in terms of their average city mileage?[10] (US DOE EPA, 2021)

CITY	Mean	SD	n
Automatic	17.4	3.44	25
Manual	22.7	4.58	25

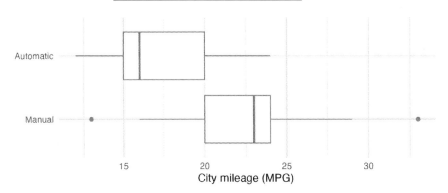

20.17. **Chicken diet: casein vs. soybean.** Casein is a common weight gain supplement for humans. Does it have an effect on chickens? An experiment was conducted to measure and compare the effectiveness of various feed supplements on the growth rate of chickens. Newly hatched chicks were randomly allocated into six groups, and each group was given a different feed supplement. In this exercise we consider chicks that were fed casein and soybean. Assume that the conditions for conducting inference using mathematical models are met, and using the data provided below, test the hypothesis that the average weight of chickens that were fed casein is different than the average weight of chickens that were fed soybean. If your hypothesis test yields a statistically significant result, discuss whether or not the higher average weight of chickens can be attributed to the casein diet. (McNeil, 1977)

Feed type	Mean	SD	n
casein	323.58	64.43	12
soybean	246.43	54.13	14

---

[10] The epa2021 data used in this exercise can be found in the **openintro** R package.

20.18. **Fuel efficiency on the highway.** Each year the US Environmental Protection Agency (EPA) releases fuel economy data on cars manufactured in that year. Below are summary statistics on fuel efficiency (in miles/gallon) from random samples of cars with manual and automatic transmissions manufactured in 2021. Do these data provide strong evidence of a difference between the average fuel efficiency of cars with manual and automatic transmissions in terms of their average highway mileage? (US DOE EPA, 2021)

HIGHWAY	Mean	SD	n
Automatic	23.7	3.90	25
Manual	30.9	5.13	25

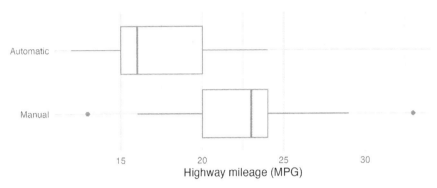

20.19. **Gaming, distracted eating, and intake.** A group of researchers who are interested in the possible effects of distracting stimuli during eating, such as an increase or decrease in the amount of food consumption, monitored food intake for a group of 44 patients who were randomized into two equal groups. The treatment group ate lunch while playing solitaire, and the control group ate lunch without any added distractions. Patients in the treatment group ate 52.1 grams of biscuits, with a standard deviation of 45.1 grams, and patients in the control group ate 27.1 grams of biscuits, with a standard deviation of 26.4 grams. Do these data provide convincing evidence that the average food intake (measured in amount of biscuits consumed) is different for the patients in the treatment group compared to the control group? Assume that conditions for conducting inference using mathematical models are satisfied. (Oldham-Cooper et al., 2011)

20.20. **Gaming, distracted eating, and recall.** A group of researchers who are interested in the possible effects of distracting stimuli during eating, such as an increase or decrease in the amount of food consumption, monitored food intake for a group of 44 patients who were randomized into two equal groups. The 22 patients in the treatment group who ate their lunch while playing solitaire were asked to do a serial-order recall of the food lunch items they ate. The average number of items recalled by the patients in this group was 4.9, with a standard deviation of 1.8. The average number of items recalled by the patients in the control group (no distraction) was 6.1, with a standard deviation of 1.8. Do these data provide strong evidence that the average numbers of food items recalled by the patients in the treatment and control groups are different? Assume that conditions for conducting inference using mathematical models are satisfied. (Oldham-Cooper et al., 2011)

Chapter 21

Inference for comparing paired means

In Chapter 20 analysis was done to compare the average population value across two different groups. Recall that one of the important conditions in doing a two-sample analysis is that the two groups are independent. Here, independence across groups means that knowledge of the observations in one group doesn't change what we'd expect to happen in the other group. But what happens if the groups are **dependent**? Sometimes dependency is not something that can be addressed through a statistical method. However, a particular dependency, **pairing**, can be modeled quite effectively using many of the same tools we have already covered in this text.

Paired data represent a particular type of experimental structure where the analysis is somewhat akin to a one-sample analysis (see Chapter 19) but has other features that resemble a two-sample analysis (see Chapter 20). As with a two-sample analysis, quantitative measurements are made on each of two different levels of the explanatory variable. However, because the observational unit is **paired** across the two groups, the two measurements are subtracted such that only the difference is retained. Table 21.1 presents some examples of studies where paired designs were implemented.

Table 21.1: Examples of studies where a paired design is used to measure the difference in the measurement over two conditions.

Observational unit	Comparison groups	Measurement	Value of interest
Car	Smooth Turn vs Quick Spin	amount of tire tread after 1,000 miles	difference in tread
Textbook	UCLA vs Amazon	price of new text	difference in price
Individual person	Pre-course vs Post-course	exam score	difference in score

Paired data.

Two sets of observations are *paired* if each observation in one set has a special correspondence or connection with exactly one observation in the other dataset.

It is worth noting that if mathematical modeling is chosen as the analysis tool, paired data inference on the difference in measurements will be identical to the one-sample mathematical techniques described in Chapter 19. However, recall from Chapter 19 that with pure one-sample data, the computational tools for hypothesis testing are not easy to implement and were not presented (although the bootstrap was presented as a computational approach for constructing a one sample confidence interval). With paired data, the randomization test fits nicely with the structure of the experiment and is presented here.

21.1 Randomization test for the mean paired difference

Consider an experiment done to measure whether tire brand Smooth Turn or tire brand Quick Spin has longer tread wear (in cm). That is, after 1,000 miles on a car, which brand of tires has more tread, on average?

21.1.1 Observed data

The observed data represent 25 tread measurements (in cm) taken on 25 tires of Smooth Turn and 25 tires of Quick Spin. The study used a total of 25 cars, so on each car, one tire was of Smooth Turn and one was of Quick Spin. Figure 21.1 presents the observed data, calculations on tread remaining (in cm).

The Smooth Turn manufacturer looks at the box plot and says:

> *Clearly the tread on Smooth Turn tires is higher, on average, than the tread on Quick Spin tires after 1,000 miles of driving.*

The Quick Spin manufacturer is skeptical and retorts:

> *But with only 25 cars, it seems that the variability in road conditions (sometimes one tire hits a pothole, etc.) could be what leads to the small difference in average tread amount.*

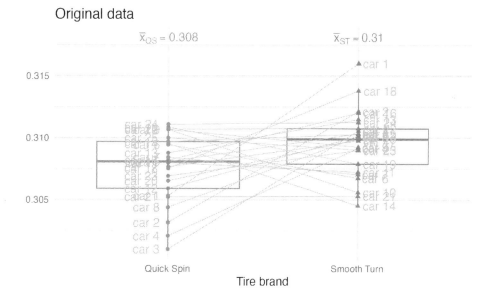

Figure 21.1: Boxplots of the tire tread data (in cm) and the brand of tire from which the original measurements came.

We'd like to be able to systematically distinguish between what the Smooth Turn manufacturer sees in the plot and what the Quick Spin manufacturer sees in the plot. Fortunately for us, we have an excellent way to simulate the natural variability (from road conditions, etc.) that can lead to tires being worn at different rates.

21.1.2 Variability of the statistic

A randomization test will identify whether the differences seen in the box plot of the original data in Figure 21.1 could have happened just by chance variability. As before, we will simulate the variability in the study under the assumption that the null hypothesis is true. In this study, the null hypothesis is that average tire tread wear is the same across Smooth Turn and Quick Spin tires.

- $H_0 : \mu_{diff} = 0$, the average tread wear is the same for the two tire brands.
- $H_A : \mu_{diff} \ne 0$, the average tread wear is different across the two tire brands.

When observations are paired, the randomization process randomly assigns the tire brand to each of the observed tread values. Note that in the randomization test for the two-sample mean setting (see Section 20.1) the explanatory variable was *also* randomly assigned to the responses. The change in the paired setting, however, is that the assignment happens *within* an observational unit (here, a car). Remember, if the null hypothesis is true, it will not matter which brand is put on which tire because the overall tread wear will be the same across pairs.

Figures 21.2 and 21.3 show that the random assignment of group (tire brand) happens within a single car. That is, every single car will still have one tire of each type. In the first randomization, it just so happens that the 4th car's tire brands were swapped and the 5th car's tire brands were not swapped.

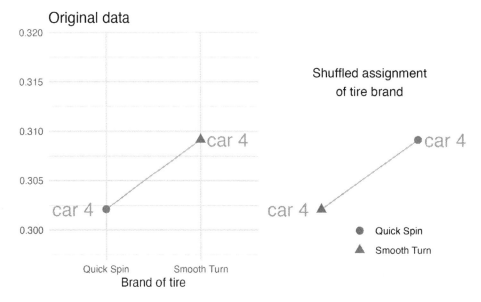

Figure 21.2: The 4th car: the tire brand was randomly permuted, and in the randomization calculation, the measurements (in cm) ended up in different groups.

We can put the shuffled assignments for all the cars into one plot as seen in Figure 21.4.

The next step in the randomization test is to sort the brands so that the assigned brand value on the x-axis aligns with the assigned group from the randomization. See Figure 21.5 which has the same randomized groups (right image in Figure 21.4 and left image in Figure 21.5) as seen previously. However, the right image in Figure 21.5 sorts the randomized groups so that we can measure the variability across groups as compared to the variability within groups.

Figure 21.6 presents a second randomization of the data. Notice how the two observations from the same car are linked by a grey line; some of the tread values have been randomly assigned to the opposite tire brand than they were originally (while some are still connected to their original tire brands).

Figure 21.7 presents yet another randomization of the data. Again, the same observations are linked by a grey line, and some of the tread values have been randomly assigned to the opposite tire brand than they were originally (while some are still connected to their original tire brands).

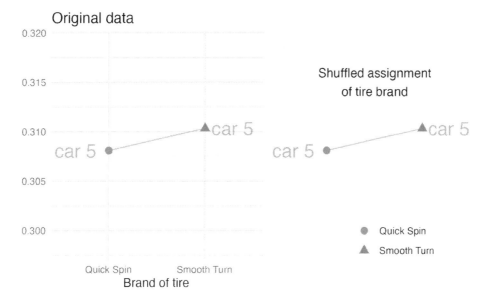

Figure 21.3: The 5th car: the tire brand was randomly permuted to stay the same! In the randomization calculation, the measurements (in cm) ended up in the original groups.

21.1.3 Observed statistic vs. null statistics

By repeating the randomization process, we can create a distribution of the average of the differences in tire treads, as seen in Figure 21.8. As expected (because the differences were generated under the null hypothesis), the center of the histogram is zero. A line has been drawn at the observed difference which is well outside the majority of the null differences simulated from natural variability by mixing up which the tire received Smooth Turn and which received Quick Spin. Because the observed statistic is so far away from the natural variability of the randomized differences, we are convinced that there is a difference between Smooth Turn and Quick Spin. Our conclusion is that the extra amount of average tire tread in Smooth Turn is due to more than just natural variability: we reject H_0 and conclude that $\mu_{ST} \neq \mu_{QS}$.

21.2 Bootstrap confidence interval for the mean paired difference

For both the bootstrap and the mathematical models applied to paired data, the analysis is virtually identical to the one-sample approach given in Chapter 19. The key to working with paired data (for bootstrapping and mathematical approaches) is to consider the measurement of interest to be the difference in measured values across the pair of observations.

21.2.1 Observed data

In an earlier edition of this textbook, we found that Amazon prices were, on average, lower than those of the UCLA Bookstore for UCLA courses in 2010. It's been several years, and many stores have adapted to the online market, so we wondered, how is the UCLA Bookstore doing today?

We sampled 201 UCLA courses. Of those, 68 required books could be found on Amazon. A portion of the dataset from these courses is shown in Figure 21.2, where prices are in US dollars.

 The ucla_textbooks_f18 data can be found in the **openintro** R package.

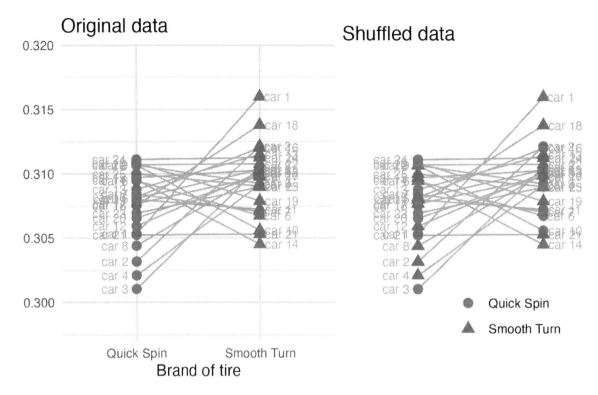

Figure 21.4: Tire tread data (in cm) with: the brand of tire from which the original measurements came (left) and shuffled brand assignment (right). As evidenced by the colors, some of the cars kept their original tire assignments and some cars swapped the tire assignments.

Table 21.2: Four cases from the ucla_textbooks_f18 dataset.

subject	course_num	bookstore_new	amazon_new	price_diff
American Indian Studies	M10	48.0	47.5	0.52
Anthropology	2	14.3	13.6	0.71
Arts and Architecture	10	13.5	12.5	0.97
Asian	M60W	49.3	55.0	-5.69

Each textbook has two corresponding prices in the dataset: one for the UCLA Bookstore and one for Amazon. When two sets of observations have this special correspondence, they are said to be **paired**.

21.2.2 Variability of the statistic

Following the example of bootstrapping the one-sample statistic, the observed *differences* can be bootstrapped in order to understand the variability of the average difference from sample to sample. Remember, the differences act as a single value to bootstrap. That is, the original dataset would include the list of 68 price differences, and each resample will also include 68 price differences (some repeated through the bootstrap resampling process). The bootstrap procedure for paired differences is quite similar to the procedure applied to the one-sample statistic case in Section 19.1.

In Figure 21.9, two 99% confidence intervals for the difference in the cost of a new book at the UCLA bookstore compared with Amazon have been calculated. The bootstrap percentile confidence interval is computing using the 0.5 percentile and 99.5 percentile bootstrapped differences and is found to be ($0.25, $7.87).

21.2. BOOTSTRAP CONFIDENCE INTERVAL FOR THE MEAN PAIRED DIFFERENCE

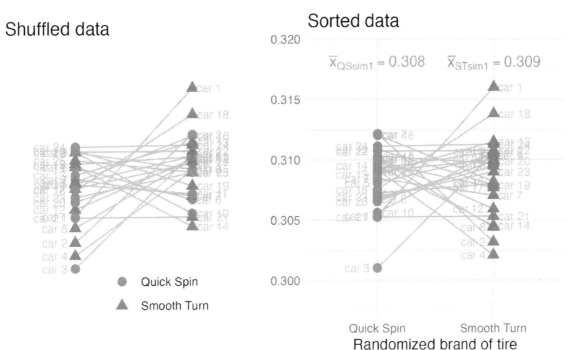

Figure 21.5: Tire tread from (left) randomized brand assignment, (right) sorted by randomized brand.

 GUIDED PRACTICE

Using the histogram of bootstrapped difference in means, estimate the standard error of the mean of the sample differences, $\bar{x}_{diff}$.[1]

The bootstrap SE interval is found by computing the SE of the bootstrapped differences ($SE_{\bar{x}_{diff}} = \$1.64$) and the normal multiplier of $z^* = 2.58$. The averaged difference is $\bar{x} = \$3.58$. The 99% confidence interval is: $\$3.58 \pm 2.58 \times \$1.64 = (\$-0.65, \$7.81)$.

The confidence intervals seem to indicate that the UCLA bookstore price is, on average, higher than the Amazon price, as the majority of the confidence interval is positive. However, if the analysis required a strong degree of certainty (e.g., 99% confidence), and the bootstrap SE interval was most appropriate (given a second course in statistics the nuances of the methods can be investigated), the results of which book seller is higher is not well determined (because the bootstrap SE interval overlaps zero). That is, the 99% bootstrap SE interval gives potential for UCLA to be lower, on average, than Amazon (because of the possible negative values for the true mean difference in price).

---

[1]The bootstrapped differences in sample means vary roughly from 0.7 to 7.5, a range of $6.80. Although the bootstrap distribution is not symmetric, we use the empirical rule (that with bell-shaped distributions, most observations are within two standard errors of the center), the standard error of the mean differences is approximately $1.70. You might note that the standard error calculation given in Section 21.3 is $SE(\bar{x}_{diff}) = \sqrt{s^2_{diff}/n_{diff}}$
$= \sqrt{13.4^2/68} = \$1.62$ (values from Section 21.3), very close to the bootstrap approximation.

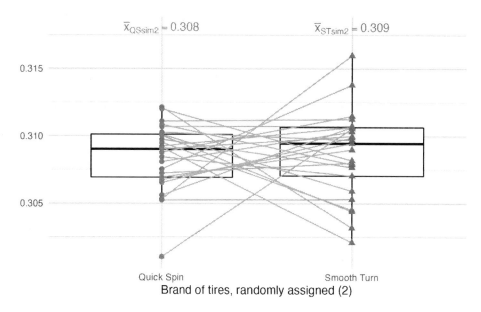

Figure 21.6: A second randomization where the brand is randomly swapped (or not) across the two tread wear measurements (in cm) from the same car.

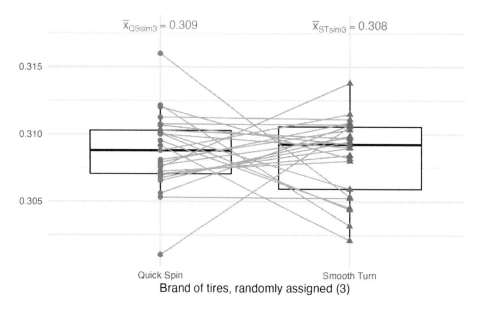

Figure 21.7: An additional randomization where the brand is randomly swapped (or not) across the two tread wear measurements (in cm) from the same car.

21.2. BOOTSTRAP CONFIDENCE INTERVAL FOR THE MEAN PAIRED DIFFERENCE 405

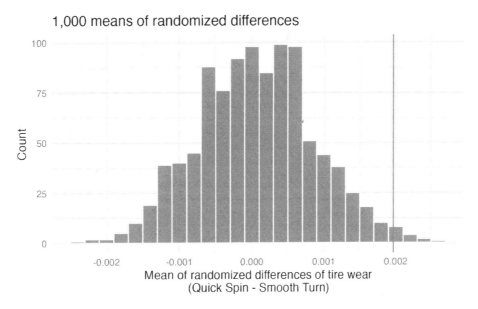

Figure 21.8: Histogram of 1,000 mean differences with tire brand randomly assigned across the two tread measurements (in cm) per pair.

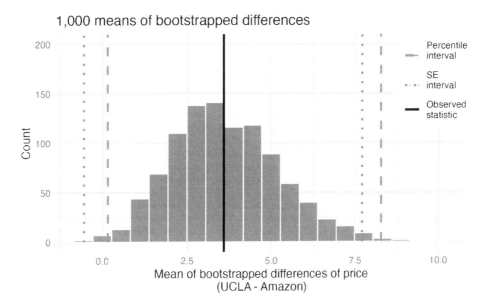

Figure 21.9: Bootstrap distribution for the average difference in new book price at the UCLA bookstore versus Amazon. 99% confidence intervals are superimposed using blue dashed (bootstrap percentile interval) and red dotted (bootstrap SE interval) lines.

21.3 Mathematical model for the mean paired difference

Thinking about the differences as a single observation on an observational unit changes the paired setting into the one-sample setting. The mathematical model for the one-sample case is covered in Section 19.2.

21.3.1 Observed data

To analyze paired data, it is often useful to look at the difference in outcomes of each pair of observations. In the textbook data, we look at the differences in prices, which is represented as the `price_difference` variable in the dataset. Here the differences are taken as

$$\text{UCLA Bookstore price} - \text{Amazon price}$$

It is important that we always subtract using a consistent order; here Amazon prices are always subtracted from UCLA prices. The first difference shown in Table 21.2 is computed as $47.97 - 47.45 = 0.52$. Similarly, the second difference is computed as $14.26 - 13.55 = 0.71$, and the third is $13.50 - 12.53 = 0.97$. A histogram of the differences is shown in Figure 21.10.

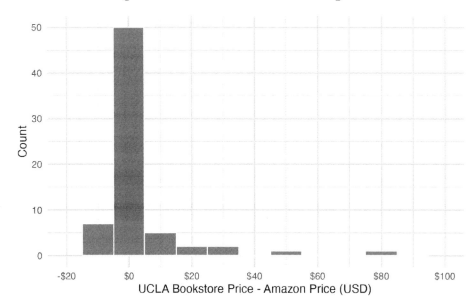

Figure 21.10: Histogram of the difference in price for each book sampled.

21.3.2 Variability of the statistic

To analyze a paired dataset, we simply analyze the differences. Table 21.3 provides the data summaries from the textbook data. Note that instead of reporting the prices separately for UCLA and Amazon, the summary statistics are given by the mean of the differences, the standard deviation of the differences, and the total number of pairs (i.e., differences). The parameter of interest is also a single value, μ_{diff}, so we can use the same t-distribution techniques we applied in Section 19.2 directly onto the observed differences.

Table 21.3: Summary statistics for the 68 price differences.

n	Mean	SD
68	3.58	13.4

21.3. MATHEMATICAL MODEL FOR THE MEAN PAIRED DIFFERENCE

EXAMPLE

Set up a hypothesis test to determine whether, on average, there is a difference between Amazon's price for a book and the UCLA bookstore's price. Also, check the conditions for whether we can move forward with the test using the t-distribution.

We are considering two scenarios: there is no difference or there is some difference in average prices.

- $H_0 : \mu_{diff} = 0$. There is no difference in the average textbook price.

- $H_A : \mu_{diff} \neq 0$. There is a difference in average prices.

Next, we check the independence and normality conditions. The observations are based on a simple random sample, so assuming the textbooks are independent seems reasonable. While there are some outliers, $n = 68$ and none of the outliers are particularly extreme, so the normality of $\bar{x}$ is satisfied. With these conditions satisfied, we can move forward with the t-distribution.

21.3.3 Observed statistic vs. null statistics

As mentioned previously, the methods applied to a difference will be identical to the one-sample techniques. Therefore, the full hypothesis test framework is presented as guided practices.

The test statistic for assessing a paired mean is a T.

The T score is a ratio of how the sample mean difference varies from zero as compared to how the observations vary.

$$T = \frac{\bar{x}_{diff} - 0}{s_{diff}/\sqrt{n_{diff}}}$$

When the null hypothesis is true and the conditions are met, T has a t-distribution with $df = n_{diff} - 1$.

Conditions:

- Independently sampled pairs.
- Large samples and no extreme outliers.

EXAMPLE

Complete the hypothesis test started in the previous Example.

To compute the test compute the standard error associated with $\bar{x}_{diff}$ using the standard deviation of the differences ($s_{diff} = 13.42$) and the number of differences ($n_{diff} = 68$):

$$SE_{\bar{x}_{diff}} = \frac{s_{diff}}{\sqrt{n_{diff}}} = \frac{13.42}{\sqrt{68}} = 1.63$$

The test statistic is the T score of $\bar{x}_{diff}$ under the null condition that the actual mean difference is 0:

$$T = \frac{\bar{x}_{diff} - 0}{SE_{\bar{x}_{diff}}} = \frac{3.58 - 0}{1.63} = 2.20$$

To visualize the p-value, the sampling distribution of $\bar{x}_{diff}$ is drawn as though H_0 is true, and the p-value is represented by the two shaded tails in the figure below. The degrees of freedom is $df = 68 - 1 = 67$. Using statistical software, we find the one-tail area of 0.0156.

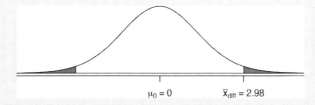

Doubling this area gives the p-value: 0.0312. Because the p-value is less than 0.05, we reject the null hypothesis. Amazon prices are, on average, lower than the UCLA Bookstore prices for UCLA courses.

Recall that the margin of error is defined by the standard error. The margin of error for $\bar{x}_{diff}$ can be directly obtained from $SE(\bar{x}_{diff})$.

Margin of error for $\bar{x}_{diff}$.

The margin of error is $t^\star_{df} \times s_{diff}/\sqrt{n_{diff}}$ where $t^\star_{df}$ is calculated from a specified percentile on the t-distribution with df degrees of freedom.

21.3. MATHEMATICAL MODEL FOR THE MEAN PAIRED DIFFERENCE

EXAMPLE

Create a 95% confidence interval for the average price difference between books at the UCLA bookstore and books on Amazon.

Conditions have already verified and the standard error computed in a previous Example.

To find the confidence interval, identify $t^\star_{67}$ using statistical software or the t-table ($t^\star_{67} = 2.00$), and plug it, the point estimate, and the standard error into the confidence interval formula:

$$\text{point estimate } \pm \ z^\star \ \times \ SE$$
$$3.58 \ \pm \ 2.00 \ \times \ 1.63$$
$$(0.32 \ , \ 6.84)$$

We are 95% confident that the UCLA Bookstore is, on average, between $0.32 and $6.84 more expensive than Amazon for UCLA course books.

GUIDED PRACTICE

We have convincing evidence that Amazon is, on average, less expensive. How should this conclusion affect UCLA student buying habits? Should UCLA students always buy their books on Amazon?[2]

---

[2]The average price difference is only mildly useful for this question. Examine the distribution shown in Figure 21.10. There are certainly a handful of cases where Amazon prices are far below the UCLA Bookstore's, which suggests it is worth checking Amazon (and probably other online sites) before purchasing. However, in many cases the Amazon price is above what the UCLA Bookstore charges, and most of the time the price isn't that different. Ultimately, if getting a book immediately from the bookstore is notably more convenient, e.g., to get started on reading or homework, it's likely a good idea to go with the UCLA Bookstore unless the price difference on a specific book happens to be quite large. For reference, this is a very different result from what we (the authors) had seen in a similar dataset from 2010. At that time, Amazon prices were almost uniformly lower than those of the UCLA Bookstore's and by a large margin, making the case to use Amazon over the UCLA Bookstore quite compelling at that time. Now we frequently check multiple websites to find the best price.

21.4 Chapter review

21.4.1 Summary

Like the two independent sample procedures in Chapter 20, the paired difference analysis can be done using a t-distribution. The randomization test applied to the paired differences is slightly different, however. Note that when randomizing under the paired setting, each null statistic is created by randomly assigning the group to a numerical outcome **within** the individual observational unit. The procedure for creating a confidence interval for the paired difference is almost identical to the confidence intervals created in Chapter 19 for a single mean.

21.4.2 Terms

We introduced the following terms in the chapter. If you're not sure what some of these terms mean, we recommend you go back in the text and review their definitions. We are purposefully presenting them in alphabetical order, instead of in order of appearance, so they will be a little more challenging to locate. However you should be able to easily spot them as **bolded text**.

bootstrap CI paired difference	paired difference CI	T score paired difference
paired data	paired difference t-test	

21.5 Exercises

Answers to odd numbered exercises can be found in Appendix A.21.

21.1. **Air quality.** Air quality measurements were collected in a random sample of 25 country capitals in 2013, and then again in the same cities in 2014. We would like to use these data to compare average air quality between the two years. Should we use a paired or non-paired test? Explain your reasoning.

21.2. **True / False: paired.** Determine if the following statements are true or false. If false, explain.

   a. In a paired analysis we first take the difference of each pair of observations, and then we do inference on these differences.

   b. Two datasets of different sizes cannot be analyzed as paired data.

   c. Consider two sets of data that are paired with each other. Each observation in one dataset has a natural correspondence with exactly one observation from the other dataset.

   d. Consider two sets of data that are paired with each other. Each observation in one dataset is subtracted from the average of the other dataset's observations.

21.3. **Paired or not? I.** In each of the following scenarios, determine if the data are paired.

   a. Compare pre- (beginning of semester) and post-test (end of semester) scores of students.

   b. Assess gender-related salary gap by comparing salaries of randomly sampled men and women.

   c. Compare artery thicknesses at the beginning of a study and after 2 years of taking Vitamin E for the same group of patients.

   d. Assess effectiveness of a diet regimen by comparing the before and after weights of subjects.

21.4. **Paired or not? II.** In each of the following scenarios, determine if the data are paired.

   a. We would like to know if Intel's stock and Southwest Airlines' stock have similar rates of return. To find out, we take a random sample of 50 days, and record Intel's and Southwest's stock on those same days.

   b. We randomly sample 50 items from Target stores and note the price for each. Then we visit Walmart and collect the price for each of those same 50 items.

   c. A school board would like to determine whether there is a difference in average SAT scores for students at one high school versus another high school in the district. To check, they take a simple random sample of 100 students from each high school.

21.5. **Sample size and pairing.** Determine if the following statement is true or false, and if false, explain your reasoning: If comparing means of two groups with equal sample sizes, always use a paired test.

21.6. **High School and Beyond, randomization test.** The National Center of Education Statistics conducted a survey of high school seniors, collecting test data on reading, writing, and several other subjects. Here we examine a simple random sample of 200 students from this survey. Side-by-side box plots of reading and writing scores as well as a histogram of the differences in scores are shown below. Also provided below is a histogram of randomized averages of paired differences of scores (read - write), with the observed difference ($\bar{x}_{read-write} = -0.545$) marked with a red vertical line. The randomization distribution was produced by doing the following 1000 times: for each student, the two scores were randomly assigned to either read or write, and the average was taken across all students in the sample.[3]

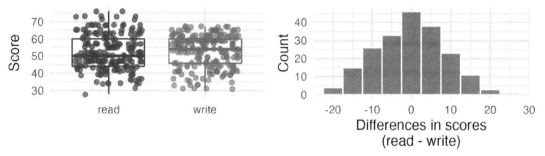

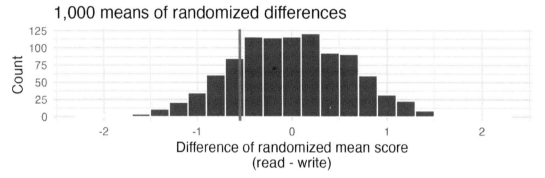

a. Is there a clear difference in the average reading and writing scores?

b. Are the reading and writing scores of each student independent of each other?

c. Create hypotheses appropriate for the following research question: is there an evident difference in the average scores of students in the reading and writing exam?

d. Is the average of the observed difference in scores ($\bar{x}_{read-write} = -0.545$) consistent with the distribution of randomized average differences? Explain.

e. Do these data provide convincing evidence of a difference between the average scores on the two exams? Estimate the p-value from the randomization test, and conclude the hypothesis test using words like "score on reading test" and "score on writing test."

21.7. **Forest management.** Forest rangers wanted to better understand the rate of growth for younger trees in the park. They took measurements of a random sample of 50 young trees in 2009 and again measured those same trees in 2019. The data below summarize their measurements, where the heights are in feet.

Year	Mean	SD	n
2009	12.0	3.5	50
2019	24.5	9.5	50
Difference	12.5	7.2	50

Construct a 99% confidence interval for the average growth of (what had been) younger trees in the park over 2009-2019.

---

[3]The `hsb2` data used in this exercise can be found in the **openintro** R package.

21.5. EXERCISES

21.8. **High School and Beyond, bootstrap interval.** We considered the differences between the reading and writing scores of a random sample of 200 students who took the High School and Beyond Survey. The mean and standard deviation of the differences are $\bar{x}_{read-write} = -0.545$ and $s_{read-write} = 8.887$ points. The bootstrap distribution below was produced by bootstrapping from the sample of differences in reading and writing scores 1,000 times.

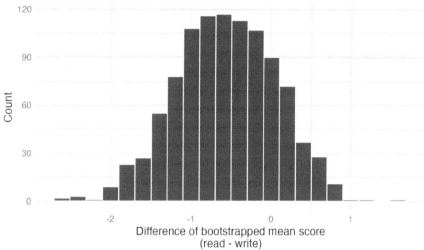

a. Find an approximate 95% bootstrap percentile confidence interval for the true average difference in scores (read - write).

b. Find an approximate 95% bootstrap SE confidence interval for the true average difference in scores (read - write).

c. Interpret both confidence intervals using words like "population" and "score".

d. From the confidence intervals calculated above, does it appear that there is a significant difference in reading and writing scores, on average?

21.9. **Possible paired randomized differences.** Data were collected on five people.

	Person 1	Person 2	Person 3	Person 4	Person 5
Observation 1	3	14	4	5	10
Observation 2	7	3	6	5	9
Difference	-4	11	-2	0	1

Which of the following could be a possible randomization of the paired differences given above? If the set of values could not be a randomized set of differences, indicate why not.

a. -2, 1, 1, 11, -2

b. -4 11 -2 0 1

c. -2, 2, -11, 11, -2, 2, 0, 1, -1

d. 0, -1, 2, -4, 11

e. 4, -11, 2, 0, -1

21.10. **Study environment.** In order to test the effects of listening to music while studying versus studying in silence, students agree to be randomized to two treatments (i.e., study with music or study in silence). There are two exams during the semester, so the researchers can either randomize the students to have one exam with music and one with silence (randomly selecting which exam corresponds to which study environment) or the researchers can randomize the students to one study habit for both exams.

The researchers are interested in estimating the true population difference of exam score for those who listen to music while studying as compared to those who study in silence.

   a. Describe the experiment which is consistent with a paired designed experiment. How is the treatment assigned, and how are the data collected such that the observations are paired?

   b. Describe the experiment which is consistent with an indpenedent samples experiment. How is the treatment assigned, and how are the data collected such that the observations are independent?

21.11. **Global warming, randomization test.** Let's consider a limited set of climate data, examining temperature differences in 1948 vs 2018. We sampled 197 locations from the National Oceanic and Atmospheric Administration's (NOAA) historical data, where the data was available for both years of interest. We want to know: were there more days with temperatures exceeding 90F in 2018 or in 1948? (NOAA, 2018) The difference in number of days exceeding 90F (number of days in 2018 - number of days in 1948) was calculated for each of the 197 locations. The average of these differences was 2.9 days with a standard deviation of 17.2 days. We are interested in determining whether these data provide strong evidence that there were more days in 2018 that exceeded 90F from NOAA's weather stations.[4]

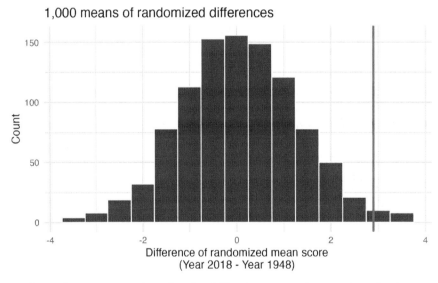

   a. Create hypotheses appropriate for the following research question: is there an evident difference in the average number of days greater than 90F across the two years (1948 and 2018)?

   b. Is the average of the observed difference in scores ($\bar{x}_{2018-1948} = 2.9$) consistent with the distribution of randomized average differences? Explain.

   c. Do these data provide convincing evidence of a difference between the average number of days? Estimate the p-value from the randomization test, and conclude the hypothesis test using words like "number of days in 1948" and "number of days in 2018."

---

[4]The `climate70` data used in this exercise can be found in the **openintro** R package.

21.5. EXERCISES

21.12. **Global warming, bootstrap interval.** We considered the change in the number of days exceeding 90F from 1948 and 2018 at 197 randomly sampled locations from the NOAA database. The mean and standard deviation of the reported differences are 2.9 days and 17.2 days.

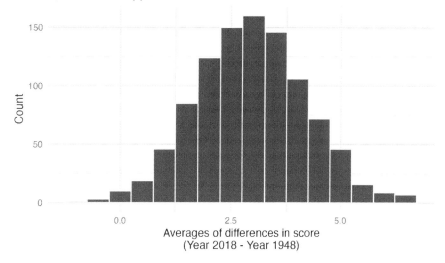

a. Calculate a 90% bootstrap percentile confidence interval for the average difference between number of days exceeding 90F between 1948 and 2018.

b. Calculate a 90% bootstrap SE confidence interval for the average difference between number of days exceeding 90F between 1948 and 2018.

c. Interpret both intervals in context.

d. Do the confidence intervals provide convincing evidence that there were more days exceeding 90F in 2018 than in 1948 at NOAA stations? Explain your reasoning.

21.13. **Global warming, mathematical test.** We considered the change in the number of days exceeding 90F from 1948 and 2018 at 197 randomly sampled locations from the NOAA database. The mean and standard deviation of the reported differences are 2.9 days and 17.2 days.

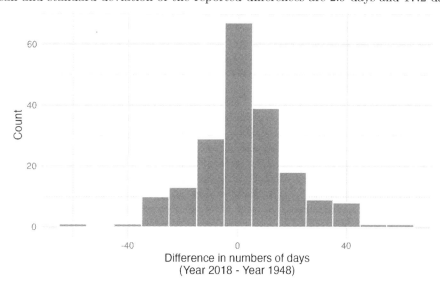

a. Is there a relationship between the observations collected in 1948 and 2018? Or are the observations in the two groups independent? Explain.

b. Write hypotheses for this research in symbols and in words.

c. Check the conditions required to complete this test. A histogram of the differences is given to the right.

See next page for parts d to g.

d. Calculate the test statistic and find the p-value.

e. Use $\alpha = 0.05$ to evaluate the test, and interpret your conclusion in context.

f. What type of error might we have made? Explain in context what the error means.

g. Based on the results of this hypothesis test, would you expect a confidence interval for the average difference between the number of days exceeding 90F from 1948 and 2018 to include 0? Explain your reasoning.

21.14. High School and Beyond, mathematical test. We considered the differences between the reading and writing scores of a random sample of 200 students who took the High School and Beyond Survey.

a. Create hypotheses appropriate for the following research question: is there an evident difference in the average scores of students in the reading and writing exam?

b. Check the conditions required to complete this test.

c. The average observed difference in scores is $\bar{x}_{read-write} = -0.545$, and the standard deviation of the differences is $s_{read-write} = 8.887$ points. Do these data provide convincing evidence of a difference between the average scores on the two exams?

d. What type of error might we have made? Explain what the error means in the context of the application.

e. Based on the results of this hypothesis test, would you expect a confidence interval for the average difference between the reading and writing scores to include 0? Explain your reasoning.

21.15. Global warming, mathematical interval. We considered the change in the number of days exceeding 90F from 1948 and 2018 at 197 randomly sampled locations from the NOAA database. The mean and standard deviation of the reported differences are 2.9 days and 17.2 days.

a. Calculate a 90% confidence interval for the average difference between number of days exceeding 90F between 1948 and 2018. We've already checked the conditions for you.

b. Interpret the interval in context.

c. Does the confidence interval provide convincing evidence that there were more days exceeding 90F in 2018 than in 1948 at NOAA stations? Explain your reasoning.

21.16. High school and beyond, mathematical interval. We considered the differences between the reading and writing scores of a random sample of 200 students who took the High School and Beyond Survey. The mean and standard deviation of the differences are $\bar{x}_{read-write} = -0.545$ and $s_{read-write} = 8.887$ points.

a. Calculate a 95% confidence interval for the average difference between the reading and writing scores of all students.

b. Interpret this interval in context.

c. Does the confidence interval provide convincing evidence that there is a real difference in the average scores? Explain.

21.5. EXERCISES

21.17. **Friday the 13th, traffic.** In the early 1990's, researchers in the UK collected data on traffic flow on Friday the 13th with the goal of addressing issues of how superstitions regarding Friday the 13th affect human behavior and and whether Friday the 13th is an unlucky day. The histograms below show the distributions of numbers of cars passing by a specific intersection on Friday the 6th and Friday the 13th for many such date pairs. Also provided are some sample statistics, where the difference is the number of cars on the 6th minus the number of cars on the 13th.[5] (Scanlon et al., 1993)

	n	Mean	SD
sixth	10	128,385	7,259
thirteenth	10	126,550	7,664
diff	10	1,836	1,176

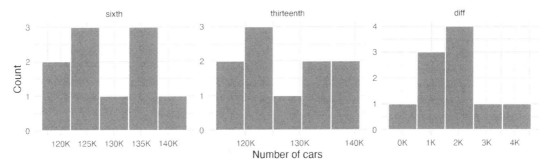

a. Are there any underlying structures in these data that should be considered in an analysis? Explain.

b. What are the hypotheses for evaluating whether the number of people out on Friday the 6th is different than the number out on Friday the 13th?

c. Check conditions to carry out the hypothesis test from part (b) using mathematical models.

d. Calculate the test statistic and the p-value.

e. What is the conclusion of the hypothesis test?

f. Interpret the p-value in this context.

g. What type of error might have been made in the conclusion of your test? Explain.

---

[5] The `friday` data used in this exercise can be found in the **openintro** R package.

21.18. **Friday the 13th, accidents.** In the early 1990's, researchers in the UK collected data the number of traffic accident related emergency room (ER) admissions on Friday the 13th with the goal of addressing issues of how superstitions regarding Friday the 13th affect human behavior and and whether Friday the 13th is an unlucky day. The histograms below show the distributions of numbers of ER admissions at specific emergency rooms on Friday the 6th and Friday the 13th for many such date pairs. Also provided are some sample statistics, where the difference is the ER admissions on the 6th minus the ER admissions on the 13th.(Scanlon et al., 1993)

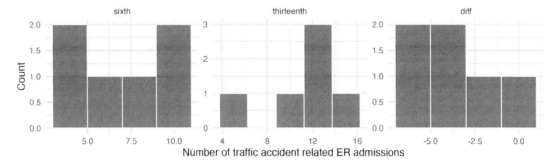

a. Conduct a hypothesis test using mathematical models to evaluate if there is a difference between the average numbers of traffic accident related emergency room admissions between Friday the 6th and Friday the 13th.

b. Calculate a 95% confidence interval using mathematical models for the difference between the average numbers of traffic accident related emergency room admissions between Friday the 6th and Friday the 13th.

c. The conclusion of the original study states, "Friday 13th is unlucky for some. The risk of hospital admission as a result of a transport accident may be increased by as much as 52%. Staying at home is recommended." Do you agree with this statement? Explain your reasoning.

Chapter 22

Inference for comparing many means

In Chapter 20 analysis was done to compare the average population value across two different groups. An important aspect of the analysis was to look at the difference in sample means as an estimate for the difference in population means. When comparing more than two groups, the difference (i.e., subtraction) will not fully capture the nuance in variation across the three or more groups. As with two groups, the research question will focus on whether the group membership is independent of the numerical response variable. Here, independence across groups means that knowledge of the observations in one group doesn't change what we'd expect to happen in the other group. But what happens if the groups are **dependent**? In this section we focus on a new statistic which incorporates differences in means across more than two groups. Although the ideas in this chapter are quite similar to the t-test, they have earned themselves their own name: **AN**alysis **O**f **VA**riance, or ANOVA.

Sometimes we want to compare means across many groups. We might initially think to do pairwise comparisons. For example, if there were three groups, we might be tempted to compare the first mean with the second, then with the third, and then finally compare the second and third means for a total of three comparisons. However, this strategy can be treacherous. If we have many groups and do many comparisons, it is likely that we will eventually find a difference just by chance, even if there is no difference in the populations. Instead, we should apply a holistic test to check whether there is evidence that at least one pair groups are in fact different, and this is where **ANOVA** saves the day.

In this section, we will learn a new method called **analysis of variance (ANOVA)** and a new test statistic called an F-statistic (which we will introduce in our discussion of mathematical models). ANOVA uses a single hypothesis test to check whether the means across many groups are equal:

- H_0 : The mean outcome is the same across all groups. In statistical notation, $\mu_1 = \mu_2 = \cdots = \mu_k$ where μ_i represents the mean of the outcome for observations in category i.

- H_A : At least one mean is different.

Generally we must check three conditions on the data before performing ANOVA:

- the observations are independent within and between groups,
- the responses within each group are nearly normal, and
- the variability across the groups is about equal.

When these three conditions are met, we may perform an ANOVA to determine whether the data provide convincing evidence against the null hypothesis that all the μ_i are equal.

EXAMPLE

College departments commonly run multiple sections of the same introductory course each semester because of high demand. Consider a statistics department that runs three sections of an introductory statistics course. We might like to determine whether there are substantial differences in first exam scores in these three classes (Section A, Section B, and Section C). Describe appropriate hypotheses to determine whether there are any differences between the three classes.

The hypotheses may be written in the following form:

- H_0 : The average score is identical in all sections, $\mu_A = \mu_B = \mu_C$. Assuming each class is equally difficult, the observed difference in the exam scores is due to chance.
- H_A : The average score varies by class. We would reject the null hypothesis in favor of the alternative hypothesis if there were larger differences among the class averages than what we might expect from chance alone.

Strong evidence favoring the alternative hypothesis in ANOVA is described by unusually large differences among the group means. We will soon learn that assessing the variability of the group means relative to the variability among individual observations within each group is key to ANOVA's success.

EXAMPLE

Examine Figure 22.1. Compare groups I, II, and III. Can you visually determine if the differences in the group centers is unlikely to have occurred if there were no differences in the groups? Now compare groups IV, V, and VI. Do these differences appear to be unlikely to have occurred if there were no differences in the groups?

Any real difference in the means of groups I, II, and III is difficult to discern, because the data within each group are very volatile relative to any differences in the average outcome. On the other hand, it appears there are differences in the centers of groups IV, V, and VI. For instance, group V appears to have a higher mean than that of the other two groups. Investigating groups IV, V, and VI, we see the differences in the groups' centers are noticeable because those differences are large *relative to the variability in the individual observations within each group.*

22.1 Case study: Batting

We would like to discern whether there are real differences between the batting performance of baseball players according to their position: outfielder (OF), infielder (IF), and catcher (C). We will use a dataset called `mlb_players_18`, which includes batting records of 429 Major League Baseball (MLB) players from the 2018 season who had at least 100 at bats. Six of the 429 cases represented in `mlb_players_18` are shown in Figure 22.1, and descriptions for each variable are provided in Figure 22.2. The measure we will use for the player batting performance (the outcome variable) is on-base percentage (`OBP`). The on-base percentage roughly represents the fraction of the time a

22.1. CASE STUDY: BATTING

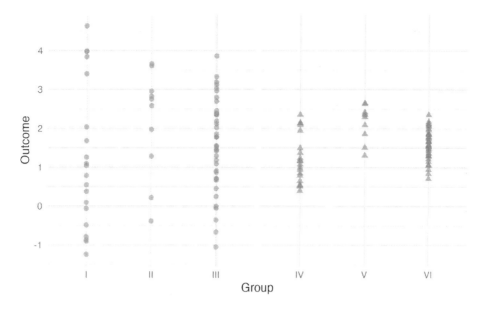

Figure 22.1: Side-by-side dot plot for the outcomes for six groups. Two sets of groups: first set is comprised of Groups I, II, and III, the second set is comprised of Groups IV, V, and VI.

player successfully gets on base or hits a home run.

 The `mlb_players_18` data can be found in the **openintro** R package.

Table 22.1: Six cases and some of the variables from the `mlb_players_18` data frame.

name	team	position	AB	H	HR	RBI	AVG	OBP
Abreu, J	CWS	IF	499	132	22	78	0.265	0.325
Acuna Jr., R	ATL	OF	433	127	26	64	0.293	0.366
Adames, W	TB	IF	288	80	10	34	0.278	0.348
Adams, M	STL	IF	306	73	21	57	0.239	0.309
Adduci, J	DET	IF	176	47	3	21	0.267	0.290
Adrianza, E	MIN	IF	335	84	6	39	0.251	0.301

Table 22.2: Variables and their descriptions for the `mlb_players_18` dataset.

Variable	Description
`name`	Player name
`team`	The abbreviated name of the player's team
`position`	The player's primary field position (OF, IF, C)
`AB`	Number of opportunities at bat
`H`	Number of hits
`HR`	Number of home runs
`RBI`	Number of runs batted in
`AVG`	Batting average, which is equal to H/AB
`OBP`	On-base percentage, which is roughly equal to the fraction of times a player gets on base or hits a home run

GUIDED PRACTICE

The null hypothesis under consideration is the following: $\mu_{OF} = \mu_{IF} = \mu_C$. Write the null and corresponding alternative hypotheses in plain language.[1]

EXAMPLE

The player positions have been divided into three groups: outfield (OF), infield (IF), and catcher (C). What would be an appropriate point estimate of the on-base percentage by outfielders, μ_{OF}?

A good estimate of the on-base percentage by outfielders would be the sample average of OBP for just those players whose position is outfield: $\bar{x}_{OF} = 0.320$.

22.2 Randomization test for comparing many means

Table 22.3 provides summary statistics for each group. A side-by-side box plot for the on-base percentage is shown in Figure 22.2. Notice that the variability appears to be approximately constant across groups; nearly constant variance across groups is an important assumption that must be satisfied before we consider the ANOVA approach.

Table 22.3: Summary statistics of on-base percentage, split by player position.

Position	n	Mean	SD
OF	160	0.320	0.043
IF	205	0.318	0.038
C	64	0.302	0.038

---

[1] H_0 : The average on-base percentage is equal across the four positions. H_A : The average on-base percentage varies across some (or all) groups.

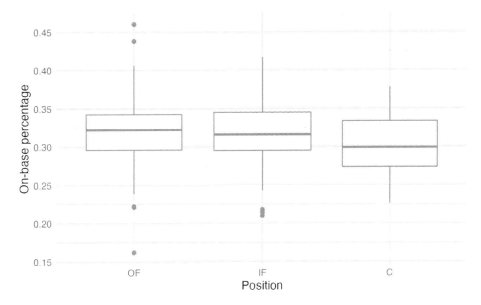

Figure 22.2: Side-by-side box plot of the on-base percentage for 429 players across three groups. There is one prominent outlier visible in the infield group, but with 205 observations in the infield group, this outlier is not extreme enough to have an impact on the calculations, so it is not a concern for moving forward with the analysis.

 EXAMPLE

The largest difference between the sample means is between the catcher and the outfielder positions. Consider again the original hypotheses:

- $H_0 : \mu_{OF} = \mu_{IF} = \mu_C$
- H_A : The average on-base percentage (μ_i) varies across some (or all) groups.

Why might it be inappropriate to run the test by simply estimating whether the difference of μ_C and μ_{OF} is "statistically significant" at a 0.05 significance level?

The primary issue here is that we are inspecting the data before picking the groups that will be compared. It is inappropriate to examine all data by eye (informal testing) and only afterwards decide which parts to formally test. This is called **data snooping** or **data fishing**. Naturally, we would pick the groups with the large differences for the formal test, and this would leading to an inflation in the Type 1 Error rate. To understand this better, let's consider a slightly different problem.

Suppose we are to measure the aptitude for students in 20 classes in a large elementary school at the beginning of the year. In this school, all students are randomly assigned to classrooms, so any differences we observe between the classes at the start of the year are completely due to chance. However, with so many groups, we will probably observe a few groups that look rather different from each other. If we select only these classes that look so different and then perform a formal test, we will probably make the wrong conclusion that the assignment wasn't random. While we might only formally test differences for a few pairs of classes, we informally evaluated the other classes by eye before choosing the most extreme cases for a comparison.

For additional information on the ideas expressed above, we recommend reading about the **prosecutor's fallacy**.[2]

---

[2] See, for example, this blog post.

22.2.1 Observed data

In the next section we will learn how to use the F statistic to test whether observed differences in sample means could have happened just by chance even if there was no difference in the respective population means.

The method of analysis of variance in this context focuses on answering one question: is the variability in the sample means so large that it seems unlikely to be from chance alone? This question is different from earlier testing procedures since we will *simultaneously* consider many groups, and evaluate whether their sample means differ more than we would expect from natural variation. We call this variability the **mean square between groups (MSG)**, and it has an associated degrees of freedom, $df_G = k - 1$ when there are k groups. The MSG can be thought of as a scaled variance formula for means. If the null hypothesis is true, any variation in the sample means is due to chance and shouldn't be too large. Details of MSG calculations are provided in the footnote.[3] However, we typically use software for these computations.

The mean square between the groups is, on its own, quite useless in a hypothesis test. We need a benchmark value for how much variability should be expected among the sample means if the null hypothesis is true. To this end, we compute a pooled variance estimate, often abbreviated as the **mean square error** (MSE), which has an associated degrees of freedom value $df_E = n - k$. It is helpful to think of MSE as a measure of the variability within the groups. Details of the computations of the MSE and a link to an extra online section for ANOVA calculations are provided in the footnote.[4]

When the null hypothesis is true, any differences among the sample means are only due to chance, and the MSG and MSE should be about equal. As a test statistic for ANOVA, we examine the fraction of MSG and MSE:

$$F = \frac{MSG}{MSE}$$

The MSG represents a measure of the between-group variability, and MSE measures the variability within each of the groups.

---

[3] Let $\bar{x}$ represent the mean of outcomes across all groups. Then the mean square between groups is computed as $MSG = \frac{1}{df_G} SSG = \frac{1}{k-1} \sum_{i=1}^{k} n_i (\bar{x}_i - \bar{x})^2$ where SSG is called the **sum of squares between groups** and n_i is the sample size of group i.

[4] See additional details on ANOVA calculations for interested readers. Let $\bar{x}$ represent the mean of outcomes across all groups. Then the **sum of squares total** (SST) is computed as

$$SST = \sum_{i=1}^{n} (x_i - \bar{x})^2$$

where the sum is over all observations in the dataset. Then we compute the **sum of squared errors** (SSE) in one of two equivalent ways: $SSE = SST - SSG = (n_1 - 1)s_1^2 + (n_2 - 1)s_2^2 + \cdots + (n_k - 1)s_k^2$ where s_i^2 is the sample variance (square of the standard deviation) of the residuals in group i. Then the MSE is the standardized form of SSE: $MSE = \frac{1}{df_E} SSE$.

22.2. RANDOMIZATION TEST FOR COMPARING MANY MEANS

The test statistic for three or more means is an F.

The F statistic is a ratio of how the groups differ (MSG) as compared to how the observations within a group vary (MSE).

$$F = \frac{MSG}{MSE}$$

When the null hypothesis is true and the conditions are met, F has an F-distribution with $df_1 = k - 1$ and $df_2 = n - k$.

Conditions:

- independent observations, both within and across groups
- large samples and no extreme outliers

22.2.2 Variability of the statistic

We recall the exams from Section 20.1 which demonstrated a two-sample randomization test for a comparison of means. Suppose now that the teacher had had such an extremely large class that three different exams were given: A, B, and C. Table 22.4 and Figure 22.3 provide a summary of the data including exam C. Again, we'd like to investigate whether or not the difficulty of the exams is the same across the three exams, so the test is

- $H_0: \mu_A = \mu_B = \mu_C$. The inherent average difficulty is the same across the three exams.
- $H_A:$ not H_0. At least one of the exams is inherently more (or less) difficult than the others.

The `classdata` data can be found in the **openintro** R package.

Table 22.4: Summary statistics of scores for each exam version.

Exam	n	Mean	SD	Min	Max
A	58	75.1	13.9	44	100
B	55	72.0	13.8	38	100
C	51	78.9	13.1	45	100

Figure 22.4 shows the process of randomizing the three different exams to the observed exam scores. If the null hypothesis is true, then the score on each exam should represent the true student ability on that material. It shouldn't matter whether they were given exam A or exam B or exam C. By reallocating which student got which exam, we are able to understand how the difference in average exam scores changes due only to natural variability. There is only one iteration of the randomization process in Figure 22.4, leading to three different randomized sample means (computed assuming the null hypothesis is true).

In the two-sample case, the null hypothesis was investigated using the difference in the sample means. However, as noted above, with three groups (three different exams), the comparison of the three sample means gets slightly more complicated. We have already derived the F-statistic which is exactly the way to compare the averages across three or more groups! Recall, the F statistic is a ratio of how the groups differ (MSG) as compared to how the observations within a group vary (MSE).

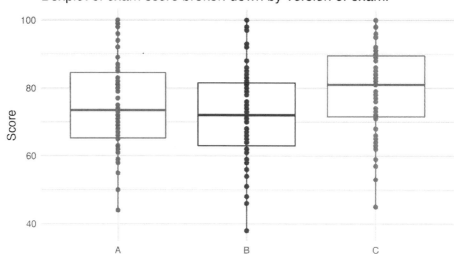

Figure 22.3: Exam scores for students given one of three different exams.

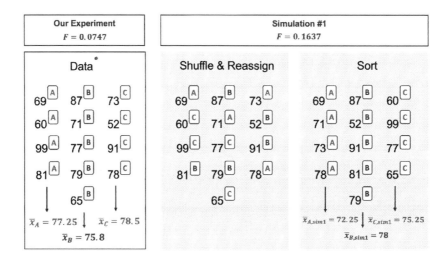

Figure 22.4: The version of the test (A or B or C) is randomly allocated to the test scores, under the null assumption that the tests are equally difficult.

Building on Figure 22.4, Figure 22.5 shows the values of the simulated F statistics over 1,000 random simulations. We see that, just by chance, the F statistic can be as large as 7.

22.2.3 Observed statistic vs. null statistic

Using statistical software, we can calculate that 3.6% of the randomized F test statistics were at or above the observed test statistic of $F = 3.48$. That is, the p-value of the test is 0.036. Assuming that we had set the level of significance to be $\alpha = 0.05$, the p-value is smaller than the level of significance which would lead us to reject the null hypothesis. We claim that the difficulty level (i.e., the true average score, μ) is different for at least one of the exams.

While it is temping to say that exam C is harder than the other two (given the inability to differentiate between exam A and exam B in Section 20.1), we must be very careful about conclusions made using different techniques on the same data.

When the null hypothesis is true, random variability that exists in nature produces data with p-values less than 0.05. How often does that happen? 5% of the time. That is to say, if you use 20 different models applied to the same data where there is no signal (i.e., the null hypothesis is

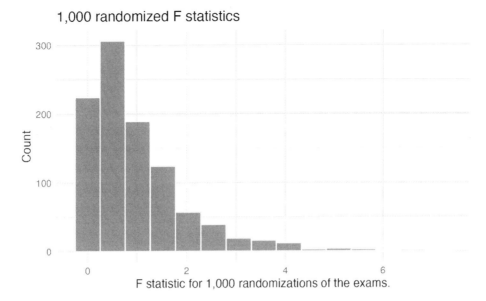

Figure 22.5: Histogram of F statistics calculated from 1,000 different randomizations of the exam type.

true), you are reasonably likely to to get a p-value less than 0.05 in one of the tests you run. The details surrounding the ideas of this problem, called a **multiple comparisons test** or **multiple comparisons problem**, are outside the scope of this textbook, but should be something that you keep in the back of your head. To best mitigate any extra type I errors, we suggest that you set up your hypotheses and testing protocol before running any analyses. Once the conclusions have been reached, you should report your findings instead of running a different type of test on the same data.

22.3 Mathematical model for test for comparing many means

As seen with many of the tests and statistics from previous sections, the randomization test on the F statistic has mathematical theory to describe the distribution without using a computational approach.

We return to the baseball example from Table 22.3 to demonstrate the mathematical model applied to the ANOVA setting.

22.3.1 Variability of the statistic

The larger the observed variability in the sample means (MSG) relative to the within-group observations (MSE), the larger F-statistic will be and the stronger the evidence against the null hypothesis. Because larger F-statistics represent stronger evidence against the null hypothesis, we use the upper tail of the distribution to compute a p-value.

 The F statistic and the F-test.

Analysis of variance (ANOVA) is used to test whether the mean outcome differs across two or more groups. ANOVA uses a test statistic, the F-statistic, which represents a standardized ratio of variability in the sample means relative to the variability within the groups. If H_0 is true and the model conditions are satisfied, an F-statistic follows an F distribution with parameters $df_1 = k - 1$ and $df_2 = n - k$. The upper tail of the F distribution is used to represent the p-value.

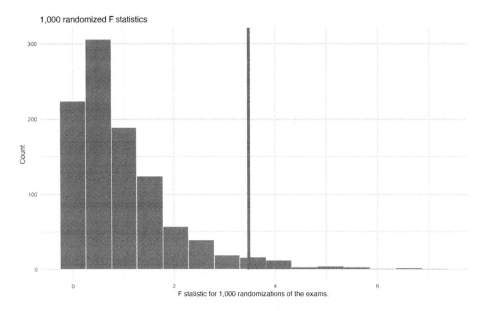

Figure 22.6: Histogram of F statistics calculated from 1000 different randomizations of the exam type. The observed F statistic is given as a red vertical line 3.48. The area to the right is more extreme than the observed value and represents the p-value.

GUIDED PRACTICE

For the baseball data, $MSG = 0.00803$ and $MSE = 0.00158$. Identify the degrees of freedom associated with MSG and MSE and verify the F-statistic is approximately 5.077.[5]

22.3.2 Observed statistic vs. null statistics

We can use the F-statistic to evaluate the hypotheses in what is called an F-test. A p-value can be computed from the F statistic using an F distribution, which has two associated parameters: df_1 and df_2. For the F-statistic in ANOVA, $df_1 = df_G$ and $df_2 = df_E$. An F distribution with 2 and 426 degrees of freedom, corresponding to the F statistic for the baseball hypothesis test, is shown in Figure 22.7.

EXAMPLE

The p-value corresponding to the shaded area in Figure 22.7 is equal to about 0.0066. Does this provide strong evidence against the null hypothesis?

The p-value is smaller than 0.05, indicating the evidence is strong enough to reject the null hypothesis at a significance level of 0.05. That is, the data provide strong evidence that the average on-base percentage varies by player's primary field position.

Note that the small p-value indicates that there is a notable difference between the mean batting averages of the different positions. However, the ANOVA test does not provide a mechanism for knowing *which* group is driving the differences. If we move forward with all possible two mean

---

[5]There are $k = 3$ groups, so $df_G = k - 1 = 2$. There are $n = n_1 + n_2 + n_3 = 429$ total observations, so $df_E = n - k = 426$. Then the F-statistic is computed as the ratio of MSG and MSE: $F = \frac{MSG}{MSE} = \frac{0.00803}{0.00158} = 5.082 \approx 5.077$. ($F = 5.077$ was computed by using values for MSG and MSE that were not rounded.)

22.3. MATHEMATICAL MODEL FOR TEST FOR COMPARING MANY MEANS

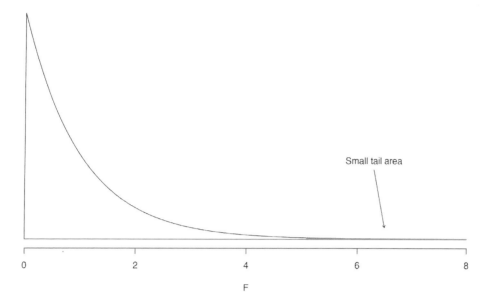

Figure 22.7: An F distribution with $df_1 = 2$ and $df_2 = 426$.

comparisons, we run the risk of a high type I error rate. As we saw at the end of Section 22.2, the follow-up questions surrounding individual group comparisons is called a problem of **multiple comparisons** and is outside the scope of this text. We encourage you to learn more about multiple comparisons, however, so that additional comparisons, after you have rejected the null hypothesis in an ANOVA test, do not lead to undue false positive conclusions.

22.3.3 Reading an ANOVA table from software

The calculations required to perform an ANOVA by hand are tedious and prone to human error. For these reasons, it is common to use statistical software to calculate the F-statistic and p-value.

An ANOVA can be summarized in a table very similar to that of a regression summary, which we saw in Chapters 7 and 8. Table 22.5 shows an ANOVA summary to test whether the mean of on-base percentage varies by player positions in the MLB. Many of these values should look familiar; in particular, the F-statistic and p-value can be retrieved from the last two columns.

Table 22.5: ANOVA summary for testing whether the average on-base percentage differs across player positions.

term	df	sumsq	meansq	statistic	p.value
position	2	0.0161	0.0080	5.08	0.0066
Residuals	426	0.6740	0.0016		

22.3.4 Conditions for an ANOVA analysis

There are three conditions we must check for an ANOVA analysis: all observations must be independent, the data in each group must be nearly normal, and the variance within each group must be approximately equal.

- **Independence.** If the data are a simple random sample, this condition can be assumed to be satisfied. For processes and experiments, carefully consider whether the data may be independent (e.g., no pairing). For example, in the MLB data, the data were not sampled. However, there are not obvious reasons why independence would not hold for most or all observations.

- **Approximately normal.** As with one- and two-sample testing for means, the normality assumption is especially important when the sample size is quite small when it is ironically difficult to check for non-normality. A histogram of the observations from each group is shown in Figure 22.8. Since each of the groups we're considering have relatively large sample sizes, what we're looking for are major outliers. None are apparent, so this conditions is reasonably met.

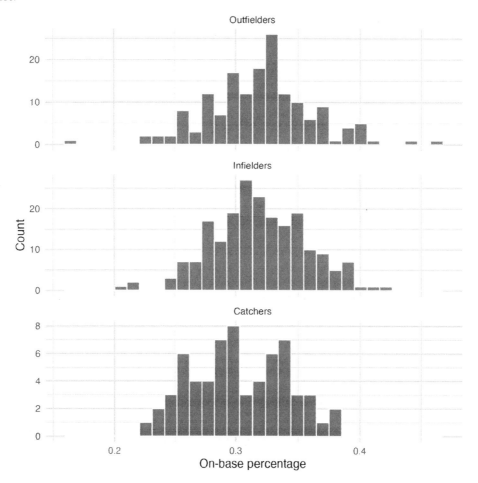

Figure 22.8: Histograms of OBP for each field position.

- **Constant variance.** The last assumption is that the variance in the groups is about equal from one group to the next. This assumption can be checked by examining a side-by-side box plot of the outcomes across the groups, as in Figure 22.2. In this case, the variability is similar in the four groups but not identical. We see in Table 22.3 that the standard deviation doesn't vary much from one group to the next.

 Diagnostics for an ANOVA analysis.

Independence is always important to an ANOVA analysis. The normality condition is very important when the sample sizes for each group are relatively small. The constant variance condition is especially important when the sample sizes differ between groups.

22.4 Chapter review

22.4.1 Summary

In this chapter we have provided both the randomization test and the mathematical model appropriate for addressing questions of equality of means across two or more groups. Note that there were important technical conditions required for confirming that the F distribution appropriately modeled the ANOVA test statistic. Also, you may have noticed that there was no discussion of creating confidence intervals. That is because the ANOVA statistic does not have a direct analogue parameter to estimate. If there is interest in comparisons of mean differences (across each set of two groups), then the methods from Chapter 20 comparing two independent means should be applied.

22.4.2 Terms

We introduced the following terms in the chapter. If you're not sure what some of these terms mean, we recommend you go back in the text and review their definitions. We are purposefully presenting them in alphabetical order, instead of in order of appearance, so they will be a little more challenging to locate. However you should be able to easily spot them as **bolded text**.

analysis of variance	F-test	sum of squared error (SSE)
ANOVA	mean square between groups (MSG)	sum of squares between groups (SSG)
data fishing	mean square error (MSE)	sum of squares total (SST)
data snooping	multiple comparisons	
degrees of freedom	prosecutor's fallacy	

22.5 Exercises

Answers to odd numbered exercises can be found in Appendix A.22.

22.1. **Fill in the blank.** When doing an ANOVA, you observe large differences in means between groups. Within the ANOVA framework, this would most likely be interpreted as evidence strongly favoring the _____ hypothesis.

22.2. **Which test?** We would like to test if students who are in the social sciences, natural sciences, arts and humanities, and other fields spend the same amount of time, on average, studying for a course. What type of test should we use? Explain your reasoning.

22.3. **Cuckoo bird egg lengths, randomize once.** Cuckoo birds lay their eggs in other birds' nests, making them known as brood parasites. One question relates to whether the size of the cuckoo egg differs depending on the species of the host bird.[6] (Latter, 1902)

Consider the following plots, one represents the original data, the second represents data where the host species has been randomly assigned to the egg length.

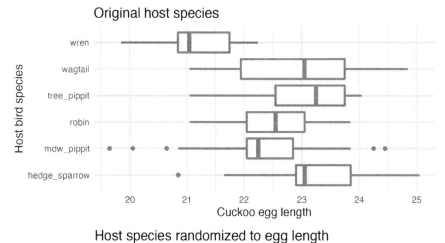

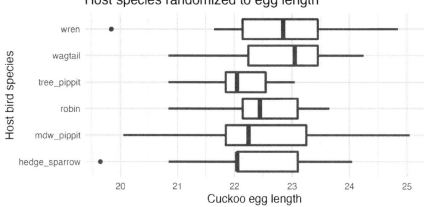

a. Consider the average length of the eggs for each species. Is the average length for the original data: more variable, less variable, or about the same as the randomized species? Describe what you see in the plots.

b. Consider the standard deviation of the lengths of the eggs within each species. Is the within species standard deviation of the length for the original data: bigger, smaller, or about the same as the randomized species?

c. Recall that the F statistic's numerator measures how much the groups vary (MSG) with the denominator measuring how much the within species values vary (MSE), which of the plots above would have a larger F statistic, the original data or the randomized data? Explain.

---
[6] The Cuckoo data used in this exercise can be found in the **Stat2Data** R package.

22.5. EXERCISES

22.4. Cuckoo bird egg lengths, randomization test. Cuckoo birds lay their eggs in other birds' nests, making them known as brood parasites. One question relates to whether the size of the cuckoo egg differs depending on the species of the host bird.[7] (Latter, 1902)

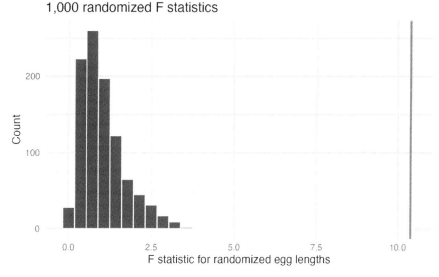

Using the randomization distribution of the F statistic (host species randomized to egg length), conduct a hypothesis test to evaluate if there is a difference, in the population, between the average egg lengths for different host bird species. Make sure to state your hypotheses clearly and interpret your results in context of the data.

22.5. Chicken diet and weight, many groups. An experiment was conducted to measure and compare the effectiveness of various feed supplements on the growth rate of chickens. Newly hatched chicks were randomly allocated into six groups, and each group was given a different feed supplement. Sample statistics and a visualization of the observed data are shown below. (McNeil, 1977)

Feed type	Mean	SD	n
casein	323.58	64.43	12
horsebean	160.20	38.63	10
linseed	218.75	52.24	12
meatmeal	276.91	64.90	11
soybean	246.43	54.13	14
sunflower	328.92	48.84	12

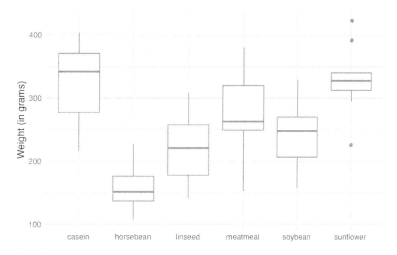

See next page for the rest of the exercise.

---

[7]The data `Cuckoo` used in this exercise can be found in the **Stat2Data** R package.

The ANOVA output below can be used to test for differences between the average weights of chicks on different diets. Conduct a hypothesis test to determine if these data provide convincing evidence that the average weight of chicks varies across some (or all) groups. Make sure to check relevant conditions.

term	df	sumsq	meansq	statistic	p.value
feed	5	231,129	46,226	15.4	<0.0001
Residuals	65	195,556	3,009		

22.6. **Teaching descriptive statistics.** A study compared five different methods for teaching descriptive statistics. The five methods were traditional lecture and discussion, programmed textbook instruction, programmed text with lectures, computer instruction, and computer instruction with lectures. 45 students were randomly assigned, 9 to each method. After completing the course, students took a 1-hour exam.

   a. What are the hypotheses for evaluating if the average test scores are different for the different teaching methods?

   b. What are the degrees of freedom associated with the F-test for evaluating these hypotheses?

   c. Suppose the p-value for this test is 0.0168. What is the conclusion?

22.7. **Coffee, depression, and physical activity.** Caffeine is the world's most widely used stimulant, with approximately 80% consumed in the form of coffee. Participants in a study investigating the relationship between coffee consumption and exercise were asked to report the number of hours they spent per week on moderate (e.g., brisk walking) and vigorous (e.g., strenuous sports and jogging) exercise. Based on these data the researchers estimated the total hours of metabolic equivalent tasks (MET) per week, a value always greater than 0. The table below gives summary statistics of MET for women in this study based on the amount of coffee consumed. (Lucas et al., 2011)

	Caffeinated coffee consumption				
	1 cup / week or fewer	2-6 cups / week	1 cups / day	2-3 cups / day	4 cups / day or more
Mean	18.7	19.6	19.3	18.9	17.5
SD	21.1	25.5	22.5	22.0	22.0
n	12,215.0	6,617.0	17,234.0	12,290.0	2,383.0

   a. Write the hypotheses for evaluating if the average physical activity level varies among the different levels of coffee consumption.

   b. Check conditions and describe any assumptions you must make to proceed with the test.

   c. Below is the output associated with this test. What is the conclusion of the test?

	df	sumsq	meansq	statistic	p.value
cofee	4	10,508	2,627	5.2	0
Residuals	50,734	25,564,819	504		
Total	50,738	25,575,327			

22.5. EXERCISES

22.8. **Student performance across discussion sections.** A professor who teaches a large introductory statistics class (197 students) with eight discussion sections would like to test if student performance differs by discussion section, where each discussion section has a different teaching assistant. The summary table below shows the average final exam score for each discussion section as well as the standard deviation of scores and the number of students in each section.

	Sec 1	Sec 2	Sec 3	Sec 4	Sec 5	Sec 6	Sec 7	Sec 8
Mean	92.94	91.11	91.80	92.45	89.30	88.30	90.12	93.35
SD	4.21	5.58	3.43	5.92	9.32	7.27	6.93	4.57
n	33.00	19.00	10.00	29.00	33.00	10.00	32.00	31.00

The ANOVA output below can be used to test for differences between the average scores from the different discussion sections.

	df	sumsq	meansq	statistic	p.value
section	7	525	75.0	1.87	0.077
Residuals	189	7,584	40.1		
Total	196	8,109			

Conduct a hypothesis test to determine if these data provide convincing evidence that the average score varies across some (or all) groups. Check conditions and describe any assumptions you must make to proceed with the test.

22.9. **GPA and major.** Undergraduate students taking an introductory statistics course at Duke University conducted a survey about GPA and major. The side-by-side box plots show the distribution of GPA among three groups of majors. Also provided is the ANOVA output.

term	df	sumsq	meansq	statistic	p.value
major	2	0.03	0.02	0.21	0.81
Residuals	195	15.77	0.08		

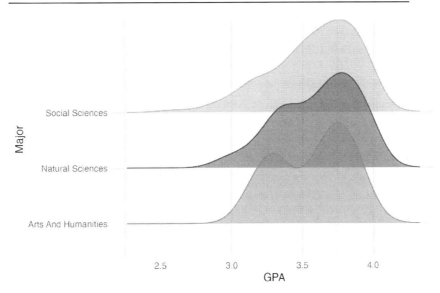

a. Write the hypotheses for testing for a difference between average GPA across majors.

b. What is the conclusion of the hypothesis test?

c. How many students answered these questions on the survey, i.e. what is the sample size?

22.10. **Work hours and education.** The General Social Survey collects data on demographics, education, and work, among many other characteristics of US residents. (NORC, 2010) Using ANOVA, we can consider educational attainment levels for all 1,172 respondents at once. Below are the distributions of hours worked by educational attainment and relevant summary statistics that will be helpful in carrying out this analysis.

Educational attainment	Mean	SD	n
Lt High School	38.7	15.8	121
High School	39.6	15.0	546
Junior College	41.4	18.1	97
Bachelor	42.5	13.6	253
Graduate	40.8	15.5	155

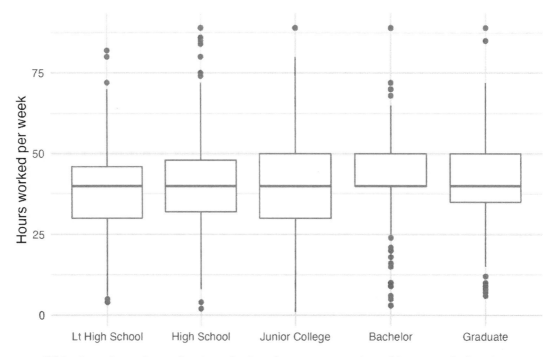

a. Write hypotheses for evaluating whether the average number of hours worked varies across the five groups.

b. Check conditions and describe any assumptions you must make to proceed with the test.

c. Below is the output associated with this test. What is the conclusion of the test?

term	df	sumsq	meansq	statistic	p.value
degree	4	2,006	502	2.19	0.07
Residuals	1,167	267,382	229		

22.5. EXERCISES

22.11. **True / False: ANOVA, I.** Determine if the following statements are true or false in ANOVA, and explain your reasoning for statements you identify as false.

  a. As the number of groups increases, the modified significance level for pairwise tests increases as well.

  b. As the total sample size increases, the degrees of freedom for the residuals increases as well.

  c. The constant variance condition can be somewhat relaxed when the sample sizes are relatively consistent across groups.

  d. The independence assumption can be relaxed when the total sample size is large.

22.12. **True / False: ANOVA, II.** Determine if the following statements are true or false, and explain your reasoning for statements you identify as false.

  If the null hypothesis that the means of four groups are all the same is rejected using ANOVA at a 5% significance level, then...

  a. we can then conclude that all the means are different from one another.

  b. the standardized variability between groups is higher than the standardized variability within groups.

  c. the pairwise analysis will identify at least one pair of means that are significantly different.

  d. the appropriate α to be used in pairwise comparisons is 0.05 / 4 = 0.0125 since there are four groups.

22.13. **Matching observed data with randomized F statistics.** Consider the following two datasets. The response variable is the score and the explanatory variable is whether the individual is in one of four groups.

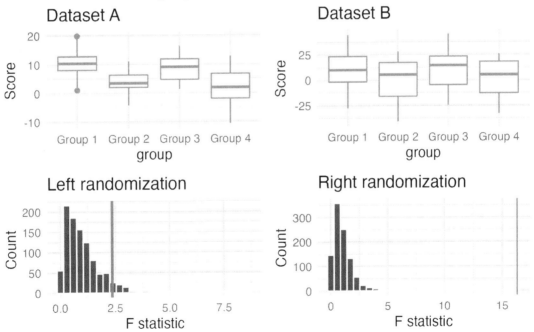

The randomizations (randomly assigning group to the score, calculating a randomization F statistic) were done 1000 times for each of Dataset A and B. The red line on each plot indicates the observed F statistic for the original (unrandomized) data.

  a. Does the randomization distribution on the left correspond to Dataset A or B? Explain.

  b. Does the randomization distribution on the right correspond to Dataset A or B? Explain.

22.14. **Child care hours.** The China Health and Nutrition Survey aims to examine the effects of the health, nutrition, and family planning policies and programs implemented by national and local governments. (Center, 2006) It, for example, collects information on number of hours Chinese parents spend taking care of their children under age 6. The side-by-side box plots below show the distribution of this variable by educational attainment of the parent. Also provided below is the ANOVA output for comparing average hours across educational attainment categories.

term	df	sumsq	meansq	statistic	p.value
edu	4	4,142	1,036	1.26	0.28
Residuals	794	653,048	822		

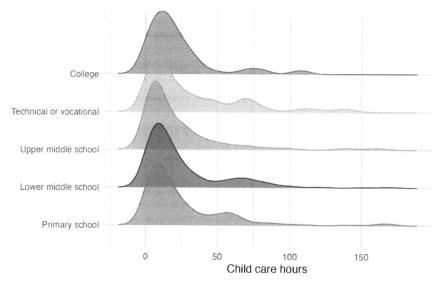

a. Write the hypotheses for testing for a difference between the average number of hours spent on child care across educational attainment levels.

b. What is the conclusion of the hypothesis test?

Chapter 23

Applications: Infer

23.1 Recap: Computational methods

The computational methods we have presented are used in two settings. First, in many real life applications (as in those covered here), the mathematical model and computational model give identical conclusions. When there are no differences in conclusions, the advantage of the computational method is that it gives the analyst a good sense for the logic of the statistical inference process. Second, when there is a difference in the conclusions (seen primarily in methods beyond the scope of this text), it is often the case that the computational method relies on fewer technical conditions and is therefore more appropriate to use.

23.1.1 Randomization

The important feature of randomization tests is that the data is permuted in such a way that the null hypothesis is true. The randomization distribution provides a distribution of the statistic of interest under the null hypothesis, which is exactly the information needed to calculate a p-value — where the p-value is the probability of obtaining the observed data or more extreme when the null hypothesis is true. Although there are ways to adjust the randomization for settings other than the null hypothesis being true, they are not covered in this book and they are not used widely. In approaching research questions with a randomization test, be sure to ask yourself what the null hypothesis represents and how it is that permuting the data is creating different possible null data representations.

Hypothesis tests. When using a randomization test, we proceed as follows:

- Write appropriate hypotheses.
- Compute the observed statistic of interest.
- Permute the data repeatedly, each time, recalculating the statistic of interest.
- Compute the proportion of times the permuted statistics are as extreme as or more extreme than the observed statistic, this is the p-value.
- Make a conclusion based on the p-value, and write the conclusion in context and in plain language so anyone can understand the result.

23.1.2 Bootstrapping

Bootstrapping, in contrast to randomization tests, represents a proxy sampling of the original population. With bootstrapping, the analyst is not forcing the null hypothesis to be true (or false, for that matter), but instead, they are replicating the variability seen in taking repeated samples from a population. Because there is no underlying true (or false) null hypothesis, bootstrapping is typically used for creating confidence intervals for the parameter of interest. Bootstrapping can be used to test particular values of a parameter (e.g., by evaluating whether a particular value of interest is contained in the confidence interval), but generally, bootstrapping is used for interval estimation instead of testing.

Confidence intervals. The following is how we generally computed a confidence interval using bootstrapping:

- Repeatedly resample the original data, with replacement, using the same sample size as the original data.
- For each resample, calculate the statistic of interest.
- Calculate the confidence interval using one of the following methods:
    - Bootstrap percentile interval: Obtain the endpoints representing the middle (e.g., 95%) of the bootstrapped statistics. The endpoints will be the confidence interval.
    - Bootstrap standard error (SE) interval: Find the SE of the bootstrapped statistics. The confidence interval will be given by the original observed statistic plus or minus some multiple (e.g., 2) of SEs.
- Put the conclusions in context and in plain language so even non-statisticians and data scientists can understand the results.

23.2 Recap: Mathematical models

The mathematical models which have been used to produce inferential analyses follow a consistent framework for different parameters of interest. As a way to contrast and compare the mathematical approach, we offer the following summaries in Tables 23.1 and 23.2.

23.2.1 z-procedures

Generally, when the response variable is categorical (or binary), the summary statistic is a proportion and the model used to describe the proportion is the standard normal curve (also referred to as a z-curve or a z-distribution). We provide Table 23.1 partly as a mechanism for understanding z-procedures and partly to highlight the extremely common usage of the z-distribution in practice.

Table 23.1: Similarities of z-methods across one and two independent samples analysis of a binary response variable.

	One sample	Two independent samples
Response variable	Binary	Binary
Parameter of interest	Proportion: p	Difference in proportions: $p_1 - p_2$
Statistic of interest	Proportion: $\hat{p}$	Difference in proportions: $\hat{p}_1 - \hat{p}_2$
Standard error: HT	$\sqrt{\frac{p_0(1-p_0)}{n}}$	$\sqrt{\hat{p}_{pool}\left(1-\hat{p}_{pool}\right)\left(\frac{1}{n_1}+\frac{1}{n_2}\right)}$
Standard error: CI	$\sqrt{\frac{\hat{p}(1-\hat{p})}{n}}$	$\sqrt{\frac{\hat{p}_1(1-\hat{p}_1)}{n_1}+\frac{\hat{p}_2(1-\hat{p}_2)}{n_2}}$
Conditions	1. Independence, 2. Success-failure	1. Independence, 2. Success-failure

Hypothesis tests. When applying the z-distribution for a hypothesis test, we proceed as follows:

- Write appropriate hypotheses.
- Verify conditions for using the z-distribution.
    - One-sample: the observations (or differences) must be independent. The success-failure condition of at least 10 success and at least 10 failures should hold.
    - For a difference of proportions: each sample must separately satisfy the success-failure conditions, and the data in the groups must also be independent.
- Compute the point estimate of interest and the standard error.
- Compute the Z score and p-value.
- Make a conclusion based on the p-value, and write a conclusion in context and in plain language so anyone can understand the result.

Confidence intervals. Similarly, the following is how we generally computed a confidence interval using a z-distribution:

- Verify conditions for using the z-distribution. (See above.)
- Compute the point estimate of interest, the standard error, and $z^\star$.
- Calculate the confidence interval using the general formula:
  point estimate $\pm\ z^\star SE$.
- Put the conclusions in context and in plain language so even non-statisticians and data scientists can understand the results.

23.2.2 t-procedures

With quantitative response variables, the t-distribution was applied as the appropriate mathematical model in three distinct settings. Although the three data structures are different, their similarities and differences are worth pointing out. We provide Table 23.2 partly as a mechanism for understanding t-procedures and partly to highlight the extremely common usage of the t-distribution in practice.

Table 23.2: Similarities of t-methods across one sample, paired sample, and two independent samples analysis of a numeric response variable.

	One sample	Paired sample	Two independent samples
Response variable	Numeric	Numeric	Numeric
Parameter of interest	Mean: μ	Paired mean: μ_{diff}	Difference in means: $\mu_1 - \mu_2$
Statistic of interest	Mean: $\bar{x}$	Paired mean: $\bar{x}_{diff}$	Difference in means: $\bar{x}_1 - \bar{x}_2$
Standard error	$\frac{s}{\sqrt{n}}$	$\frac{s_{diff}}{\sqrt{n_{diff}}}$	$\sqrt{\frac{s_1^2}{n_1} + \frac{s_2^2}{n_2}}$
Degrees of freedom	$n-1$	$n_{diff} - 1$	$\min(n_1 - 1, n_2 - 1)$
Conditions	1. Independence, 2. Normality or large samples	1. Independence, 2. Normality or large samples	1. Independence, 2. Normality or large samples

Hypothesis tests. When applying the t-distribution for a hypothesis test, we proceed as follows:

- Write appropriate hypotheses.
- Verify conditions for using the t-distribution.
    - One-sample or differences from paired data: the observations (or differences) must be independent and nearly normal. For larger sample sizes, we can relax the nearly normal requirement, e.g., slight skew is okay for sample sizes of 15, moderate skew for sample sizes of 30, and strong skew for sample sizes of 60.
    - For a difference of means when the data are not paired: each sample mean must separately satisfy the one-sample conditions for the t-distribution, and the data in the groups must also be independent.
- Compute the point estimate of interest, the standard error, and the degrees of freedom For df, use $n-1$ for one sample, and for two samples use either statistical software or the smaller of $n_1 - 1$ and $n_2 - 1$.
- Compute the T score and p-value.
- Make a conclusion based on the p-value, and write a conclusion in context and in plain language so anyone can understand the result.

Confidence intervals. Similarly, the following is how we generally computed a confidence interval using a t-distribution:

- Verify conditions for using the t-distribution. (See above.)
- Compute the point estimate of interest, the standard error, the degrees of freedom, and $t^\star_{df}$.
- Calculate the confidence interval using the general formula:
point estimate $\pm\ t^\star_{df} SE$.
- Put the conclusions in context and in plain language so even non-statisticians and data scientists can understand the results.

23.3 Case study: Redundant adjectives

Take a look at the images in Figure 23.1. How would you describe the circled item in the top image (A)? Would you call it "the triangle"? Or "the blue triangle"? How about in the bottom image (B)? Does your answer change?

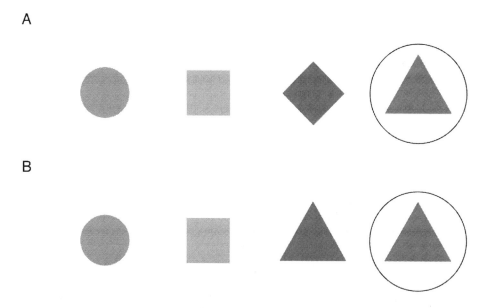

Figure 23.1: Two sets of four shapes. In A, the circled triangle is the only triangle. In B, the circled triangle is the only blue triangle.

23.3. CASE STUDY: REDUNDANT ADJECTIVES

In the top image in Figure 23.1 the circled item is the only triangle, while in the bottom image the circled item is one of two triangles. While in the top image "the triangle" is a sufficient description for the circled item, many of us might choose to refer to it as the "blue triangle" anyway. In the bottom image there are two triangles, so "the triangle" is no longer sufficient, and to describe the circled item we must qualify it with the color as well, as "the blue triangle".

Your answers to the above questions might be different if you're answering in a different language than English. For example, in Spanish, the adjective comes after the noun (e.g., "el triángulo azul") therefore the incremental value of the additional adjective might be different for the top image.

Researchers studying frequent use of redundant adjectives (e.g., referring to a single triangle as "the blue triangle") and incrementality of language processing designed an experiment where they showed the following two images to 22 native English speakers (undergraduates from University College London) and 22 native Spanish speakers (undergraduates from the Universidad de las Islas Baleares). They found that in both languages, the subjects used more redundant color adjectives in denser displays where it would be more efficient. (Rubio-Fernandez et al., 2021)

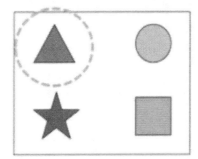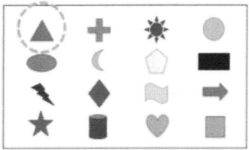

Figure 23.2: Images used in one of the experiments described in [@rubio-fernandez2021].

In this case study we will examine data from redundant adjective study, which the authors have made available on Open Science Framework at osf.io/9hw68.

Table 23.3 shows the top six rows of the data. The full dataset has 88 rows. Remember that there are a total of 44 subjects in the study (22 English and 22 Spanish speakers). There are two rows in the dataset for each of the subjects: one representing data from when they were shown an image with 4 items on it and the other with 16 items on it. Each subject was asked 10 questions for each type of image (with a different layout of items on the image for each question). The variable of interest to us is `redundant_perc`, which gives the percentage of questions the subject used a redundant adjective to identify "the blue triangle".

Table 23.3: Top six rows of the data collected in the study.

language	subject	items	n_questions	redundant_perc
English	1	4	10	100
English	1	16	10	100
English	2	4	10	0
English	2	16	10	0
English	3	4	10	100
English	3	16	10	100

23.3.1 Exploratory analysis

In one of the images shown to the subjects, there are 4 items, and in the other, there are 16 items. In each of the images the circled item is the only triangle, therefore referring to it as "the blue triangle" or as "el triángulo azul" is considered redundant. If the subject's response was "the triangle", they were recorded to have not used a redundant adjective. If the response was "the blue triangle", they were recorded to have used a redundant adjective. Figure 23.3 shows the results of the experiment. We can see that English speakers are more likely than Spanish speakers to use redundant adjectives, and also that in both languages, subjects are more likely to use a redundant adjective when there are more items in the image (i.e. in a denser display).

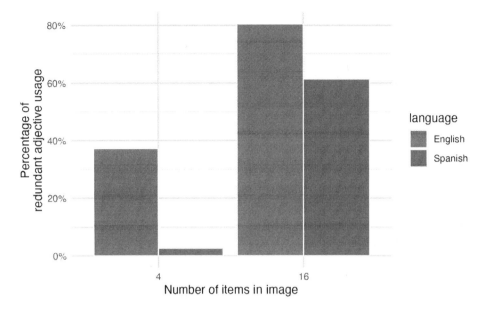

Figure 23.3: Results of redundant adjective usage experiment from [@rubio-fernandez2021]. English speakers are more likely than Spanish speakers to use redundant adjectives, regardless of number of items in image. For both images, respondents are more likely to use a redundant adjective when there are more items in the image.

These values are also shown in Table 23.4.

Table 23.4: Summary of redundant adjective usage experiment from study.

Language	Number of items	Percentage redundant
English	4	37.27
English	16	80.45
Spanish	4	2.73
Spanish	16	61.36

23.3.2 Confidence interval for a single mean

In this experiment, the average percentage of redundant adjective usage among subjects who responded in English when presented with an image with 4 items in it is 37.27. Along with the sample average as a point estimate, however, we can construct a confidence interval for the true mean redundant adjective usage of English speakers who use redundant color adjectives when describing items in an image that is not very dense.

Using a computational method, we can construct the interval via bootstrapping. Figure 23.4 shows the distribution of 1,000 bootstrapped means from this sample. The 95% confidence interval (that is calculated by taking the 2.5th and 97.5th percentile of the bootstrap distribution is 19.1% to 56.4%. Note that this interval for the true population parameter is only valid if we can assume that the sample of English speakers are representative of the population of all English speakers.

Using a similar technique, we can also construct confidence intervals for the true mean redundant adjective usage percentage for English speakers who are shown dense (16 item) displays and for Spanish speakers with both types (4 and 16 items) displays. However these confidence intervals are

23.3. CASE STUDY: REDUNDANT ADJECTIVES

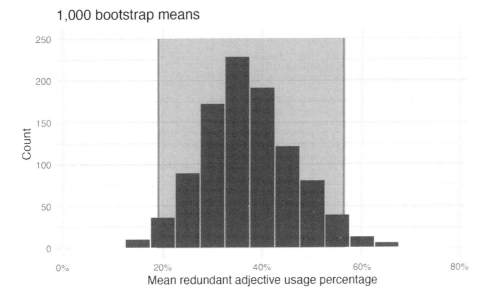

Figure 23.4: Distribution of 1,000 bootstrapped means of redundant adjective usage percentage among English speakers who were shown four items in images. Overlaid on the distribution is the 95% bootstrap percentile interval that ranges from 19.1% to 56.4%.

not very meaningful to compare to one another as the interpretation of the "true mean redundant adjective usage percentage" is quite an abstract concept. Instead, we might be more interested in comparative questions such as "Does redundant adjective usage differ between dense and sparse displays among English speakers and among Spanish speakers?" or "Does redundant adjective usage differ between English speakers and Spanish speakers?" To answer either of these questions we need to conduct a hypothesis test.

23.3.3 Paired mean test

Let's start with the following question: "Do the data provide convincing evidence of a difference in mean redundant adjective usage percentages between sparse (4 item) and dense (16 item) displays for English speakers?" Note that the English speaking participants were each evaluated on both the 4 item and the 16 item displays. Therefore, the variable of interest is the difference in redundant percentage. The statistic of interest will be the average of the differences, here $\bar{x}_{diff} = 43.18$.

Data from the first six English speaking participants are seen in Table 23.5. Although the redundancy percentages seem higher in the 16 item task, a hypothesis test will tell us whether the differences observed in the data could be due to natural variability.

Table 23.5: Six participants who speak English with redundancy difference.

subject	redundant_perc_4	redundant_perc_16	diff_redundant_perc
1	100	100	0
2	0	0	0
3	100	100	0
4	10	80	70
5	0	90	90
6	0	70	70

We can answer the research question using a hypothesis test with the following hypotheses:

$$H_0 : \mu_{diff} = 0$$
$$H_A : \mu_{diff} \neq 0$$

where μ_{diff} is the true difference in redundancy percentages when comparing a 16 item display with a 4 item display. Recall that the computational method used to assess a hypothesis pertaining to the true average of a paired difference shuffles the observed percentage across the two groups (4 item vs 16 item) but **within** a single participant. The shuffling process allows for repeated calculations of potential sample differences under the condition that the null hypothesis is true.

Figure 23.5 shows the distribution of 1,000 mean differences from redundancy percentages permuted across the two conditions. Note that the distribution is centered at 0, since the structure of randomly assigning redundancy percentages to each item display will balance the data out such that the average of any differences will be zero.

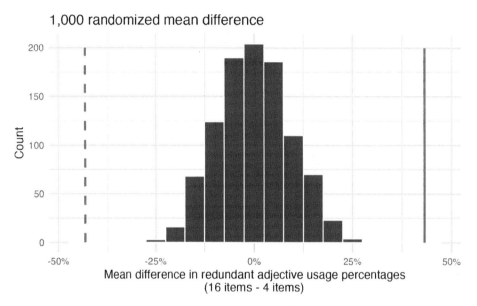

Figure 23.5: Distribution of 1,000 mean differences of redundant adjective usage percentage among English speakers who were shown images with 4 and 16 items. Overlaid on the distribution is the observed average difference in the sample (solid line) as well as the difference in the other direction (dashed line), which is far out in the tail, yielding a p-value that is approximately 0.

With such a small p-value, we reject the null hypothesis and conclude that the data provide convincing evidence of a difference in mean redundant adjective usage percentages across different displays for English speakers.

23.3.4 Two independent means test

Finally, let's consider the question "How does redundant adjective usage differ between English speakers and Spanish speakers?" The English speakers are independent from the Spanish speakers, but since the same subjects were shown the two types of displays, we can't combine data from the two display types (4 objects and 16 objects) together while maintaining independence of observations. Therefore, to answer questions about language differences, we will need to conduct two hypothesis tests, one for sparse displays and the other for dense displays. In each of the tests, the hypotheses are as follows:

$$H_0 : \mu_{English} = \mu_{Spanish}$$
$$H_A : \mu_{English} \neq \mu_{Spanish}$$

Here, the randomization process is slightly different than the paired setting (because the English and Spanish speakers do not have a natural pairing across the two groups). To answer the research question using a computational method, we can use a randomization test where we permute the data across all participants under the assumption that the null hypothesis is true (no difference in mean redundant adjective usage percentages across English vs Spanish speakers).

Figure 23.6 shows the null distributions for each of these hypothesis tests. The p-value for the 4 item display comparison is very small (0.002) while the p-value for the 16 item display is much larger (0.114).

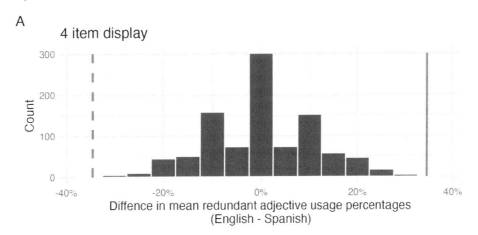

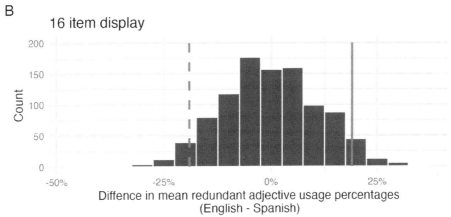

Figure 23.6: Distributions of 1,000 differences in randomized means of redundant adjective usage percentage between English and Spanish speakers. Plot A shows the differences in 4 item displays and Plot B shows the differences in 16 item displays. In each plot, the observed differences in the sample (solid line) as well as the differences in the other direction (dashed line) are overlaid.

Based on the p-values (a measure of deviation from the null claim), we can conclude that the data provide convincing evidence of a difference in mean redundant adjective usage percentages between languages in 4 item displays (small p-value) but not in 16 item displays (not small p-value). The results suggests that language patterns around redundant adjective usage might be more similar for denser displays than sparser displays.

23.4 Interactive R tutorials

Navigate the concepts you've learned in this chapter in R using the following self-paced tutorials. All you need is your browser to get started!

Tutorial 5: Statistical inference
https://openintrostat.github.io/ims-tutorials/05-infer

Tutorial 5 - Lesson 1: Inference for a single proportion
https://openintro.shinyapps.io/ims-05-infer-01

Tutorial 5 - Lesson 2: Hypothesis tests to compare proportions
https://openintro.shinyapps.io/ims-05-infer-02

Tutorial 5 - Lesson 3: Chi-squared test of independence
https://openintro.shinyapps.io/ims-05-infer-03

Tutorial 5 - Lesson 4: Chi-squared goodness of fit Test
https://openintro.shinyapps.io/ims-05-infer-04

Tutorial 5 - Lesson 5: Bootstrapping for estimating a parameter
https://openintro.shinyapps.io/ims-05-infer-05

Tutorial 5 - Lesson 6: Introducing the t-distribution
https://openintro.shinyapps.io/ims-05-infer-06

Tutorial 5 - Lesson 7: Inference for difference in two means
https://openintro.shinyapps.io/ims-05-infer-07

Tutorial 5 - Lesson 8: Comparing many means
https://openintro.shinyapps.io/ims-05-infer-08

You can also access the full list of tutorials supporting this book at https://openintrostat.github.io/ims-tutorials.

23.5 R labs

Further apply the concepts you've learned in this part in R with computational labs that walk you through a data analysis case study.

Inference for categorical responses - Texting while driving
https://www.openintro.org/go?id=ims-r-lab-infer-1

Inference for numerical responses - Youth Risk Behavior Surveillance System
https://www.openintro.org/go?id=ims-r-lab-infer-2

You can also access the full list of labs supporting this book at https://www.openintro.org/go?id=ims-r-labs.

PART VI

INFERENTIAL MODELING

Chapter 24

Inference for linear regression with a single predictor

 We now bring together ideas of inferential analyses with the descriptive models seen in Chapters 7. In particular, we will use the least squares regression line to test whether or not there is a relationship between two continuous variables. Additionally, we will build confidence intervals which quantify the slope of the linear regression line. The setting is now focused on predicting a numeric response variable (for linear models) or a binary response variable (for logistic models), we continue to ask questions about the variability of the model from sample to sample. The sampling variability will inform the conclusions about the population that can be drawn.

Many of the inferential ideas are remarkably similar to those covered in previous chapters. The technical conditions for linear models are typically assessed graphically, although independence of observations continues to be of utmost importance.

We encourage the reader to think broadly about the models at hand without putting too much dependence on the exact p-values that are reported from the statistical software. Inference on models with multiple explanatory variables can suffer from data snooping which result in false positive claims. We provide some guidance and hope the reader will further their statistical learning after working through the material in this text.

24.1 Case study: Sandwich store

24.1.1 Observed data

We start the chapter with a hypothetical example describing the linear relationship between dollars spent advertising for a chain sandwich restaurant and monthly revenue. The hypothetical example serves the purpose of illustrating how a linear model varies from sample to sample. Because we have made up the example and the data (and the entire population), we can take many many samples from the population to visualize the variability. Note that in real life, we always have exactly one sample (that is, one dataset), and through the inference process, we imagine what might have happened had we taken a different sample. The change from sample to sample leads to an understanding of how the single observed dataset is different from the population of values, which is typically the fundamental goal of inference.

Consider the following hypothetical population of all of the sandwich stores of a particular chain seen in Figure 24.1. In this made-up world, the CEO actually has all the relevant data, which is why they can plot it here. The CEO is omniscient and can write down the population model which describes the true population relationship between the advertising dollars and revenue. There appears to be a linear relationship between advertising dollars and revenue (both in $1,000).

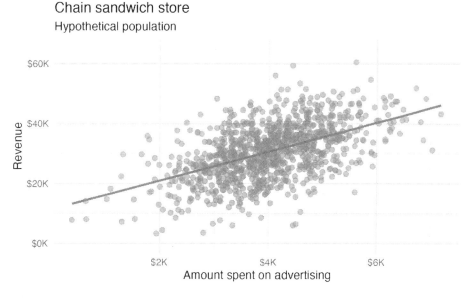

Figure 24.1: Revenue as a linear model of advertising dollars for a population of sandwich stores, in thousands of dollars.

You may remember from Chapter 7 that the population model is:

$$y = \beta_0 + \beta_1 x + \varepsilon.$$

Again, the omniscient CEO (with the full population information) can write down the true population model as:

```
expected revenue = 11.23 + 4.8 × advertising.
```

24.1.2 Variability of the statistic

Unfortunately, in our scenario, the CEO is not willing to part with the full set of data, but they will allow potential franchise buyers to see a small sample of the data in order to help the potential buyer decide whether or not set up a new franchise. The CEO is willing to give each potential franchise buyer a random sample of data from 20 stores.

As with any numerical characteristic which describes a subset of the population, the estimated slope of a sample will vary from sample to sample. Consider the linear model which describes revenue (in $1,000) based on advertising dollars (in $1,000).

The least squares regression model uses the data to find a sample linear fit:

$$\hat{y} = b_0 + b_1 x.$$

A random sample of 20 stores shows a different least square regression line depending on which observations are selected. A subset of size 20 stores shows a similar positive trend between advertising and revenue (to what we saw in Figure 24.1 which described the population) despite having fewer observations on the plot.

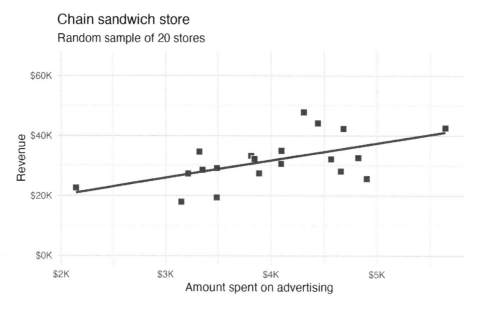

Figure 24.2: A random sample of 20 stores from the entire population. A linear trend between advertising and revenue continues to be observed.

A second sample of size 20 also shows a positive trend!

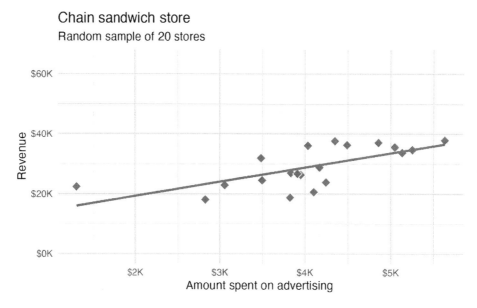

Figure 24.3: A different random sample of 20 stores from the entire population. Again, a linear trend between advertising and revenue is observed.

But the lines are slightly different!

That is, there is **variability** in the regression line from sample to sample. The concept of the sampling variability is something you've seen before, but in this lesson, you will focus on the variability of the line often measured through the variability of a single statistic: **the slope of the line**.

You might notice in Figure 24.5 that the $\hat{y}$ values given by the lines are much more consistent in the middle of the dataset than at the ends. The reason is that the data itself anchors the lines in such a way that the line must pass through the center of the data cloud. The effect of the fan-shaped lines is that predicted revenue for advertising close to $4,000 will be much more precise than the revenue predictions made for $1,000 or $7,000 of advertising.

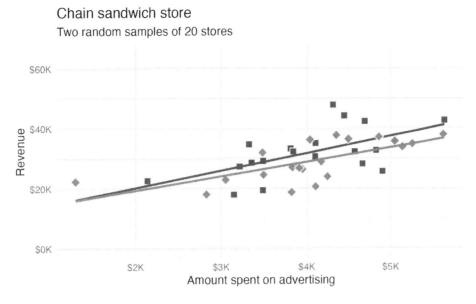

Figure 24.4: The linear models from the two different random samples are quite similar, but they are not the same line.

The distribution of slopes (for samples of size $n = 20$) can be seen in a histogram, as in Figure 24.6.

Recall, the example described in this introduction is hypothetical. That is, we created an entire population in order demonstrate how the slope of a line would vary from sample to sample. The tools in this textbook are designed to evaluate only one single sample of data. With actual studies, we do not have repeated samples, so we are not able to use repeated samples to visualize the variability in slopes. We have seen variability in samples throughout this text, so it should not come as a surprise that different samples will produce different linear models. However, it is nice to visually consider the linear models produced by different slopes. Additionally, as with measuring the variability of previous statistics (e.g., $\overline{X}_1 - \overline{X}_2$ or $\hat{p}_1 - \hat{p}_2$), the histogram of the sample statistics can provide information related to inferential considerations.

In the following sections, the distribution (i.e., histogram) of b_1 (the estimated slope coefficient) will be constructed in the same three ways that, by now, may be familiar to you. First (in Section 24.2), the distribution of b_1 when $\beta_1 = 0$ is constructed by randomizing (permuting) the response variable. Next (in Section 24.3), we can bootstrap the data by taking random samples of size n from the original dataset. And last (in Section 24.4), we use mathematical tools to describe the variability using the t-distribution that was first encountered in Section 19.2.

24.2 Randomization test for the slope

Consider data on 100 randomly selected births gathered originally from the US Department of Health and Human Services. Some of the variables are plotted in Figure 24.7.

The scientific research interest at hand will be in determining the linear relationship between weight of baby at birth (in lbs) and number of weeks of gestation. The dataset is quite rich and deserves exploring, but for this example, we will focus only on the weight of the baby.

 The `births14` data can be found in the **openintro** R package. We will work with a random sample of 100 observations from these data.

As you have seen previously, statistical inference typically relies on setting a null hypothesis which is hoped to be subsequently rejected. In the linear model setting, we might hope to have a linear relationship between `weeks` and `weight` in settings where `weeks` gestation is known and `weight` of

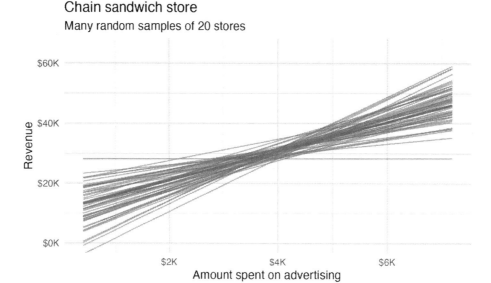

Figure 24.5: If repeated samples of size 20 are taken from the entire population, each linear model will be slightly different. The red line provides the linear fit to the entire population.

baby needs to be predicted.

The relevant hypotheses for the linear model setting can be written in terms of the population slope parameter. Here the population refers to a larger population of births in the US.

- $H_0 : \beta_1 = 0$, there is no linear relationship between weight and weeks.
- $H_A : \beta_1 \neq 0$, there is some linear relationship between weight and weeks.

Recall that for the randomization test, we permute one variable to eliminate any existing relationship between the variables. That is, we set the null hypothesis to be true, and we measure the natural variability in the data due to sampling but **not** due to variables being correlated. Figure 24.8 shows the observed data and a scatterplot of one permutation of the weight variable. The careful observer can see that each of the observed values for weight (and for weeks) exist in both the original data plot as well as the permuted weight plot, but the weight and weeks gestation are no longer matched for a given birth. That is, each weight value is randomly assigned to a new weeks gestation.

By repeatedly permuting the response variable, any pattern in the linear model that is observed is due only to random chance (and not an underlying relationship). The randomization test compares the slopes calculated from the permuted response variable with the observed slope. If the observed slope is inconsistent with the slopes from permuting, we can conclude that there is some underlying relationship (and that the slope is not merely due to random chance).

24.2.1 Observed data

We will continue to use the births data to investigate the linear relationship between weight and weeks gestation. Note that the least squares model (see Chapter 7) describing the relationship is given in Table 24.1. The columns in Table 24.1 are further described in Section 24.4.

Table 24.1: The least squares estimates of the intercept and slope are given in the estimate column. The observed slope is 0.335.

term	estimate	std.error	statistic	p.value
(Intercept)	-5.72	1.61	-3.54	6e-04
weeks	0.34	0.04	8.07	<0.0001

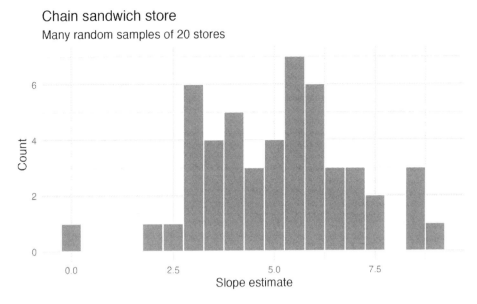

Figure 24.6: Variability of slope estimates taken from many different samples of stores, each of size 20.

24.2.2 Variability of the statistic

After permuting the data, the least squares estimate of the line can be computed. Repeated permutations and slope calculations describe the variability in the line (i.e., in the slope) due only to the natural variability and not due to a relationship between `weight` and `weeks` gestation. Figure 24.9 shows two different permutations of `weight` and the resulting linear models.

As you can see, sometimes the slope of the permuted data is positive, sometimes it is negative. Because the randomization happens under the condition of no underlying relationship (because the response variable is completely mixed with the explanatory variable), we expect to see the center of the randomized slope distribution to be zero.

24.2.3 Observed statistic vs. null statistics

As we can see from Figure 24.10, a slope estimate as extreme as the observed slope estimate (the red line) never happened in many repeated permutations of the `weight` variable. That is, if indeed there were no linear relationship between `weight` and `weeks`, the natural variability of the slopes would produce estimates between approximately -0.15 and +0.15. We reject the null hypothesis. Therefore, we believe that the slope observed on the original data is not just due to natural variability and indeed, there is a linear relationship between `weight` of baby and `weeks` gestation for births in the US.

24.3 Bootstrap confidence interval for the slope

As we have seen in previous chapters, we can use bootstrapping to estimate the sampling distribution of the statistic of interest (here, the slope) without the null assumption of no relationship (which was the condition in the randomization test). Because interest is now in creating a CI, there is no null hypothesis, so there won't be any reason to permute either of the variables.

24.3.1 Observed data

Returning to the births data, we may want to consider the relationship between `mage` (mother's age) and `weight`. Is `mage` a good predictor of `weight`? And if so, what is the relationship? That is, what

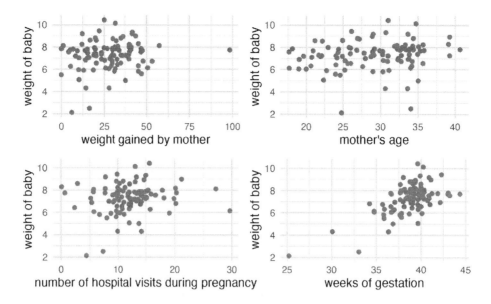

Figure 24.7: Weight of baby at birth (in lbs) as plotted by four other birth variables (mother's weight gain, mother's age, number of hospital visits, and weeks gestation).

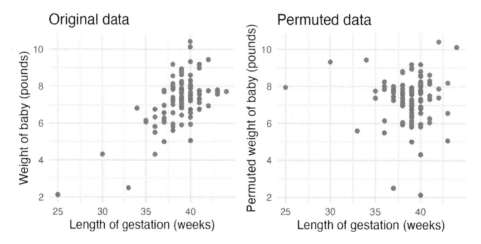

Figure 24.8: Original (left) and permuted (right) data. The permutation removes the linear relationship between `weight` and `weeks`. Repeated permutations allow for quantifying the variability in the slope under the condition that there is no linear relationship (i.e., that the null hypothesis is true).

is the slope that models average `weight` of baby as a function of `mage` (mother's age)? The linear model regressing `weight` on `mage` is provided in Table 24.2.

Table 24.2: The least squares estimates of the intercept and slope are given in the estimate column. The observed slope is 0.036

term	estimate	std.error	statistic	p.value
(Intercept)	6.23	0.71	8.79	<0.0001
mage	0.04	0.02	1.50	0.1362

24.3.2 Variability of the statistic

Because the focused is not on a null distribution, sample with replacement $n = 100$ observations from the original dataset. Recall that with bootstrapping the resample always has the same number of observations as the original dataset in order to mimic the process of taking a sample from the

24.3. BOOTSTRAP CONFIDENCE INTERVAL FOR THE SLOPE

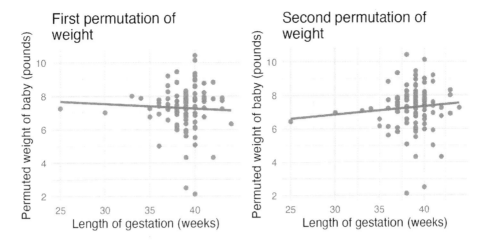

Figure 24.9: Two different permutations of the `weight` variable with slightly different least squares regression lines.

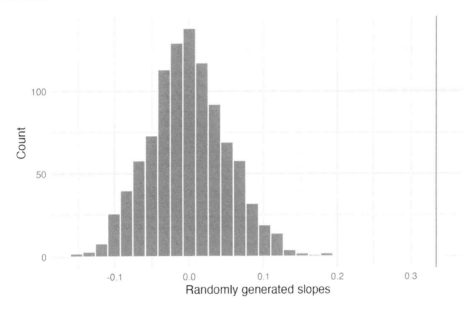

Figure 24.10: Histogram of slopes given different permutations of the `weight` variable. The vertical red line is at the observed value of the slope, 0.335.

population. When sampling in the linear model case, consider each observation to be a single dot. If the dot is resampled, both the `weight` and the `mage` measurement are observed. The measurements are linked to the dot (i.e., to the birth in the sample).

Figure 24.12 shows the original data as compared with a single bootstrap sample, resulting in (slightly) different linear models. The red circles represent points in the original data which were not included in the bootstrap sample. The blue circles represents a point that was repeatedly resampled (and is therefore darker) in the bootstrap sample. The green circles represents a particular structure to the data which is observed in both the original and bootstrap samples. By repeatedly resampling, we can see dozens of bootstrapped slopes on the same plot in Figure 24.13.

Recall that in order to create a confidence interval for the slope, we need to find the range of values that the statistic (here the slope) takes on from different bootstrap samples. Figure 24.14 is a histogram of the relevant bootstrapped slopes. We can see that a 95% bootstrap percentile interval for the true population slope is given by (-0.01, 0.081). We are 95% confident that for the model describing the population of births, described by mother's age and `weight` of baby, a one unit increase in `mage` (in years) will be associated with an increase in predicted average baby `weight` of between -0.01 and 0.081 pounds (notice that the CI overlaps zero!).

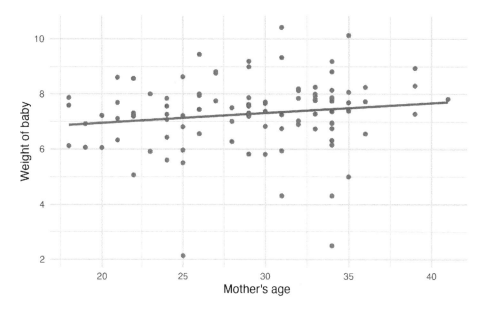

Figure 24.11: Original data: `weight` of baby as a linear model of mother's age. Notice that the relationship between `mage` and `weight` is not as strong as the relationship we saw previously between `weeks` and `weight`.

EXAMPLE

Using Figure 24.14, calculate the bootstrap estimate for the standard error of the slope. Using the bootstrap standard error, find a 95% bootstrap SE confidence interval for the true population slope, and interpret the interval in context.

Notice that most of the bootstrapped slopes fall between -0.01 and +0.08 (a range of 0.09). Using the empirical rule (that with bell-shaped distributions, most observations are within two standard errors of the center), the standard error of the slopes is approximately 0.0225. The normal cutoff for a 95% confidence interval is $z^* = 1.96$ which leads to a confidence interval of $b_1 \pm 1.96 \cdot SE \rightarrow 0.036 \pm 1.96 \cdot 0.0225 \rightarrow (-0.0081, 0.0801)$. The bootstrap SE confidence interval is almost identical to the bootstrap percentile interval. In context, we are 95% confident that for the model describing the population of births, described by mother's age and `weight` of baby, a one unit increase in `mage` (in years) will be associated with an increase in predicted average baby `weight` of between -0.0081 and 0.0801 pounds

24.4 Mathematical model for testing the slope

When certain technical conditions apply, it is convenient to use mathematical approximations to test and estimate the slope parameter. The approximations will build on the t-distribution which were described in Chapter 19. The mathematical model is often correct and is usually easy to implement computationally. The validity of the technical conditions will be considered in detail in Section 24.6.

In this section, we discuss uncertainty in the estimates of the slope and y-intercept for a regression line. Just as we identified standard errors for point estimates in previous chapters, we first discuss standard errors for these new estimates.

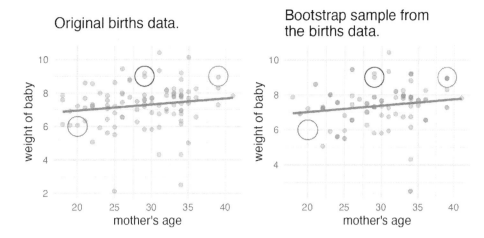

Figure 24.12: Original and one bootstrap sample of the births data. Note that it is difficult to differentiate the two plots, as (within a single bootstrap sample) the observations which have been resampled twice are plotted as points on top of one another. The red circles represent points in the original data which were not included in the bootstrap sample. The blue circles represents a data point that was repeatedly resampled (and is therefore darker) in the bootstrap sample. The green circles represents a particular structure to the data which is observed in both the original and bootstrap samples.

24.4.1 Observed data

Midterm elections and unemployment

Elections for members of the United States House of Representatives occur every two years, coinciding every four years with the U.S. Presidential election. The set of House elections occurring during the middle of a Presidential term are called midterm elections. In America's two-party system (the vast majority of House members through history have been either Republicans or Democrats), one political theory suggests the higher the unemployment rate, the worse the President's party will do in the midterm elections. In 2020 there were 232 Democrats, 198 Republicans, and 1 Libertarian in the House.

To assess the validity of this claim, we can compile historical data and look for a connection. We consider every midterm election from 1898 to 2018, with the exception of those elections during the Great Depression. The House of Representatives is made up of 435 voting members.

 The `midterms_house` data can be found in the **openintro** R package.

Figure 24.15 shows these data and the least-squares regression line:

$$\texttt{percent change in House seats for President's party}$$
$$= -7.36 - 0.89 \times \texttt{(unemployment rate)}$$

We consider the percent change in the number of seats of the President's party (e.g., percent change in the number of seats for Republicans in 2018) against the unemployment rate.

Examining the data, there are no clear deviations from linearity or substantial outliers (see Section 7.1.3 for a discussion on using residuals to visualize how well a linear model fits the data). While the data are collected sequentially, a separate analysis was used to check for any apparent correlation between successive observations; no such correlation was found.

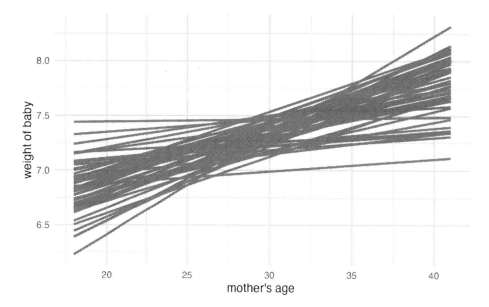

Figure 24.13: Repeated bootstrap resamples of size 100 are taken from the original data. Each of the bootstrapped linear model is slightly different.

 GUIDED PRACTICE

The data for the Great Depression (1934 and 1938) were removed because the unemployment rate was 21% and 18%, respectively. Do you agree that they should be removed for this investigation? Why or why not?[1]

There is a negative slope in the line shown in Figure 24.15. However, this slope (and the y-intercept) are only estimates of the parameter values. We might wonder, is this convincing evidence that the "true" linear model has a negative slope? That is, do the data provide strong evidence that the political theory is accurate, where the unemployment rate is a useful predictor of the midterm election? We can frame this investigation into a statistical hypothesis test:

- H_0: $\beta_1 = 0$. The true linear model has slope zero.
- H_A: $\beta_1 \neq 0$. The true linear model has a slope different than zero. The unemployment is predictive of whether the President's party wins or loses seats in the House of Representatives.

We would reject H_0 in favor of H_A if the data provide strong evidence that the true slope parameter is different than zero. To assess the hypotheses, we identify a standard error for the estimate, compute an appropriate test statistic, and identify the p-value.

24.4.2 Variability of the statistic

Just like other point estimates we have seen before, we can compute a standard error and test statistic for b_1. We will generally label the test statistic using a T, since it follows the t-distribution.

We will rely on statistical software to compute the standard error and leave the explanation of how this standard error is determined to a second or third statistics course. Table 24.3 shows software output for the least squares regression line in Figure 24.15. The row labeled unemp includes all relevant information about the slope estimate (i.e., the coefficient of the unemployment variable).

---

[1] The answer to this question relies on the idea that statistical data analysis is somewhat of an art. That is, in many situations, there is no "right" answer. As you do more and more analyses on your own, you will come to recognize the nuanced understanding which is needed for a particular dataset. In terms of the Great Depression, we will provide two contrasting considerations. Each of these points would have very high leverage on any least-squares regression line, and years with such high unemployment may not help us understand what would happen in other years where the unemployment is only modestly high. On the other hand, these are exceptional cases, and we would be discarding important information if we exclude them from a final analysis.

24.4. MATHEMATICAL MODEL FOR TESTING THE SLOPE

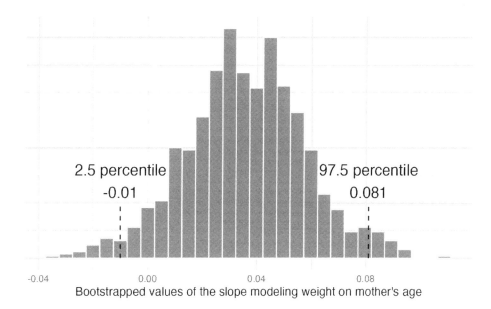

Figure 24.14: The original births data on `weight` and `mage` is bootstrapped 1,000 times. The histogram provides a sense for the variability of the slope of the linear model slope from sample to sample.

Table 24.3: Output from statistical software for the regression line modeling the midterm election losses for the President's party as a response to unemployment.

term	estimate	std.error	statistic	p.value
(Intercept)	-7.36	5.16	-1.43	0.16
unemp	-0.89	0.83	-1.07	0.30

EXAMPLE

What do the first and second columns of Table 24.3 represent?

The entries in the first column represent the least squares estimates, b_0 and b_1, and the values in the second column correspond to the standard errors of each estimate. Using the estimates, we could write the equation for the least square regression line as

$$\hat{y} = -7.36 - 0.89x$$

where $\hat{y}$ in this case represents the predicted change in the number of seats for the president's party, and x represents the unemployment rate.

We previously used a t-test statistic for hypothesis testing in the context of numerical data. Regression is very similar. In the hypotheses we consider, the null value for the slope is 0, so we can compute the test statistic using the T score formula:

$$T = \frac{\text{estimate} - \text{null value}}{\text{SE}} = \frac{-0.89 - 0}{0.835} = -1.07$$

This corresponds to the third column of Table 24.3 .

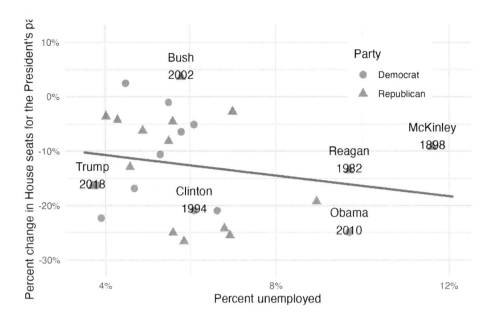

Figure 24.15: The percent change in House seats for the President's party in each election from 1898 to 2010 plotted against the unemployment rate. The two points for the Great Depression have been removed, and a least squares regression line has been fit to the data.

EXAMPLE

Use Table 24.3 to determine the p-value for the hypothesis test

---

The last column of the table gives the p-value for the two-sided hypothesis test for the coefficient of the unemployment rate 0.2961 That is, the data do not provide convincing evidence that a higher unemployment rate has any correspondence with smaller or larger losses for the President's party in the House of Representatives in midterm elections.

24.4.3 Observed statistic vs. null statistics

As the final step in a mathematical hypothesis test for the slope, we use the information provided to make a conclusion about whether or not the data could have come from a population where the true slope was zero (i.e., $\beta_1 = 0$). Before evaluating the formal hypothesis claim, sometimes it is important to check your intuition. Based on everything we've seen in the examples above describing the variability of a line from sample to sample, ask yourself if the linear relationship given by the data could have come from a population in which the slope was truly zero.

EXAMPLE

Examine Figure ??, which relates the Elmhurst College aid and student family income. Are you convinced that the slope is meaningfully different from zero? That is, do you think a formal hypothesis test would reject the claim that the true slope of the line should be zero?

While the relationship between the variables is not perfect, there is an evident decreasing trend in the data. This suggests the hypothesis test will reject the null claim that the slope is zero.

24.4. MATHEMATICAL MODEL FOR TESTING THE SLOPE

The `elmhurst` data can be found in the **openintro** R package.

The tools in this section help you go beyond a visual interpretation of the linear relationship toward a formal mathematical claim about whether the slope estimate is meaningfully different from 0 to suggest that the true population slope is different from 0.

Table 24.4: Summary of least squares fit for the Elmhurst College data, where we are predicting the gift aid by the university based on the family income of students.

term	estimate	std.error	statistic	p.value
(Intercept)	24319.33	1291.45	18.83	<0.0001
family_income	-0.04	0.01	-3.98	2e-04

GUIDED PRACTICE

Table 24.4 shows statistical software output from fitting the least squares regression line shown in Figure ??. Use the output to formally evaluate the following hypotheses.[2]

- H_0: The true coefficient for family income is zero.
- H_A: The true coefficient for family income is not zero.

Inference for regression.

We usually rely on statistical software to identify point estimates, standard errors, test statistics, and p-values in practice. However, be aware that software will not generally check whether the method is appropriate, meaning we must still verify conditions are met. See Section 24.6.

---

[2]We look in the second row corresponding to the family income variable. We see the point estimate of the slope of the line is -0.0431, the standard error of this estimate is 0.0108, and the t-test statistic is $T = -3.98$. The p-value corresponds exactly to the two-sided test we are interested in: 0.0002. The p-value is so small that we reject the null hypothesis and conclude that family income and financial aid at Elmhurst College for freshman entering in the year 2011 are negatively correlated and the true slope parameter is indeed less than 0, just as we believed in our analysis of Figure ??.

24.5 Mathematical model, interval for the slope

24.5.1 Observed data

Similar to how we can conduct a hypothesis test for a model coefficient using regression output, we can also construct a confidence interval for that coefficient.

EXAMPLE

Compute the 95% confidence interval for the coefficient using the regression output from Table 24.4.

The point estimate is -0.0431 and the standard error is $SE = 0.0108$. When constructing a confidence interval for a model coefficient, we generally use a t-distribution. The degrees of freedom for the distribution are noted in the regression output, $df = 48$, allowing us to identify $t^\star_{48} = 2.01$ for use in the confidence interval.

We can now construct the confidence interval in the usual way:

$$\text{point estimate} \pm t^\star_{48} \times SE$$
$$-0.0431 \pm 2.01 \times 0.0108$$
$$(-0.0648, -0.0214)$$

We are 95% confident that with each dollar increase in , the university's gift aid is predicted to decrease on average by \$0.0214 to \$0.0648.

24.5.2 Variability of the statistic

Confidence intervals for coefficients.

Confidence intervals for model coefficients (e.g., the intercept or the slope) can be computed using the t-distribution:

$$b_i \pm t^\star_{df} \times SE_{b_i}$$

where $t^\star_{df}$ is the appropriate t-value corresponding to the confidence level with the model's degrees of freedom.

On the topic of intervals in this book, we've focused exclusively on confidence intervals for model parameters. However, there are other types of intervals that may be of interest, including prediction intervals for a response value and also confidence intervals for a mean response value in the context of regression.

24.6 Checking model conditions

In the previous sections, we used randomization and bootstrapping to perform inference when the mathematical model was not valid due to violations of the technical conditions. In this section, we'll provide details for when the mathematical model is appropriate and a discussion of technical conditions needed for the randomization and bootstrapping procedures.

24.6.1 What are the technical conditions for the mathematical model?

When fitting a least squares line, we generally require

- **Linearity.** The data should show a linear trend. If there is a nonlinear trend (e.g., first panel of Figure 24.16) an advanced regression method from another book or later course should be applied.

- **Independent observations.** Be cautious about applying regression to data, which are sequential observations in time such as a stock price each day. Such data may have an underlying structure that should be considered in a model and analysis. An example of a dataset where successive observations are not independent is shown in the fourth panel of Figure 24.16. There are also other instances where correlations within the data are important, which is further discussed in Chapter 25.

- **Nearly normal residuals.** Generally, the residuals must be nearly normal. When this condition is found to be unreasonable, it is usually because of outliers or concerns about influential points, which we'll talk about more in Section 7.3. An example of a residual that would be a potentially concern is shown in the second panel of Figure 24.16, where one observation is clearly much further from the regression line than the others.

- **Constant or equal variability.** The variability of points around the least squares line remains roughly constant. An example of non-constant variability is shown in the third panel of Figure 24.16, which represents the most common pattern observed when this condition fails: the variability of y is larger when x is larger.

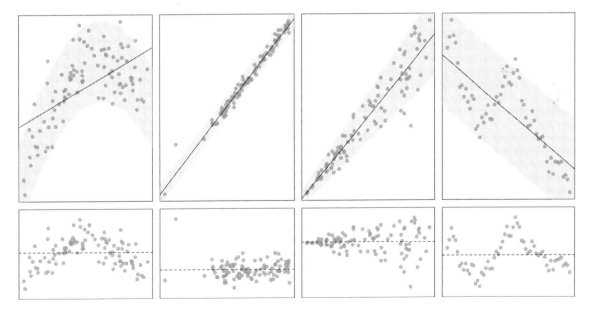

Figure 24.16: Four examples showing when the methods in this chapter are insufficient to apply to the data. The top set of graphs represents the x and y relationship. The bottom set of graphs is a residual plot. First panel: linearity fails. Second panel: there are outliers, most especially one point that is very far away from the line. Third panel: the variability of the errors is related to the value of x. Fourth panel: a time series dataset is shown, where successive observations are highly correlated.

GUIDED PRACTICE

Should we have concerns about applying least squares regression to the Elmhurst data in Figure 7.15?[3]

The technical conditions are often remembered using the **LINE** mnemonic. The linearity, normality, and equality of variance conditions usually can be assessed through residual plots, as seen in Figure 24.16. A careful consideration of the experimental design should be undertaken to confirm that the observed values are indeed independent.

- L: **linear** model
- I: **independent** observations
- N: points are **normally** distributed around the line
- E: **equal** variability around the line for all values of the explanatory variable

24.6.2 Why do we need technical conditions?

As with other inferential techniques we have covered in this text, if the technical conditions above don't hold, then it is not possible to make concluding claims about the population. That is, without the technical conditions, the T score (or Z score) will not have the assumed t-distribution (or standard normal z-distribution). That said, it is almost always impossible to check the conditions precisely, so we look for large deviations from the conditions. If there are large deviations, we will be unable to trust the calculated p-value or the endpoints of the resulting confidence interval.

The model based on Linearity

The linearity condition is among the most important if your goal is to understand a linear model between x and y. For example, the value of the slope will not be at all meaningful if the true relationship between x and y is quadratic, as in Figure 7.3. Not only should we be cautious about the inference, but the model *itself* is also not an accurate portrayal of the relationship between the variables.

In Section 25 we discuss model modifications that can often lead to an excellent fit of strong relationships other than linear ones. However, an extended discussion on the different methods for modeling functional forms other than linear is outside the scope of this text.

The importance of Independence

The technical condition describing the independence of the observations is often the most crucial but also the most difficult to diagnose. It is also extremely difficult to gather a dataset which is a true random sample from the population of interest. (Note: a true randomized experiment from a fixed set of individuals is much easier to implement, and indeed, randomized experiments are done in most medical studies these days.)

Dependent observations can bias results in ways that produce fundamentally flawed analyses. That is, if you hang out at the gym measuring height and weight, your linear model is surely not a representation of all students at your university. At best it is a model describing students who use the gym (but also who are willing to talk to you, that use the gym at the times you were there measuring, etc.).

In lieu of trying to answer whether or not your observations are a true random sample, you might instead focus on whether or not you believe your observations are representative of the populations. Humans are notoriously bad at implementing random procedures, so you should be wary of any process that used human intuition to balance the data with respect to, for example, the demographics of the individuals in the sample.

---

[3]The trend appears to be linear, the data fall around the line with no obvious outliers, the variance is roughly constant. These are also not time series observations. Least squares regression can be applied to these data.

24.6. CHECKING MODEL CONDITIONS

Some thoughts on Normality

The normality condition requires that points vary symmetrically around the line, spreading out in a bell-shaped fashion. You should consider the "bell" of the normal distribution as sitting on top of the line (coming off the paper in a 3-D sense) so as to indicate that the points are dense close to the line and disperse gradually as they get farther from the line.

The normality condition is less important than linearity or independence for a few reasons. First, the linear model fit with least squares will still be an unbiased estimate of the true population model. However, the standard errors associated with variability of the line will not be well estimated. Fortunately the Central Limit Theorem tells us that most of the analyses (e.g., SEs, p-values, confidence intervals) done using the mathematical model will still hold (even if the data are not normally distributed around the line) as long as the sample size is large enough. One analysis method that *does* require normality, regardless of sample size, is creating intervals which predict the response of individual outcomes at a given x value, using the linear model. One additional reason to worry slightly less about normality is that neither the randomization test nor the bootstrapping procedures require the data to be normal around the line.

Equal variability for prediction in particular

As with normality, the equal variability condition (that points are spread out in similar ways around the line for all values of x) will not cause problems for the estimate of the linear model. That said, the **inference** on the model (e.g., computing p-values) will be incorrect if the variability around the line is heterogeneous. Data that exhibit non-equal variance across the range of x-values will have the potential to seriously mis-estimate the variability of the slope which will have consequences for the inference results (i.e., hypothesis tests and confidence intervals).

The inference results for both a randomization test or a bootstrap confidence interval are robust to the equal variability condition, so they give the analyst methods to use when the data are heteroskedastic (that is, exhibit unequal variability around the regression line). Although randomization tests and bootstrapping allow us to analyze data using fewer conditions, some technical conditions are required for all methods described in this text (e.g., independent observation). When the equal variability condition is violated and a mathematical analysis (e.g., p-value from T score) is needed, there are other existing methods (outside the scope of this text) which can easily handle the unequal variance (e.g., weighted least squares analysis).

24.6.3 What if all the technical conditions are met?

When the technical conditions are met, the least squares regression model and inference is provided by virtually all statistical software. In addition to being ubiquitous, however, an additional advantage to the least squares regression model (and related inference) is that the linear model has important extensions (which are not trivial to implement with bootstrapping and randomization tests). In particular, random effects models, repeated measures, and interaction are all linear model extensions which require the above technical conditions. When the technical conditions hold, the extensions to the linear model can provide important insight into the data and research question at hand. We will discuss some of the extended modeling and associated inference in Chapter 25 and Section 26. Many of the techniques used to deal with technical condition violations are outside the scope of this text, but they are taught in universities in the very next class after this one. If you are working with linear models or curious to learn more, we recommend that you continue learning about statistical methods applicable to a larger class of datasets.

24.7 Chapter review

24.7.1 Summary

Recall that early in the text we presented graphical techniques which communicated relationships across multiple variables. We also used modeling to formalize the relationships. Many chapters were dedicated to inferential methods which allowed claims about the population to be made based on samples of data. Not only did we present the mathematical model for each of the inferential techniques, but when appropriate, we also presented bootstrapping and permutation methods.

Here in Chapter 24 we brought all of those ideas together by considering inferential claims on linear models through randomization tests, bootstrapping, and mathematical modeling. We continue to emphasize the importance of experimental design in making conclusions about research claims. In particular, recall that variability can come from different sources (e.g., random sampling vs. random allocation, see Figure 2.8).

24.7.2 Terms

We introduced the following terms in the chapter. If you're not sure what some of these terms mean, we recommend you go back in the text and review their definitions. We are purposefully presenting them in alphabetical order, instead of in order of appearance, so they will be a little more challenging to locate. However you should be able to easily spot them as **bolded text**.

bootstrap CI for the slope	randomization test for the slope	techinical conditions linear regression
inference with single precictor regression	t-distribution for slope	variability of the slope

24.8 Exercises

Answers to odd numbered exercises can be found in Appendix A.24.

24.1. **Body measurements, randomization test.** Researchers studying anthropometry collected body and skeletal diameter measurements, as well as age, weight, height and sex for 507 physically active individuals. A linear model is built to predict height based on shoulder girth (circumference of shoulders measured over deltoid muscles), both measured in centimeters.[4] (Heinz et al., 2003)

Below are two items. The first is the standard linear model output for predicting height from shoulder girth. The second is a histogram of slopes from 1,000 randomized datasets (1,000 times, hgt was permuted and regressed against sho_gi). The red vertical line is drawn at the observed slope value which was produced in the linear model output.

term	estimate	std.error	statistic	p.value
(Intercept)	105.832	3.27	32.3	<0.0001
sho_gi	0.604	0.03	20.0	<0.0001

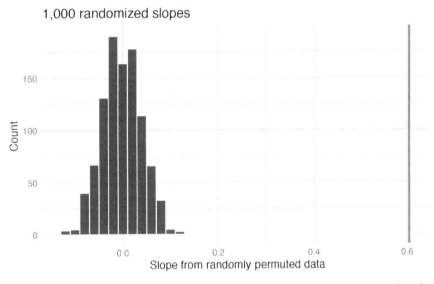

a. What are the null and alternative hypotheses for evaluating whether the slope of the model predicting height from shoulder girth is differen than 0.

b. Using the histogram which describes the distribution of slopes when the null hypothesis is true, find the p-value and conclude the hypothesis test in the context of the problem (use words like shoulder girth and height).

c. Is the conclusion based on the histogram of randomized slopes consistent with the conclusion which would have been obtained using the mathematical model? Explain.

---

[4]The bdims data used in this exercise can be found in the **openintro** R package.

24.2. **Body measurements, mathematical test.** The scatterplot and least squares summary below show the relationship between weight measured in kilograms and height measured in centimeters of 507 physically active individuals. (Heinz et al., 2003)

term	estimate	std.error	statistic	p.value
(Intercept)	-105.01	7.54	-13.9	<0.0001
hgt	1.02	0.04	23.1	<0.0001

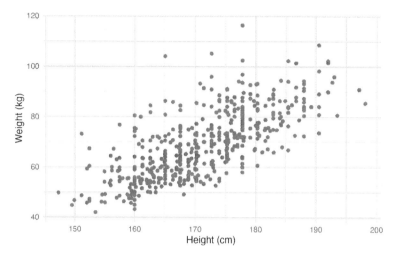

a. Describe the relationship between height and weight.

b. Write the equation of the regression line. Interpret the slope and intercept in context.

c. Do the data provide convincing evidence that the true slope parameter is different than 0? State the null and alternative hypotheses, report the p-value (using a mathematical model), and state your conclusion.

d. The correlation coefficient for height and weight is 0.72. Calculate R^2 and interpret it.

24.3. **Body measurements, bootstrap percentile interval.** In order to estimate the slope of the model predicting height based on shoulder girth (circumference of shoulders measured over deltoid muscles), 1,000 bootstrap samples are taken from a dataset of body measurements from 507 people. A linear model predicting height from shoulder girth is fit to each bootstrap sample, and the slope is estimated. A histogram of these slopes is shown below. (Heinz et al., 2003)

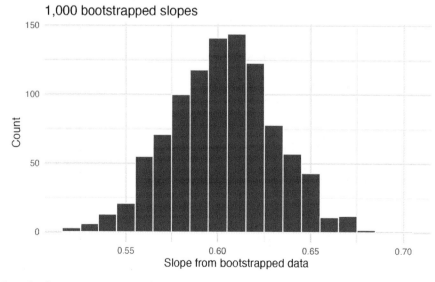

a. Using the bootstrap percentile method and the histogram above, find a 98% confidence interval for the slope parameter.

b. Interpret the confidence interval in the context of the problem.

24.4. **Body measurements, standard error bootstrap interval.** A linear model is built to predict height based on shoulder girth (circumference of shoulders measured over deltoid muscles), both measured in centimeters. (Heinz et al., 2003)

Below are two items. The first is the standard linear model output for predicting height from shoulder girth. The second is the bootstrap distribution of the slope statistic from 1,000 different bootstrap samples of the data.

term	estimate	std.error	statistic	p.value
(Intercept)	105.832	3.27	32.3	<0.0001
sho_gi	0.604	0.03	20.0	<0.0001

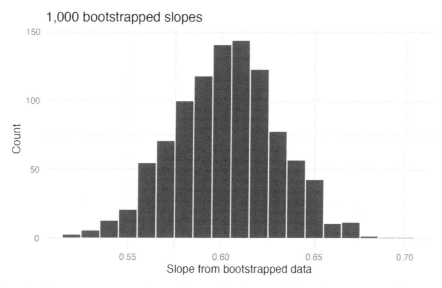

a. Using the histogram, approximate the standard error of the slope statistic (that is, quantify the variability of the slope statistic from sample to sample).

b. Find a 98% bootstrap SE confidence interval for the slope parameter.

c. Interpret the confidence interval in the context of the problem.

24.5. **Murders and poverty, randomization test.** The following regression output is for predicting annual murders per million (annual_murders_per_mil) from percentage living in poverty (perc_pov) in a random sample of 20 metropolitan areas.

Below are two items. The first is the standard linear model output for predicting annual murders per million from percentage living in poverty for metropolitan areas. The second is a histogram of slopes from 1000 randomized datasets (1000 times, annual_murders_per_mil was permuted and regressed against perc_pov). The red vertical line is drawn at the observed slope value which was produced in the linear model output.

term	estimate	std.error	statistic	p.value
(Intercept)	-29.90	7.79	-3.84	0.0012
perc_pov	2.56	0.39	6.56	<0.0001

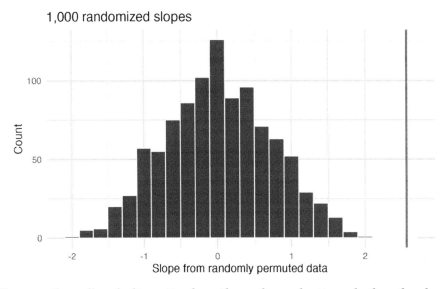

a. What are the null and alternative hypotheses for evaluating whether the slope of the model for predicting annual murder rate from poverty percentage is different than 0?

b. Using the histogram which describes the distribution of slopes when the null hypothesis is true, find the p-value and conclude the hypothesis test in the context of the problem (use words like murder rate and poverty).

c. Is the conclusion based on the histogram of randomized slopes consistent with the conclusion which would have been obtained using the mathematical model? Explain.

24.6. **Murders and poverty, mathematical test.** The table below shows the output of a linear model annual murders per million (`annual_murders_per_mil`) from percentage living in poverty (`perc_pov`) in a random sample of 20 metropolitan areas.

term	estimate	std.error	statistic	p.value
(Intercept)	-29.90	7.79	-3.84	0.0012
perc_pov	2.56	0.39	6.56	<0.0001

a. What are the hypotheses for evaluating whether the slope of the model predicting annual murder rate from poverty percentage is different than 0?

b. State the conclusion of the hypothesis test from part (a) in context of the data. What does this say about whether poverty percentage is a useful predictor of annual murder rate?

c. Calculate a 95% confidence interval for the slope of poverty percentage, and interpret it in context of the data.

d. Do your results from the hypothesis test and the confidence interval agree? Explain.

24.7. **Murders and poverty, bootstrap percentile interval.** Data on annual murders per million (`annual_murders_per_mil`) and percentage living in poverty (`perc_pov`) is collected from a random sample of 20 metropolitan areas. Using these data we want to estimate the slope of the model predicting `annual_murders_per_mil` from `perc_pov`. We take 1,000 bootstrap samples of the data and fit a linear model predicting `annual_murders_per_mil` from `perc_pov` to each bootstrap sample. A histogram of these slopes is shown below.

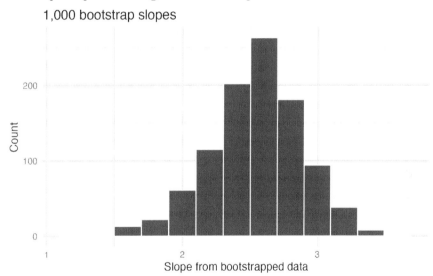

a. Using the percentile bootstrap method and the histogram above, find a 90% confidence interval for the slope parameter.

b. Interpret the confidence interval in the context of the problem.

24.8. **Murders and poverty, standard error bootstrap interval.** A linear model is built to predict annual murders per million (annual_murders_per_mil) from percentage living in poverty (perc_pov) in a random sample of 20 metropolitan areas.

Below are two items. The first is the standard linear model output for predicting annual murders per million from percentage living in poverty for metropolitan areas. The second is the bootstrap distribution of the slope statistic from 1000 different bootstrap samples of the data.

term	estimate	std.error	statistic	p.value
(Intercept)	-29.90	7.79	-3.84	0.0012
perc_pov	2.56	0.39	6.56	<0.0001

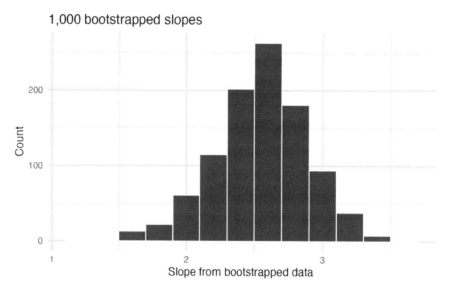

a. Using the histogram, approximate the standard error of the slope statistic (that is, quantify the variability of the slope statistic from sample to sample).

b. Find a 90% bootstrap SE confidence interval for the slope parameter.

c. Interpret the confidence interval in the context of the problem.

24.9. **Baby's weight and father's age, randomization test.** US Department of Health and Human Services, Centers for Disease Control and Prevention collect information on births recorded in the country. The data used here are a random sample of 1000 births from 2014. Here, we study the relationship between the father's age and the weight of the baby.[5] (ICPSR, 2014)

Below are two items. The first is the standard linear model output for predicting baby's weight (in pounds) from father's age (in years). The second is a histogram of slopes from 1000 randomized datasets (1000 times, `weight` was permuted and regressed against `fage`). The red vertical line is drawn at the observed slope value which was produced in the linear model output.

term	estimate	std.error	statistic	p.value
(Intercept)	7.101	0.199	35.674	<0.0001
fage	0.005	0.006	0.757	0.4495

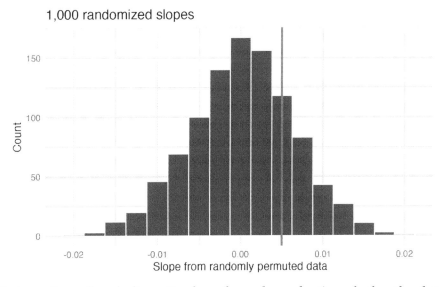

a. What are the null and alternative hypotheses for evaluating whether the slope of the model for predicting baby's weight from father's age is different than 0?

b. Using the histogram which describes the distribution of slopes when the null hypothesis is true, find the p-value and conclude the hypothesis test in the context of the problem (use words like father's age and weight of baby). What does the conclusion of your test say about whether the father's age is a useful predictor of baby's weight?

c. Is the conclusion based on the histogram of randomized slopes consistent with the conclusion which would have been obtained using the mathematical model? Explain.

---

[5]The `births14` data used in this exercise can be found in the **openintro** R package.

24.10. **Baby's weight and father's age, mathematical test.** Is the father's age useful in predicting the baby's weight? The scatterplot and least squares summary below show the relationship between baby's weight (measured in pounds) and father's age for a random sample of babies. (ICPSR, 2014)

term	estimate	std.error	statistic	p.value
(Intercept)	7.1042	0.1936	36.698	<0.0001
fage	0.0047	0.0061	0.779	0.4359

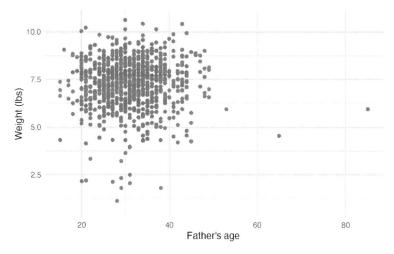

a. What is the predicted weight of a baby whose father is 30 years old.

b. Do the data provide convincing evidence that the model for predicting baby weights from father's age has a slope different than 0? State the null and alternative hypotheses, report the p-value (using a mathematical model), and state your conclusion.

c. Based on your conclusion, is father's age a useful predictor of baby's weight?

24.11. **Baby's weight and father's age, bootstrap percentile interval.** US Department of Health and Human Services, Centers for Disease Control and Prevention collect information on births recorded in the country. The data used here are a random sample of 1000 births from 2014. Here, we study the relationship between the father's age and the weight of the baby. Below is the bootstrap distribution of the slope statistic from 1,000 different bootstrap samples of the data. (ICPSR, 2014)

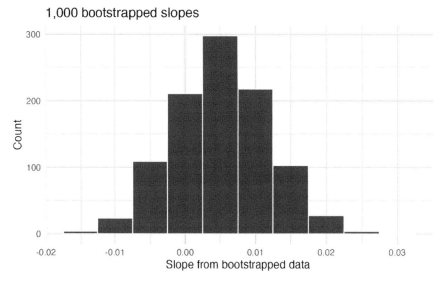

a. Using the bootstrap percentile method and the histogram above, find a 95% confidence interval for the slope parameter.

b. Interpret the confidence interval in the context of the problem.

24.12. **Baby's weight and father's age, standard error bootstrap interval.** US Department of Health and Human Services, Centers for Disease Control and Prevention collect information on births recorded in the country. The data used here are a random sample of 1000 births from 2014. Here, we study the relationship between the father's age and the weight of the baby. (ICPSR, 2014)

Below are two items. The first is the standard linear model output for predicting baby's weight (in pounds) from father's age (in years). The second is the bootstrap distribution of the slope statistic from 1000 different bootstrap samples of the data.

term	estimate	std.error	statistic	p.value
(Intercept)	7.101	0.199	35.674	<0.0001
fage	0.005	0.006	0.757	0.4495

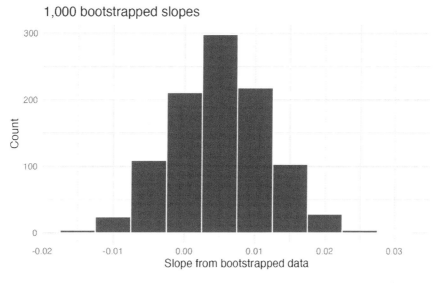

a. Using the histogram, approximate the standard error of the slope statistic (that is, quantify the variability of the slope statistic from sample to sample).

b. Find a 95% bootstrap SE confidence interval for the slope parameter.

c. Interpret the confidence interval in the context of the problem.

24.13. **I heart cats.** Researchers collected data on heart and body weights of 144 domestic adult cats. The table below shows the output of a linear model predicting heat weight (measured in grams) from body weight (measured in kilograms) of these cats.[6]

term	estimate	std.error	statistic	p.value
(Intercept)	-0.357	0.692	-0.515	0.6072
Bwt	4.034	0.250	16.119	<0.0001

a. What are the hypotheses for evaluating whether body weight is positively associated with heart weight in cats?

b. State the conclusion of the hypothesis test from part (a) in context of the data.

c. Calculate a 95% confidence interval for the slope of body weight, and interpret it in context of the data.

d. Do your results from the hypothesis test and the confidence interval agree? Explain.

---

[6] The cats data used in this exercise can be found in the **MASS** R package.

24.14. **Beer and blood alcohol content** Many people believe that weight, drinking habits, and many other factors are much more important in predicting blood alcohol content (BAC) than simply considering the number of drinks a person consumed. Here we examine data from sixteen student volunteers at Ohio State University who each drank a randomly assigned number of cans of beer. These students were evenly divided between men and women, and they differed in weight and drinking habits. Thirty minutes later, a police officer measured their blood alcohol content (BAC) in grams of alcohol per deciliter of blood. The scatterplot and regression table summarize the findings. [7] (?)

term	estimate	std.error	statistic	p.value
(Intercept)	-0.0127	0.0126	-1.00	0.332
beers	0.0180	0.0024	7.48	<0.0001

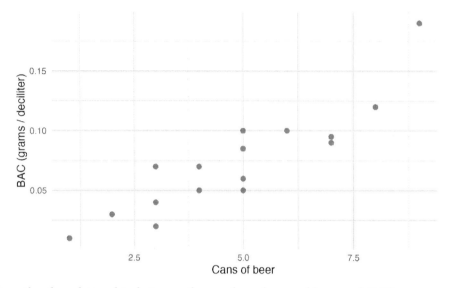

a. Describe the relationship between the number of cans of beer and BAC.

b. Write the equation of the regression line. Interpret the slope and intercept in context.

c. Do the data provide convincing evidence that drinking more cans of beer is associated with an increase in blood alcohol? State the null and alternative hypotheses, report the p-value, and state your conclusion.

d. The correlation coefficient for number of cans of beer and BAC is 0.89. Calculate R^2 and interpret it in context.

e. Suppose we visit a bar, ask people how many drinks they have had, and also take their BAC. Do you think the relationship between number of drinks and BAC would be as strong as the relationship found in the Ohio State study?

---

[7] The bac data used in this exercise can be found in the **openintro** R package.

24.15. **Urban homeowners, conditions.** The scatterplot below shows the percent of families who own their home vs. the percent of the population living in urban areas. (Bureau, 2010) There are 52 observations, each corresponding to a state in the US. Puerto Rico and District of Columbia are also included.

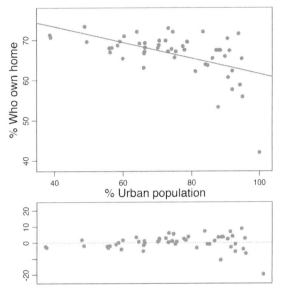

a. For these data, R^2 is 29.16%. What is the value of the correlation coefficient? How can you tell if it is positive or negative?

b. Examine the residual plot. What do you observe? Is a simple least squares fit appropriate for these data? Which of the LINE conditions are met or not met?

Chapter 25

Inference for linear regression with multiple predictors

In Chapter 8, the least squares regression method was used to estimate linear models which predicted a particular response variable given more than one explanatory variable. Here, we discuss whether each of the variables individually is a statistically significant predictor of the outcome or whether the model might be just as strong without that variable. That is, as before, we apply inferential methods to ask whether a variable could have come from a population where the particular coefficient at hand was zero. If one of the linear model coefficients is truly zero (in the population), then the estimate of the coefficient (using least squares) will vary around zero. The inference task at hand is to decide whether the coefficient's difference from zero is large enough to decide that the data cannot possibly have come from a model where the true population coefficient is zero. Both the derivations from the mathematical model and the randomization model are beyond the scope of this book, but we are able to calculate p-values using statistical software. We will discuss interpreting p-values in the multiple regression setting and note some scenarios where careful understanding of the context and the relationship between variables is important. We use cross-validation as a method for independent assessment of the multiple linear regression model.

25.1 Multiple regression output from software

Recall the `loans` data from Chapter 8.

The `loans_full_schema` data can be found in the **openintro** R package. Based on the data in this dataset we have created two new variables: `credit_util` which is calculated as the total credit utilized divided by the total credit limit and `bankruptcy` which turns the number of bankruptcies to an indicator variable (0 for no bankruptcies and 1 for at least 1 bankruptcies). We will refer to this modified dataset as `loans`.

25.1. MULTIPLE REGRESSION OUTPUT FROM SOFTWARE

Now, our goal is to create a model where `interest_rate` can be predicted using the variables `debt_to_income`, `term`, and `credit_checks`. As you learned in Chapter 8, least squares can be used to find the coefficient estimates for the linear model. The unknown population model can be written as:

$$E[\texttt{interest\_rate}] = \beta_0 + \beta_1 \times \texttt{debt\_to\_income}$$
$$+ \beta_2 \times \texttt{term}$$
$$+ \beta_3 \times \texttt{credit\_checks}$$

Table 25.1: Summary of a linear model for predicting interest rate based on `debt_to_income`, `term`, and `credit_checks`. Each of the variables has its own coefficient estimate as well as p-value significance.

term	estimate	std.error	statistic	p.value
(Intercept)	4.31	0.20	22.1	<0.0001
debt_to_income	0.04	0.00	13.3	<0.0001
term	0.16	0.00	37.9	<0.0001
credit_checks	0.25	0.02	12.8	<0.0001

The estimated equation for the regression model may be written as a model with three predictor variables:

$$\widehat{\texttt{interest\_rate}} = 4.31 + 0.041 \times \texttt{debt\_to\_income}$$
$$+ 0.16 \times \texttt{term}$$
$$+ 0.25 \times \texttt{credit\_checks}$$

Not only does Table 25.1 provide the estimates for the coefficients, it also provides information on the inference analysis (i.e., hypothesis testing) which are the focus of this chapter.

In Section 24, you learned that the hypothesis test for a linear model with **one predictor**[1] can be written as:

if only one predictor, $H_0 : \beta_1 = 0$.

That is, if the true population slope is zero, the p-value measures how likely it would be to select data which produced the observed slope (b_1) value.

With **multiple predictors**, the hypothesis is similar, however, it is now conditioned on each of the other variables remaining in the model.

if multiple predictors, $H_0 : \beta_i = 0$ given other variables in the model

Using the example above and focusing on each of the variable p-values (here we won't discuss the p-value associated with the intercept), we can write out the three different hypotheses:

- $H_0 : \beta_1 = 0$, given `term` and `credit_checks` are included in the model
- $H_0 : \beta_2 = 0$, given `debt_to_income` and `credit_checks` are included in the model
- $H_0 : \beta_3 = 0$, given `debt_to_income` and `term` are included in the model

The very low p-values from the software output tell us that each of the variables acts as an important predictor in the model, despite the inclusion of the other two. Consider the p-value on $H_0 : \beta_1 = 0$. The low p-value says that it would be extremely unlikely to see data that produce a coefficient on `debt_to_income` as large as 0.041 if the true relationship between `debt_to_income`and `interest_-rate` was non-existent (i.e., if $\beta_1 = 0$) and the model also included `term` and `credit_checks`. You might have thought that the value 0.041 is a small number (i.e., close to zero), but in the units of

---

[1]In previous sections, the term **explanatory variable** was used instead of **predictor**. The words are synonymous and are used separately in the different sections to be consistent with how most analysts use them: explanatory variable for testing, predictor for modeling.

the problem, 0.041 turns out to be far away from zero, it's all about context! The p-values on `term` and on `credit_checks` are interpreted similarly.

Sometimes a set of predictor variables can impact the model in unusual ways, often due to the predictor variables themselves being correlated.

25.2 Multicollinearity

In practice, there will almost always be some degree of correlation between the explanatory variables in a multiple regression model. For regression models, it is important to understand the entire context of the model, particularly for correlated variables. Our discussion will focus on interpreting coefficients (and their signs) in relationship to other variables as well as the significance (i.e., the p-value) of each coefficient.

Consider an example where we'd like to predict how much money is in a coin dish based only on the number of coins in the dish. We ask 26 students to tell us about their individual coin dishes, collecting data on the total dollar amount, the total number of coins, and the total number of low coins.[2] The number of low coins is the number of coins minus the number of quarters (a quarter is the largest commonly used US coin, at US$0.25). Figure 25.1 illustrates a sample of U.S. coins, their total worth (`total_amount`), the total `number of coins`, and the `number of low coins`.

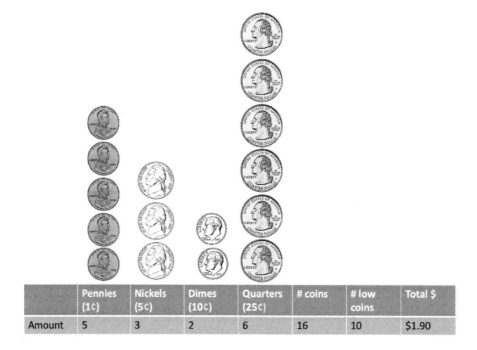

Figure 25.1: A sample of coins with 16 total coins, 10 low coins, and a net worth of $1.90.

The collected data is given in Figure 25.2 and shows that the `total_amount` of money is more highly correlated with the total `number of coins` than it is with the `number of low coins`. We also note that the total `number of coins` and the `number of low coins` are positively correlated.

Using the total `number of coins` as the predictor variable, Table 25.2 provides the least squares estimate of the coefficient is 0.13. For every additional coin in the dish, we would predict that the student had US$0.13 more. The $b_1 = 0.13$ coefficient has a small p-value associated with it, suggesting we would not have seen data like this if `number of coins` and `total_amount` of money were not linearly related.

$$\widehat{\texttt{total\_amount}} = 0.55 + 0.13 \times \texttt{number\_of\_coins}$$

---

[2]In all honesty, this particular dataset is fabricated, and the original idea for the problem comes from Jeff Witmer at Oberlin College.

25.2. MULTICOLLINEARITY

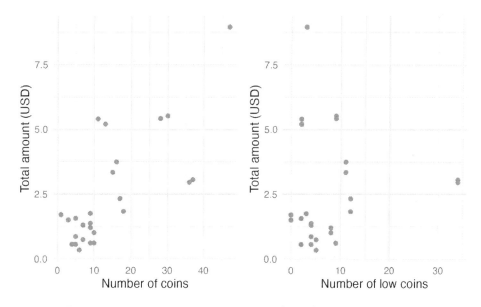

Figure 25.2: Plot describing the total amount of money (USD) as a function of the number of coins and the number of low coins. As you might expect, the total amount of money is more highly postively correlated with the total number of coins than with the number of low coins.

Table 25.2: Linear model output predicting the total amount of money based on the total number of coins.

term	estimate	std.error	statistic	p.value
(Intercept)	0.55	0.44	1.23	0.2301
number_of_coins	0.13	0.02	5.54	<0.0001

Using the `number of low coins` as the predictor variable, Table 25.3 provides the least squares estimate of the coefficient is 0.02. For every additional low coin in the dish, we would predict that the student had US$0.02 more. The $b_1 = 0.02$ coefficient has a large p-value associated with it, suggesting we could easily have seen data like ours even if the `number of low coins` and `total_amount` of money are not at all linearly related.

$$\widehat{\texttt{total\_amount}} = 2.28 + 0.02 \times \texttt{number\_of\_low\_coins}$$

Table 25.3: Linear model output predicting the total amount of money based on the number of low coins.

term	estimate	std.error	statistic	p.value
(Intercept)	2.28	0.58	3.9	0.0
number_of_low_coins	0.02	0.05	0.4	0.7

EXAMPLE

Come up with an example of two observations that have the same number of low coins but the number of total coins differs by one. What is the difference in total amount?

Two samples of coins with the same number of low coins (3), but a different number of total coins (4 vs 5) and a different total total amount ($0.41 vs $0.66).

 EXAMPLE

Come up with an example of two observations that have the same total number of coins but a different number of low coins. What is the difference in total amount?

Two samples of coins with the same total number of coins (4), but a different number of low coins (3 vs 4) and a different total total amount ($0.41 vs $0.17).

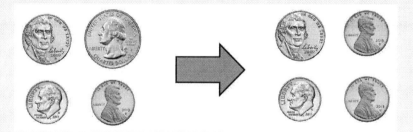

Using both the total `number of coins` and the `number of low coins` as predictor variables, Table 25.4 provides the least squares estimates of both coefficients as 0.21 and -0.16. Now, with two variables in the model, the interpretation is more nuanced.

- The coefficient indicates a change in one variable while keeping the other variable constant. For every additional coin in the dish **while** the `number of low coins` stays constant, we would predict that the student had US$0.21 more. Re-considering the phrase "every additional coin in the dish **while** the number of low coins stays constant" makes us realize that each increase is a single additional quarter (larger samples sizes would have led to a b_1 coefficient closer to 0.25 because of the deterministic relationship described here).

- For every additional low coin in the dish **while** the total `number of coins` stays constant, we would predict that the student had US$0.16 less. Re-considering the phrase "every additional low coin in the dish **while** the number of total coins stays constant" makes us realize that a quarter is being swapped out for a penny, nickel, or dime.

25.3 Cross-validation for prediction error

Considering the coefficients across Tables 25.2, 25.3, and 25.4 within the context and knowledge we have of US coins allows us to understand the correlation between variables and why the signs of the coefficients would change depending on the model. Note also, however, that the p-value for the `number of low coins` coefficient changed from Table 25.3 to Table 25.4. It makes sense that the variable describing the `number of low coins` provides more information about the `total_amount` of money when it is part of a model which also includes the total `number of coins` than it does when it is used as a single variable in a simple linear regression model.

$$\widehat{\texttt{total\_amount}} = 0.80 + 0.21 \times \texttt{number\_of\_coins} - 0.16 \times \texttt{number of low coins}$$

Table 25.4: Linear model output predicting the total amount of money based on both the total number of coins and the number of low coins.

term	estimate	std.error	statistic	p.value
(Intercept)	0.80	0.30	2.65	0.0142
number_of_coins	0.21	0.02	9.89	<0.0001
number_of_low_coins	-0.16	0.03	-5.51	<0.0001

When working with multiple regression models, interpreting the model coefficient is mot always as straightforward as it was with the coin example. However, we encourage you to always think carefully about the variables in the model, consider how they might be correlated among themselves, and work through different models to see how using different sets of variables might produce different relationships for predicting the response variable of interest.

Multicollinearity.

Multicollinearity happens when the predictor variables are correlated within themselves. When the predictor variables themselves are correlated, the coefficients in a multiple regression model can be difficult to interpret.

Although diving into the details are beyond the scope of this text, we will provide one more reflection about multicollinearity. If the predictor variables have some degree of correlation, it can be quite difficult to interpret the value of the coefficient or evaluate whether the variable is a statistically significant predictor of the outcome. However, even a model that suffers from high multicollinearity will likely lead to unbiased predictions of the response variable. So if the task at hand is only to do prediction, multicollinearity is likely to not cause you substantial problems.

25.3 Cross-validation for prediction error

In Section 25.1, p-values were calculated on each of the model coefficients. The p-value gives a sense of which variables are important to the model; however, a more extensive treatment of variable selection is warranted in a follow-up course or textbook. Here, we use cross-validation prediction error to focus on which variable(s) are important for predicting the response variable of interest. In general, linear models are also used to make predictions of individual observations. In addition to model building, cross-validation provides a method for generating predictions that are not overfit to the particular dataset at hand. We continue to encourage you to take up further study on the topic of cross-validation, as it is among the most important ideas in modern data analysis, and we are only able to scratch the surface here.

Cross-validation is a computational technique which removes some observations before a model is run, then assesses the model accuracy on the held-out sample. By removing some observations, we provide ourselves with an independent evaluation of the model (that is, the removed observations do not contribute to finding the parameters which minimize the least squares equation). Cross-validation can be used in many different ways (as an independent assessment), and here we will just scratch the surface with respect to one way the technique can be used to compare models. See Figure 25.3 for a visual representation of the cross-validation process.

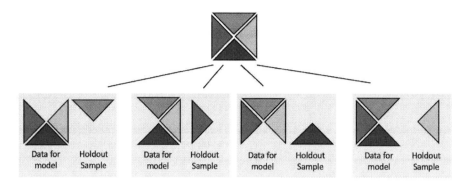

Figure 25.3: The dataset is broken into k folds (here k = 4). One at a time, a model is built using k-1 of the folds, and predictions are calculated on the single held out sample which will be completely independent of the model estimation.

 The penguins data can be found in the **palmerpenguings** R package.

Our goal in this section is to compare two different regression models which both seek to predict the mass of an individual penguin in grams. The observations of three different penguin species include measurements on body size and sex. The data were collected by Dr. Kristen Gorman and the Palmer Station, Antarctica LTER as part of the Long Term Ecological Research Network. (Gorman et al., 2014b) Although not exactly aligned with this research project, you might be able to imagine a setting where the dimensions of the penguin are known (through, for example, aerial photographs) but the mass is not known. The first model will predict `body_mass_g` by using only the `bill_length_mm`, a variable denoting the length of a penguin's bill, in mm. The second model will predict `body_mass_g` by using `bill_length_mm`, `bill_depth_mm`, `flipper_length_mm`, `sex`, and `species`.

 Prediction error.

The predicted error (also previously called the **residual**) is the difference between the observed value and the predicted value (from the regression model).

$$\text{prediction error}_i = e_i = y_i - \hat{y}_i$$

The presentation below (see the comparison of Figures 25.5 and 25.7) shows that the model with more variables predicts `body_mass_g` with much smaller errors (predicted minus actual body mass) than the model which uses only `bill_length_g`. We have deliberately used a model that intuitively makes sense (the more body measurements, the more predictable mass is). However, in many settings, it is not obvious which variables or which models contribute most to accurate predictions. Cross-validation is one way to get accurate independent predictions with which to compare different models.

25.3.1 Comparing two models to predict body mass in penguins

The question we will seek to answer is whether the predictions of `body_mass_g` are substantially better when `bill_length_mm`, `bill_depth_mm`, `flipper_length_mm`, `sex`, and `species` are used in the model, as compared with a model on `bill_length_mm` only.

25.3. CROSS-VALIDATION FOR PREDICTION ERROR

We refer to the model given with only `bill_lengh_mm` as the **smaller** model. It is seen in Table 25.5 with coefficient estimates of the parameters as well as standard errors and p-values. We refer to the model given with `bill_lengh_mm`, `bill_depth_mm`, `flipper_length_mm`, `sex`, and `species` as the **larger** model. It is seen in Table 25.6 with coefficient estimates of the parameters as well as standard errors and p-values. Given what we know about high correlations between body measurements, it is somewhat unsurprising that all of the variables have low p-values, suggesting that each variable is a statistically significant predictor of `body_mass_g`, given all other variables in the model. However, in this section, we will go beyond the use of p-values to consider independent predictions of `body_mass_g` as a way to compare the smaller and larger models.

The smaller model:

$$E[\texttt{body\_mass\_g}] = \beta_0 + \beta_1 \times \texttt{bill\_length\_mm}$$
$$\widehat{\texttt{body\_mass\_g}} = 362.31 + 87.42 \times \texttt{bill\_length\_mm}$$

Table 25.5: The smalller model: least squares estimates of the regression model predicting `body_mass_g` from `bill_length_mm`.

term	estimate	std.error	statistic	p.value
(Intercept)	362.3	283.4	1.28	0.2019
bill_length_mm	87.4	6.4	13.65	<0.0001

The larger model:

$$\begin{aligned}E[\texttt{body\_mass\_g}] = \beta_0 &+ \beta_1 \times \texttt{bill\_length\_mm} \\ &+ \beta_2 \times \texttt{bill\_depth\_mm} \\ &+ \beta_3 \times \texttt{flipper\_length\_mm} \\ &+ \beta_4 \times \texttt{sex}_{male} \\ &+ \beta_5 \times \texttt{species}_{Chinstrap} \\ &+ \beta_6 \times \texttt{species}_{Gentoo} \end{aligned}$$

$$\begin{aligned}\widehat{\texttt{body\_mass\_g}} = -1460.99 &+ 18.20 \times \texttt{bill\_length\_mm} \\ &+ 67.22 \times \texttt{bill\_depth\_mm} \\ &+ 15.95 \times \texttt{flipper\_length\_mm} \\ &+ 389.89 \times \texttt{sex}_{male} \\ &- 251.48 \times \texttt{species}_{Chinstrap} \\ &+ 1014.63 \times \texttt{species}_{Gentoo} \end{aligned}$$

Table 25.6: The larger model: least squares estimates of the regression model predicting `body_mass_g` from `bill_length_mm`, `bill_depth_mm`, `flipper_length_mm`, `sex`, and `species`.

term	estimate	std.error	statistic	p.value
(Intercept)	-1461.0	571.31	-2.56	0.011
bill_length_mm	18.2	7.11	2.56	0.0109
bill_depth_mm	67.2	19.74	3.40	7e-04
flipper_length_mm	15.9	2.91	5.48	<0.0001
sexmale	389.9	47.85	8.15	<0.0001
speciesChinstrap	-251.5	81.08	-3.10	0.0021
speciesGentoo	1014.6	129.56	7.83	<0.0001

In order to compare the smaller and larger models in terms of their **ability to predict penguin mass**, we need to build models that can provide independent predictions based on the penguins in the holdout samples created by cross-validation. To reiterate, each of the predictions that (when combined together) will allow us to distinguish between the smaller and larger are independent of the data which were used to build the model. In this example, using cross-validation, we remove one quarter of the data before running the least squares calculations. Then the least squares model is used to predict the `body_mass_g` of the penguins in the holdout sample. Here we use a 4-fold cross-validation (meaning that one quarter of the data is removed each time) to produce four different versions of each model (other times it might be more appropriate to use 2-fold or 10-fold or even run the model separately after removing each individual data point one at a time).

Figure 25.4 displays how a model is fit to 3/4 of the data (note the slight differences in coefficients as compared to Table 25.5), and then predictions are made on the holdout sample.

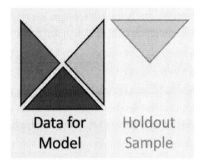
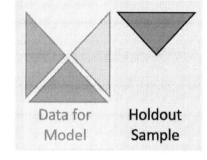
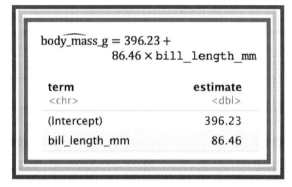
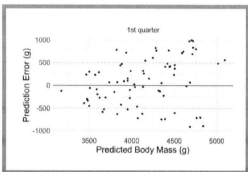

Figure 25.4: The coefficients are estimated using the least squares model on 3/4 of the dataset with only a single predictor variable. Predictions are made on the remaining 1/4 of the observations. The y-axis in the scatterplot represents the residual: true observed value minus the predicted value. Note that the predictions are independent of the estimated model coefficients.

By repeating the process for each holdout quarter sample, the residuals from the model can be plotted against the predicted values. We see that the predictions are scattered which shows a good model fit but that the prediction errors vary ± 1000g of the true body mass.

The cross-validation SSE is the sum of squared error associated with the predictions. Let $\hat{y}_{cv,i}$ be the prediction for the ith observation where the ith observation was in the hold-out fold and the other three folds were used to create the linear model. For the model using only `bill_length_mm` to predict `body_mass_g`, the CV SSE is 141,552,822.

Cross-validation SSE.

The prediction error from the cross-validated model can be used to calculate a single numerical summary of the model. The cross-validation SSE is the sum of squared cross-validation prediction errors.

$$\text{CV SSE} = \sum_{i=1}^{n}(\hat{y}_{cv,i} - y_i)^2$$

25.3. CROSS-VALIDATION FOR PREDICTION ERROR

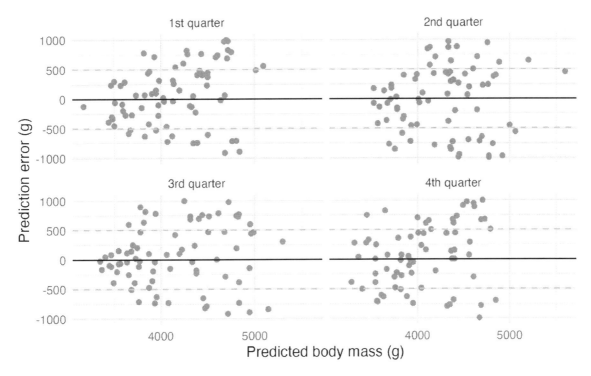

Figure 25.5: One quarter at a time, the data were removed from the model building, and the body mass of the removed penguins was predicted. The least squares regression model was fit independently of the removed penguins. The predictions of body mass are based on bill length only. The x-axis represents the predicted value, the y-axis represents the error (difference between predicted value and actual value).

The same process is repeated for the larger number of explanatory variables. Note that the coefficients estimated for the first cross-validation model (in Figure 25.6) are slightly different from the estimates computed on the entire dataset (seen in Table 25.6). Figure 25.6 displays the cross-validation process for the multivariable model with a full set of residual plots given in Figure 25.7. Note that the residuals are mostly within ± 500g, providing much more precise predictions for the independent body mass values of the individual penguins.

Figure 25.5 shows that the independent predictions are centered around the true values (i.e., errors are centered around zero), but that the predictions can be as much as 1000g off when using only `bill_length_mm` to predict `body_mass_g`. On the other hand, when using `bill_length_mm`, `bill_depth_mm`, `flipper_length_mm`, `sex`, and `species` to predict `body_mass_g`, the prediction errors seem to be about half as big, as seen in Figure 25.7. For the model using `bill_length_mm`, `bill_depth_mm`, `flipper_length_mm`, `sex`, and `species` to predict `body_mass_g`, the CV SSE is 27,728,698. Consistent with visually comparing the two sets of residual plots, the sum of squared prediction errors is smaller for the model which uses more predictor variables. The model with more predictor variables seems like the better model (according to the cross-validated prediction errors criteria).

We have provided a very brief overview to and example using cross-validation. Cross-validation is a computational approach to model building and model validation as an alternative to reliance on p-values. While p-values have a role to play in understanding model coefficients, throughout this text, we have continued to present computational methods that broaden statistical approaches to data analysis. Cross-validation will be used again in Section 26 with logistic regression. We encourage you to consider both standard inferential methods (such as p-values) and computational approaches (such as cross-validation) as you build and use multivariable models of all varieties.

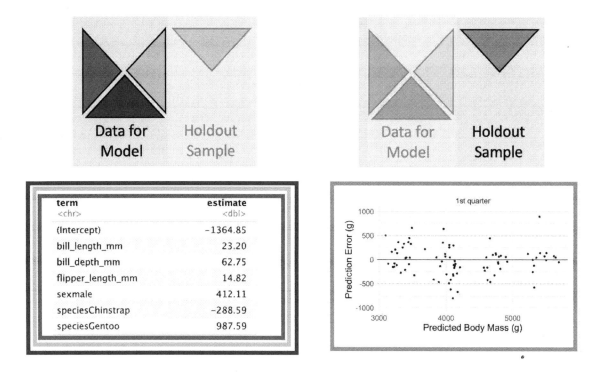

Figure 25.6: The coefficients are estimated using the least squares model on 3/4 of the dataset with the five specified predictor variables. Predictions are made on the remaining 1/4 of the observations. The y-axis in the scatterplot represents the residual: true observed value minus the predicted value. Note that the predictions are independent of the estimated model coefficients.

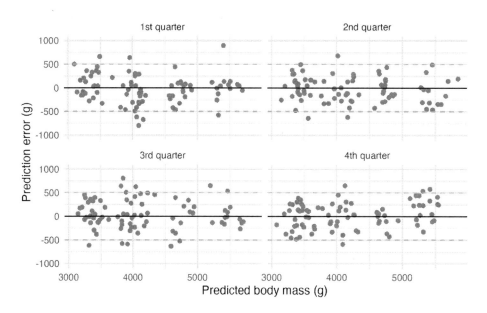

Figure 25.7: One quarter at a time, the data were removed from the model building, and the body mass of the removed penguins was predicted. The least squares regression model was fit independently of the removed penguins. The predictions of body mass are based on bill length only. The x-axis represents the predicted value, the y-axis represents the error (difference between predicted value and actual value).

25.4 Chapter review

25.4.1 Summary

Building on the modeling ideas from Chapter 8, we have now introduced methods for evaluating coefficients (based on p-values) and evaluating models (cross-validation). There are many important aspects to consider when working with multiple variables in a single model, and we have only glanced at a few topics. Remember, multicollinearity can make coefficient interpretation difficult. A topic not covered in this text but important for multiple regression models is interaction, and we hope that you learn more about how variables work together as you continue to build up your modeling skills.

25.4.2 Terms

We introduced the following terms in the chapter. If you're not sure what some of these terms mean, we recommend you go back in the text and review their definitions. We are purposefully presenting them in alphabetical order, instead of in order of appearance, so they will be a little more challenging to locate. However you should be able to easily spot them as **bolded text**.

cross-validation	multicollinearity	prediction error
inference on multiple linear regression	multiple predictors	predictor

25.5 Exercises

Answers to odd numbered exercises can be found in Appendix A.25.

25.1. **GPA, mathematical interval.** A survey of 55 Duke University students asked about their GPA (gpa), number of hours they study weekly (studyweek), number of hours they sleep nightly (sleepnight), and whether they go out more than two nights a week (out_mt2). We use these data to build a model predicting GPA from the other variables. Summary of the model is shown below. Note that out_mt2 is 1 if the student goes out more than two nights a week, and 0 otherwise.[3]

term	estimate	std.error	statistic	p.value
(Intercept)	3.508	0.347	10.114	<0.0001
studyweek	0.002	0.004	0.400	0.6908
sleepnight	0.000	0.047	0.008	0.994
out_mt2	0.151	0.097	1.551	0.127

a. Calculate a 95% confidence interval for the coefficient of out_mt2 (go out more than two night a week) in the model, and interpret it in the context of the data.

2. Would you expect a 95% confidence interval for the slope of the remaining variables to include 0? Explain.

25.2. **GPA, collinear predictors.** In this exercise we work with data from a survey of 55 Duke University students who were asked about their GPA, number of hours they sleep nightly, and number of nights they go out each week.

The plots below describe the show the distribution of each of these variables (on the diagonal) as well as provide information on the pairwise correlations between them.

Also provided below are three regression model outputs: gpa vs. out, gpa vs. sleepnight, and gpa vs. out + sleepnight.

term	estimate	std.error	statistic	p.value
(Intercept)	3.504	0.106	33.011	<0.0001
out	0.045	0.046	0.998	0.3229

term	estimate	std.error	statistic	p.value
(Intercept)	3.46	0.318	10.874	<0.0001
sleepnight	0.02	0.045	0.445	0.6583

term	estimate	std.error	statistic	p.value
(Intercept)	3.483	0.320	10.888	<0.0001
out	0.044	0.050	0.886	0.3796
sleepnight	0.003	0.048	0.072	0.9432

See next page for the plots and parts a to c.

---

[3]The gpa data used in this exercise can be found in the **openintro** R package.

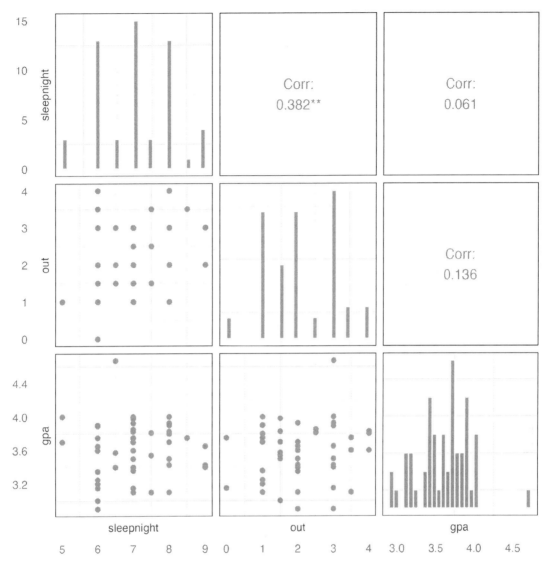

a. There are three variables described in the figure, and each is paired with each other to create three different scatterplots. Rate the pairwise relationships from most correlated to least correlated.

b. When using only one variable to model gpa, is out a significant predictor variable? Is sleepnight a significant predictor variable? Explain.

c. When using both out and sleepnight to predict gpa in a multiple regression model, are either of the variables significant? Explain.

25.3. **Tourism spending.** The Association of Turkish Travel Agencies reports the number of foreign tourists visiting Turkey and tourist spending by year. Three plots are provided: scatterplot showing the relationship between these two variables along with the least squares fit, residuals plot, and histogram of residuals.[4]

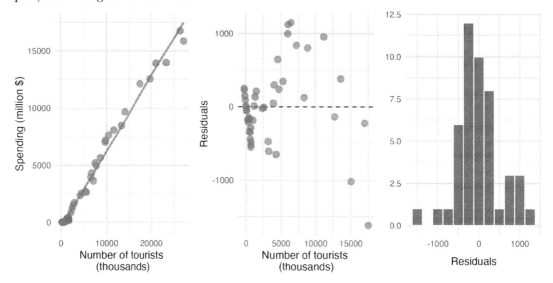

a. Describe the relationship between number of tourists and spending.

b. What are the predictor and the outcome variables?

c. Why might we want to fit a regression line to these data?

d. Do the data meet the LINE conditions required for fitting a least squares line? In addition to the scatterplot, use the residual plot and histogram to answer this question.

---

[4]The `tourism` data used in this exercise can be found in the **openintro** R package.

25.4. **Cherry trees with collinear predictors.** Timber yield is approximately equal to the volume of a tree, however, this value is difficult to measure without first cutting the tree down. Instead, other variables, such as height and diameter, may be used to predict a tree's volume and yield. Researchers wanting to understand the relationship between these variables for black cherry trees collected data from 31 such trees in the Allegheny National Forest, Pennsylvania. Height is measured in feet, diameter in inches (at 54 inches above ground), and volume in cubic feet.[5] (Hand, 1994)

The plots below describe the show the distribution of each of these variables (on the diagonal) as well as provide information on the pairwise correlations between them.

Also provided below are three regression model outputs: volume vs. diam, volume vs. height, and volume vs. height + diam.

term	estimate	std.error	statistic	p.value
(Intercept)	-36.94	3.365	-11.0	<0.0001
diam	5.07	0.247	20.5	<0.0001

term	estimate	std.error	statistic	p.value
(Intercept)	-87.12	29.273	-2.98	0.006
height	1.54	0.384	4.02	0.000

term	estimate	std.error	statistic	p.value
(Intercept)	-57.988	8.638	-6.71	<0.0001
height	0.339	0.130	2.61	0.0145
diam	4.708	0.264	17.82	<0.0001

See next page for the plots and parts a to c.

---

[5] The cherry data used in this exercise can be found in the **openintro** R package.

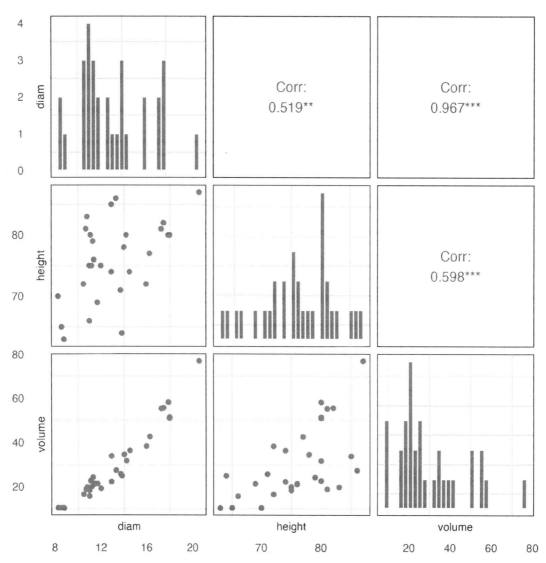

a. There are three variables described in the figure, and each is paired with each other to create three different scatterplots. Rate the pairwise relationships from most correlated to least correlated.

b. When using only one variable to model a tree's `volume`, is `diameter` a significant predictor variable? Is `height` a significant predictor variable? Explain.

c. When using both `diameter` and `height` to predict a tree's `volume`, are both predictor variables still significant? Explain.

25.5. **Movie returns.** A FiveThirtyEight.com article reports that "Horror movies get nowhere near as much draw at the box office as the big-time summer blockbusters or action/adventure movies, but there's a huge incentive for studios to continue pushing them out. The return-on-investment potential for horror movies is absurd." To investigate how the return-on-investment (ROI) compares between genres and how this relationship has changed over time, an introductory statistics student fit a linear regression model to predict the ratio of gross revenue of movies to the production costs from genre and release year for 1,070 movies released between 2000 and 2018. Using the plots given below, determine if this regression model is appropriate for these data. In particular, use the residual plot to check the LINE conditons. (FiveThirtyEight, 2015)

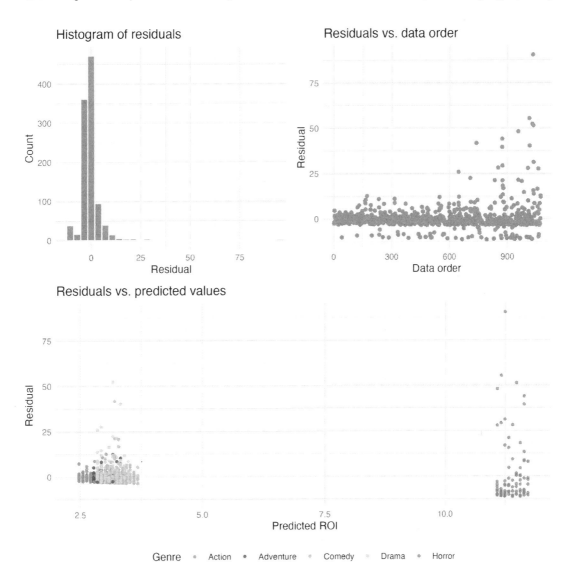

25.6. **Difficult encounters.** A study was conducted at a university outpatient primary care clinic in Switzerland to identify factors associated with difficult doctor-patient encounters. The data consist of 527 patient encounters, conducted by the 27 medical residents employed at the clinic. After each encounter, the attending physician completed two questionnaires: the Difficult Doctor Patient Relationship Questionnaire (DDPRQ-10) and the patient's vulnerability grid (PVG).

A higher score on the DDPRQ-10 indicates a more difficult encounter. The maximum possible score is 60 and encounters with score 30 and higher are considered difficult.

A model was fit for the association of DDPRQ-10 score with features of the attending physician: age, sex (identify as male or not), and years of training.

term	estimate	std.error	statistic	p.value
(Intercept)	30.594	2.886	10.601	<0.0001
age	-0.016	0.104	-0.157	0.876
sexMale	-0.535	0.781	-0.686	0.494
yrs_train	0.096	0.215	0.445	0.656

a. The intercept of the model is 30.594. What is the age, sex, and years of training of a physician whom this model would predict to have a DDPRQ-10 score of 30.594.

b. Is there evidence of a significant association between DDPRQ-10 score and any of the physician features?

25.7. **Baby's weight, mathematical test.** US Department of Health and Human Services, Centers for Disease Control and Prevention collect information on births recorded in the country. The data used here are a random sample of 1,000 births from 2014. Here, we study the relationship between smoking and weight of the baby. The variable `smoke` is coded 1 if the mother is a smoker, and 0 if not. The summary table below shows the results of a linear regression model for predicting the average birth weight of babies, measured in pounds, based on the smoking status of the mother.[6] (ICPSR, 2014)

term	estimate	std.error	statistic	p.value
(Intercept)	-3.82	0.57	-6.73	<0.0001
weeks	0.26	0.01	18.93	<0.0001
mage	0.02	0.01	2.53	0.0115
sexmale	0.37	0.07	5.30	<0.0001
visits	0.02	0.01	2.09	0.0373
habitsmoker	-0.43	0.13	-3.41	7e-04

Also shown below are a series of diagnostics plots.

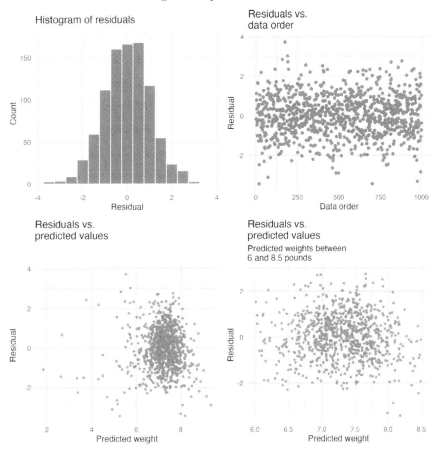

a. Determine if the conditions for doing inference based on mathematical models with these data are met using the diagnostic plots above. If not, describe how to proceed with the analysis.

b. Using the regression output, evaluate whether the true slope of `habit` (i.e. whether the mother is a smoker) is different than 0, given the other variables in the model. State the null and alternative hypotheses, report the p-value (using a mathematical model), and state your conclusion.

---

[6]The `births14` data used in this exercise can be found in the **openintro** R package.

25.8. **Baby's weight with collinear predictors.** In this exercise we study the relationship between the weight of the baby and two explanatory variables: number of `weeks` of gestation and number of pregnancy hospital `visits`. (ICPSR, 2014)

The plots below describe the show the distribution of each of these variables (on the diagonal) as well as provide information on the pairwise correlations between them.

Also provided below are three regression model outputs: `weight` vs. `weeks`, `weight` vs. `visits`, and `weight` vs. `weeks + visits`.

term	estimate	std.error	statistic	p.value
(Intercept)	-0.83	1.76	-0.47	0.6395
weeks	0.21	0.05	4.70	<0.0001

term	estimate	std.error	statistic	p.value
(Intercept)	6.56	0.36	18.36	<0.0001
visits	0.08	0.03	2.61	0.0105

term	estimate	std.error	statistic	p.value
(Intercept)	-0.44	1.78	-0.25	0.81
weeks	0.19	0.05	4.00	0.00
visits	0.04	0.03	1.26	0.21

See next page for the plots and parts a to c.

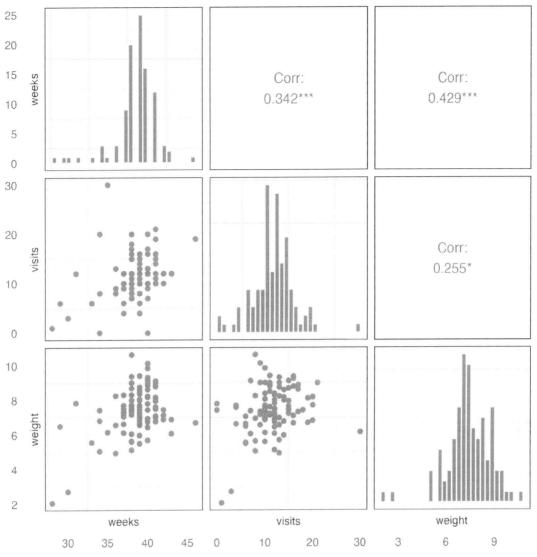

a. There are three variables described in the figure, and each is paired with each other to create three different scatterplots. Rate the pairwise relationships from most correlated to least correlated.

b. When using only one variable to model the baby's `weight`, is `weeks` a significant predictor variable? Is `visits` a significant predictor variable? Explain.

c. When using both `visits` and `weeks` to predict the baby's `weight`, are both predictor variables still significant? Explain.

25.9. **Baby's weight, cross-validation.** Using a random sample of 1,000 US births from 2014, we study the relationship between the weight of the baby and various explanatory variables. (ICPSR, 2014)

The plots below describe prediction errors associated with two different models designed to predict weight of baby at birth; one model uses 7 predictors, one model uses 2 predictors. Using 4-fold cross-validation, the data were split into 4 folds. Three of the folds estimate the β_i parameters using b_i, and the model is applied to the held out fold for prediction. The process was repeated 4 times (each time holding out one of the folds).

$$E[\text{weight}] = \beta_0 + \beta_1 \times \text{fage} + \beta_2 \times \text{mage}$$
$$+ \beta_3 \times \text{mature} + \beta_4 \times \text{weeks}$$
$$+ \beta_5 \times \text{visits} + \beta_6 \times \text{gained}$$
$$+ \beta_7 \times \text{habit}$$

$$E[\text{weight}] = \beta_0 + \beta_1 \times \text{weeks} + \beta_2 \times \text{mature}$$

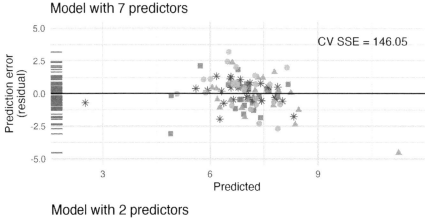

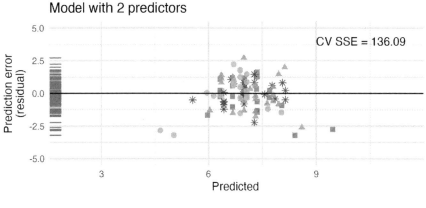

a. In the first graph, note the point at roughly (predicted = 11 and error = -4). Estimate the observed and predcited value for that observation.

b. Using the same point, describe which cross-validation fold(s) were used to build its prediction model.

c. For the plot on the left, for one of the cross-validation folds, how many coefficients were estimated in the linear model? For the plot on the right, for one of the cross-validation folds, how many coefficients were estimated in the linear model?

d. Do the values of the residuals (along the y-axis, not the x-axis) seem markedly different for the two models? Explain.

25.10. **Baby's weight, cross-validation to choose model.** Using a random sample of 1,000 US births from 2014, we study the relationship between the weight of the baby and various explanatory variables. (ICPSR, 2014)

The plots below describe prediction errors associated with two different models designed to predict weight of baby at birth; one model uses 7 predictors, one model uses 2 predictors. Using 4-fold cross-validation, the data were split into 4 folds. Three of the folds estimate the β_i parameters using b_i, and the model is applied to the held out fold for prediction. The process was repeated 4 times (each time holding out one of the folds).

$$E[\text{weight}] = \beta_0 + \beta_1 \times \text{fage} + \beta_2 \times \text{mage}$$
$$+ \beta_3 \times \text{mature} + \beta_4 \times \text{weeks}$$
$$+ \beta_5 \times \text{visits} + \beta_6 \times \text{gained}$$
$$+ \beta_7 \times \text{habit}$$

$$E[\text{weight}] = \beta_0 + \beta_1 \times \text{weeks} + \beta_2 \times \text{mature}$$

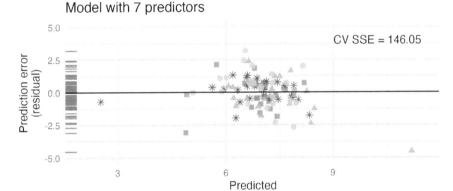

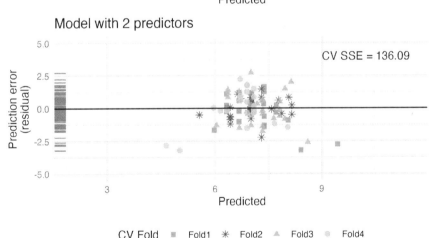

a. Using the spread of the points, which model should be chosen for a final report on these data? Explain.

b. Using the summary statistic (CV SSE), which model should be chosen for a final report on these data? Explain.

c. Why would the model with more predictors fit the data less closely than the model with only two predictors?

25.11. **RailTrail, cross-validation.** The Pioneer Valley Planning Commission (PVPC) collected data north of Chestnut Street in Florence, MA for ninety days from April 5, 2005 to November 15, 2005. Data collectors set up a laser sensor, with breaks in the laser beam recording when a rail-trail user passed the data collection station.[7]

The plots below describe prediction errors associated with two different models designed to predict the volume of riders on the RailTrail; one model uses 6 predictors, one model uses 2 predictors. Using 3-fold cross-validation, the data were split into 3 folds. Three of the folds estimate the β_i parameters using b_i, and the model is applied to the held out fold for prediction. The process was repeated 4 times (each time holding out one of the folds).

$$E[\text{volume}] = \beta_0 + \beta_1 \times \text{hightemp} + \beta_2 \times \text{lowtemp}$$
$$+ \beta_3 \times \text{spring} + \beta_4 \times \text{weekday}$$
$$+ \beta_5 \times \text{cloudcover} + \beta_6 \times \text{precip}$$

$$E[\text{volume}] = \beta_0 + \beta_1 \times \text{hightemp} + \beta_2 \times \text{precip}$$

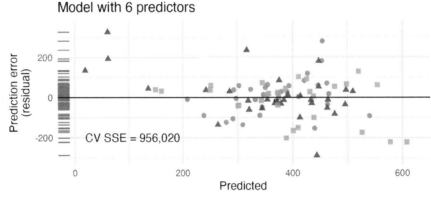

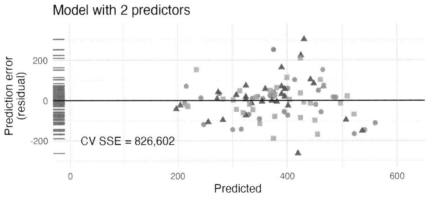

a. In the second graph, note the point at roughly (predicted = 400 and error = 100). Estimate the observed and predcited value for that observation.

b. Using the same point, describe which cross-validation fold(s) were used to build its prediction model.

c. For the plot on the left, for one of the cross-validation folds, how many coefficients were estimated in the linear model? For the plot on the right, for one of the cross-validation folds, how many coefficients were estimated in the linear model?

d. Do the values of the residuals (along the y-axis, not the x-axis) seem markedly different for the two models? Explain.

---
[7] The RailTrail data used in this exercise can be found in the **mosaicData** R package.

25.12. **RailTrail, cross-validation to choose model.** The Pioneer Valley Planning Commission (PVPC) collected data north of Chestnut Street in Florence, MA for ninety days from April 5, 2005 to November 15, 2005. Data collectors set up a laser sensor, with breaks in the laser beam recording when a rail-trail user passed the data collection station.

The plots below describe prediction errors associated with two different models designed to predict the volume of riders on the RailTrail; one model uses 6 predictors, one model uses 2 predictors. Using 3-fold cross-validation, the data were split into 3 folds. Three of the folds estimate the β_i parameters using b_i, and the model is applied to the held out fold for prediction. The process was repeated 4 times (each time holding out one of the folds).

$$E[\text{volume}] = \beta_0 + \beta_1 \times \text{hightemp} + \beta_2 \times \text{lowtemp}$$
$$+ \beta_3 \times \text{spring} + \beta_4 \times \text{weekday}$$
$$+ \beta_5 \times \text{cloudcover} + \beta_6 \times \text{precip}$$

$$E[\text{volume}] = \beta_0 + \beta_1 \times \text{hightemp} + \beta_2 \times \text{precip}$$

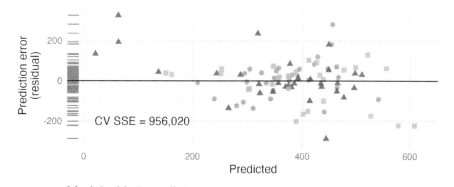

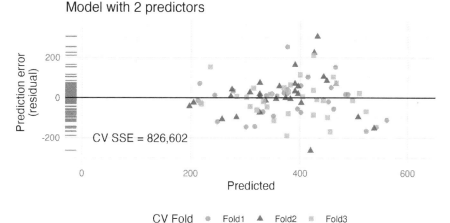

a. Using the spread of the points, which model should be chosen for a final report on these data? Explain.

b. Using the summary statistic (CV SSE), which model should be chosen for a final report on these data? Explain.

c. Why would the model with more predictors fit the data less closely than the model with only two predictors?

Chapter 26

Inference for logistic regression

Combining ideas from Chapter 9 on logistic regression, Chapter 13 on inference with mathematical models, and Chapters 24 and 25 which apply inferential techniques to the linear model, we wrap up the book by considering inferential methods applied to a logistic regression model. Additionally, we use cross-validation as a method for independent assessment of the logistic regression model.

As with multiple linear regression, the inference aspect for logistic regression will focus on interpretation of coefficients and relationships between explanatory variables. Both p-values and cross-validation will be used for assessing a logistic regression model.

Consider the `email` data which describes email characteristics which can be used to predict whether a particular incoming email is (unsolicited bulk email). Without reading every incoming message, it might be nice to have an automated way to identify spam emails. Which of the variables describing each email are important for predicting the status of the email?

The `email` data can be found in the **openintro** R package.

26.1 Model diagnostics

Before looking at the hypothesis tests associated with the coefficients (turns out they are very similar to those in linear regression!), it is valuable to understand the technical conditions that underlie the inference applied to the logistic regression model. Generally, as you've seen in the logistic regression modeling examples, it is imperative that the response variable in binary. Additionally, the key technical condition for logistic regression has to do with the relationship between the predictor variables (x_i values) and the probability the outcome will be a success. It turns out, the relationship is a specific functional form called a logit function, where $\text{logit}(p) = \log_e(\frac{p}{1-p})$. The function may feel complicated, and memorizing the formula of the logit is not necessary for understanding logistic regression. What you do need to remember is that the probability of the outcome being a success is a function of a linear combination of the explanatory variables.

26.1. MODEL DIAGNOSTICS

Table 26.1: Variables and their descriptions for the `email` dataset. Many of the variables are indicator variables, meaning they take the value 1 if the specified characteristic is present and 0 otherwise.

Variable	Description
spam	Indicator for whether the email was spam.
to_multiple	Indicator for whether the email was addressed to more than one recipient.
from	Whether the message was listed as from anyone (this is usually set by default for regular outgoing email).
cc	Number of people cc'ed.
sent_email	Indicator for whether the sender had been sent an email in the last 30 days.
attach	The number of attached files.
dollar	The number of times a dollar sign or the word "dollar" appeared in the email.
winner	Indicates whether "winner" appeared in the email.
format	Indicates whether the email was written using HTML (e.g., may have included bolding or active links).
re_subj	Whether the subject started with "Re:", "RE:", "re:", or "rE:"
exclaim_subj	Whether there was an exclamation point in the subject.
urgent_subj	Whether the word "urgent" was in the email subject.
exclaim_mess	The number of exclamation points in the email message.
number	Factor variable saying whether there was no number, a small number (under 1 million), or a big number.

Logistic regression conditions.

There are two key conditions for fitting a logistic regression model:

1. Each outcome Y_i is independent of the other outcomes.
2. Each predictor x_i is linearly related to logit(p_i) if all other predictors are held constant.

The first logistic regression model condition — independence of the outcomes — is reasonable if we can assume that the emails that arrive in an inbox within a few months are independent of each other with respect to whether they're spam or not.

The second condition of the logistic regression model is not easily checked without a fairly sizable amount of data. Luckily, we have 3921 emails in the dataset! Let's first visualize these data by plotting the true classification of the emails against the model's fitted probabilities, as shown in Figure 26.1.

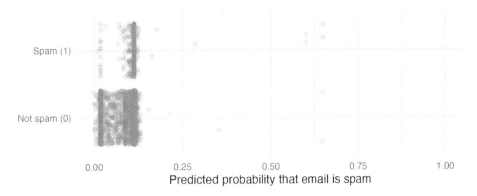

Figure 26.1: The predicted probability that each of the 3921 emails that are spam. Points have been jittered so that those with nearly identical values aren't plotted exactly on top of one another.

We'd like to assess the quality of the model. For example, we might ask: if we look at emails that we modeled as having 10% chance of being spam, do we find out 10% of the actually are spam? We can check this for groups of the data by constructing a plot as follows:

1. Bucket the data into groups based on their predicted probabilities.
2. Compute the average predicted probability for each group.
3. Compute the observed probability for each group, along with a 95% confidence interval for the true probability of success for those individuals.
4. Plot the observed probabilities (with 95% confidence intervals) against the average predicted probabilities for each group.

If the model does a good job describing the data, the plotted points should fall close to the line $y = x$, since the predicted probabilities should be similar to the observed probabilities. We can use the confidence intervals to roughly gauge whether anything might be amiss. Such a plot is shown in Figure 26.2.

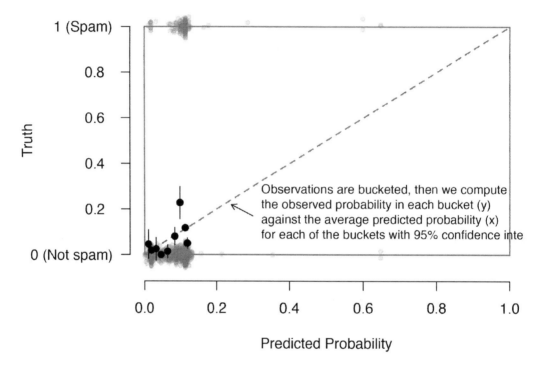

Figure 26.2: The dashed line is within the confidence bound of the 95% confidence intervals of each of the buckets, suggesting the logistic fit is reasonable.

A plot like Figure 26.2 helps to better understand the deviations. Additional diagnostics may be created that are similar to those featured in Section 24.6. For instance, we could compute residuals as the observed outcome minus the expected outcome ($e_i = Y_i - \hat{p}_i$), and then we could create plots of these residuals against each predictor.

26.2 Multiple logistic regression output from software

As you learned in Chapter 8, optimization can be used to find the coefficient estimates for the logistic model. The unknown population model can be written as:

$$\log_e\left(\frac{p}{1-p}\right) = \beta_0 + \beta_1 \times \texttt{to\_multiple} + \beta_2 \times \texttt{cc}$$
$$+ \beta_3 \times \texttt{dollar} + \beta_4 \times \texttt{urgent\_subj}$$

The estimated equation for the regression model may be written as a model with four predictor variables (where $\hat{p}$ is the estimated probability of being a spam email message):

$$\log_e\left(\frac{\hat{p}}{1-\hat{p}}\right) = -2.05 + -1.91 \times \texttt{to\_multiple} + 0.02 \times \texttt{cc}$$
$$- 0.07 \times \texttt{dollar} + 2.66 \times \texttt{urgent\_subj}$$

Table 26.2: Summary of a logistic model for predicting whether an email is based on the variables to_multiple, cc, dollar, and urgent_subj. Each of the variables has its own coefficient estimate and p-value.

term	estimate	std.error	statistic	p.value
(Intercept)	-2.05	0.06	-34.67	<0.0001
to_multiple1	-1.91	0.30	-6.37	<0.0001
cc	0.02	0.02	1.16	0.245
dollar	-0.07	0.02	-3.38	7e-04
urgent_subj1	2.66	0.80	3.32	9e-04

Not only does Table 26.2 provide the estimates for the coefficients, it also provides information on the inference analysis (i.e., hypothesis testing) which are the focus of this chapter.

As in Section 25, with **multiple predictors**, each hypothesis test (for each of the explanatory variables) is conditioned on each of the other variables remaining in the model.

if multiple predictors $H_0: \beta_i = 0$ given other variables in the model

Using the example above and focusing on each of the variable p-values (here we won't discuss the p-value associated with the intercept), we can write out the four different hypotheses (associated with the p-value corresponding to each of the coefficients / row in Table 26.2):

- $H_0: \beta_1 = 0$ given cc, dollar, and urgent_subj are included in the model
- $H_0: \beta_2 = 0$ given to_multiple, dollar, and urgent_subj are included in the model
- $H_0: \beta_3 = 0$ given to_multiple, cc, and urgent_subj are included in the model
- $H_0: \beta_4 = 0$ given to_multiple, dollar, and dollar are included in the model

The very low p-values from the software output tell us that three of the variables (that is, not cc) act as statistically significant predictors in the model at the significance level of 0.05, despite the inclusion of any of the other variables. Consider the p-value on $H_0: \beta_1 = 0$. The low p-value says that it would be extremely unlikely to observe data that yield a coefficient on to_multiple at least as far from 0 as -1.91 (i.e. $|b_1| > 1.91$) if the true relationship between to_multiple and spam was non-existent (i.e., if $\beta_1 = 0$) **and** the model also included cc and dollar and urgent_subj. Note also that the coefficient on dollar has a small associated p-value, but the magnitude of the coefficient is also seemingly small (0.07). It turns out that in units of standard errors (0.02 here), 0.07 is actually quite far from zero, it's all about context! The p-values on the remaining variables are interpreted similarly. From the initial output (p-values) in Table 26.2, it seems as though to_multiple, dollar, and urgent_subj are important variables for modeling whether an email is spam. We remind you that although p-values provide some information about the importance of each of the predictors in the model, there are many, arguably more important, aspects to consider when choosing the best model.

As with linear regression (see Section 25.2), existence of predictors that are correlated with each other can affect both the coefficient estimates and the associated p-values. However, investigating multicollinearity in a logistic regression model is saved for a text which provides more detail about logistic regression. Next, as a model building alternative (or enhancement) to p-values, we revisit cross-validation within the context of predicting status for each of the individual emails.

26.3 Cross-validation for prediction error

The p-value is a probability measure under a setting of no relationship. That p-value provides information about the degree of the relationship (e.g., above we measure the relationship between spam and `to_multiple` using a p-value), but the p-value does not measure how well the model will predict the individual emails (e.g., the accuracy of the model). Depending on the goal of the research project, you might be inclined to focus on variable importance (through p-values) or you might be inclined to focus on prediction accuracy (through cross-validation).

Here we present a method for using cross-validation accuracy to determine which variables (if any) should be used in a model which predicts whether an email is . A full treatment of cross-validation and logistic regression models is beyond the scope of this text. Using cross-validation, we can build k different models which are used to predict the observations in each of the k holdout samples. The smaller model uses only the `to_multiple` variable, see the complete dataset (not cross-validated) model output in Table 26.3. The logistic regression model can be written as (where $\hat{p}$ is the estimated probability of being a spam email message):

$$\log_e\left(\frac{\hat{p}}{1-\hat{p}}\right) = -2.12 + -1.81 \times \texttt{to\_multiple}$$

Table 26.3: Summary of a logistic model for predicting whether an email is based on only the predictor variable `to_multiple`. Each of the variables has its own coefficient estimate and p-value.

term	estimate	std.error	statistic	p.value
(Intercept)	-2.12	0.06	-37.67	<0.0001
to_multiple1	-1.81	0.30	-6.09	<0.0001

For each cross-validated model, the coefficients change slightly, and the model is used to make independent predictions on the holdout sample. The model from the first cross-validation sample is given in Table 26.3 and can be compared to the coefficients in Table 26.3.

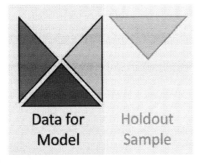
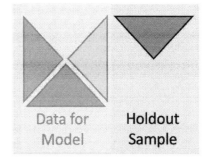
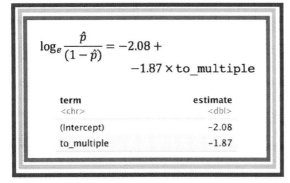

Figure 26.3: The coefficients are estimated using the least squares model on 3/4 of the dataset with a single predictor variable. Predictions are made on the remaining 1/4 of the observations. Note that the predictions are independent of the estimated model coefficients, and the prediction error rate is quite high.

Because the `email` dataset has a ratio of roughly 90% non-spam and 10% spam emails, a model which randomly guessed all non- would have an overall accuracy of 90%! Clearly, we'd like to capture

26.3. CROSS-VALIDATION FOR PREDICTION ERROR

Table 26.4: One quarter at a time, the data were removed from the model building, and whether the email was spam (TRUE) or not (FALSE) was predicted. The logistic regression model was fit independently of the removed emails. Only `to_multiple` is used to predict whether the email is spam. Because we used a cutoff designed to identify spam emails, the accuracy of the non-spam email predictions is very low.

fold	count	accuracy	notspamTP	spamTP
1st quarter	980	0.26	0.19	0.98
2nd quarter	981	0.23	0.15	0.96
3rd quarter	979	0.25	0.18	0.96
4th quarter	981	0.24	0.17	0.98

the information with the spam emails, so our interest is in the percent of spam emails which are identified as (see Table 26.4). Additionally, in the logistic regression model, we use a 10% cutoff to predict whether or not the email is . Fortunately, we've done a great job of predicting ! However, the trade-off was that most of the non-spam emails are now predicted to be which is not acceptable for a prediction algorithm. Adding more variables to the model may help with both the and not- predictions.

The larger model uses `to_multiple`, `attach`, `winner`, `format`, `re_subj`, `exclaim_mess`, and `number` as the set of seven predictor variables, see the complete dataset (not cross-validated) model output in Table 26.5. The logistic regression model can be written as (where $\hat{p}$ is the estimated probability of being a spam email message):

$$\log_e\left(\frac{\hat{p}}{1-\hat{p}}\right) = -0.34 - 2.56 \times \texttt{to\_multiple} + 0.20 \times \texttt{attach}$$
$$+ 1.73 \times \texttt{winner}_{yes} - 1.28 \times \texttt{format}$$
$$- 2.86 \times \texttt{re\_subj} + 0 \times \texttt{exclaim\_mess}$$
$$- 1.07 \times \texttt{number}_{small} - 0.42 \times \texttt{number}_{big}$$

Table 26.5: Summary of a logistic model for predicting whether an email is based on only the predictor variable `to_multiple`. Each of the variables has its own coefficient estimate and p-value.

term	estimate	std.error	statistic	p.value
(Intercept)	-0.34	0.11	-3.02	0.0025
to_multiple1	-2.56	0.31	-8.28	<0.0001
attach	0.20	0.06	3.29	0.001
winneryes	1.73	0.33	5.33	<0.0001
format1	-1.28	0.13	-9.80	<0.0001
re_subj1	-2.86	0.37	-7.83	<0.0001
exclaim_mess	0.00	0.00	0.26	0.7925
numbersmall	-1.07	0.14	-7.54	<0.0001
numberbig	-0.42	0.20	-2.10	0.0357

Somewhat expected, the larger model (see Table 26.6) was able to capture more nuance in the emails which lead to better predictions. However, it is not true that adding variables will always lead to better predictions, as correlated or noise variables may dampen the signal from those variables which truly predict the status. We encourage you to learn more about multiple variable models and cross-validation in your future exploration of statistical topics.

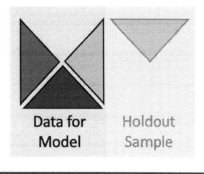
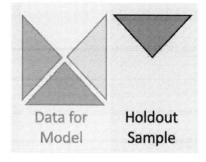

Figure 26.4: The coefficients are estimated using the least squares model on 3/4 of the dataset with the seven specified predictor variables. Predictions are made on the remaining 1/4 of the observations. Note that the predictions are independent of the estimated model coefficients. The predictions are now much better for both the and the non-spam emails (than they were with a single predictor variable).

Table 26.6: One quarter at a time, the data were removed from the model building, and whether the email was spam (TRUE) or not (FALSE) was predicted. The logistic regression model was fit independently of the removed emails. Now, the variables to_multiple, attach, winner, format, re_subj, exclaim_mess, and number are used to predict whether the email is spam.

fold	count	accuracy	notspamTP	spamTP
1st quarter	980	0.77	0.77	0.71
2nd quarter	981	0.80	0.81	0.70
3rd quarter	979	0.76	0.77	0.65
4th quarter	981	0.78	0.79	0.75

26.4 Chapter review

26.4.1 Summary

Throughout the text, we have presented a modern view to introduction to statistics. Early we presented graphical techniques which communicated relationships across multiple variables. We also used modeling to formalize the relationships. In Chapter 26 we considered inferential claims on models which include many variables used to predict the probability of the outcome being a success. We continue to emphasize the importance of experimental design in making conclusions about research claims. In particular, recall that variability can come from different sources (e.g., random sampling vs. random allocation, see Figure 2.8).

As you might guess, this text has only scratched the surface of the world of statistical analyses that can be applied to different datasets. In particular, to do justice to the topic, the linear models and generalized linear models we have introduced can each be covered with their own course or book. Hierarchical models, alternative methods for fitting parameters (e.g., Ridge Regression or LASSO), and advanced computational methods applied to models (e.g., permuting the response variable? one explanatory variable? all the explanatory variables?) are all beyond the scope of this book. However, your successful understanding of the ideas we have covered has set you up perfectly to move on to a higher level of statistical modeling and inference. Enjoy!

26.4.2 Terms

We introduced the following terms in the chapter. If you're not sure what some of these terms mean, we recommend you go back in the text and review their definitions. We are purposefully presenting them in alphabetical order, instead of in order of appearance, so they will be a little more challenging to locate. However you should be able to easily spot them as **bolded text**.

cross-validation	multiple predictors
inference on logistic regression	technical conditions

26.5 Exercises

Answers to odd numbered exercises can be found in Appendix A.26.

26.1. **Marijuana use in college.** Researchers studying whether the value systems adolescents conflict with those of their children asked 445 college students if they use marijuana. They also asked the students' parents if they've used marijuana when they were in college. The following model was fit to predict student drug use from parent drug use.[1] (Ellis and Stone, 1979)

term	estimate	std.error	statistic	p.value
(Intercept)	-0.405	0.133	-3.04	0.0023
parentsused	0.791	0.194	4.09	<0.0001

a. State the hypotheses for evaluating whether parents' marijuana usage is a significant predictor of their kids' marijuana usage.

b. Based on the regression output, state the sample statistic and the p-value of the test.

c. State the conclusion of the hypothesis test in context of the data and the research question.

26.2. **Treating heart attacks.** Researchers studying the effectiveness of Sulfinpyrazone in the prevention of sudden death after a heart attack conducted a randomized experiment on 1,475 patients. The following model was fit to predict experiment outcome (`died` or `lived`, where success is defined as `lived`) from the experimental group (`control` and `treatment`).[2] (Group, 1980)

term	estimate	std.error	statistic	p.value
(Intercept)	2.431	0.135	18.05	<0.0001
grouptreatment	0.395	0.210	1.89	0.0594

a. State the hypotheses for evaluating whether experimental group is a significant predictor of treatment outcome.

b. Based on the regression output, state the sample statistic and the p-value of the test.

c. State the conclusion of the hypothesis test in context of the data and the research question.

---

[1] The `drug_use` data used in this exercise can be found in the **openintro** R package.
[2] The `sulphinpyrazone` data used in this exercise can be found in the **openintro** R package.

26.5. EXERCISES

26.3. **Possum classification, cross-validation.** The common brushtail possum of the Australia region is a bit cuter than its distant cousin, the American opossum. We consider 104 brushtail possums from two regions in Australia, where the possums may be considered a random sample from the population. The first region is Victoria, which is in the eastern half of Australia and traverses the southern coast. The second region consists of New South Wales and Queensland, which make up eastern and northeastern Australia. We use logistic regression to differentiate between possums in these two regions. The outcome variable, called pop, takes value 1 when a possum is from Victoria and 0 when it is from New South Wales or Queensland.[3]

$$\log_e\left(\frac{p}{1-p}\right) = \beta_0 + \beta_1 \times \texttt{tail\_l}$$

$$\log_e\left(\frac{p}{1-p}\right) = \beta_0 + \beta_1 \times \texttt{total\_l} + \beta_2 \times \texttt{sex}$$

Model with tail length predictor

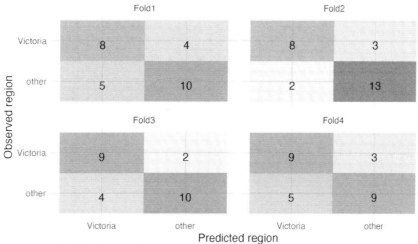

Model with total length and sex as predictors

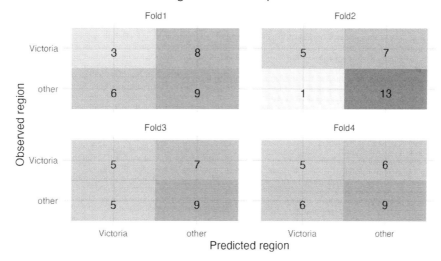

a. How many observations are in Fold2? Use the model with only tail length as a predictor variable. Of the observations in Fold2, how many of them were correctly predicted to be from Vicotria? How many of them were incorrectly predicted to be from Victoria?

b. How many observations are used to build the model which predicts for the observations in Fold2?

c. For one of the cross-validation folds, how many coefficients were estimated for the model which uses tail length as a predictor? For one of the cross-validation folds, how many coefficients were estimated for the model which uses total length and sex as predictors?

---
[3]The possum data used in this exercise can be found in the **openintro** R package.

26.4. **Possum classification, cross-validation to choose model.** In this exercise we consider 104 brushtail possums from two regions in Australia (the first region is Victoria and the second is New South Wales and Queensland), where the possums may be considered a random sample from the population. We use logistic regression to classify the possums into the two regions. The outcome variable, called pop, takes value 1 when a possum is from Victoria and 0 when it is from New South Wales or Queensland.

$$\log_e\left(\frac{p}{1-p}\right) = \beta_0 + \beta_1 \times \text{tail\_l}$$

$$\log_e\left(\frac{p}{1-p}\right) = \beta_0 + \beta_1 \times \text{total\_l} + \beta_2 \times \text{sex}$$

Model with tail length predictor

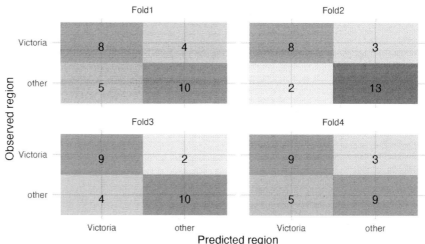

Model with total length and sex as predictors

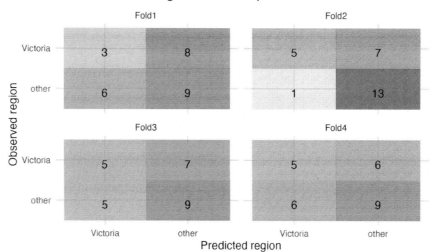

a. For the model with tail length, how many of the observations were correctly classified? What proportion of the observations were correctly classified?

b. For the model with total length and sex, how many of the observations were correctly classified? What proportion of the observations were correctly classified?

c. If you have to choose between using only tail length as a predictor versus using total length and sex as predictors (for classification into region), which model would you choose? Explain.

d. Given the predictions provided, what model might be preferable to either of the models given above?

26.5. **Premature babies, cross-validation.** US Department of Health and Human Services, Centers for Disease Control and Prevention collect information on births recorded in the country. The data used here are a random sample of 1000 births from 2014 (with some rows removed due to missing data). Here, we use logistic regression to model whether or not the baby is premature from various explanatory variables.[4] (ICPSR, 2014)

$$\log_e\left(\frac{p}{1-p}\right) = \beta_0 + \beta_1 \times \texttt{mage} + \beta_2 \times \texttt{weight}$$
$$+ \beta_3 \times \texttt{mature} + \beta_4 \times \texttt{visits}$$
$$+ \beta_5 \times \texttt{gained} + \beta_6 \times \texttt{habit}$$

$$\log_e\left(\frac{p}{1-p}\right) = \beta_0 + \beta_1 \times \texttt{weight} + \beta_2 \times \texttt{mature}$$

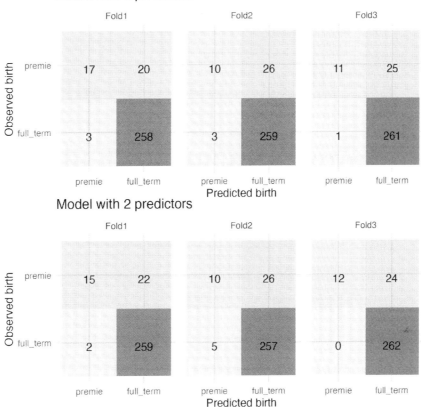

a. How many observations are in Fold2? Use the model with only `weight` and `mature` as predictor variables. Of the observations in Fold2, how many of them were correctly predicted to be premature? How many of them were incorrectly predicted to be premature?

b. How many observations are used to build the model which predicts for the observations in Fold2?

c. In the original dataset, are most of the births premature or full term? Explain.

d. For one of the cross-validation folds, how many coefficients were estimated for the model which uses `mage`, `weight`, `mature`, `visits`, `gained`, and `habit` as predictors? For one of the cross-validation folds, how many coefficients were estimated for the model which uses `weight` and `mature` as predictors?

---

[4]The `births14` data used in this exercise can be found in the **openintro** R package.

26.6. **Premature babies, cross-validation to choose model.** US Department of Health and Human Services, Centers for Disease Control and Prevention collect information on births recorded in the country. The data used here are a random sample of 1000 births from 2014 (with some rows removed due to missing data). Here, we use logistic regression to model whether or not the baby is premature from various explanatory variables. (ICPSR, 2014)

$$\log_e\left(\frac{p}{1-p}\right) = \beta_0 + \beta_1 \times \texttt{mage} + \beta_2 \times \texttt{weight}$$
$$+ \beta_3 \times \texttt{mature} + \beta_4 \times \texttt{visits}$$
$$+ \beta_5 \times \texttt{gained} + \beta_6 \times \texttt{habit}$$

$$\log_e\left(\frac{p}{1-p}\right) = \beta_0 + \beta_1 \times \texttt{weight} + \beta_2 \times \texttt{mature}$$

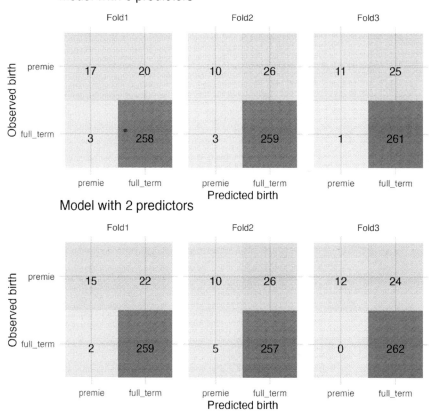

a. For the model with 6 predictors, how many of the observations were correctly classified? What proportion of the observations were correctly classified?

b. For the model with 2 predictors, how many of the observations were correctly classified? What proportion of the observations were correctly classified?

c. If you have to choose between the model with 6 predictors and the model with 2 predictors (for predicting whether a baby will be premature), which model would you choose? Explain.

Chapter 27

Applications: Model and infer

27.1 Case study: Mario Kart

In this case study, we consider Ebay auctions of a video game called *Mario Kart* for the Nintendo Wii. The outcome variable of interest is the total price of an auction, which is the highest bid plus the shipping cost. We will try to determine how total price is related to each characteristic in an auction while simultaneously controlling for other variables. For instance, all other characteristics held constant, are longer auctions associated with higher or lower prices? And, on average, how much more do buyers tend to pay for additional Wii wheels (plastic steering wheels that attach to the Wii controller) in auctions? Multiple regression will help us answer these and other questions.

 The mariokart data can be found in the **openintro** R package.

The mariokart data set includes results from 141 auctions. Four observations from this data set are shown in Table 27.1, and descriptions for each variable are shown in Table 27.2. Notice that the condition and stock photo variables are indicator variables, similar to bankruptcy in the loans data set from Chapter 25.

Table 27.1: Top four rows of the mariokart dataset.

price	cond_new	stock_photo	duration	wheels
51.5	new	yes	3	1
37.0	used	yes	7	1
45.5	new	no	3	1
44.0	new	yes	3	1

Table 27.2: Variables and their descriptions for the mariokart dataset.

Variable	Description
price	Final auction price plus shipping costs, in US dollars.
cond_new	Indicator variable for if the game is new (1) or used (0).
stock_photo	Indicator variable for if the auction's main photo is a stock photo.
duration	The length of the auction, in days, taking values from 1 to 10.
wheels	The number of Wii wheels included with the auction. A Wii wheel is an optional steering wheel accessory that holds the Wii controller.

27.1.1 Mathematical approach to linear models

In Table 27.3 we fit a mathematical linear regression model with the game's condition as a predictor of auction price.

$$E[\text{price}] = \beta_0 + \beta_1 \times \text{cond\_new}$$

Results of the model are summarized below:

Table 27.3: Summary of a linear model for predicting price based on cond_new.

term	estimate	std.error	statistic	p.value
(Intercept)	42.9	0.81	52.67	<0.0001
cond_new	10.9	1.26	8.66	<0.0001

GUIDED PRACTICE

Write down the equation for the model, note whether the slope is statistically different from zero, and interpret the coefficient.[1]

Sometimes there are underlying structures or relationships between predictor variables. For instance, new games sold on Ebay tend to come with more Wii wheels, which may have led to higher prices for those auctions. We would like to fit a model that includes all potentially important variables simultaneously, which would help us evaluate the relationship between a predictor variable and the outcome while controlling for the potential influence of other variables.

We want to construct a model that accounts for not only the game condition but simultaneously accounts for three other variables:

$$E[\text{price}] = \beta_0 + \beta_1 \times \text{cond\_new} + \beta_2 \times \text{stock\_photo}$$
$$+ \beta_3 \times \text{duration} + \beta_4 \times \text{wheels}$$

Table 27.4 summarizes the full model. Using the output, we identify the point estimates of each coefficient and the corresponding impact (measured with information for the standard error to compute the p-value).

Table 27.4: Summary of a linear model for predicting price based on cond_new, stock_photo, duration, and wheels.

term	estimate	std.error	statistic	p.value
(Intercept)	36.21	1.51	23.92	<0.0001
cond_new	5.13	1.05	4.88	<0.0001
stock_photo	1.08	1.06	1.02	0.3085
duration	-0.03	0.19	-0.14	0.8882
wheels	7.29	0.55	13.13	<0.0001

---

[1] The equation for the line may be written as $\widehat{\text{price}} = 47.15 + 10.90 \times \text{cond\_new}$. Examining the regression output in we can see that the p-value for cond_new is very close to zero, indicating there is strong evidence that the coefficient is different from zero when using this one-variable model. The variable cond_new is a two-level categorical variable that takes value 1 when the game is new and value 0 when the game is used. This means the 10.90 model coefficient predicts a price of an extra $10.90 for those games that are new versus those that are used.

27.1. CASE STUDY: MARIO KART

GUIDED PRACTICE

Write out the model's equation using the point estimates from Table 27.4. How many predictors are there in the model? How many coefficients are estimated?[2]

GUIDED PRACTICE

What does β_4, the coefficient of variable x_4 (Wii wheels), represent? What is the point estimate of β_4?[3]

GUIDED PRACTICE

Compute the residual of the first observation in Table 27.1 using the equation identified in Table 27.4.[4]

EXAMPLE

In Table 27.3, we estimated a coefficient for cond_new in of $b_1 = 10.90$ with a standard error of $SE_{b_1} = 1.26$ when using simple linear regression. Why might there be a difference between that estimate and the one in the multiple regression setting?

If we examined the data carefully, we would see that there is multicollinearity among some predictors. For instance, when we estimated the connection of the outcome price and predictor cond_new using simple linear regression, we were unable to control for other variables like the number of Wii wheels included in the auction. That model was biased by the confounding variable wheels. When we use both variables, this particular underlying and unintentional bias is reduced or eliminated (though bias from other confounding variables may still remain).

---

[2]$\widehat{price} = 36.21 + 5.13 \times$ cond_new $+ 1.08 \times$ stock_photo $- 0.03 \times$ duration $+ 7.29 \times$ *textttwheels*, with the $k = 4$ predictors but 5 coefficients (including the intercept).

[3]In the population of all auctions, it is the average difference in auction price for each additional Wii wheel included when holding the other variables constant. The point estimate is $b_4 = 7.29$

[4]$e_i = y_i - \hat{y}_i = 51.55 - 49.62 = 1.93$.

27.1.2 Computational approach to linear models

Previously, using a mathematical model, we investigated the coefficients associated with `cond_new` when predicting `price` in a linear model.

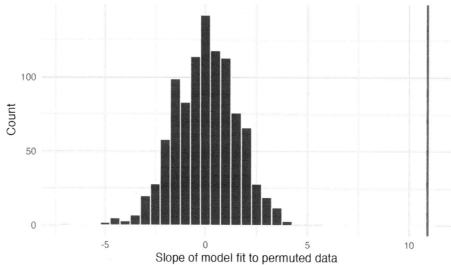

Figure 27.1: Estimated slopes from linear models (`price` regressed on `cond_new`) built on 1,000 randomized datasets. Each dataset was permuted under the null hypothesis.

EXAMPLE

In Figure 27.1, the red line (the observed slope) is far from the bulk of the histogram. Explain why the randomly permuted datasets produce slopes that are quite different from the observed slope.

The null hypothesis is that, in the population, there is no linear relationship between the `price` and the `cond_new` of the *Mario Kart* games. When the data are randomly permuted, prices are randomly assigned to a condition (new or used), so that the null hypothesis is forced to be true, i.e. permutation is done under the assumption that no relationship between the two variables exists. In the actual study, the new *Mario Kart* games do actually cost more (on average) than the used games! So the slope describing the actual observed relationship is not one that is likely to have happened in a randomly dataset permuted under the assumption that the null hypothesis is true.

GUIDED PRACTICE

Using the histogram in Figure 27.1, find the p-value and conclude the hypothesis test in the context of the problem (use words like price of the game and whether or not it is new).[5]

---

[5]The observed slope is 10.9 which is nowhere near the range of values for the permuted slopes (roughly -5 to 5). Because the observed slope is not a plausible value under the null distribution, the p-value is essentially zero. We reject the null hypothesis and claim that there is a relationship between whether the game is new (or not) and the average predicted price of the game.

GUIDED PRACTICE

Is the conclusion based on the histogram of randomized slopes consistent with the conclusion obtained using the mathematical model? Explain.[6]

Although knowing there is a relationship between the condition of the game and its price, we might be more interested in the difference in price, here given by the slope of the linear regression line. That is, β_1 represents the population value for the difference in price between new *Mario Kart* games and used games.

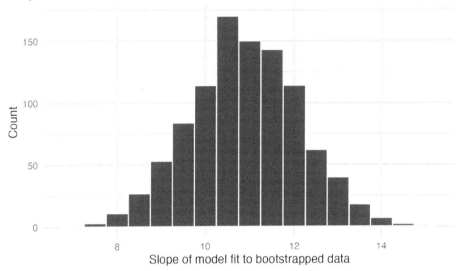

Figure 27.2: Estimated slopes from linear models (`price` regressed on `cond_new`) built on 1000 bootstrapped datasets. Each bootstrap dataset was a resample taken from the original Mario Kart auction data.

EXAMPLE

Figure 27.2 displays the slope estimates taken from bootstrap samples of the original data. Using the histogram, estimate the standard error of the slope. Is your estimate similar to the value of the standard error of the slope provided in the output of the mathematical linear model?

The slopes seem to vary from approximately 8 to 14. Using the empirical rule, we know that if a variable has a bell-shaped distribution, most of the observations will be with 2 standard errors of the center. Therefore, a rough approximation of the standard error is 1.5. The standard error given in Table 27.3 is 1.26 which is not too different from the value computed using the bootstrap approach.

GUIDED PRACTICE

Use Figure 27.2 to create a 90% standard error bootstrap confidence interval for the true slope. Interpret the interval in context.[7]

---

[6]The p-value in Table 27.3 is also essentially zero, so the null hypothesis is also rejected when the mathematical model approach is taken. Often, the mathematical and computational approaches to inference will give quite similar answers.

GUIDED PRACTICE

Use Figure 27.2 to create a 90% bootstrap percentile confidence interval for the true slope. Interpret the interval in context.[8]

27.1.3 Cross-validation

In Chapter 8, models were compared using R^2_{adj}. In Chapter 25, however, a computational approach was introduced to compare models by removing chunks of data one at a time and assessing how well the variables predicted the observations that had been held out.

Figure 27.3 was created by cross-validating models with the same variables as in Table 27.3 and Table 27.4. We applied 3-fold cross-validation, so 1/3 of the data was removed while 2/3 of the observations were used to build each model (first cond_new only and then cond_new, stock_photo, duration, and wheels). Note that each time 1/3 of the data is removed, the resulting model will produce slightly different coefficients.

The points in Figure 27.3 represent the prediction (x-axis) and residual (y-axis) for each observation run through the cross-validated model. In other words, the model is built (using the other 2/3) without the observation (which is in the 1/3 being used. The residuals give us a sense for how well the model will do at predicting observations which were not a part of the original dataset (e.g., future studies).

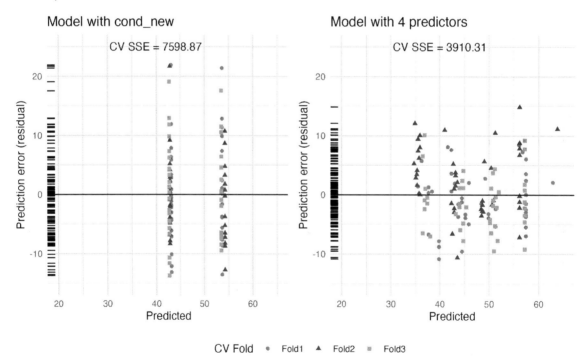

Figure 27.3: Cross-validation predictions and errors from linear models built on two different sets of variables. Left regressed price on cond_new; right regressed price on cond_new, stock_photo, duration, and wheels.

---

[7]Using the bootstrap SE method, we know the normal percentile is $z^\star = 1.645$, which gives a CI of $b_1 \pm 1.645 \cdot SE \rightarrow 10.9 \pm 1.645 \cdot 1.5 \rightarrow (8.43, 13.37)$. For games that are new, the average price is higher by between \$8.43 and \$13.37, with 90% confidence.

[8]Because there were 1000 bootstrap resamples, we look for the cutoffs which provide 50 bootstrap slopes on the left, 900 in the middle, and 50 on the right. Looking at the bootstrap histogram, the rough 95% confidence interval is \$9 to \$13.10. For games that are new, the average price is higher by between \$9.00 and \$13.10, with 90% confidence.

27.1. CASE STUDY: MARIO KART

GUIDED PRACTICE

In the second graph in Figure 27.3, note the point at roughly (predicted = 50 and error = 10). Estimate the observed and predicted value for that observation.[9]

GUIDED PRACTICE

In the second graph in Figure 27.3, for the same point at roughly (predicted = 50 and error = 10), describe which cross-validation fold(s) were used to build its prediction model.[10]

GUIDED PRACTICE

By noting the spread of the cross-validated prediction errors (on the y-axis) in Figure 27.3, which model should be chosen for a final report on these data?[11]

GUIDED PRACTICE

Using the summary statistic cross-validation sum of squared errors (CV SSE), which model should be chosen for a final report on these data?[12]

---

[9]The predicted value is roughly $\widehat{\text{price}} = \$50$. The observed value is roughly $\text{price}_i = \$60$ riders (using $e_i = y_i - \hat{y}_i$).

[10]The point appears to be in fold 2, so folds 1 and 3 were used to build the prediction model.

[11]The cross-validated residuals on cond_new only vary roughly from -15 to 15, while the cross-validated residuals on the four predictor model vary less, roughly from -10 to 10. Given the smaller residuals from the four predictor model, it seems as though the larger model is better.

[12]The CV SSE is smaller (by a factor of almost two!) for the model with four predictors. Using a single valued criterion (CV SSE) allows us to make a decision to choose the model with four predictors.

27.2 Interactive R tutorials

Navigate the concepts you've learned in this chapter in R using the following self-paced tutorials. All you need is your browser to get started!

Tutorial 6: Inferential modeling
https://openintrostat.github.io/ims-tutorials/06-model-infer

Tutorial 6 - Lesson 1: Inference in regression
https://openintro.shinyapps.io/ims-06-model-infer-01

Tutorial 6 - Lesson 2: Randomization test for slope
https://openintro.shinyapps.io/ims-06-model-infer-02

Tutorial 6 - Lesson 3: t-test for slope
https://openintro.shinyapps.io/ims-06-model-infer-03

Tutorial 6 - Lesson 4: Checking technical conditions for slope inference
https://openintro.shinyapps.io/ims-06-model-infer-04

Tutorial 6 - Lesson 5: Inference beyond the simple linear regression model
https://openintro.shinyapps.io/ims-06-model-infer-05

You can also access the full list of tutorials supporting this book at
https://openintrostat.github.io/ims-tutorials.

27.3 R labs

Further apply the concepts you've learned in this part in R with computational labs that walk you through a data analysis case study.

Multiple linear regression - Grading the professor
https://www.openintro.org/go?id=ims-r-lab-model-infer

You can also access the full list of labs supporting this book at
https://www.openintro.org/go?id=ims-r-labs.

Chapter A

Exercise solutions

A.1 Chapter 1

1. 23 observations and 7 variables.
3. (a) "Is there an association between air pollution exposure and preterm births?" (b) 143,196 births in Southern California between 1989 and 1993. (c) Measurements of carbon monoxide, nitrogen dioxide, ozone, and particulate matter less than $10\mu g/m^3$ (PM_{10}) collected at air-quality-monitoring stations as well as length of gestation. Continuous numerical variables.
5. (a) "What is the effect of gamification on learning outcomes compared to traditional teaching methods?" (b) 365 college students taking a statistics course (c) Gender (categorical), level of studies (categorical, ordinal), academic major (categorical), expertise in English language (categorical, ordinal), us of personal computers and games (categorical, ordinal), treatment group (categorical), score (numerical, discrete).
7. (a) Treatment: $10/43 = 0.23 \to 23\%$. (b) Control: $2/46 = 0.04 \to 4\%$. (c) A higher percentage of patients in the treatment group were pain free 24 hours after receiving acupuncture. (d) It is possible that the observed difference between the two group percentages is due to chance. (e) Explanatory: acupuncture or not. Response: if the patient was pain free or not.
9. (a) Experiment; researchers are evaluating the effect of fines on parents' behavior related to picking up their children late from daycare. (b) 10 cases: the daycare centers. (c) Number of late pickups (discrete numerical). (d) Week (numerical, discrete), group (categorical, nominal), number of late pickups (numerical discrete), and study period (categorical, ordinal).
11. (a) 344 cases (penguins) are included in the data. (b) There are 4 numerical variables in the data: bill length, bill depth, and flipper length (measured in millimeters) and body mass (measured in grams). They are all continuous. (c) There are 3 categorical variables in the data: species (Adelie, Chinstrap, Gentoo), island (Torgersen, Biscoe, and Dream), and sex (female and male).
13. (a) Airport ownership status (public/private), airport usage status (public/private), region (Central, Eastern, Great Lakes, New England, Northwest Mountain, Southern, Southwest, Western Pacific), latitude, and longitude. (b) Airport ownership status: categorical, not ordinal. Airport usage status: categorical, not ordinal. Region: categorical, not ordinal. Latitude: numerical, continuous. Longitude: numerical, continuous.
15. (a) Year, number of baby girls named Fiona born in that year, nation. (b) Year (numerical, discrete), number of baby girls named Fiona born in that year (numerical, discrete), nation (categorical, nominal).
17. (a) County, state, driver's race, whether the car was searched or not, and whether the driver was arrested or not. (b) All categorical, non-ordinal. (c) Response: whether the car was searched or not. Explanatory: race of the driver.
19. (a) Observational study. (b) Dog: Lucy. Cat: Luna. (c) Oliver and Lily. (d) Positive, as the popularity of a name for dogs increases, so does the popularity of that name for cats.

A.2 Chapter 2

1. (a) Population mean, $\mu_{2007} = 52$; sample mean, $\bar{x}_{2008} = 58$. (b) Population mean, $\mu_{2001} = 3.37$; sample mean, $\bar{x}_{2012} = 3.59$.

3. (a) Population: all births, sample: 143,196 births between 1989 and 1993 in Southern California. (b) If births in this time span at the geography can be considered to be representative of all births, then the results are generalizable to the population of Southern California. However, since the study is observational the findings cannot be used to establish causal relationships.

5. (a) The population of interest is all college students studying statistics. The sample consists of 365 such students. (b) If the students in this sample, who are likely not randomly sampled, can be considered to be representative of all college students studying statistics, then the results are generalizable to the population defined above. This is probably not a reasonable assumption since these students are from two specific majors only. Additionally, since the study is experimental, the findings can be used to establish causal relationships.

7. (a) Observation. (b) Variable. (c) Sample statistic (mean). (d) Population parameter (mean).

9. (a) Observational. (b) Use stratified sampling to randomly sample a fixed number of students, say 10, from each section for a total sample size of 40 students.

11. (a) Positive, non-linear, somewhat strong. Countries in which a higher percentage of the population have access to the internet also tend to have higher average life expectancies, however rise in life expectancy trails off before around 80 years old. (b) Observational. (c) Wealth: countries with individuals who can widely afford the internet can probably also afford basic medical care. (Note: Answers may vary.)

13. (a) Simple random sampling is okay. In fact, it's rare for simple random sampling to not be a reasonable sampling method! (b) The student opinions may vary by field of study, so the stratifying by this variable makes sense and would be reasonable. (c) Students of similar ages are probably going to have more similar opinions, and we want clusters to be diverse with respect to the outcome of interest, so this would **not** be a good approach. (Additional thought: the clusters in this case may also have very different numbers of people, which can also create unexpected sample sizes.)

15. (a) The cases are 200 randomly sampled men and women. (b) The response variable is attitude towards a fictional microwave oven. (c) The explanatory variable is dispositional attitude. (d) Yes, the cases are sampled randomly, recruited online using Amazon's Mechanical Turk. (e) This is an observational study since there is no random assignment to treatments. (f) No, we cannot establish a causal link between the explanatory and response variables since the study is observational. (g) Yes, the results of the study can be generalized to the population at large since the sample is random.

17. (a) Simple random sample. Non-response bias, if only those people who have strong opinions about the survey responds their sample may not be representative of the population. (b) Convenience sample. Under coverage bias, their sample may not be representative of the population since it consists only of their friends. It is also possible that the study will have non-response bias if some choose to not bring back the survey. (c) Convenience sample. This will have a similar issues to handing out surveys to friends. (d) Multi-stage sampling. If the classes are similar to each other with respect to student composition this approach should not introduce bias, other than potential non-response bias.

19. (a) Exam performance. (b) Light level: fluorescent overhead lighting, yellow overhead lighting, no overhead lighting (only desk lamps). (c) Wearing glasses or not.

21. (a) Exam performance. (b) Light level (overhead lighting, yellow overhead lighting, no overhead lighting) and noise level (no noise, construction noise, and human chatter noise). (c) Since the researchers want to ensure equal representation of those wearing glasses and not wearing glasses, wearing glasses is a blocking variable.

23. Need randomization and blinding. One possible outline: (1) Prepare two cups for each participant, one containing regular Coke and the other containing Diet Coke. Make sure the cups ar identical and contain equal amounts of soda. Label the cups (regular) and B (diet). (Be sure to randomize A and B for each trial!) (2) Give each participant the two cups, one cup at a time, in random order, and ask the participant to record a value that indicates ho much she liked the beverage. Be sure that neither the participant nor the person handing out the cups knows the identity of th beverage to make this a double-blind experiment. (Answers may vary.)

A.3. CHAPTER 3

25. (a) Experiment. (b) Treatment: 25 grams of chia seeds twice a day, control: placebo. (c) Yes, gender. (d) Yes, single blind since the patients were blinded to the treatment they received. (e) Since this is an experiment, we can make a causal statement. However, since the sample is not random, the causal statement cannot be generalized to the population at large.

27. (a) Non-responders may have a different response to this question, e.g., parents who returned the surveys likely don't have difficulty spending time with their children. (b) It is unlikely that the women who were reached at the same address 3 years later are a random sample. These missing responders are probably renters (as opposed to homeowners) which means that they might have a lower socio-economic status than the respondents. (c) There is no control group in this study, this is an observational study, and there may be confounding variables, e.g., these people may go running because they are generally healthier and/or do other exercises.

29. (a) Randomized controlled experiment. (b) Explanatory: treatment group (categorical, with 3 levels). Response variable: Psychological well-being. (c) No, because the participants were volunteers. (d) Yes, because it was an experiment. (e) The statement should say "evidence" instead of "proof".

A.3 Chapter 3

Application chapter, no exercises.

A.4 Chapter 4

1. (a) We see the order of the categories and the relative frequencies in the bar plot. (b) There are no features that are apparent in the pie chart but not in the bar plot. (c) We usually prefer to use a bar plot as we can also see the relative frequencies of the categories in this graph.

3. (a) The horizontal locations at which the age groups break into the various opinion levels differ, which indicates that likelihood of supporting protests varies by age group. Two variables may be associated. (b) Answers may vary. Political ideology/leaning and education level.

5. (a) Number of participants in each group. (b) Proportion of survival. (c) The standardized bar plot should be displayed as a way to visualize the survival improvement in the treatment versus the control group.

A.5 Chapter 5

1. (a) Positive association: mammals with longer gestation periods tend to live longer as well. (b) Association would still be positive. (c) No, they are not independent. See part (a).

3. The graph below shows a ramp up period. There may also be a period of exponential growth at the start before the size of the petri dish becomes a factor in slowing growth.

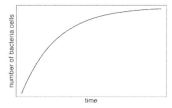

5. (a) Decrease: the new score is smaller than the mean of the 24 previous scores. (b) Calculate a weighted mean. Use a weight of 24 for the old mean and 1 for the new mean: $(24 \times 74 + 1 \times 64)/(24 + 1) = 73.6$. (c) The new score is more than 1 standard deviation away from the previous mean, so increase.

7. Any 10 employees whose average number of days off is between the minimum and the mean number of days off for the entire workforce at this plant.

9. (a) Dist B has a higher mean since 20 > 13, and a higher standard deviation since 20 is further from the rest of the data than 13. (b) Dist A has a higher mean since $-20 > -40$, and Dist B has a higher standard deviation since -40 is farther away from the rest of the data than -20. (c) Dist B has a higher mean since all values in this Dist Are higher than those in Dist A, but both distribution have the same standard deviation since they are equally variable around their respective means. (d) Both distributions have the same mean since they're both centered at 300, but Dist B has a higher standard deviation since the observations are farther from the mean than in Dist A.

11. (a) About 30. (b) Since the distribution is right skewed the mean is higher than the median. (c) Q1: between 15 and 20, Q3: between 35 and 40, IQR: about 20. (d) Values that are considered to be unusually low or high lie more than 1.5×IQR away from the quartiles. Upper fence: Q3 + 1.5 × IQR = 37.5 + 1.5 × 20 = 67.5; Lower fence: Q1 - 1.5 × IQR = 17.5 + 1.5 × 20 = −12.5; The lowest AQI recorded is not lower than 5 and the highest AQI recorded is not higher than 65, which are both within the fences. Therefore none of the days in this sample would be considered to have an unusually low or high AQI.

13. The histogram shows that the distribution is bimodal, which is not apparent in the box plot. The box plot makes it easy to identify more precise values of observations outside of the whiskers.

15. (a) Right skewed, there is a natural boundary at 0 and only a few people have many pets. Center: median, variability: IQR. (b) Right skewed, there is a natural boundary at 0 and only a few people live a very long distance from work. Center: median, variability: IQR. (c) Symmetric. Center: mean, variability: standard deviation. (d) Left skewed. Center: median, variability: IQR. (e) Left skewed. Center: median, variability: IQR.

17. No, we would expect this distribution to be right skewed. There are two reasons for this: there is a natural boundary at 0 (it is not possible to watch less than 0 hours of TV) and the standard deviation of the distribution is very large compared to the mean.

19. No, the outliers are likely the maximum and the minimum of the distribution so a statistic based on these values cannot be robust to outliers.

21. The 75th percentile is 82.5, so 5 students will get an A. Also, by definition 25% of students will be above the 75th percentile.

23. (a) If $\frac{\bar{x}}{median} = 1$, then $\bar{x} = median$. This is most likely to be the case for symmetric distributions. (b) If $\frac{\bar{x}}{median} < 1$, then $\bar{x} < median$. This is most likely to be the case for left skewed distributions, since the mean is affected (and pulled down) by the lower values more so than the median. (c) If $\frac{\bar{x}}{median} > 1$, then $\bar{x} > median$. This is most likely to be the case for right skewed distributions, since the mean is affected (and pulled up) by the higher values more so than the median.

25. (a) The distribution of percentage of population that is Hispanic is extremely right skewed with majority of counties with less than 10% Hispanic residents. However there are a few counties that have more than 90% Hispanic population. It might be preferable to, in certain analyses, to use the log-transformed values since this distribution is much less skewed. (b) The map reveals that counties with higher proportions of Hispanic residents are clustered along the Southwest border, all of New Mexico, a large swath of Southwest Texas, the bottom two-thirds of California, and in Southern Florida. In the map all counties with more than 40% of Hispanic residents are indicated by the darker shading, so it is impossible to discern the how high Hispanic percentages go. The histogram reveals that there are counties with over 90% Hispanic residents. The histogram is also useful for estimating measures of center and spread. (c) Both visualizations are useful, but if we could only examine one, we should examine the map since it explicitly ties geographic data to each county's percentage.

A.6 Chapter 6

Application chapter, no exercises.

A.7 Chapter 7

1. (a) The residual plot will show randomly distributed residuals around 0. The variance is also approximately constant. (b) The residuals will show a fan shape, with higher variability for smaller x. There will also be many points on the right above the line. There is trouble with the model being fit here.
3. (a) Strong relationship, but a straight line would not fit the data. (b) Strong relationship, and a linear fit would be reasonable. (c) Weak relationship, and trying a linear fit would be reasonable. (d) Moderate relationship, but a straight line would not fit the data. (e) Strong relationship, and a linear fit would be reasonable. (f) Weak relationship, and trying a linear fit would be reasonable.
5. (a) Exam 2 since there is less of a scatter in the plot of course grade versus exam 2. Notice that the relationship between Exam 1 and the course grade appears to be slightly nonlinear. (b) (Answers may vary.) If Exam 2 is cumulative it might be a better indicator of how a student is doing in the class.
7. (a) $r = -0.7 \to$ (4). (b) $r = 0.45 \to$ (3). (c) $r = 0.06 \to$ (1). (d) $r = 0.92 \to$ (2).
9. (a) There is a moderate, positive, and linear relationship between shoulder girth and height. (b) Changing the units, even if just for one of the variables, will not change the form, direction or strength of the relationship between the two variables.
11. (a) There is a somewhat weak, positive, possibly linear relationship between the distance traveled and travel time. There is clustering near the lower left corner that we should take special note of. (b) Changing the units will not change the form, direction or strength of the relationship between the two variables. If longer distances measured in miles are associated with longer travel time measured in minutes, longer distances measured in kilometers will be associated with longer travel time measured in hours. (c) Changing units doesn't affect correlation: $r = 0.636$.
13. In each part, we can write the age of one partner as a linear function of the other. (a) $age_{P1} = age_{P2} + 3$. (b) $age_{P1} = age_{P2} - 2$. (c) $age_{P1} = 2 \times age_{P2}$. Since the slopes are positive and these are perfect linear relationships, the correlation will be exactly 1 in all three parts. An alternative way to gain insight into this solution is to create a mock dataset, e.g., 5 women aged 26, 27, 28, 29, and 30, then find the husband ages for each wife in each part and create a scatterplot.
15. Correlation: no units. Intercept: cal. Slope: cal/cm.
17. Over-estimate. Since the residual is calculated as *observed* − *predicted*, a negative residual means that the predicted value is higher than the observed value.
19. (a) There is a positive, moderate, linear association between number of calories and amount of carbohydrates. In addition, the amount of carbohydrates is more variable for menu items with higher calories, indicating non-constant variance. There also appear to be two clusters of data: a patch of about a dozen observations in the lower left and a larger patch on the right side. (b) Explanatory: number of calories. Response: amount of carbohydrates (in grams). (c) With a regression line, we can predict the amount of carbohydrates for a given number of calories. This may be useful if only calorie counts for the food items are posted but the amount of carbohydrates in each food item is not readily available. (d) Food menu items with higher predicted protein are predicted with higher variability than those without, suggesting that the model is doing a better job predicting protein amount for food menu items with lower predicted proteins.
21. (a) First calculate the slope: $b_1 = R \times s_y/s_x = 0.636 \times 113/99 = 0.726$. Next, make use of the fact that the regression line passes through the point $(\bar{x}, \bar{y})$: $\bar{y} = b_0 + b_1 \times \bar{x}$. Plug in $\bar{x}$, $\bar{y}$, and b_1, and solve for b_0: 51. Solution: $\widehat{travel\ time} = 51 + 0.726 \times distance$. (b) b_1: For each additional mile in distance, the model predicts an additional 0.726 minutes in travel time. b_0: When the distance travelled is 0 miles, the travel time is expected to be 51 minutes. It does not make sense to have a travel distance of 0 miles in this context. Here, the y-intercept serves only to adjust the height of the line and is meaningless by itself. (c) $R^2 = 0.636^2 = 0.40$. About 40% of the variability in travel time is accounted for by the model, i.e., explained by the distance travelled. (d) $\widehat{travel\ time} = 51 + 0.726 \times distance = 51 + 0.726 \times 103 \approx 126$ minutes. (Note: we should be cautious in our predictions with this model since we have not yet evaluated whether it is a well-fit model.) (e) $e_i = y_i - \hat{y}_i = 168 - 126 = 42$ minutes. A positive residual

means that the model underestimates the travel time. (f) No, this calculation would require extrapolation.

23. (a) $\widehat{\text{poverty}} = 4.60 + 2.05 \times \text{unemployment\_rate}$. (b) The model predicts a poverty rate of 4.60% for counties with 0% unemployment, on average. This is not a meaningful value as no counties have such low unexmployment, it just serves to adjust the height of the regression line. (c) For each additional percentage increase in unemployment rate, poverty rate is predicted to be higher, on average, by 2.05%. (d) Poverty level explains 46% of the variability in poverty rates in US counties. (e) $\sqrt{0.46} = 0.678$.

25. (a) There is an outlier in the bottom right. Since it is far from the center of the data, it is a point with high leverage. It is also an influential point since, without that observation, the regression line would have a very different slope. (b) There is an outlier in the bottom right. Since it is far from the center of the data, it is a point with high leverage. However, it does not appear to be affecting the line much, so it is not an influential point. (c) The observation is in the center of the data (in the x-axis direction), so this point does *not* have high leverage. This means the point won't have much effect on the slope of the line and so is not an influential point.

27. (a) There is a negative, moderate-to-strong, somewhat linear relationship between percent of families who own their home and the percent of the population living in urban areas in 2010. There is one outlier: a state where 100% of the population is urban. The variability in the percent of homeownership also increases as we move from left to right in the plot. (b) The outlier is located in the bottom right corner, horizontally far from the center of the other points, so it is a point with high leverage. It is an influential point since excluding this point from the analysis would greatly affect the slope of the regression line.

29. (a) True. (b) False, correlation is a measure of the linear association between any two numerical variables.

31. (a) $r = 0.7 \to (1)$ (b) $r = 0.09 \to (4)$ (c) $r = -0.91 \to (2)$ (d) $r = 0.96 \to (3)$.

A.8 Chapter 8

1. Annika is right. All variables being highly correlated, including the predictor variables being highly correlated with each other, is not desirable as this would result in multicollinearity.

3. No, they shouldn't include all variables as days_since_start and days_since_race are perfectly correlated with each other. They should only include one of them.

5. (a) $\widehat{\text{weight}} = 7.270 - 0.593 \times \text{habit}_{\text{smoker}}$. (b) The estimated body weight of babies born to smoking mothers is 0.593 ounds lower than those who are born to non-smoking mothers. Smoker: $\widehat{\text{weight}} = 7.270 - 0.593 \times 1 = 6.68$ pounds. Non-smoker: $\widehat{\text{weight}} = 7.270 - 0.593 \times 0 = 7.270$ pounds.

7. (a) Horror movies. (b) Not necessarily, the change in adjusted R^2 is quite small.

9. (a) $\widehat{\text{weight}} = -3.82 + 0.26 \times \text{weeks} + 0.02 \times \text{mage} + 0.37 \times \text{sex\_male} + 0.02 \times \text{visits} - 0.43 \times \text{habit\_smoker}$. (b) b_{weeks}: The model predicts a 0.26 pound increase in the birth weight of the baby for each additional week in length of pregnancy, all else held constant. $b_{\text{habit}_{\text{smoker}}}$: The model predicts a 0.43 pound decrease in the birth weight of the babies born to smoker mothers compared to non-smokers, all else held constant. (c) Habit might be correlated with one of the other variables in the model, which introduces multicollinearity and complicates model estimation. (d) 7.13~lbs.

11. Remove age.

13. Add weeks.

A.9 Chapter 9

1. (a) False. The line is fit to predict the probability of success, not the binary outcome. (b) False. Residuals are not used in logistic regression like they are in linear regression because the observed value is always either zero or one (and the predicted value is a probability). The goal of the logistic regression is not to get a perfect prediction (of zero or one), so minimizing residuals is not part of the modeling process. (c) True.

3. (a) There are a few potential outliers, e.g., on the left in the variable, but nothing that will be of serious concern in a dataset this large. (b) When coefficient estimates are sensitive to which variables are included in the model, this typically indicates that some variables are collinear. For example, a possum's gender may be related to its head length, which would explain why the coefficient (and p-value) changed when we removed the variable. Likewise, a possum's skull width is likely to be related to its head length, probably even much more closely related than the head length was to gender.

5. (a) The logistic model relating $\hat{p}$ to the predictors may be written as $\log\left(\frac{\hat{p}}{1-\hat{p}}\right) = 33.5095 - 1.4207 \times \texttt{sex}_{\texttt{male}} - 0.2787 \times \texttt{skull\_w} + 0.5687 \times \texttt{total\_l} - 1.8057 \times \texttt{tail\_l}$. Only `total_l` has a positive association with a possum being from Victoria. (b) $\hat{p} = 0.0062$. While the probability is very near zero, we have not run diagnostics on the model. We might also be a little skeptical that the model will remain accurate for a possum found in a US zoo. For example, perhaps the zoo selected a possum with specific characteristics but only looked in one region. On the other hand, it is encouraging that the possum was caught in the wild. (Answers regarding the reliability of the model probability will vary.)

7. (a) The variable `exclaim_subj` should be removed, since it's removal reduces AIC the most (and the resulting model has lower AIC than the None Dropped model). (b) The variable `cc` should be removed. (c) Removing any variable will increase AIC, so we should not remove any variables from this set.

A.10 Chapter 10

Application chapter, no exercises.

A.11 Chapter 11

1. (a) Mean. Each student reports a numerical value: a number of hours. (b) Mean. Each student reports a number, which is a percentage, and we can average over these percentages. (c) Proportion. Each student reports Yes or No, so this is a categorical variable and we use a proportion. (d) Mean. Each student reports a number, which is a percentage like in part (b). (e) Proportion. Each student reports whether or not s/he expects to get a job, so this is a categorical variable and we use a proportion.

3. (a) Alternative. (b) Null. (c) Alternative. (d) Alternative. (e) Null. (f) Alternative. (g) Null.

5. (a) $H_0: \mu = 8$ (On average, New Yorkers sleep 8 hours a night.) $H_A: \mu < 8$ (On average, New Yorkers sleep less than 8 hours a night.) (b) $H_0: \mu = 15$ (The average amount of company time each employee spends not working is 15 minutes for March Madness.) $H_A: \mu > 15$ (The average amount of company time each employee spends not working is greater than 15 minutes for March Madness.)

7. (a) (i) False. Instead of comparing counts, we should compare percentages of people in each group who suffered cardiovascular problems. (ii) True. (iii) False. Association does not imply causation. We cannot infer a causal relationship based on an observational study. The difference from part (ii) is subtle. (iv) True. (b) Proportion of all patients who had cardiovascular problems: $\frac{7,979}{227,571} \approx 0.035$ (c) The expected number of heart attacks in the Rosiglitazone group, if having cardiovascular problems and treatment were independent, can be calculated as the number of patients in that group multiplied by the overall cardiovascular problem rate in the study: $67,593 * \frac{7,979}{227,571} \approx 2370$. (d) (i) H_0: The treatment and cardiovascular problems are independent. They have no relationship, and the difference in incidence rates between the Rosiglitazone and Pioglitazone groups is due to chance. H_A: The treatment and cardiovascular problems are not independent. The difference in the incidence rates between the Rosiglitazone and Pioglitazone groups is not due to chance and Rosiglitazone is associated with an increased risk of serious cardiovascular problems. (ii) A higher number of patients with cardiovascular problems than expected under the assumption of independence would provide support for the alternative hypothesis as this would suggest that Rosiglitazone increases the

risk of such problems. (iii) In the actual study, we observed 2,593 cardiovascular events in the Rosiglitazone group. In the 100 simulations under the independence model, the simulated differences were never so high, which suggests that the actual results did not come from the independence model. That is, the variables do not appear to be independent, and we reject the independence model in favor of the alternative. The study's results provide convincing evidence that Rosiglitazone is associated with an increased risk of cardiovascular problems.

A.12 Chapter 12

1. (a) The statistic is the sample proportion (0.289); the parameter is the population proportion (unknown). (b) $\hat{p}$ and p. (c) Bootstrap sample proportion. (d) 0.289. (e) Roughly (0.22, 0.35). (f) We can be 95% confident that between 22% and 35% of all YouTube videos take place outdoors.
3. With 98% confidence, the true proportion of all US adult Twitter users (in 2013) who get at least some of the news from Twitter is between 0.48 and 0.56.
5. (a) A or perhaps D. (b) A, B, C, or D. (c) B or C. (d) B. (e) None.
7. (a) This claim is reasonable, since the entire interval lies above 50%. (b) The value of 70% lies outside of the interval, so we have convincing evidence that the researcher's conjecture is wrong. (c) A 90% confidence interval will be narrower than a 95% confidence interval. Even without calculating the interval, we can tell that 70% would not fall in the interval, and we would reject the researcher's conjecture based on a 90% confidence level as well.

A.13 Chapter 13

1. (a) Recall that the general formula is *point estimate* $\pm z^\star \times SE$. First, identify the three different values. The point estimate is 45%, $z^\star = 1.96$ for a 95% confidence level, and $SE = 1.2\%$. Then, plug the values into the formula: $45\% \pm 1.96 \times 1.2\%$ → (42.6%, 47.4%) We are 95% confident that the proportion of US adults who live with one or more chronic conditions is between 42.6% and 47.4%. (b) (i) False. Confidence intervals provide a range of plausible values, and sometimes the truth is missed. A 95% confidence interval "misses" about 5% of the time. (ii) True. Notice that the description focuses on the true population value. (iii) True. If we examine the 95% confidence interval, we can see that 50% is not included in this interval. This means that in a hypothesis test, we would reject the null hypothesis that the proportion is 0.5. (iv) False. The standard error describes the uncertainty in the overall estimate from natural fluctuations due to randomness, not the uncertainty corresponding to individuals' responses.
3. A Z score of 0.47 denotes that the sample proportion is 0.47 standard errors greater than the hypothesized value of the population proportion.
5. (a) Sampling distribution. (b) To know whether the distribution is skewed, we need to know the proportion. We've been told the proportion is likely above 5% and below 30%, and the success-failure condition would be satisfied for any of these values. If the population proportion is in this range, the sampling distribution will be symmetric. (c) Standard error. (d) The distribution will tend to be more variable when we have fewer observations per sample.

A.14 Chapter 14

1. (a) H_0: Anti-depressants do not affect the symptoms of Fibromyalgia. H_A: Anti-depressants do affect the symptoms of Fibromyalgia (either helping or harming). (b) Concluding that anti-depressants either help or worsen Fibromyalgia symptoms when they actually do neither. (c) Concluding that anti-depressants do not affect Fibromyalgia symptoms when they actually do.
3. (a) H_0: The restaurant meets food safety and sanitation regulations. H_A: The restaurant does not meet food safety and sanitation regulations. (b) The food safety inspector concludes that the restaurant does not meet food safety and sanitation regulations and shuts down the

restaurant when the restaurant is actually safe. (c) The food safety inspector concludes that the restaurant meets food safety and sanitation regulations and the restaurant stays open when the restaurant is actually not safe. (d) A Type 1 Error may be more problematic for the restaurant owner since his restaurant gets shut down even though it meets the food safety and sanitation regulations. (e) A Type 2 Error may be more problematic for diners since the restaurant deemed safe by the inspector is actually not. (f) Strong evidence. Diners would rather a restaurant that meet the regulations get shut down than a restaurant that doesn't meet the regulations not get shut down.
5. The hypotheses should be about the population proportion (p), not the sample proportion. The null hypothesis should have an equal sign. The alternative hypothesis should have a not-equals sign, and it should reference the null value, $p_0 = 0.6$, not the observed sample proportion. The correct way to set up these hypotheses is: $H_0 : p = 0.6$ and $H_A : p \neq 0.6$.

A.15 Chapter 15

Application chapter, no exercises.

A.16 Chapter 16

1. First, the hypotheses should be about the population proportion (p), not the sample proportion. Second, the null value should be what we are testing (0.25), not the observed value (0.29). The correct way to set up these hypotheses is: $H_0 : p = 0.25$ and $H_A : p > 0.25$.
3. (a) $H_0 : p = 0.20$, $H_A : p > 0.20$. (b) $\hat{p} = 159/650 = 0.245$. (c) Answers will vary. Each student can be represented with a card. Take 100 cards, 20 black cards representing those who support proposals to defund police departments and 80 red cards representing those who do not. Shuffle the cards and draw with replacement (shuffling each time in between draws) 650 cards representing the 650 respondents to the poll. Calculate the proportion of black cards in this sample, $\hat{p}_{sim}$, i.e., the proportion of those who upport proposals to defund police departments. The p-value will be the proportion of simulations where $\hat{p}_{sim} \geq 0.245$. (Note: We would generally use a computer to perform the simulations.) (d) There 1 only one simulated proportion that is at least 0.245, therefore the approximate p-value is 0.001. Your p-value may vary slightly since it is based on a visual estimate. Since the p-value is smaller than 0.05, we reject H_0. The data provide convincing evidence that the proportion of Seattle adults who support proposals to defund police departments is greater than 0.20, i.e. more than one in five.
5. (a) $H_0 : p = 0.5$, $H_A : p \neq 0.5$. (b) The p-value is roughly 0.4, There is not evidence in the data (possibly because there are only 7 cats being measured!) to conclude that the cats have a preference one way or the other between the two shapes.
7. (a) $SE(\hat{p}) = 0.189$. (c) Roughly 0.188. (c) Yes. (d) No. (e) The parametric bootstrap is discrete (only a few distinct options) and the mathematical model is continuous (infinite options on a continuum).
9. (a) The parametric bootstrap simulation was done with $p = 0.7$, and the data bootstrap simulation was done with $p = 0.6$. (b) The parametric bootstrap is centered at 0.7; the data bootstrap is centered at 0.6. (c) The standard error of the sample proportion is given to be roughly 0.1 for both histograms. (d) Both histograms are reasonably symmetric. Note that histograms which describe the variability of proportions become more skewed as the center of the distribution gets closer to 1 (or zero) because the boundary of 1.0 restricts the symmetry of the tail of the distribution. For this reason, the parametric bootstrap histogram is slightly more skewed (left).
11. (a) The parametric bootstrap for testing. The data bootstrap distribution for confidence intervals. (b) $H_0 : p = 0.7$; $H_A : p \neq 0.7$. p-value > 0.05. There is no evidence that the proportion of full-time statistics majors who work is different from 70%. (c) We are 98% confident that the true proportion of all full-time student statistics majors who work at least 5 hours per week is between 35% and 80%. (d) Using $z^\star = 2.33$, the 98% confidence interval is 0.367 to 0.833.

13. (a) False. Doesn't satisfy success-failure condition. (b) True. The success-failure condition is not satisfied. In most samples we would expect $\hat{p}$ to be close to 0.08, the true population proportion. While $\hat{p}$ can be much above 0.08, it is bound below by 0, suggesting it would take on a right skewed shape. Plotting the sampling distribution would confirm this suspicion. (c) False. $SE_{\hat{p}} = 0.0243$, and $\hat{p} = 0.12$ is only $\frac{0.12-0.08}{0.0243} = 1.65$ SEs away from the mean, which would not be considered unusual. (d) True. $\hat{p} = 0.12$ is 2.32 standard errors away from the mean, which is often considered unusual. (e) False. Decreases the SE by a factor of $1/\sqrt{2}$.

15. (a) True. See the reasoning of 6.1(b). (b) True. We take the square root of the sample size in the SE formula. (c) True. The independence and success-failure conditions are satisfied. (d) True. The independence and success-failure conditions are satisfied.

17. (a) False. A confidence interval is constructed to estimate the population proportion, not the sample proportion. (b) True. 95% CI: 82% ± 2%. (c) True. By the definition of the confidence level. (d) True. Quadrupling the sample size decreases the SE and ME by a factor of $1/\sqrt{4}$. (e) True. The 95% CI is entirely above 50%.

19. With a random sample, independence is satisfied. The success-failure condition is also satisfied. $ME = z^{\star}\sqrt{\frac{\hat{p}(1-\hat{p})}{n}} = 1.96\sqrt{\frac{0.56 \times 0.44}{600}} = 0.0397 \approx 4\%$.

21. (a) No. The sample only represents students who took the SAT, and this was also an online survey. (b) (0.5289, 0.5711). We are 90% confident that 53% to 57% of high school seniors who took the SAT are fairly certain that they will participate in a study abroad program in college. (c) 90% of such random samples would produce a 90% confidence interval that includes the true proportion. (d) Yes. The interval lies entirely above 50%.

23. (a) We want to check for a majority (or minority), so we use the following hypotheses: $H_0 : p = 0.5$ and $H_A : p \neq 0.5$. We have a sample proportion of $\hat{p} = 0.55$ and a sample size of $n = 617$ independents. Since this is a random sample, independence is satisfied. The success-failure condition is also satisfied: 617×0.5 and $617 \times (1 - 0.5)$ are both at least 10 (we use the null proportion $p_0 = 0.5$ for this check in a one-proportion hypothesis test). Therefore, we can model $\hat{p}$ using a normal distribution with a standard error of $SE = \sqrt{\frac{p(1-p)}{n}} = 0.02$. (We use the null proportion $p_0 = 0.5$ to compute the standard error for a one-proportion hypothesis test.) Next, we compute the test statistic: $Z = \frac{0.55-0.5}{0.02} = 2.5$. This yields a one-tail area of 0.0062, and a p-value of $2 \times 0.0062 = 0.0124$. Because the p-value is smaller than 0.05, we reject the null hypothesis. We have strong evidence that the support is different from 0.5, and since the data provide a point estimate above 0.5, we have strong evidence to support this claim by the TV pundit. (b) No. Generally we expect a hypothesis test and a confidence interval to align, so we would expect the confidence interval to show a range of plausible values entirely above 0.5. However, if the confidence level is misaligned (e.g., a 99% confidence level and a $\alpha = 0.05$ significance level), then this is no longer generally true.

25. (a) $H_0 : p = 0.5$. $H_A : p > 0.5$. Independence (random sample, < 10% of population) is satisfied, as is the success-failure conditions (using $p_0 = 0.5$, we expect 40 successes and 40 failures). $Z = 2.91 \rightarrow$ p-value = 0.0018. Since the p-value < 0.05, we reject the null hypothesis. The data provide strong evidence that the rate of correctly identifying a soda for these people is significantly better than just by random guessing. (b) If in fact people cannot tell the difference between diet and regular soda and they randomly guess, the probability of getting a random sample of 80 people where 53 or more identify a soda correctly would be 0.0018.

27. (a) The sample is from all computer chips manufactured at the factory during the week of production. We might be tempted to generalize the population to represent all weeks, but we should exercise caution here since the rate of defects may change over time. (b) The fraction of computer chips manufactured at the factory during the week of production that had defects. (c) Estimate the parameter using the data: $\hat{p} = \frac{27}{212} = 0.127$. (d) *Standard error* (or *SE*). (e) Compute the SE using $\hat{p} = 0.127$ in place of p: $SE \approx \sqrt{\frac{\hat{p}(1-\hat{p})}{n}} = \sqrt{\frac{0.127(1-0.127)}{212}} = 0.023$. (f) The standard error is the standard deviation of $\hat{p}$. A value of 0.10 would be about one standard error away from the observed value, which would not represent a very uncommon deviation. (Usually beyond about 2 standard errors is a good rule of thumb.) The engineer should not be surprised. (g) Recomputed standard error using $p = 0.1$: $SE = \sqrt{\frac{0.1(1-0.1)}{212}} = 0.021$. This value isn't very different, which is typical when the standard error is computed using relatively similar proportions (and even sometimes when those proportions are quite different!).

29. (a) The visitors are from a simple random sample, so independence is satisfied. The success-

failure condition is also satisfied, with both 64 and 752 − 64 = 688 above 10. Therefore, we can use a normal distribution to model $\hat{p}$ and construct a confidence interval. (b) The sample proportion is $\hat{p} = \frac{64}{752} = 0.085$. The standard error is $SE = \sqrt{\frac{0.085(1-0.085)}{752}} = 0.010$. (c) For a 90% confidence interval, use $z^\star = 1.65$. The confidence interval is $0.085 \pm 1.65 \times 0.010 \to (0.0685, 0.1015)$. We are 90% confident that 6.85% to 10.15% of first-time site visitors will register using the new design.

A.17 Chapter 17

1. (a) The parameter is $p_{Asican-Indian} - p_{Chinese}$. The statistic is $\hat{p}_{Asian-Indian} - \hat{p}_{Chinese} = 223/4373 - 279/4736 = -0.008$ (b) Roughly 0.005. (c) $H_0 : p_{Asian-Indian} - p_{Chinese} = 0$;, $H_A : p_{Asian-Indian} - p_{Chinese} \neq 0$. The evidence is borderline but worth further study. There is not strong evidence that the true difference in proportion of current smokers is different across the two ethnic groups.

3. (a) Roughly 0.00625. (b) We are 95% confident that the true proportion of Filipino Americans who are current smokers is between 5.28 and 7.72 percentage points higher in the control vaccine group than the proportion of Chinese Americans who smoke. (c) We are 95% confident that the true proportion of Filipino Americans who are current smokers is between 5.2 and 7.7 percentage points higher in the control vaccine group than the proportion of Chinese Americans who smoke.

5. (a) While the standard errors of the difference in proportion across the two graphs are roughly the same (approximately 0.012), the centers are not. Computational method A is centered at 0.07 (the difference in the observed sample proportions) and Computational method B is centered at 0. (b) What is the difference between the proportions of Bachelor's and Associate's students who believe that the COVID-19 pandemic will negatively impact their ability to complete the degree? (c) Is the proportion of Bachelor's students who believe that their ability to complete the degree will be negatively impacted by the COVID-19 pandemic different than that of Associate's students?

7. (a) 26 Yes and 94 No in Nevaripine and 10 Yes and 110 No in Lopinavir group. (b) $H_0 : p_N = p_L$. There is no difference in virologic failure rates between the Nevaripine and Lopinavir groups. $H_A : p_N \neq p_L$. There is some difference in virologic failure rates between the Nevaripine and Lopinavir groups. (c) Random assignment was used, so the observations in each group are independent. If the patients in the study are representative of those in the general population (something impossible to check with the given information), then we can also confidently generalize the findings to the population. The success-failure condition, which we would check using the pooled proportion ($\hat{p}_{pool} = 36/240 = 0.15$), is satisfied. $Z = 2.89 \to$ p-value $= 0.0039$. Since the p-value is low, we reject H_0. There is strong evidence of a difference in virologic failure rates between the Nevaripine and Lopinavir groups. Treatment and virologic failure do not appear to be independent.

9. (a) Standard error: $SE = \sqrt{\frac{0.79(1-0.79)}{347} + \frac{0.55(1-0.55)}{617}} = 0.03$. Using $z^\star = 1.96$, we get: $0.79 - 0.55 \pm 1.96 \times 0.03 \to (0.181, 0.299)$. We are 95% confident that the proportion of Democrats who support the plan is 18.1% to 29.9% higher than the proportion of Independents who support the plan. (b) True.

11. (a) In effect, we're checking whether men are paid more than women (or vice-versa), and we'd expect these outcomes with either chance under the null hypothesis: $H_0 : p = 0.5$ and $H_A : p \neq 0.5$. We'll use p to represent the fraction of cases where men are paid more than women. (b) There isn't a good way to check independence here since the jobs are not a simple random sample. However, independence doesn't seem unreasonable, since the individuals in each job are different from each other. The success-failure condition is met since we check it using the null proportion: $p_0 n = (1 - p_0)n = 10.5$ is greater than 10. We can compute the sample proportion, SE, and test statistic: $\hat{p} = 19/21 = 0.905$ and $SE = \sqrt{\frac{0.5 \times (1-0.5)}{21}} = 0.109$ and $Z = \frac{0.905 - 0.5}{0.109} = 3.72$. The test statistic Z corresponds to an upper tail area of about 0.0001, so the p-value is 2 times this value: 0.0002. Because the p-value is smaller than 0.05, we reject the notion that all these gender pay disparities are due to chance. Because we observe that men are paid more in a higher proportion of cases and we have rejected H_0, we can conclude that

men are being paid higher amounts in ways not explainable by chance alone. If you're curious for more info around this topic, including a discussion about adjusting for additional factors that affect pay, please see the following video by Healthcare Triage: youtu.be/aVhgKSULNQA.

13. (a) $H_0 : p = 0.5$. $H_A : p \neq 0.5$. Independence (random sample) is satisfied, as is the success-failure conditions (using $p_0 = 0.5$, we expect 40 successes and 40 failures). $Z = 2.91 \to$ the one tail area is 0.0018, so the p-value is 0.0036. Since the p-value < 0.05, we reject the null hypothesis. Since we rejected H_0 and the point estimate suggests people are better than random guessing, we can conclude the rate of correctly identifying a soda for these people is significantly better than just by random guessing. (b) If in fact people cannot tell the difference between diet and regular soda and they were randomly guessing, the probability of getting a random sample of 80 people where 53 or more identify a soda correctly (or 53 or more identify a soda incorrectly) would be 0.0036.

15. Before we can calculate a confidence interval, we must first check that the conditions are met. There aren't at least 10 successes and 10 failures in each of the four groups (treatment/control and yawn/not yawn), $(\hat{p}_C - \hat{p}_T)$ is not expected to be approximately normal and therefore cannot calculate a confidence interval for the difference between the proportions of participants who yawned in the treatment and control groups using large sample techniques and a critical Z score.

17. (a) False. The confidence interval includes 0. (b) False. We are 95% confident that 16% fewer to 2% Americans who make less than $40,000 per year are not at all personally affected by the government shutdown compared to those who make $40,000 or more per year. (c) False. As the confidence level decreases the width of the confidence level decreases as well. (d) True.

19. (a) Type 1. (b) Type 2. (c) Type 2.

21. No. The samples at the beginning and at the end of the semester are not independent since the survey is conducted on the same students.

23. (a) The proportion of the normal curve centered at -0.1 with a standard deviation of 0.15 that is less than -2 * standard error is 0.09. (b) The proportion of the normal curve centered at -0.4 with a standard deviation of 0.145 that is less than 2 * standard error is 0.78. (c) The proportion of the normal curve centered at -0.1 with a standard deviation of 0.0671 that is less than 2 * standard error is 0.31. (d) The proportion of the normal curve centered at -0.4 with a standard deviation of 0.0678 that is less than 2 * standard error is 1. (e) The larger the value of δ and the larger the sample size, the more likely that the future study will lead to sample proportions which are able to reject the null hypothesis.

A.18 Chapter 18

1. (a) Two-way table is shown below. (b-i) $E_{row_1, col_1} = \frac{(row\ 1\ total) \times (col\ 1\ total)}{table\ total} = 35$. This is lower than the observed value. (b-ii) $E_{row_2, col_2} = \frac{(row\ 2\ total) \times (col\ 2\ total)}{table\ total} = 115$. This is lower than the observed value.

	Quit		
Treatment	Yes	No	Total
Patch + support group	40	110	150
Only patch	30	120	150
Total	70	230	300

3. (a) Sun = 0.343, Partial = 0.325, Shade = 0.331. (b) For each, the numbers are listed in the order sun, partial, and shade: Desert (40,9, 38,7, 39.4), Mountain (36.7, 34.8, 35.5), Valley (36.4, 34.5, 35.1). (c) Yes. (d) We can't evaluate the association without a formal test.

5. The original dataset will have a higher Chi-squared statistic than the randomized dataset.

7. (a) The two variables are independent. (b) The randomized Chi-squared values range from zero to approximately 15. (c) The null hypothesis is that the variables are independent; the

alternative hypothesis is that the variables are associated. The p-value is extremely small. The habitat provides information about the likelihood of being in the different sunshine states.

9. (a) The two variables are independent. (b) The randomized Chi-squared values range from zero to approximately 25. (c) The null hypothesis is that the variables are independent; the alternative hypothesis is that the variables are associated. The p-value is around 0. There is convincing evidence to claim that site and sunlight preference are associated. (d) With larger sample sizes, the power (the probability of rejecting H_0 when H_A is true) is higher.

11. (a) False. The Chi-square distribution has one parameter called degrees of freedom. (b) True. (c) True. (d) False. As the degrees of freedom increases, the shape of the Chi-square distribution becomes more symmetric.

13. The hypotheses are H_0 : Sleep levels and profession are independent. H_A : Sleep levels and profession are associated. The observations are independent and the sample sizes are large enough to conduct a Chi-square test of independence. The Chi-square statistic is 1 with 2 degrees of freedom. The p-value is 0.6. Since the p-value is high (default to alpha = 0.05), we fail to reject H_0. The data do not provide convincing evidence of an association between sleep levels and profession.

15. (a) H_0: The age of Los Angeles residents is independent of shipping carrier preference variable. H_A: The age of Los Angeles residents is associated with the shipping carrier preference variable. (b) The conditions are not satisfied since some expected counts are below 5.

A.19 Chapter 19

1. (a) Average sleep of 20 in sample vs. all New Yorkers. (b) Average height of students in study vs all undergraduates.

3. (a) Use the sample mean to estimate the population mean: 171.1. Likewise, use the sample median to estimate the population median: 170.3. (b) Use the sample standard deviation (9.4) and sample IQR (177.8 − 163.8 = 14). (c) $Z_{180} = 0.95$ and $Z_{155} = -1.71$. Neither of these observations is more than two standard deviations away from the mean, so neither would be considered unusual. (d) No, sample point estimates only estimate the population parameter, and they vary from one sample to another. Therefore we cannot expect to get the same mean and standard deviation with each random sample. (e) We use the standard error of the mean to measure the variability in means of random samples of same size taken from a population. The variability in the means of random samples is quantified by the standard error. Based on this sample, $SE_{\bar{x}} = \frac{9.4}{\sqrt{507}} = 0.417$.

5. (a) The kindergartners will have a smaller standard deviation of heights. We would expect their heights to be more similar to each other compared to a group of adults' heights. (b) The standard error of the mean will depend on the variability of individual heights. The standard error of the adult sample averages will be around $9.4/\sqrt{100} = 0.94$cm. The standard error of the kindergartner sample averages will be smaller.

7. (a) $df = 6 - 1 = 5$, $t_5^\star = 2.02$. (b) $df = 21 - 1 = 20$, $t_{20}^\star = 2.53$. (c) $df = 28$, $t_{28}^\star = 2.05$. (d) $df = 11$, $t_{11}^\star = 3.11$.

9. (a) 0.085, do not reject H_0. (b) 0.003, reject H_0. (c) 0.438, do not reject H_0. (d) 0.042, reject H_0.

11. (a) Roughly 0.1 weeks. (b) Roughly (38.45 weeks, 38.85 weeks). (c) Roughly (38.49 weeks, 38.91 weeks).

13. (a) False (b) False. (c) True. (d) False.

15. The mean is the midpoint: $\bar{x} = 20$. Identify the margin of error: $ME = 1.015$, then use $t_{35}^\star = 2.03$ and $SE = s/\sqrt{n}$ in the formula for margin of error to identify $s = 3$.

17. (a) H_0: $\mu = 8$ (New Yorkers sleep 8 hrs per night on average.) H_A: $\mu \neq 8$ (New Yorkers sleep less or more than 8 hrs per night on average.) (b) Independence: The sample is random. The min/max suggest there are no concerning outliers. $T = -1.75$. $df = 25 - 1 = 24$. (c) p-value = 0.093. If in fact the true population mean of the amount New Yorkers sleep per night was 8 hours, the probability of getting a random sample of 25 New Yorkers where the average amount of sleep is 7.73 hours per night or less (or 8.27 hours or more) is 0.093. (d) Since p-value >

0.05, do not reject H_0. The data do not provide strong evidence that New Yorkers sleep more or less than 8 hours per night on average. (e) Yes, since we did not rejected H_0.

19. With a larger critical value, the confidence interval ends up being wider. This makes intuitive sense as when we have a small sample size and the population standard deviation is unknown, we should have a wider interval than if we knew the population standard deviation, or if we had a large enough sample size.

21. (a) We will conduct a 1-sample t-test. H_0: $\mu = 5$. H_A: $\mu \neq 5$. We'll use $\alpha = 0.05$. This is a random sample, so the observations are independent. To proceed, we assume the distribution of years of piano lessons is approximately normal. $SE = 2.2/\sqrt{20} = 0.4919$. The test statistic is $T = (4.6 - 5)/SE = -0.81$. $df = 20 - 1 = 19$. The one-tail area is about 0.21, so the p-value is about 0.42, which is bigger than $\alpha = 0.05$ and we do not reject H_0. That is, we do not have sufficiently strong evidence to reject the notion that the average is 5 years. (b) Using $SE = 0.4919$ and $t^\star_{df=19} = 2.093$, the confidence interval is (3.57, 5.63). We are 95% confident that the average number of years a child takes piano lessons in this city is 3.57 to 5.63 years. (c) They agree, since we did not reject the null hypothesis and the null value of 5 was in the t-interval.

A.20 Chapter 20

1. The hypotheses should use population means (μ) not sample means ($\bar{x}$), the null hypothesis should set the two population means equal to each other, the alternative hypothesis should be two-tailed and use a not equal to sign.

3. $H_0 : \mu_{0.99} = \mu_1$ and $H_A : \mu_{0.99} \neq \mu_1$. p-value < 0.05, reject H_0. The data provide convincing evidence that the difference in population averages of price per carat of 0.99 carats and 1 carat diamonds are different.

5. (a) We are 95% confident that the population average price per carat of 0.99 carat diamonds is $2 to $23 lower than the population average price per carat of 1 carat diamonds. (b) We are 95% confident that the population average price per carat of 0.99 carat diamonds is $2.20 to $21.80 lower than the population average price per carat of 1 carat diamonds.

7. The difference is not zero (statistically significant), but there is no evidence that the difference is large (practically significant), because the interval provides values as low as 1 lb.

9. $H_0 : \mu_{0.99} = \mu_1$ and $H_A : \mu_{0.99} \neq \mu_1$. Independence: Both samples are random and represent less than 10% of their respective populations. Also, we have no reason to think that the 0.99 carats are not independent of the 1 carat diamonds since they are both sampled randomly. Normality: The distributions are not extremely skewed, hence we can assume that the distribution of the average differences will be nearly normal as well. $T_{22} = 2.23$, p-value = 0.0131. Since p-value less than 0.05, reject H_0. The data provide convincing evidence that the difference in population averages of price per carat of 0.99 carats and 1 carat diamonds are different.

11. We are 95% confident that the population average price per carat of 0.99 carat diamonds is $2.96 to $22.42 lower than the population average price per carat of 1 carat diamonds.

13. (a) $\mu_{\bar{x}_1} = 15$, $\sigma_{\bar{x}_1} = 20/\sqrt{50} = 2.8284$. (b) $\mu_{\bar{x}_2} = 20$, $\sigma_{\bar{x}_1} = 10/\sqrt{30} = 1.8257$. (c) $\mu_{\bar{x}_2 - \bar{x}_1} = 20 - 15 = 5$, $\sigma_{\bar{x}_2 - \bar{x}_1} = \sqrt{(20/\sqrt{50})^2 + (10/\sqrt{30})^2} = 3.3665$. (d) Think of $\bar{x}_1$ and $\bar{x}_2$ as being random variables, and we are considering the standard deviation of the difference of these two random variables, so we square each standard deviation, add them together, and then take the square root of the sum: $SD_{\bar{x}_2 - \bar{x}_1} = \sqrt{SD^2_{\bar{x}_2} + SD^2_{\bar{x}_1}}$.

15. (a) Chicken fed linseed weighed an average of 218.75 grams while those fed horsebean weighed an average of 160.20 grams. Both distributions are relatively symmetric with no apparent outliers. There is more variability in the weights of chicken fed linseed. (b) $H_0 : \mu_{ls} = \mu_{hb}$. $H_A : \mu_{ls} \neq \mu_{hb}$. We leave the conditions to you to consider. $T = 3.02$, $df = min(11, 9) = 9 \rightarrow$ p-value = 0.014. Since p-value < 0.05, reject H_0. The data provide strong evidence that there is a significant difference between the average weights of chickens that were fed linseed and horsebean. (c) Type 1 Error, since we rejected H_0. (d) Yes, since p-value > 0.01, we would not have rejected H_0.

17. $H_0 : \mu_C = \mu_S$. $H_A : \mu_C \neq \mu_S$. $T = 3.27$, $df = 11 \rightarrow$ p-value = 0.007. Since p-value < 0.05, reject H_0. The data provide strong evidence that the average weight of chickens that were fed casein is different than the average weight of chickens that were fed soybean (with weights from

casein being higher). Since this is a randomized experiment, the observed difference can be attributed to the diet.

19. $H_0 : \mu_T = \mu_C$. $H_A : \mu_T \neq \mu_C$. $T = 2.24$, $df = 21 \rightarrow$ p-value = 0.036. Since p-value < 0.05, reject H_0. The data provide strong evidence that the average food consumption by the patients in the treatment and control groups are different. Furthermore, the data indicate patients in the distracted eating (treatment) group consume more food than patients in the control group.

A.21 Chapter 21

1. Paired, data are recorded in the same cities at two different time points. The temperature in a city at one point is not independent of the temperature in the same city at another time point

3. (a) Since it's the same students at the beginning and the end of the semester, there is a pairing between the datasets, for a given student their beginning and end of semester grades are dependent. (b) Since the subjects were sampled randomly, each observation in the men's group does not have a special correspondence with exactly one observation in the other (women's) group. (c) Since it's the same subjects at the beginning and the end of the study, there is a pairing between the datasets, for a subject student their beginning and end of semester artery thickness are dependent. (d) Since it's the same subjects at the beginning and the end of the study, there is a pairing between the datasets, for a subject student their beginning and end of semester weights are dependent.

5. False. While it is true that paired analysis requires equal sample sizes, only having the equal sample sizes isn't, on its own, sufficient for doing a paired test. Paired tests require that there be a special correspondence between each pair of observations in the two groups.

7. The data are paired, since this is a before-after measurement of the same trees, so we will construct a confidence interval using the differences summary statistics. But before we proceed with a confidence interval, we must first check conditions: Independent: this is satisfied since the trees were randomly sampled. Normality: since $n = 50 \geq 30$, we only need consider whether there are any particularly extreme outliers. None are mentioned, and it doesn't seem like we'd expect to observe any such cases from data of this type, so we'll consider this condition to be satisfied. With the conditions satisfied, we can proceed with calculations. First, compute the standard error and degrees of freedom: $SE = \frac{7.2}{\sqrt{50}} = 1.02$ and $df = 50 - 1 = 49$. Next, we find $t^\star = 2.68$ for a 99% confidence interval using a t-distribution with 49 degrees of freedom, and then we construct the confidence interval: $\bar{x} \pm t^\star \times SE = 12.5 \pm 2.68 \times 1.02 = (9.77, 15.23)$. We are 99% confident that the average growth of young trees in this area during the 10-year period was 9.77 to 15.23 feet.

9. (a) No. (b) Yes. (c) No. (d) No and yes. (e) Yes.

11. (a) Let $diff = 2018 - 1948$. Then, $H_0 : \mu_{diff} = 0$ and $H_A : \mu_{diff} \neq 0$. (b) The observed average of difference is just outside the randomized differences. (c) Since the p-value < 0.05, reject H_0. The data provide convincing evidence of a difference between the average number of 90F degree days in 2018 and the average number of 90F degree days in 1948.

13. (a) For each observation in one dataset, there is exactly one specially corresponding observation in the other dataset for the same geographic location. The data are paired. (b) $H_0 : \mu_{\text{diff}} = 0$ (There is no difference in average number of days exceeding 90F in 1948 and 2018 for NOAA stations.) $H_A : \mu_{\text{diff}} \neq 0$ (There is a difference.) (c) Locations were randomly sampled, so independence is reasonable. The sample size is at least 30, so we're just looking for particularly extreme outliers: none are present (the observation off left in the histogram would be considered a clear outlier, but not a particularly extreme one). Therefore, the conditions are satisfied. (d) $SE = 17.2/\sqrt{197} = 1.23$. $T = \frac{2.9-0}{1.23} = 2.36$ with degrees of freedom $df = 197 - 1 = 196$. This leads to a one-tail area of 0.0096 and a p-value of about 0.019. (e) Since the p-value is less than 0.05, we reject H_0. The data provide strong evidence that NOAA stations observed more 90F days in 2018 than in 1948. (f) Type 1 Error, since we may have incorrectly rejected H_0. This error would mean that NOAA stations did not actually observe a decrease, but the sample we took just so happened to make it appear that this was the case. (g) No, since we rejected H_0, which had a null value of 0.

15. (a) $SE = 1.23$ and $z^\star = 1.65$. $2.9 \pm 1.65 \times 1.23 \rightarrow (0.87, 4.93)$. (b) We are 90% confident that there was an increase of 0.87 to 4.93 in the average number of days that hit 90F in 2018 relative

to 1948 for NOAA stations. (c) Yes, since the interval lies entirely above 0.

17. (a) These data are paired. For example, the Friday the 13th in say, September 1991, would probably be more similar to the Friday the 6th in September 1991 than to Friday the 6th in another month or year. (b) Let $\mu_{diff} = \mu_{sixth} - \mu_{thirteenth}$. $H_0 : \mu_{diff} = 0$. $H_A : \mu_{diff} \neq 0$. (c) Independence: The months selected are not random. However, if we think these dates are roughly equivalent to a simple random sample of all such Friday 6th/13th date pairs, then independence is reasonable. To proceed, we must make this strong assumption, though we should note this assumption in any reported results. Normality: With fewer than 10 observations, we would need to see clear outliers to be concerned. There is a borderline outlier on the right of the histogram of the differences, so we would want to report this in formal analysis results. (d) $T = 4.93$ for $df = 10 - 1 = 9 \to$ p-value = 0.001. (e) Since p-value < 0.05, reject H_0. The data provide strong evidence that the average number of cars at the intersection is higher on Friday the 6th than on Friday the 13th. (We should exercise caution about generalizing the interpetation to all intersections or roads.) (f) If the average number of cars passing the intersection actually was the same on Friday the 6th and 13th, then the probability that we would observe a test statistic so far from zero is less than 0.01. (g) We might have made a Type 1 Error, i.e., incorrectly rejected the null hypothesis.

A.22 Chapter 22

1. Alternative.
3. (a) Means across original data are more variable. (b) Standard deviation of egg lengths are about the same for both plots. (c) F statistic is bigger for the original data.
5. H_0: $\mu_1 = \mu_2 = \cdots = \mu_6$. H_A: The average weight varies across some (or all) groups. Independence: Chicks are randomly assigned to feed types (presumably kept separate from one another), therefore independence of observations is reasonable. Approx. normal: the distributions of weights within each feed type appear to be fairly symmetric. Constant variance: Based on the side-by-side box plots, the constant variance assumption appears to be reasonable. There are differences in the actual computed standard deviations, but these might be due to chance as these are quite small samples. $F_{5,65} = 15.36$ and the p-value is approximately 0. With such a small p-value, we reject H_0. The data provide convincing evidence that the average weight of chicks varies across some (or all) feed supplement groups.
7. (a) H_0: The population mean of MET for each group is equal to the others. H_A: At least one pair of means is different. (b) Independence: We don't have any information on how the data were collected, so we cannot assess independence. To proceed, we must assume the subjects in each group are independent. In practice, we would inquire for more details. Normality: The data are bound below by zero and the standard deviations are larger than the means, indicating very strong skew. However, since the sample sizes are extremely large, even extreme skew is acceptable. Constant variance: This condition is sufficiently met, as the standard deviations are reasonably consistent across groups. (c) Since p-value is very small, reject H_0. The data provide convincing evidence that the average MET differs between at least one pair of groups.
9. (a) H_0: Average GPA is the same for all majors. H_A: At least one pair of means are different. (b) Since p-value > 0.05, fail to reject H_0. The data do not provide convincing evidence of a difference between the average GPAs across three groups of majors. (c) The total degrees of freedom is $195 + 2 = 197$, so the sample size is $197 + 1 = 198$.
11. (a) False. As the number of groups increases, so does the number of comparisons and hence the modified significance level decreases. (b) True. (c) True. (d) False. We need observations to be independent regardless of sample size.
13. (a) Left is Dataset B. (b) Right is Dataset A.

A.23 Chapter 23

Application chapter, no exercises.

A.24 Chapter 24

1. (a) $H_0 : \beta_1 = 0$, $H_A : \beta_1 \neq 0$. (b) The observed slope of 0.604 is not a plausible value, the p-value is extremely small, and the null hypothesis can be rejected. c. The p-value is also extremely small.
3. (a) Roughly 0.53 to 0.67. (b) For individuals with one cm larger shoulder girth, their average height is predicted to be between 0.53 and 0.67 cm taller, with 98% confidence.
5. (a) $H_0 : \beta_1 = 0$, $H_A : \beta_1 \neq 0$. (b) The observed slope of 2.559 is not a plausible value, the p-value is extremely small, and the null hypothesis can be rejected. (c) The p-value is also extremely small.
7. (a) Rough 90% confidence interval is 1.9 to 3.1. (b) For a one unit (one percentage point) increase in poverty across given metropolitan areas, the predicted average annual murder rate will be between 1.9 and 3.1 persons per million larger, with 90% confidence.
9. (a) $H_0 : \beta_1 = 0$, $H_A : \beta_1 \neq 0$. (b) The p-value is roughly 0.45 which is much bigger than 0.05. The null hypothesis cannot be rejected. There is no evidence with these data that there is a linear relationship between a father's age and the baby's weight. (c) The p-value of 0.449 is quite similar. The hypothesis test conclusion is the same, the data do not support a linear model.
11. (a) Rough 95% confidence interval is (-.008, 0.016). (b) 95% confident that for individuals with fathers who are one year older, their average weight is predicted to be between -0.008 and 0.016 pounds heavier.
13. (a) H_0: The true slope coefficient of body weight is zero ($\beta_1 = 0$). H_A: The true slope coefficient of body weight is different than zero ($\beta_1 \neq 0$). (b) The p-value is extremely small (zero to 4 decimal places), which is lower than the significance level of 0.05. With such a low p-value, we reject H_0. The data provide strong evidence that the true slope coefficient of body weight is greater than zero and that body weight is positively associated with heart weight in cats. (c) (3.539, 4.529). We are 95% confident that for each additional kilogram in cats' weights, we expect their hearts to be heavier by 3.539 to 4.529 grams, on average. (d) Yes, we rejected the null hypothesis and the confidence interval lies above 0.
15. (a) $r = \sqrt{0.292} \approx -0.54$. We know the correlation is negative due to the negative association shown in the scatterplot. (b) The residuals appear to be fan shaped, indicating non-constant variance. Therefore a simple least squares fit is not appropriate for these data.

A.25 Chapter 25

1. (a) (-0.044, 0.346). We are 95% confident that student who go out more than two nights a week on average have GPAs 0.044 points lower to 0.346 points higher than those who do not go out more than two nights a week, when controlling for the other variables in the model. (b) Yes, since the p-value is larger than 0.05 in all cases (not including the intercept).
3. (a) There is a positive, very strong, linear association between the number of tourists and spending. (b) Explanatory: number of tourists (in thousands). Response: spending (in millions of US dollars). (c) We can predict spending for a given number of tourists using a regression line. This may be useful information for determining how much the country may want to spend in advertising abroad, or to forecast expected revenues from tourism. (d) Even though the relationship appears linear in the scatterplot, the residual plot actually shows a nonlinear relationship (**Linearity** is violated). This is not a contradiction: residual plots can show divergences from linearity that can be difficult to see in a scatterplot. A simple linear model is inadequate for modeling these data. It is also important to consider that these data are observed sequentially, which means there may be a hidden structure not evident in the current plots but that is important to consider (and might lead to a violation of the **Independence** condition).
5. (a) **Linearity**: Horror movies seem to show a much different pattern than the other genres. While the residuals plots show a random scatter over years and in order of data collection, there is a clear pattern in residuals for various genres, which signals that this regression model is not appropriate for these data. **Independent observations**: The variability of the residuals is higher for data that comes later in the dataset. We don't know if the data are sorted by

year, but if so, there may be a temporal pattern in the data that voilates the independence condition. **N**ormality: The residuals are right skewed (skewed to the high end). Constant or **E**qual variability: The residuals vs. predicted values plot reveals some outliers. This plot for only babies with predicted birth weights between 6 and 8.5 pounds looks a lot better, suggesting that for bulk of the data the constant variance condition is met.

7. (a) Linearity: With so many observations in the dataset, we look for particularly extreme outliers in the histogram of residuals and do not see any. We also don't see a non-linear pattern emerging in the residuals vs. predicted plot. Independent observations: The sample is random and there does not seem ti be a trend in the residuals vs. order of data collection plot. Normality: The histogram of residuals appears to be unimodal and symmetric, centered at 0. Constant or equal variability: The residuals vs. predicted values plot reveals some outliers. This plot for only babies with predicted birth weights between 6 and 8.5 pounds looks a lot better, suggesting that for bulk of the data the constant variance condition is met. All concerns raised here are relatively mild. There are some outliers, but there is so much data that the influence of such observations will be minor. (b) H_0: The true slope coefficient of habit is zero ($\beta_5 = 0$). H_A: The true slope coefficient of height is different than zero ($\beta_5 \neq 0$). The p-value for the two-sided alternative hypothesis ($\beta_5 \neq 0$) is incredibly 0.0007 (smaller than 0.05), so we reject H_0. The data provide convincing evidence that height and weight are positively correlated, given the other variables in the model. The true slope parameter is indeed greater than 0.

9. (a) Roughly $\widehat{\texttt{weight}} = 11$ pounds and $\texttt{weight}_i = 7$ pounds. (b) Folds 1, 2, and 4 were used to build the prediction model. (c) The plot on the left estimates 8 parameters; the plot on the right estimates 3 parameters. (d) The residuals are not substantially different.

11. (a) Roughly $\widehat{\texttt{volume}} = 400$ riders and $\texttt{volume}_i = 500$ riders. (b) Folds 2 and 3 were used to build the prediction model. (c) The plot on the left estimates 7 parameters; the plot on the right estimates 3 parameters. (d) The residuals are not substantially different.

A.26 Chapter 26

1. (a) $H_0 : \beta_1 = 0$, the slope of the model predicting kids' marijuana use in college from their parents' marijuana use in college is 0. $H_A : \beta_1 \neq 0$, the slope of the model predicting kids' marijuana use in college from their parents' marijuana use in college is different than 0. (b) The test statistic is $Z = 4.09$ and the associated p-value is less than 0.0001. (c) With a small p-value we reject H_0. The data provide convincing evidence that the slope of the model predicting kids' marijuana use in college from their parents' marijuana use in college is different than 0, i.e. that parents' marijuana use in college is a significant predictor of kids' marijuana use in college.

3. (a) 26 observations are in Fold2. 8 correctly and 2 incorrectly predicted to be from Victoria. (b) 78 observations are used to build the model. (c) 2 coefficients for tail length; 3 coefficients for total length and sex.

5. (a) 298 observations are in Fold2. 10 correctly and 26 incorrectly predicted to be premature. (b) 596 observations are used to build the model. (c) The vast majority of the observations fall into the row corresponding to the observed status of full term. (d) 7 coefficients for the larger model; 3 coefficients for the smaller model.

Bibliography

Adolph, S. (1990). Influence of behavioral thermoregulation on microhabitat use by two *Sceloporus* lizards. *Ecology*, 71:315–327.

Adolph, S. C. (1987). *Physiological and behavioral ecology of the lizards* Sceloporus occidentalis *and* Sceloporus graciosus. PhD thesis, University of Washington, Seattle, Washington.

AEC (2008). College-Bound Students' Interests in Study Abroad and Other International Learning Activities.

Allais, G., Romoli, M., Rolando, S., Airola, G., Castagnoli Gabellari, I., Allais, R., and Benedetto, C. (2011). Ear acupuncture in the treatment of migraine attacks: a randomized trial on the efficacy of appropriate versus inappropriate acupoints. *Neurological Sciences*, 32(1):173–175.

Allison, T. and Cicchetti, D. (1975). Sleep in mammals: ecological and constitutional correlates. *Arch. Hydrobiol*, 75:442.

Asbury, D. and Adolph, S. C. (2007). Behavioral plasticity in an ecological generalist: microhabitat use by western fence lizards. *Evolutionary Ecology Research*, 9:801–815.

AT, L. S., MR, R., CF, C., and SE, G. (2011). The $16,819 Pay Gap For Newly Trained Physicians: The Unexplained Trend Of Men Earning More Than Women. *Health Affairs*, 30(2).

Audera, C., Patulny, R. V., Sander, B. H., and Douglas, R. M. (2001). Mega-dose vitamin C in treatment of the common cold: a randomised controlled trial. *Medical journal of Australia*, 175(7):359–362.

Backstrom, L. (2011). Anatomy of facebook. *Facebook Data Team's Notes*.

Benson, J. B. (1993). Season of birth and onset of locomotion: Theoretical and methodological implications. *Infant behavior and development*, 16(1):69–81.

Bertrand, M. and Mullainathan, S. (2003). Are Emily and Greg More Employable than Lakisha and Jamal? A Field Experiment on Labor Market Discrimination. Technical report.

Böttiger, B. W., Bode, C., Kern, S., Gries, A., Gust, R., Glätzer, R., Bauer, H., Motsch, J., and Martin, E. (2001). Efficacy and safety of thrombolytic therapy after initially unsuccessful cardiopulmonary resuscitation: a prospective clinical trial. *The Lancet*, 357(9268):1583–1585.

Bucciol, A. and Piovesan, M. (2011). Luck or cheating? A field experiment on honesty with children. *Journal of Economic Psychology*, 32(1):73–78.

Bureau, U. C. (2010). 2010 Census Urban and Rural Classification and Urban Area Criteria Housing Characteristics.

CDC (2008). Perceived Insufficient Rest or Sleep Among Adults – United States, 2008.

CDC (2018). 2018 Assisted Reproductive Technology Fertility Clinic Success Rates Report.

Center, U. C. P. (2006). China Health and Nutrition Survey, 2006.

Chance, B. and Rossman, A. (2018). *Investigating Statistical Concepts, Applications, and Methods*.

Chimowitz, M. I., Lynn, M. J., Derdeyn, C. P., Turan, T. N., Fiorella, D., Lane, B. F., Janis, L. S., Lutsep, H. L., Barnwell, S. L., Waters, M. F., et al. (2011). Stenting versus aggressive medical therapy for intracranial arterial stenosis. *New England Journal of Medicine*, 365(11):993–1003.

NY Times article reporting on the study: http://www.nytimes.com/2011/09/08/health/research/08stent.html.

Conner, T. S., Brookie, K. L., Carr, A. C., Mainvil, L. A., and Vissers, M. C. (2017). Let them eat fruit! The effect of fruit and vegetable consumption on psychological well-being in young adults: A randomized controlled trial. *PloS one*, 12(2).

Datoo, M. S., Natama, M. H., Somé, A., Traoré, O., Rouamba, T., Bellamy, D., Yameogo, P., Valia, D., Tegneri, M., Ouedraogo, F., et al. (2021). High Efficacy of a Low Dose Candidate Malaria Vaccine, R21 in 1 Adjuvant Matrix-M^{TM}, with Seasonal Administration to Children in Burkina Faso. *The Lancet*.

Demos (2011). The State of Young America: The Poll.

Ellis, G. J. and Stone, L. H. (1979). Marijuana Use in College: An Evaluation of a Modeling Explanation. *Youth and Society*, 10(4):323.

FiveThirtyEight (2015). Scary Movies Are The Best Investment In Hollywood.

Foundation, K. F. (2019). The Public On Next Steps For The ACA And Proposals To Expand Coverage, data collected January 9 - 14, 2019.

Foundation, N. S. (2012). Sleep in America Poll: Transportation Workers' Sleep.

Frederick, S., Novemsky, N., Wang, J., Dhar, R., and Nowlis, S. (2009). Opportunity cost neglect. *Journal of Consumer Research*, 36(4):553–561.

Gallup (2012). Employed Americans in Better Health Than the Unemployed, data collected January 2, 2011 - May 21, 2012.

Gallup (2021a). Half of College Students Say COVID-19 May Impact Completion.

Gallup (2021b). U.S. Support for Vaccination Proof Varies by Activity, data collected in April 2021.

Garbutt, J., Banister, C., Spitznagel, E., and Piccirillo, J. (2012). Amoxicillin for Acute Rhinosinusitis: A Randomized Controlled Trial. *JAMA: The Journal of the American Medical Association*, 307(7):685–692.

Gneezy, U. and Rustichini, A. (2000). A fine is a price. *The Journal of Legal Studies*, 29(1):1–17.

Gorman, K. B., Williams, T. D., and Fraser, W. R. (2014a). Ecological sexual dimorphism and environmental variability within a community of Antarctic penguins (genus Pygoscelis). *PloS one*, 9(3):e90081.

Gorman, K. B., Williams, T. D., and Fraser, W. R. (2014b). Ecological sexual dimorphism and environmental variability within a community of Antarctic penguins (genus Pygoscelis). *PloS one*, 9(3):e90081.

Graham, D., Ouellet-Hellstrom, R., MaCurdy, T., Ali, F., Sholley, C., Worrall, C., and Kelman, J. (2010). Risk of acute myocardial infarction, stroke, heart failure, and death in elderly medicare patients treated with rosiglitazone or pioglitazone. *JAMA*, 304(4):411.

Group, A. R. T. R. (1980). Sulfinpyrazone in the prevention of sudden death after myocardial infarction. *New England Journal of Medicine*, 302(5):250–256.

Hand, D. (1994). *A handbook of small data sets*. Chapman & Hall/CRC.

Hayden, R. W. (2019). Questionable claims for simple versions of the bootstrap. *Journal of Statistics Education*, 27(3):208–215.

Heinz, G., Peterson, L., Johnson, R., and Kerk, C. (2003). Exploring relationships in body dimensions. *Journal of Statistics Education*, 11(2).

Hepler, J. and Albarracín, D. (2013). Attitudes without objects: Evidence for a dispositional attitude, its measurement, and its consequences. *Journal of Personality and Social Psychology*, 104(6):1060.

Hesterberg, T. C. (2015). What teachers should know about the bootstrap: Resampling in the undergraduate statistics curriculum. *The American Statistician*, 69(4):371–386.

ICPSR (2014). United States Department of Health and Human Services. Centers for Disease Control and Prevention. National Center for Health Statistics. Natality Detail File, 2014 United States. Inter-university Consortium for Political and Social Research, 2016-10-07.

King, C., Suamani, J., Sanuku, N., Cheng, Y.-C., Satofan, S., Mancuso, B., Goss, C. W., Robinson, L. J., Siba, P. M., Weil, G. J., and Kazura, J. W. (2018). A Trial of a Triple-Drug Treatment for Lymphatic Filariasis. *New England Journal of Medicine*, 379:1801–1810.

Latter, O. H. (1902). The Egg of Cuculus Canorus. An Enquiry into the Dimensions of the Cuckoo's Egg and the Relation of the Variations to the Size of the Eggs of the Foster-Parent, with Notes on Coloration. *Biometrika*, 1:164–176.

Legaki, N.-Z., Xi, N., Hamari, J., Karpouzis, K., and Assimakopoulos, V. (2020). The effect of challenge-based gamification on learning: An experiment in the context of statistics education. *International Journal of Human-Computer Studies*, 144.

Lockman, S., Shapiro, R. L., Smeaton, L. M., Wester, C., Thior, I., Stevens, L., Chand, F., Makhema, J., Moffat, C., Asmelash, A., et al. (2007). Response to antiretroviral therapy after a single, peripartum dose of nevirapine. *New England Journal of Medicine*, 356(2):135–147.

Lucas, M., Mirzaei, F., Pan, A., Okereke, O., Willett, W., O'Reilly, E., Koenen, K., and Ascherio, A. (2011). Coffee, caffeine, and risk of depression among women. *Archives of Internal Medicine*, 171(17):1571.

Manson, J. E., Cook, N. R., Lee, I.-M., Christen, W., Bassuk, S. S., Mora, S., Gibson, H., Albert, C. M., Gordon, D., Copeland, T., et al. (2019). Marine n-3 fatty acids and prevention of cardiovascular disease and cancer. *New England Journal of Medicine*, 380(1):23–32.

McNeil, D. R. (1977). Interactive data analysis: a practical primer.

Ménard, C., Hagège, A. A., Agbulut, O., Barro, M., Morichetti, M. C., Brasselet, C., Bel, A., Messas, E., Bissery, A., Bruneval, P., et al. (2005). Transplantation of cardiac-committed mouse embryonic stem cells to infarcted sheep myocardium: a preclinical study. *The Lancet*, 366(9490):1005–1012.

Mortada, W., Sobh, M., El-Defrawy, M., and Farahat, S. (2000). Study of lead exposure from automobile exhaust as a risk for nephrotoxicity among traffic policemen. *American journal of nephrology*, 21(4):274–279.

Nieman, D., Cayea, E., Austin, M., Henson, D., McAnulty, S., and Jin, F. (2009). Chia seed does not promote weight loss or alter disease risk factors in overweight adults. *Nutrition Research*, 29(6):414–418.

NOAA (2018). Climate Data Online. Retrieved April 24, 2019.

NORC (2010). General Social Survey, 2010.

NORC (2016). Data Explorer - Government responsible to promote gender equality.

NORC (2018). Data Explorer - Should marijuana be made legal?

Nuñez, J. R., Anderton, C. R., and Renslow, R. S. (2018). Optimizing colormaps with consideration for color vision deficiency to enable accurate interpretation of scientific data. *PloS one*, 13(7).

O'Connor, A. (2011). Coffee Drinking Linked to Less Depression in Women. *New York Times*.

Oldham-Cooper, R., Hardman, C., Nicoll, C., Rogers, P., and Brunstrom, J. (2011). Playing a computer game during lunch affects fullness, memory for lunch, and later snack intake. *The American Journal of Clinical Nutrition*, 93(2):308.

Orben, A. and Baukney-Przybylski, A. (2018). Screens, Teens and Psychological Well-Being: Evidence from three time-use diary studies. *Psychological Science*.

Parker-Pope, T. (2011). The School Bully Is Sleepy. *New York Times*.

Pew Research Center (2011). Is College Worth It?, data collected March 15-29, 2011.

Pew Research Center (2013a). The Diagnosis Difference.

Pew Research Center (2013b). Twitter News Consumers: Young, Mobile and Educated.

Pew Research Center (2018). A Majority of Teens Have Experienced Some Form of Cyberbullying, data collected March 7 - April 10, 2018.

Pew Research Center (2021a). The State of Online Harassment, data collected September 8 - 13, 2020.

Pew Research Center (2021b). How Americans' attitudes about climate change differ by generation, party and other factors, data collected April 20 to 29, 2021.

Pfizer (2021). Pfizer-Biontech Announce Positive Topline Results Of Pivotal Covid-19 Vaccine Study In Adolescents.

Pierson, E., Simoiu, C., Overgoor, J., Corbett-Davies, S., Jenson, D., Shoemaker, A., Ramachandran, V., Barghouty, P., Phillips, C., Shroff, R., et al. (2020). A large-scale analysis of racial disparities in police stops across the united states. *Nature human behaviour*, 4(7):736–745.

Piff, P., Stancato, D., Côté, S., Mendoza-Denton, R., and Keltner, D. (2012). Higher social class predicts increased unethical behavior. *Proceedings of the National Academy of Sciences*.

Poll, M. (2011). Road Rules: Re-Testing Drivers at Age 65?, March 4, 2011.

Public Policy Polling (2015). Americans on College Degrees, Classic Literature, the Seasons, and More, data collected Feb 20-22, 2015.

Rabin, R. (2010). Risks: Smokers Found More Prone to Dementia. *New York Times*.

Ramsey, F. and Schafer, D. (2012). *The Statistical Sleuth*. Cengage Learning, 3rd edition.

Rao, M., Bar, L., Yu, Y., Srinivasan, M., Mukherjea, A., Li, J., Chung, S., Venkatraman, S., Dan, S., and Palaniappan, L. (2021). Disaggregating Asian American Cigarette and Alternative Tobacco Product Use: Results from the National Health Interview Survey (NHIS) 2006–2018. *Journal of Racial and Ethnic Health Disparities*.

Ritz, B., Yu, F., Chapa, G., and Fruin, S. (2000). Effect of air pollution on preterm birth among children born in southern california between 1989 and 1993. *Epidemiology*, 11(5):502–511.

Rosen, B. and Jerdee, T. H. (1974). Influence of sex role stereotypes on personnel decisions. *Journal of Applied Psychology*, 59(1):9.

Rubio-Fernandez, P., Mollica, F., and Jara-Ettinger, J. (2021). Speakers and listeners exploit word order for communicative efficiency: A cross-linguistic investigation. *Journal of Experimental Psychology: General*, 150(3):583–594.

Scanlon, T., Luben, R., Scanlon, and F.L., Singleton, N. (1993). Is Friday the 13th Bad For Your Health? *BMJ*, 307:1584–1586.

Smith, G. E., Chouinard, P. A., and Byosiere, S. (2021). If I fits I sits: A citizen science investigation into illusory contour susceptibility in domestic cats (*Felis silvestris catus*). *Applied Animal Behaviour Science*, 240.

Survey USA (2012). News Poll 19333, data collected on June 27, 2012.

Survey USA (2019). News Poll 24568, data collected on April 21, 2019.

Survey USA (2021). News Poll 25938, data collected April 29 - May 5, 2021.

Turnbull, B., Brown, B., and Hu, M. (1974). Survivorship of Heart Transplant Data. *Journal of the American Statistical Association*, 69:74–80.

US DOE EPA (2021). Fuel Economy, 2021 Data File.

Vaughn, A. (2011). Poll finds young adults optimistic, but not about money. *Los Angeles Times*.

Washington Post (2020). Washington Post-Schar School national poll, data collected June 2-7, 2020. Retrieved September 5, 2020.

Wickham, H. (2014). Tidy data. *Journal of Statistical Software*, 59(10).

Wickham, H. (2016). *ggplot2: Elegant Graphics for Data Analysis*. Springer-Verlag New York.

YouGov (2021). Do you think the coronavirus will end up bringing the world closer together, or leave us further apart?

Made in the USA
Las Vegas, NV
21 August 2022